ENDOCRINOLOGY AND PHYSIOLOGY OF REPRODUCTION

ENDOCRINOLOGY AND PHYSIOLOGY OF REPRODUCTION

Edited by

P. C. K. Leung

The University of British Columbia
Vancouver, British Columbia, Canada

D. T. Armstrong

University of Western Ontario
London, Ontario, Canada

K. B. Ruf and
W. H. Moger

Dalhousie University
Halifax, Nova Scotia, Canada

and

H. G. Friesen

University of Manitoba
Winnipeg, Manitoba, Canada

Springer Science+Business Media, LLC

Library of Congress Cataloging in Publication Data

Endocrinology and physiology and reproduction.

Most chapters were presented as plenary lectures or symposium talks at the 1986 30th Congress of the International Union of Physiological Sciences in Vancouver, B.C.
Includes bibliographies and indexes.
1. Human reproduction—Endocrine aspects—Congresses. 2. Human reproduction—Congresses. I. Leung, P. C. K. II. International Union of Physiological Sciences. Congress (30th: 1986: Vancouver, B.C.) [DNLM: 1. Endocrine Glands—physiology—congresses. 2. Reproduction—congresses. W3 IN84 30th 1986 / WQ 205 E565 1986]
QP252.E53 1987 612'.6 87-11280
ISBN 978-1-4899-1973-1

ISBN 978-1-4899-1973-1 ISBN 978-1-4899-1971-7 (eBook)
DOI 10.1007/978-1-4899-1971-7

Originally published by Plenum Press, New York in 1987
Softcover reprint of the hardcover 1st edition 1987

FOREWORD

Most of the following chapters were presented as plenary lectures or symposium talks at the 1986 XXXth Congress of the International Union of Physiological Sciences in Vancouver, B.C. A distinguished international group of endocrinologists and physiologists have contributed up-to-date reviews of their particular fields. The early chapters are largely concerned with the brain and neuroendocrine mechanisms controlling the secretion of gonadotropin releasing hormone (GnRH) and its action on the anterior pituitary gland. Later chapters focus on the gonads themselves and the systemic and intrinsic hormones influencing the functional cytology of ovarian and testicular cells. Such comprehensive subjects as sex differentiation, puberty, placentation and parturition are also discussed authoritatively.

According to Pfaff and Cohen and Arai et al., gonadal steroids, especially estrogen, exert multiple effects on certain hypothalamic and preoptic neurons, including growth, protein synthesis and electrical changes, which promote plasticity and facilitate synaptogenesis. The electrophysiology of the hypothalamic GnRH pulse generator in the rhesus monkey is reviewed more specifically by Knobil. In ovariectomized ewes, Clarke finds both positive and negative effects of estrogen on hypothalamic release of GnRH as well as on pituitary responsiveness to the peptide. Flerkó et al. and Motta et al. describe mechanisms by which brain and pituitary peptides can influence hypothalamic function by short and ultrashort feedback circuits to modulate neural control of pituitary gonadotropin release. An involvement of the amino acid neurotransmitter γ-aminobutyric acid in the control of GnRH release is also proposed by Wuttke et al. In a study of hypothalamic biogenic amines, Coen emphasizes the importance of adrenergic nerves in stimulating GnRH release. Admitting the stimulatory effects, Bergen and Leung demonstrate that electrical stimulation of the ascending dorsal midbrain noradrenergic pathway, but not the ventral tract, markedly inhibits pulsatile LH release in ovariectomized rats, supporting the existence of an inhibitory noradrenergic system in the modulation of GnRH discharge. In the developing female rat approaching puberty, Ojeda et al. describe the interplay of ovarian estrogen, norepinephrine (NE) and vasoactive intestinal peptide (VIP) in evoking the release of sufficient GnRH to activate a proestrous ovulatory LH surge. Estrogen enhances both NE-induced prostaglandin E_2 (PGE_2) synthesis and PGE_2-induced GnRH release. A protein kinase C-mediated pathway may also be activated in the hypothalamus and participate in stimulating GnRH discharge. Millar et al. review structural and functional characteristics of a recently discovered precursor of GnRH, as well as the biological activity of non-GnRH synthetic peptide sequences of the GnRH precursor.

Two chapters by Catt et al. and Naor are concerned with biochemical mechanisms by which GnRH stimulates LH release. They agree that GnRH first stimulates a rapid phosphodiester hydrolysis of phosphatidylinositol

4,5-biphosphate (PIP_2) to inositol-triphosphate (IP_3) and diacylglycerol (DG). IP_3 appears to mobilize cellular Ca^{2+} while DG activates protein kinase C. These findings suggest that the LH response to GnRH is mediated by two intracellular pathways involving Ca^{2+} and diacylglycerol as second messengers.

Four chapters are concerned with mechanisms by which local and circulating hormones influence ovarian cell functions such as steroidogenesis, inhibin secretion, oogenesis and luteolysis. Hsueh and colleagues have studied extensively the effects of follicle-stimulating hormone on ovarian granulosa cells in culture and find that the ability of a selected follicle(s) to become dominant while the majority of follicles become atretic cannot be explained on the basis of gonadotropin levels alone. Local modulatory factors are also important. Likewise, Armstrong et al. find that follicular steroid biosynthesis is influenced by the local action of the ovarian steroids themselves. Steroids represent one of the mechanisms by which follicular recruitment, growth, atresia, and ovulation may be influenced by local intraovarian factors. Dekel suggests that meiosis is arrested in the intrafollicular oocyte by the paracrine transfer of cAMP from the surrounding cumulus cells, and that the preovulatory surge of LH releases the meiotic inhibition by breaking down the communication between cumulus and oocyte. Cumulus-free oocytes resume meiosis and become fertilizable. Sheldrick and Flint propose that in the sheep oxytocin secreted by the corpus luteum may contribute to luteolysis by stimulating the release of the ovine luteolysin, prostaglandin $F_{2\alpha}$. The oxytocin receptors in the uterus develop at the appropriate stage in the estrous cycle, and the short circuit interchange between corpus luteum and uterus may involve countercurrent distribution of the peptide and prostaglandin in the ovarian and uterine veins and arteries.

Four chapters focus on the testis and its specialized cells involved in steroidogenesis and spermatogenesis. Moger et al. review the evidence that androgen steroidogenesis by the interstitial Leydig cells is influenced by catecholamines. Pomerantz and Jansz find that disruption of spermatogenesis by unilateral surgical cryptorchidism induces a hyperresponsiveness of the Leydig cells in both testes to treatment with LH in vitro. This suggests an intergonadal transfer of the influence of unilateral aspermatogenesis caused by the cryptorchidism. According to Bergh and Damber, treatment of the male rat with hCG/LH induces a rapid rise in testosterone secretion which in turn causes the local formation of at least two factors influencing testicular blood vessels, one affecting arteriolar and one increasing venular permeability by attracting polymorphonuclear leukocytes. The nature and physiological roles of the factors is unknown. The descent of the testis has been restudied by Wensing. The essential outgrowth of the gubernaculum appears to depend on some unknown testicular hormone other than testosterone. However, there are indications that testosterone plays a major role in the subsequent regression of the gubernaculum.

Josso et al. describe the factors responsible for sexual differentiation of the fetus. Genetic and hormonal stimuli including an early anti-Mullerian hormone secreted by Sertoli cells and testosterone produced subsequently by Leydig cells induce the male alterations in organs that would otherwise develop autonomously as female structures. Female organogenesis is disrupted by exposure to male hormones such as may occur in congenital adrenal hyperplasia.

Two placental lactogens have been demonstrated by Duckworth et al. in the rat. The first, with a molecular weight of about 40,000 daltons, peaks at 12 days of pregnancy and is inhibited by the maternal pituitary. The second, of about half the molecular weight of the first, appears at about day 11 and peaks at day 20. Its level is markedly increased by maternal hypophysectomy or ovariectomy at midpregnancy but strongly suppressed by fetectomy. The structural relations of these and other lactogen-related molecules in the rat placenta are under study with genetic coding techniques.

The volume closes with a review of hormonal influences on fetal and perinatal water metabolism, by Perks and Cassin, and Thorburn's essay on the comparative physiology of mechanisms controlling the timing of parturition. The mammalian fetus is usually successful in delaying parturition by suppressing excitatory prostaglandin release from the maternal reproductive tract until its organ systems needed for extrauterine survival are sufficiently mature. This important process requiring precisely programmed interactions between fetus and mother exemplifies many of the neuroendocrine mechanisms encountered repeatedly in the physiology of reproduction.

C.H. Sawyer

Department of Anatomy
University of California
Los Angeles, California, USA 90024

CONTENTS

SECTION I: HYPOTHALAMUS AND OTHER BRAIN AREAS

SECTION II: PITUITARY

SECTION III: GONADS

SECTION IV: FETUS AND PLACENTA

ESTROGEN ACTING ON HYPOTHALAMIC NEURONS MAY HAVE TROPHIC EFFECT ON THOSE NEURONS AND THE CELLS ON WHICH THEY SYNAPSE

Donald W. Pfaff and Rochelle S. Cohen

Neurobiology and Behavior
The Rockefeller University
1230 York Avenue
New York, New York, USA 10021

INTRODUCTION

Work on the effects of steroid hormones in classical peripheral target organs - estrogen in the uterus, progesterone in the chick oviduct, glucocorticoids in the liver and on mouse mammary tumor virus - has offered the neurobiologist analogies and hypotheses to test when investigating synthetic events related to the more complicated subject of hormone effects on nerve cells. In particular, many sex hormone actions in the reproductive tract involve growth: estrogens, androgens and progestins can have trophic consequences for the cells on which they act.

Our lab has been interested in the synthetic consequences of steroid sex hormones in nerve cells, as a consequence of two longstanding lines of research. First, steroid hormone binding in the brain (Pfaff, 1968) indicated that some of the mechanisms consequent to nuclear receptor occupation by estrogens or androgens could involve new RNA and protein synthesis. Second, having determined the circuit which activates lordosis behavior, the estrogen-progestin dependent reproductive behavior of many female quadrupeds, we saw that the interruption of hormone effects on this behavior by actinomycin D or protein synthesis inhibitors would require studies of messenger RNA and protein synthesis (Pfaff, 1980, 1983). After the two sections below summarizing the conclusions from these lines of research in their most general and abstract forms, we present new ideas regarding the trophic actions of estrogen, not only in the cells in the ventromedial hypothalamus in which they are bound, but also discuss the idea of trophic consequences for the midbrain neurons on which ventromedial hypothalamic cells synapse.

SEX HORMONE RECEPTORS IN NERVE CELLS

Using steroid hormone autoradiography, we mapped the precise locations of neurons with estrogen or androgen receptors. For these studies we used a wide range of vertebrate species covering, in our lab, all the major vertebrate classes.

Certain findings about the neuroanatomical distributions of estrogen and androgen concentrating neurons were universally true, not only across rats, but also across all the vertebrate species studied. These findings demonstrated an orderly, lawful development of such hormone concentrating cells in the vertebrate brain (Pfaff and Keiner, 1973; Morrell and Pfaff, 1978; Pfaff, 1980). First, in representative species from all major vertebrate classes, neurons specifically concentrating radioactive estrogens or androgens could be detected autoradiographically. Second, in all species, these hormone concentrating cells could be found in the medial preoptic area, in cell groups of the basomedial (tuberal) hypothalamus (in rats, these cell groups are the ventromedial nucleus, arcuate nucleus and ventral premammillary nucleus), in specific, phylogenetically ancient limbic forebrain structures (in the rat, these cell groups include the medial nucleus of the amygdala and the lateral septum), and in a specific area of the mesencephalon deep to the tectum. Thirdly, papers on the autoradiographic findings in each species (see review treatments above) include evidence from the endocrine and behavioral physiology of that species that nerve cells which concentrate estrogenic or androgenic hormones actually participate in the control of functions modulated by these hormones. These physiologic findings usually emphasize the participation of those cell groups in the regulation of gonadotropin release or mating behavior.

NEURAL CIRCUIT FOR A MAMMALIAN BEHAVIOR

Two properties of lordosis behavior were strategically important in allowing it to be analyzed, yielding the first complete neural circuit for a mammalian behavior (Pfaff, 1980). First, its sensory determinants and its motor expression are simple enough to be analyzed in a relatively straightforward way from a neuroanatomical and neurophysiological point of view. Second, its strong dependence on estrogens and progesterone can be used to experimental advantage.

Lordosis is the primary female-typical reproductive behavior of rodents and many other quadrupeds. In a variety of species it is preceded by a long chain of hormone-dependent communicative and courtship behaviors, and in other species it assumes social meanings beyond reproduction. It is a standing response, coupled with vertebral dorsiflexion which, in rodents, is extreme.

All the features of the main, complete circuit description for activating lordosis behavior (Pfaff, 1980) have been replicated and detail added (Pfaff and Schwartz-Giblin, 1986). Lordosis behavior is triggered by cutaneous input on the flanks followed by pressure on the posterior rump, tail-base and perineum. First order interneurons, deep in the dorsal horn, respond promptly and vigorously to behaviorally adequate cutaneous stimulation. However, spinal rats never do lordosis behavior, no matter how many hormone treatments or pharmacological treatments they have been given. In the obligatory supraspinal loop, ascending fibers travel in the anterolateral columns, terminating in the medullary reticular formation, the dorsal caudal part of the lateral vestibular nucleus, and the midbrain central gray. Sensory information does not have to reach the hypothalamus, on a mount-by-mount basis, for lordosis behavior to be triggered.

The main importance of the hypothalamic module for the control of this behavior is to accumulate estrogens and progesterone from the bloodstream and yield a hormone-dependent output. Neurons in the ventromedial nucleus of the hypothalamus are crucial for this behavior. Lesions of these ventromedial hypothalamic cells lead to a loss of lordosis behavior, while electrical stimulation of these neurons at low frequencies leads to lordosis facilitation. In the absence of circulating estrogens, placing an estrogenic implant next to these ventromedial hypothalamic neurons can facilitate lordosis; conversely, in the presence of circulating estrogens, placing an implant of an anti-estrogen next to these cells will decrease lordosis behavior. Axons from the ventromedial hypothalamus reach the midbrain via a medial (periventricular and a sweeping lateral route, both of which may contribute to the behavior), but quantitatively the contribution of the lateral running descending axons is more important.

Both sensory, ascending information and hypothalamic influences are received by neurons in the central gray of the midbrain and the midbrain reticular formation just lateral to the central gray, and these neurons facilitate lordosis behavior. Central gray lesions reduce lordosis, while electrical stimulation facilitates it. The time courses of behavioral actions of these midbrain neurons are markedly faster than those of medial hypothalamic neurons. An important feature of midbrain neuron action is to receive peptides synthesized in neuronal cell groups in the medial hypothalamus and the basal forebrain - for example, LHRH - and to translate this signal into altered electrical excitability. Axons descend from the central gray of the midbrain to the ventral, medial medullary reticular formation. Central gray stimulation greatly potentiates the actions of reticulospinal neurons on deep back muscles important for lordosis behavior, and also can synergize with lateral vestibulospinal actions on these muscles.

Of all the tracts descending from brainstem to spinal cord, the only ones required for lordosis behavior are the lateral vestibulospinal tract (LVST) and the medullary, lateral reticulospinal tract (RST). The physiological properties of these tracts (Pfaff and Schwartz-Giblin, 1986) fit perfectly with what is known about the motor properties of lordosis behavior.

Back at the spinal level, descending signals from the lateral vestibulospinal tract and reticulospinal tract can act through monosynaptic connections to motoneurons for deep back muscles, but it is also likely that last-order interneurons participate in these descending influences. The motoneurons for the muscles which execute lordosis behavior - lateral longissimus and transversospinalis - are on the medial side of the ventral horn at lumbar spinal cord levels (Pfaff and Schwartz-Giblin, 1986). The properties of the muscles lateral longissimus (LL) and transversospinalis, which dorsiflex the vertebral column to expose the perineal region and allow fertilization, fit perfectly with what is required for lordosis behavior.

The completion of the circuit for lordosis behavior, both in the robust body of cellular and behavioral data collected and in its orderly submission to a comprehensive and internally consistent model, prove that it is possible to achieve a detailed cellular explanation for a mammalian behavior. The cell groups and mechanisms for lordosis arrange themselves naturally into modules: a spinal cord module which receives the major impact of somatosensory input and organizes motor output; a lower brainstem module which integrates postural adaptations across spinal cord segments; a midbrain module which receives hypothalamic and preoptic peptides and proteins; and a hypothalamic module which adds the endocrine control component to this behavioral mechanism.

The neuroendocrine features of these mechanisms indicate that many steroid hormone effects on peripheral target tissues may have been conserved in steroid actions on the hypothalamus. Moreover, these hormone effects synchronize reproductive behavior with environmental conditions and gametogenic preparations for reproduction (Pfaff and Schwartz-Giblin, 1986).

RNA AND PROTEIN SYNTHESIS

Since, considering the actions of estrogens and androgens in their peripheral target tissues (McEwen et al., 1979; Pfaff, 1983), it is frequently the case that these steroids are associated with altered RNA and protein synthesis, we were stimulated to develop techniques whereby RNA levels could be measured in individual neurons (in situ hybridization), for application to rat brain tissue, and techniques for looking at proteins which are synthesized and transported under various hormonal conditions.

Likewise, since work in many laboratories has shown that estrogen and progestin effects in the hypothalamus, crucial for lordosis behavior, can be interrupted by RNA and protein synthesis inhibitors (reviewed in Pfaff, 1980), it was important to work with measurements of RNA levels in hypothalamic neurons and protein synthesis in the hypothalamus, as a function of steroid hormone treatment.

In Situ Hybridization applied to Hypothalamic Neurons

For easy quantification of RNA levels in a hypothalamic region, dot blots or slot blots are useful. For determining that a single RNA band is the source of the hybridization signal, Northern blots are useful. However, for complicated tissues with great cellular heterogeneity, cell-by-cell resolution is desired and for this feature in situ hybridization is required. We have applied in situ hybridization techniques to neural tissue, and analyzed the data in a quantitative manner (McCabe et al., 1985, 1986; Shivers et al., 1986). Optimization of in situ hybridization techniques depends on the abundance of the message studied, the size of the cell studied, the size of the radioactive DNA or riboprobe used, the amount of the enzyme RNAase, and problems with background or nonspecific binding. Methodological steps such as the type, amount and timing of fixation, the amount of deproteination, the time and temperature of the hybridization reaction and the stringency of the wash all depend upon the above-mentioned experimental requirements. For example, with the genes for oxytocin and vasopressin in supraoptic nucleus of hypothalamus, ethanol-acetic acid fixation, followed by hybridization that proceeded overnight at room temperature was adequate (McCabe et al., 1985, 1986). In situ hybridization can be combined with immunocytochemistry in the same tissue section (Shivers et al., 1986) to show that messegner RNA is being expressed in the neurons which, indeed, manufacture the appropriate protein.

Using in situ hybridization with a 59-mer that includes the coding sequence for LHRH, we have localized neurons which express the LHRH gene (Shivers et al., 1986b). Neurons with hybridizable LHRH message were discovered exactly in the locations predicted from immunocytochemistry. Thus, speculations that the immunocytochemistry had been a relatively insensitive measure of presumed LHRH neurons elsewhere in the brain with very small levels of the peptide, turned out to be incorrect - LHRH gene expressing neurons are in the medial preoptic area, amongst the fibers of the diagonal band of Broca and in the ventral septum (Shivers et al., 1986b; Rothfield et al., 1986). Background in these in situ hybridization experiments was extremely low; the cells were undoubtedly labeled. There was a very small number of cells with hybridizable LHRH message compared to what would have been predicted from immunocytochemistry (Shivers et al., 1983) - this may be due to the technical conditions of the hybridization reaction, but it may also be related to the claim that a relatively small percentage of LHRH is used even during the preovulatory period, so that any demand for new synthesis of LHRH would be correspondingly low. To a remarkable extent, LHRH gene expressing neurons appeared to be labeled to about the same degree, rather than having an even gradation of labeling from zero through heavily labeled. If this turns out not to be due to some limitation of the technical conditions, it would imply that the LHRH gene is turned on in an all-or-none, digital fashion. Finally, there was a tendency for LHRH gene expressing neurons to be found in clusters. Preliminary results indicate that seven days of estrogen treatment can stimulate LHRH gene expression in preoptic neurons, compared to seven days following ovariectomy without hormone

Table 1. LHRH Messenger RNA measured in Rat Preoptic Neurons by in situ Hybridization with Labelled 59-mer.

	No. of Labelled Cells (A)	Grains per Labelled Cell (B)	Total Content ** (A × B)	ESTROGEN MINUS CONTROL (% CHANGE)	
OVARIECTOMIZED CONTROLS (n = 6 rats)					
Probe concentration*					
low	21	26 ± 2	546		
medium	11	23 ± 2	253		
high	22	23 ± 1	506		
Total	54	24 ± 2	1296		
ESTROGEN TREATED (n = 6 rats)					
Probe concentration*					
low	22	30 ± 2	660	114	(21%)
medium	16	31 ± 3	496	243	(96%)
high	30	32 ± 2	960	454	(90%)
Total	68	31 ± 2	2108	812	(63%)

*Counts per 20 microliters: Low probe concentration 62.8 × 10^3. Medium, 129 × 10^3. High, 178 × 10^3.

**LHRH messenger RNA content measured by (Numbers of labelled cells) × (Grains per labelled cell).

(From Rothfeld et al., 1986)

Table 2. Ratio of Proenkephalin mRNA to Total Poly A^+, in Ventral Medial Hypothalamus

Ovariex Control	Estradiol-treated
.22	.85
.28	.70
.32	1.19
.32	.82
.19	.47
.30	
x .27	x .81

treatment (Table 1) (Rothfeld et al., 1986). In the male rat, as well, long-term steroid treatment may increase the levels of LHRH message: two weeks after castration there was a tendency toward a smaller number of cells detected with LHRH message, and a small number of grains per cell, compared to intact male rat controls (Rothfeld et al., 1986b).

The gene for proenkephalin is expressed in a wide variety of mammalian neurons, at different levels of the neuraxis (Harlan et al., 1985, 1986). Of great interest is the fact that enkephalin producing neurons can be found in the ventrolateral portion of the ventromedial nucleus of the hypothalamus, a strong estrogen-binding cell group. Quantitative slot blot analysis indicates an estrogen induction of the preproenkephalin gene by a factor of about 3.0 in the ventromedial nucleus of the hypothalamus (Table 2) (Romano et al., 1986).

In situ hybridization applied to nerve cells will become increasingly useful as a measure of gene expression under a variety of endocrine and environmental conditions, especially since the numbers of opportunities to apply this technique will go up rapidly as the number of fully analyzed gene sequences increases.

Hypothalamic Protein Synthesis and Transport

Proteins synthesized in hypothalamic neurons which might be relevant for a particular behavior should be altered under the functional conditions which similarly alter the behavior, and they probably should be transported to cell groups involved in that behavior. For the hormonal control of lordosis behavior, for instance, proteins synthesized in the ventromedial hypothalamus would be expected to be influenced by estrogen and progesterone, and transported to the midbrain central gray. Using HPLC, we studied hypothalamic proteins synthesized following the delivery of cocktails of five tritiated amino acids to the ventral medial hypothalamus, with or without estrogen treatment in ovariectomized female rats (Pfaff et al., 1984). Most interesting was a class of proteins transported to the midbrain from ventral medial hypothalamus - these were synthesized more rapidly following estrogen treatment, and were consistent with a molecular weight range just above 68,000 (Pfaff et al., 1984).

Much greater resolution of newly synthesized hypothalamic proteins can be gained from 2-dimensional gel analysis (Mobbs et al., 1985) with local microinjection of S35 labeled amino acids in the ventral medial hypothalamus, Mobbs discovered a protein of about 70,000 molecular weight with a PI of about 6.5 which is induced with estrogen treatment of ovariectomized female rats and is transported to midbrain (Mobbs et al., 1986). The identity and functions of this protein are under investigation.

Clearly it is not the only protein in medial hypothalamus whose synthesis could be affected by estrogen treatment - subsequent work both with in vivo and in vitro labeling protocols may be expected to reveal the full range of proteins which are hormonally sensitive. Only then will we be able to see what kinds of altered chemical signals could arrive in midbrain and other terminal zones important for the hypothalamic controls over specific behaviors. We expect that the electrical effects of steroid hormones in hypothalamus (Pfaff, 1983) have, as part of their function, the coupled release of peptides in midbrain and other target cell groups and that, together with electrical signalling, these peptides help to direct the extra-hypothalamic circuitry involved in the execution of specific behaviors.

POSSIBILITY OF TROPHIC EFFECTS

Trophic Actions on the Estrogen-binding Cell

Prominent trophic actions of steroids in other tissues, such as the growth effects in the uterus under the influence of estrogen, make it necessary to consider possible growth responses of estrogen binding nerve cells following hormone treatment. The similarity of the gene for the estrogen receptor to an oncogene (Greene et al., 1986) is a tangential reason for pursuing the same line of thought, made slightly more compelling by demonstrations that oncogenes can be expressed in normal nerve cells (Swanson et al., 1986; Furth et al., 1986). Indeed, nerve cells with the constant requirement of sending protoplasm down the axon may be manifesting a constant requirement for a growth-like response, despite the fact that in the normal case in the adult, they do not divide. Wright (1983) has shown that estrogen prevents normally programmed cell death in the cervical sympathetic ganglion. Following hypothalamic transections that allow room for axonal proliferation in the arcuate nucleus, Matsumoto and Arai (1981) found that estrogen would signficiantly increase the rate of synaptogenesis. Indeed, estrogen increases the rate of neurite proliferation (Toran-Allerand, 1980). In this regard there is the possibility that an estrogen-induced trophic factor(s) alters pre-existing synapses. We are presently examining the presynaptic terminals of ventromedial hypothalamic neurons and postsynaptic processes with which they contact in the midbrain central gray of estrogen-treated animals for ultrastructural signs indicative of changes in the circuit at the synaptic level that may, for example, reinforce it. Interestingly, other studies indicate that the induction of reproductive behavior is facilitated by repeated hormone treatments (Beach et al., 1974; Gerall and Dunlap, 1973; Parsons et al., 1979; Whalen and Nakayama, 1965). The long-term potentiation of lordosis by estradiol is not due to a change in estrogen or progestin receptor levels in the hypothalamus, preoptic area and septum (Parsons et al., 1979). Precedents for hormone related changes at the synaptic level have been seen in rat occipital cortical synapses after ovariectomy (Medosch and Diamond, 1982).

Ultrastructural results in hypothalamic nerve cells support the notion that estrogens can be followed by a growth-like response (Cohen and Pfaff, 1981). Estrogen treatment was followed by a massive increase in the amount of rough endoplasmic reticulum in certain ventromedial hypothalamic nerve cells, correlated with a stimulation of the number of dense-cored vesicles (Cohen and Pfaff, 1981). In turn, these changes in the cytoplasm were correlated with an increase in electron dense material on the surface of the nucleolus (Cohen et al., 1984), which, upon enzyme digestion, proved to be DNA (Chung et al., 1984). These findings were replicated and extended with much briefer durations of estrogen treatment (Jones et al., 1985), and since some of these effects are coupled to increases in ventromedial hypothalamic neuron cell size, one is led to think of a trophic reaction to the estrogen. The ultrastructural results led to the prediction that the amount of hybridizable ribosomal RNA would be increased by estrogen treatment. Indeed, after 6 h or 24 h of estrogen, using in situ hybridization, the amount of hybridizable ribosomal RNA was seen to be significantly elevated (Jones et al., 1986). In hypothalamic cells without significant amounts of estrogen receptor, there was no estrogen effect. After 15 days of estrogen treatment, even in the ventrolateral portion of the ventromedial nucleus of the hypothalamus, there was not a significant estrogen effect on the amount of hybridizable ribosomal RNA, even though with this duration of estrogen treatment there is an increased amount of stacked rough endoplasmic reticulum. We do not know yet whether an early wave of synthesis of ribosomal RNA led to the formation of stacked rough endoplasmic reticulum (without the absolute increase in amount after 15 days), or whether some of the ribosomal RNA in stacked rough endoplasmic reticulum form is inaccessible to the radioactive DNA probe. Further experiments with different technical conditions of hybridization should answer this question. We view the proliferation of rough endoplasmic reticulum, the nucleolar changes, the nuclear changes after short durations of estrogen treatment, and the in situ hybridization results as a part of a massive response by the hypothalamic neuron to estrogen, getting ready for new protein synthesis.

Trophic Effects of Hypothalamic Neurons on Postsynaptic Elements in the Midbrain

It is reasonable to consider possible trophic effects of peptide secreting hypothalamic and preoptic neurons on nerve cells elsewhere in the nervous system, since hypothalamically produced proteins and peptides such as prolactin and TRH are already known to have trophic effects. During the course of investigations of normal synaptic morphology in the midbrain (Chung et al. 1984b, 1986), we made electrolytic lesions of the ventromedial hypothalamus and were surprised to see not only presynaptic degeneration in the midbrain central gray, but also an obvious degeneration of postsynaptic elements. Having followed the time course of degeneration following electrolytic lesions, we considered the possibility that this is not actually transynaptic degeneration following loss of hypothalamic nerve cell bodies: it might be due to destruction of fibers passing through the hypothalamus, or it might be retrograde degeneration in midbrain central gray neurons that project to hypothalamus. To answer these questions we used excitotoxin chemical lesions of hypothalamus using kainic acid or N-methyl aspartic acid treatment. These would neither disrupt fibers passing through, nor would they induce retrograde degeneration. These chemical lesions, also, were followed by degeneration of postsynaptic elements in the midbrain central gray (Chung et al., 1986b). Transynaptic degeneration in the midbrain central gray indicates a trophic action of hypothalamic nerve cell groups which include estrogen binding neurons on the postsynaptic elements in terminal fields receiving hypothalamic axons.

One hypothesis for the basis of a trophic action is that hypothalamic axons so dominate the dendritic surface of certain midbrain cells (analogous to visual input to the lateral geniculate nucleus) that removal of the input causes the postsynaptic element to die. Two items of data in our current experiments argue against this possibility. First, we have observed degenerating postsynaptic elements which receive presynaptic terminals which are still normal. Secondly, the quantitative aspects of post- and pre-synaptic degeneration in the midbrain indicate that the effects are even more severe postsynaptically than they are presynaptically. A second hypothesis is that hypothalamic neurons are manufacturing, transporting and secreting trophic factors. Finally, it could be that removing the hypothalamic target of certain midbrain neurons has a reverse trophic effect on midbrain cells. Among these three hypotheses, the second, postulating a hypothalamic trophic factor, is testable. Since LHRH can have trophic actions in the pituitary, since TRH has widespread trophic actions, since prolactin is a well-known trophic factor and can be manufactured by hypothalamic nerve cells (Harlan et al., 1983, 1987), and since ACTH affects neuronal maturation (Strand et al., 1986), we have several compounds, and combinations of chemicals, to try in cell survival and cell growth assays. Estrogen binding cells - especially those which produce prolactin (Shivers et al., 1987) or which are POMC neurons that produce ACTH (Morrell et al., 1985) - could form part of the basis for trophic actions on hypothalamic neurons in extra-hypothalamic terminal fields.

REFERENCES

Beach, F. A., and Orndoff, R. K., 1974, Variation in the responsiveness of female rats to ovarian hormones as a function of preceding hormonal deprivation, Horm. Behav., 5:202.

Chung, S. K., Cohen, R. S., and Pfaff, D. W., 1984, Ultrastructure and enzyme digestion of nucleoli and associated structures in hypothalamic nerve cells viewed in resinless sections, Biologie Cellulaire, 51:23.

Chung, S. K., Pfaff, D. W., and Cohen, R. S., 1984b, Projections of ventromedial hypothalamic neurons to the midbrain central gray: an ultrastructural study, Soc. for Neurosci. Abstr. 10:211 (Abstr. No. 62.11).

Chung, S. K., Cohen, R. S., and Pfaff, D. W., 1986, Trans-synaptic degeneration in midbrain following hypothalamic lesion, Neuroscience, submitted.

Chung, S. K., Cohen, R. S., and Pfaff, D. W., 1986b, Effects of neurochemical hypothalamic lesion on midbrain neurons, Neuroscience, submitted.

Cohen, R. S., and Pfaff, D. W., 1981, Ultrastructure of neurons in the ventromedial nucleus of the hypothalamus of ovariectomized rats with or without estrogen treatment, Cell and Tiss. Res., 217:451.

Cohen, R. S., Chung, S. K., and Pfaff, D. W., 1984, Alteration by estrogen of the nucleoli in nerve cells of the rat hypothalamus, Cell and Tiss. Res., 235:485.

Gerall, A. A., and Dunlap, J. L., 1973, The effect of experience and hormones on the initial receptivity in female and male rats, Physiol. Behav., 10:851.

Greene, G. L., Gilna, P., Waterfield, M., Baker, A., Hort, Y., and Shine, J., 1986, Sequence and expression of human estrogen receptor complementary DNA, Science, 231:1150.

Harlan, R., Shivers, B., and Pfaff, D. W., 1983, Midbrain microinfusions of prolactin increase the estrogen-dependent behavior, lordosis, Science, 219:1451.

Harlan, R. E., Shivers, B. D., Romano, G. J., Howells, R. D., and Pfaff, D. W., 1985, Localization of cells containing preproenkephalin mRNA in the rat forebrain by in situ hybridization, Soc. for Neurosci. Abstr., 11:143 (Abstr. No. 46.10).

Harlan, R. E., Shivers, B. D., Romano, G. J., Howells, R. D., and Pfaff, D. W., 1986, Localization of cells containing preproenkephalin mRNA in the rat forebrain by in situ hybridization, J. Comp. Neurol., submitted.

Harlan, R. E., Shivers, B. D., and Pfaff, D. W., 1987, Immunocytochemical mapping of immunoreactive prolactin in female rat brain, Neuroendocrinology, submitted.

Jones, K. J., Pfaff, D. W., and McEwen, B. S., 1985, Early estrogen-induced nuclear changes in rat hypothalamic ventromedial neurons: an ultrastructural and morphometric analysis, J. Comp. Neurol., 239:255.

Jones, K. J., Chikaraishi, D. M., Harrington, C. A., McEwen, B. S., and Pfaff, D. W., 1986, Estradiol (E_2)-induced changes in rRNA levels in rat hypothalamic neurons detected by in situ hybridization, Mol. Brain Res., in press.
Matsumoto, A., and Arai, Y., 1981, Neuronal plasticity in the deafferented hypothalamic arcuate nucleus of adult female rats and its enhancement by treatment with estrogen, J. Comp. Neurol., 197:197.
McCabe, J. T., Morrell, J. I., Richter, D., and Pfaff, D. W., 1985, Localization of neuroendocrinologically relevant RNA in brain by in situ hybridization, in: "Frontiers in Neuroendocrinology," W. F. Ganong and L. Martini, eds., Vol. 9, pp. 149-167, Raven Press, New York.
McCabe, J. T., Morrell, J. I., Ivell, R., Schmale, H., Richter, D., and Pfaff, D. W., 1986, In situ hybridization technique to localize rRNA and mRNA in mammalian neurons, J. Histochem. Cytochem., 34:45.
McEwen, B. S., Davis, P. G., Parsons, B., and Pfaff, D. W., 1979, The brain as a target for steroid hormone action, in: "Annual Review of Neuro-Science," W. M. Cowan, Z. W. Hall and E. R. Kandel, eds., Vol. 2, pp. 65-112, Pao Alto.
Medosch, C. M., and Diamond, M. C., 1982, Rat occipital cortical synapses after ovariectomy, Exp. Neurol., 75:120.
Mobbs, C. V., Harlan, R. E., and Pfaff, D. W., 1985, An estradiol-induced protein synthesized in the ventral medial hypothalamus (VMN) and transported to the midbrain central gray (MCG), Soc. for Neurosci. Abstr., 11:1271 (Abstr. No. 372.10).
Mobbs, C. V., Harlan, R. E., and Pfaff, D. W., 1986, An estradiol-induced protein in the hypothalamus, J. Neurosci., submitted.
Morrell, J. I., and Pfaff, D. W., 1978, A neuroendocrine approach to brain function: localization of sex steroid concentrating cells in vertebrate brains, Amer. Zool., 18:447.
Morrell, J. I., McGinty, J. F., and Pfaff, D. W., 1985, A subset of β-endorphin- or dynorphin-containing neurons in the medial basal hypothalamus accumulates estradiol, Neuroendocrinology, 41:417.
Parsons, B., McLusky, N. J., Krieger, M. S., McEwen, B. S., and Pfaff, D. W., 1979, The effects of long-term exposure on the induction of sexual behavior and measurements of brain estrogen and progestin receptors in the female rat, Horm. Behav., 13:301.
Pfaff, D. W., 1968, Uptake of estradiol-17β-H^3 in the female rat brain: an autoradiographic study, Endocrinology, 82:1149.
Pfaff, D. W., 1980, Estrogens and Brain Function: Neural Analysis of a Hormone-controlled Mammalian Reproductive Behavior, Springer-Verlag, New York.
Pfaff, D. W., 1983, Impact of estrogens on hypothalamic nerve cells: ultrastructural, chemical, and electrical effects, Recent Prog. Horm. Res., 39:127.
Pfaff, D. W., and Keiner, M., 1973, Atlas of estradiol-concentrating cells in the central nervous system of the female rat, J. Comp. Neurol., 151:121.
Pfaff, D. W., Rosello, L., and Blackburn, P., 1984, Proteins synthesized in medial hypothalamus and transported to midbrain in estrogen-treated female rats, Exp. Brain Res., 57:204.
Pfaff, D. W., and Schwartz-Giblin, S., 1986, Physiological mechanisms of female reproductive behavior, in: "Textbook of Physiology," E. Knobil and J. Neill, eds., Raven Press, New York, in preparation.
Romano, G. J., Harlan, R. E., Shivers, B. D., Howells, R. D., and Pfaff, D. W., 1986, Estrogen increases proenkephalin mRNA levels in the mediobasal hypothalamus of the rat, Soc. for Neurosci. Abstr., 12:692 (Abstr. No. 188.17).
Rothfeld, J., Hejtmancik, J. F., and Pfaff, D. W., 1985, Effects of estrogen on LHRH gene expression, Nature, submitted.

Rothfeld, J. M., Shivers, B., Hejtmancik, J. F., Conn, P. M., and Pfaff, D. W., 1986b, Quantitation of LHRH mRNA in neurons in the intact and castrate male rat forebrain, Soc. for Neurosci. Abstr., 12:3 (Abstr. No. 4.8).

Shivers, B. D., Harlan, R., Morrell, J. I., and Pfaff, D. W., 1983, Immunocytochemical localization of luteinizing hormone-releasing hormone in male and female rat brains, Neuroendocrinology, 36:1.

Shivers, B. D., Harlan, R. E., Pfaff, D. W., and Schachter, B. S., 1986, Combination of immunocytochemistry and in situ hybridization in the same tissue section of rat pituitary, J. Histochem. Cytochem., 34:39.

Shivers, B. D., Harlan, R. E., Hejtmancik, J. F., Conn, P. M., and Pfaff, D. W., 1986b, Localization of cells containing LHRH-mRNA in rat forebrain using in situ hybridization, Endocrinology, 118:883.

Shivers, B. D., Harlan, R. E., and Pfaff, D. W., 1987, Hypothalamic immunoreactive prolactin neurons are targets for estrogenic action, Neuroendocrinology, submitted.

Strand, F., Frischer, R., King, J., and Rose, K., 1986, Neuropeptides integrate motor function and structure during development, in: "Neuropeptides and Brain Function," E. P. deKloet et al., eds., Elsevier, Amsterdam, in press.

Swanson, M. E., Elste, A. M., and Greenberg, S. M., and Schwartz, J. H., 1986, Abundant expression of RAS proteins in Aplysia neurons, J. Cell Biology, submitted.

Toran-Allerand, C. D., 1980, Sex steroids and the development of the newborn mouse hypothalamus and preoptic area in vitro. II. Morphological correlates and hormonal specificity, Brain Res., 189:413.

Whalen, R. E., and Nakayama, K., 1965, Induction of estrus behavior: facilitation by repeated hormone treatment, J. Endocrinology, 33:525.

Wright, L. L., and Smolen, A. J., 1983, Effects of 17β-estradiol on developing superior cervical ganglion neurons and synapses, Develop. Brain Res., 6:299.

GONADAL STEROID CONTROL OF SYNAPTOGENESIS

IN THE NEUROENDOCRINE BRAIN

Yasumasa Arai, Akira Matsumoto
and Masako Nishizuka

Department of Anatomy
Juntendo University School of Medicine
Tokyo, Japan 113

INTRODUCTION

Sex steroid hormones exert very complex effects on the brain. The neural circuitries which participate in neuroendocrine control of reproductive functions (especially gonadotropin secretion and sexual behaviors) are made under the direct control of gonadal steroids throughout life. The best documented of these is the long-loop feedback action of gonadal hormones on the neuroendocrine brain. In perinatal animals, sex steroids affect brain sexual differentiation to produce major sex differences in neuroendocrine and behavioral functions (Goy and McEwen, 1980; MacLusky and Naftolin, 1981). Recent studies indicate that aromatizable androgen or estrogen act on the developing brain tissues to promote neuronal growth and neural circuit formation (Toran-Allerand, 1976; Toran-Allerand et al., 1983). Synaptogenesis can be facilitated by estrogen (Matsumoto and Arai, 1976; Arai and Matsumoto, 1978; Nishizuka and Arai, 1981a). These organizational effects of gonadal steroids appear to be regionally specific and correlated with the presence and topographical localization of the sex steroid-receptor containing neurons.

In the present paper, attention is focused on synaptogenic action of sex steroids on the developing brain to study the steroidal influence on the circuit formation. In addition, synaptic plasticity to sex steroids is also discussed in the adult brain.

GONADAL STEROIDS AND SEXUALLY DIMORPHIC NEUROENDOCRINE BRAIN

Male-female structural differences have been demonstrated in certain brain regions. Most of these differences can be modified by manipulation of perinatal hormonal environment. Gorski et al. (1978) have found a marked sex difference in the medial preoptic area (POA) of the rat. The volume of an intensely staining neuron group of the POA (the sexually dimorphic nucleus of the POA) is markedly greater in the male than in the female. Its volume in the female rat is increased by perinatal treatment with androgen (Döhler et al., 1982). Sexual dimorphism in nuclear volume has also been found in the ventromedial nucleus (VMN) (Matsumoto and Arai, 1983) and the medial amygdaloid nucleus (MAN) in the rat (Mizukami et al., 1983). The volume of the

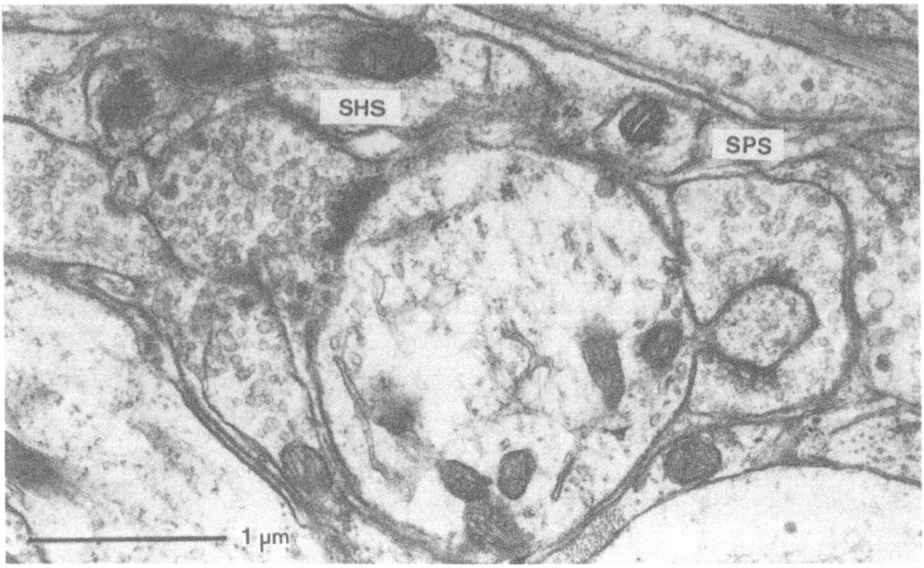

Fig. 1. Shaft synapse contacting on a dendritic shaft (SHS) and spine synapse on a dendritic spine (SPS) in the neurophil of the VMN.

VMN, for example, is greater in males than in females. Neonatal castration of males which produces functionally "feminine males" reduces the volume of the VMN to the level comparable to that of females.

At an electron microscopic level, sex difference in synaptic organization has been found in several regions. According to the site of the synaptic contact, three types of synapses are roughly classified: shaft synapses made on the dendritic shaft (Fig. 1), spine synapses on the dendritic spine (Fig. 1), and somatic synapses on the cell body. In the arcuate nucleus (ARCN), the number of spine synapses is approximately twice in females than in males, whereas somatic synapses in females are twice as many as in males (Matsumoto and Arai, 1980). Interestingly, neonatal castration of males increases the number of spine synapses to the female level. On the other hand, the female synaptic pattern can be reversed by neonatal injection of testosterone propionate (TP). In the dorsomedial POA, the situation is similar to that in the ARCN. No difference in the number of shaft synapses is found between the two sexes, whereas the number of spine synapses of nonamygdaloid origin is greater in female rats than in males (Raisman and Field, 1973). Neonatal castration of males also causes an increase in the number of spine synapses to almost the same level as in females. In the suprachiasmatic nucleus, however, the incidence of spine synapses is higher in males than in females (Guldner, 1982; Le Blond et al., 1982). A characteristic feature of the sexually dimorphic pattern of the MAN is dependent on a difference in the number of shaft synapses (Nishizuka and Arai, 1981b). These findings suggest that synaptic organization may vary according to the genomic responses of the individual nuclei to organizational action of sex steroids.

In the VMN, there is a regional difference in distribution of sex steroid receptors (Pfaff and Keiner, 1973). This may be reflected in a regional difference in synaptic pattern of the male VMN. As shown in Fig. 2,

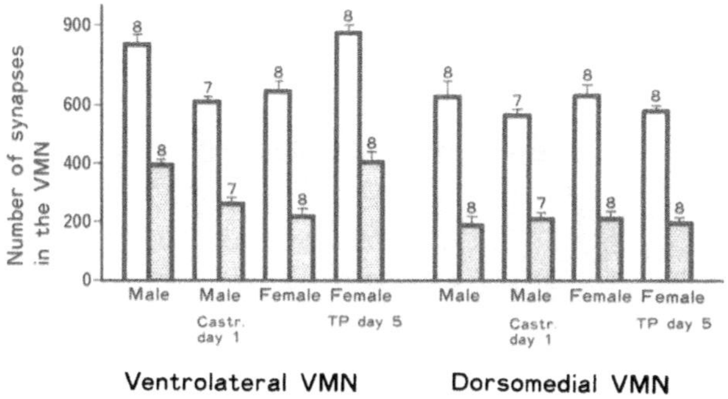

Fig. 2. The number of shaft and spine synapses in the VMN of normal, and neonatally androgenized or castrated rats. Open bars indicate the number of shaft synapses and dotted ones, the number of spine synapses. Vertical bars indicate SEM. Numbers above the vertical bars refer to the number of rats examined.

the number of shaft and spine synapses in the ventrolateral part of the VMN (VL-VMN) where sex steroid receptors are abundant is significantly greater than that in the dorsomedial part of the VMN where sex steroid receptors are almost absent. The synaptic organization of the male DM-VMN is almost identical to that of female VL-VMN. This implies that the synaptic number of the VL-VMN is greater in males than in females, suggesting the presence of sexually dimorphic synaptic organization in the VL-VMN (Matsumoto and Arai, 1986). Neonatal treatment of females with TP significantly increases the shaft and spine synapses in the VL-VMN to a level comparable to that of normal males, but not in the DM-VMN. Neonatal castration of males significantly reduces the number of shaft and spine synapses in the VL-DMN to levels comparable to those of normal females. These findings confirm the importance of the neonatal sex steroid environment for the development of the sexually dimorphic pattern of the VMN. Since the neuropil of the female VL-VMN is not exposed to organizational action of androgen during neonatal days, it develops almost the same synaptic pattern as the DM-VMN which lacks in sex steroid receptors. A high correlation between the sexually dimorphic synaptic pattern and the presence of neurons containing sex steroid receptors indicates that the occurrence of synaptic sexual differentiation is rather specific to the sex steroid sensitive neuronal system. Other regions which show a sexually dimorphic synaptic pattern are all (except the suprachiasmatic nucleus) known to be abundant in sex steroid receptors.

GONADAL STEROIDS AND SYNAPTOGENESIS IN THE DEVELOPING NEUROENDOCRINE BRAIN

The synaptic population in the hypothalamic and limbic structures is quite small at neonatal age and then increases progressively during the course of postnatal development. In the ARCN, synaptic density remains less

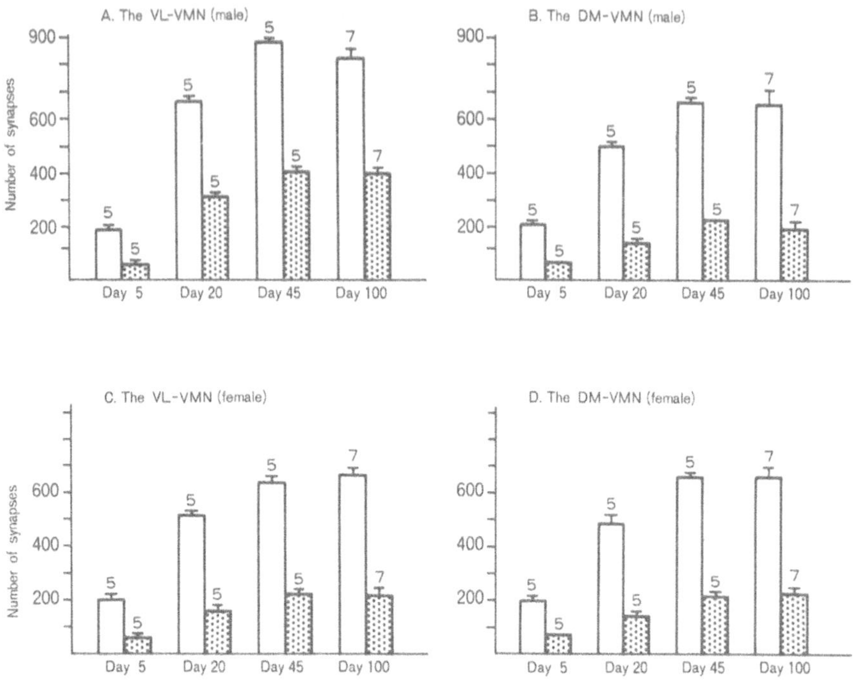

Fig. 3. The number of shaft (open bars) and spine (dotted bars) synapses in the VL-VMN and DM-VMN during postnatal development.

than 50% of the adult level at 21 days of age, and reaches a plateau around the onset of puberty (Matsumoto and Arai, 1976b; 1981a). The synaptic number of the VMN increases considerably within the first 20 days (Fig. 3). At 20 days of age (Day 20), the numerical density of shaft and spine synapses in the DM-VMN and VL-VMN is approximately 70 - 80% of that at 45 days of age (Day 45) when it reaches the maximal level. No further increase in the number of synapses occurs at 100 days of age (Day 100). This tendency has also been detected in the POA (Reier et al., 1977; Lawrence and Raisman, 1981). In the MAN, synaptogenesis is almost completed before 21 days of age (Nishizuka and Arai, 1981a). This indicates that the maturation rate of neural circuit formation is different among these brain regions.

The neuropil matrix of these areas in the hypothalamic and limbic brain is in an immature state at neonatal age and major neural circuit networks for operating postpubertal neuroendocrine and behavioral regulations are not yet established at this period. The sexually undifferentiated neuropil of the neural substrates which contain abundant steroid receptors could be subjected to organizational action of sex steroids. As a matter of fact, estrogen markedly stimulates axonal and dendritic differentiation and synapse formation during postnatal development (Arai, 1981; Toran-Allerand, 1984). In the MAN, for example, estrogen specifically promotes the formation of shaft synapses (Nishizuka and Arai, 1981a, b). This provides clear evidence for a mechanism underlying synaptic sexual differentiation of this nucleus, because the sex difference can be attributed to a significant increase in the number of shaft synapses in response to organizational action of androgen in males. In addition, shaft synaptogenesis in newborn medial amygdaloid tissue grafted into the anterior eye chamber of adult ovariectomized rats is also markedly

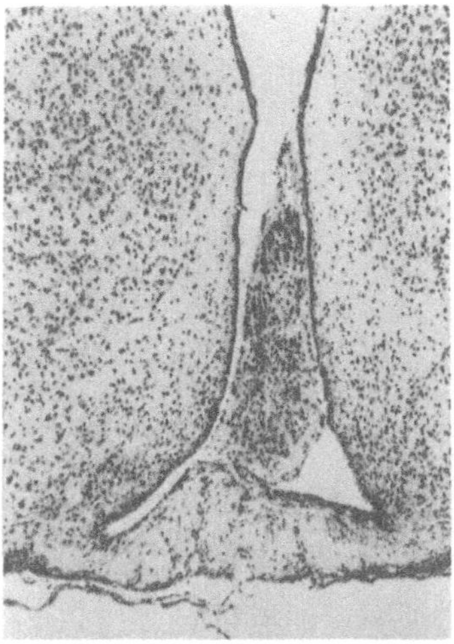

Fig. 4. POA graft in the third ventricle of an estrogen-treated female rat X 60.

Fig. 5. Effect of estrogen on volume of the POA or cortical grafts transplanted into the third ventricle of ovariectomized adult females. Black bars indicate the volume of control grafts and open bars the volume of grafts after estrogen treatment.

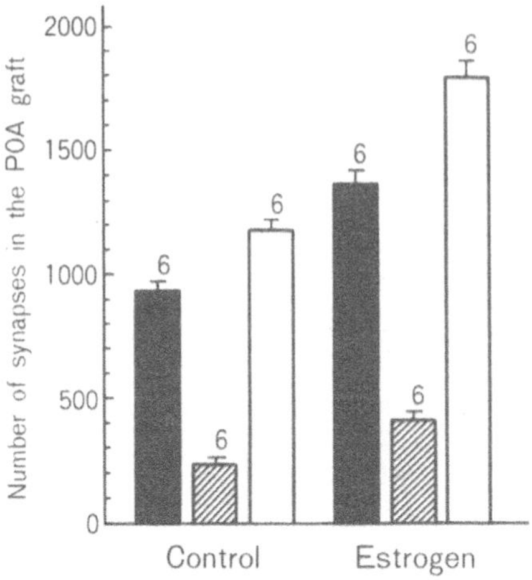

Fig. 6. Number of shaft, spine and total synapses per 10,000 μm² of the POA grafts in the control and estrogen-treated females. Black bars indicate the number of shaft synapses, hatched ones indicate that of spine synapses and open ones indicate that of total synapses.

facilitated by estrogen given via the host (Fig. 4) (Nishizuka and Arai, 1982). This may suggest that estrogen directly affects graft tissues, presumably the postsynaptic elements to form shaft synapses.

Other than the intraocular grafting, estrogen exposure of the intraventricular POA grafts through the host ovariectomized adult females induces a significant increase in the graft volume, compared to the control grafts which are not estrogenized (Fig. 5). In this case, the medial POA or parietal cortical tissues of newborn female rats was transplanted into the floor of the third ventricle. In order to study the effects of estrogen on the graft tissue, estradiol-17β in silastic tubing was placed subcutaneously in the host animals for four weeks. As for the cortical grafts, there is no significant difference in volume between estrogenized and nonestrogenized grafts. Since receptors for sex steroids are abundant in the POA but almost absent in cortical tissues, the stimulatory effects of estrogen on the growth of the grafts seems to be rather specific to estrogen-receptor containing neuronal system. The neuronal population in the POA grafts exposed to estrogen is not significantly different from that of the control grafts which are not exposed to estrogen. For the increase of the graft volume, the facilitation of growth and differentiation of neuronal components in the neuropil of the POA grafts by estrogen may be responsible. These findings are in good agreement with the finding that neuritic and dendritic growth in the explants of newborn mouse POA tissues is markedly enhanced by estrogen in culture (Toran-Allerand, 1976; Toran-Allerand et al., 1983). Furthermore, the number of shaft and spine synapses of the POA grafts exposed to estrogen is significantly greater than that of the grafts without estrogen treatment (Fig. 6). Again, this suggests that synaptogenesis as well as neuritic and dendritic growth can be stimulated by estrogen even in such ectopic sites.

SYNAPTIC PLASTICITY TO ESTROGEN IN THE ADULT BRAIN

Toran-Allerand (1984) has postulated that gonadal steroids exert organizational effects on the developing neuroendocrine brain in a cascading manner. Stimulation of axonal and dendritic growth by sex steroids could lead to sex differences in neuron size, number and synaptic pattern. Compared to steroid-insensitive neurons, neurons stimulated by sex steroids might make more effective synaptic contacts. However, the mechanisms of estrogen-stimulated synaptogenesis is still not fully understood. It has been reported that α-bungarotoxin binding capacity in the medial amygdala is sexually dimorphic and permanently modified by neonatal sex steroid manipulation (Arimatsu et al., 1981). An addition of estrogen to the culture medium of the amygdala cell aggregates causes a significant increase in toxin binding capacity (Arimatsu et al., 1985). Furthermore, there is a delay between the time of increasing toxin binding sites and the appearance of synaptic structures in vitro (Arimatsu et al., 1985). Neonatal activation of neurotransmitter receptors by estrogen might be correlated with the onset of estrogen-stimulated synaptogenesis.

As well as its organizational effect in the neonate brain, estrogen has a facilitatory effect on synapse formation in peripubertal and adult rats. When a single dose of pregnant mare's serum gonadotropin (PMSG, 20 IU) is given to 28-day-old female rats, precocious ovulation occurs three days later later. Concurrently, the synaptic number in the ARCN increases and reaches an adult level within three days (Matsumoto and Arai, 1977). Since PMSG fails to increase the number of synapses in the ARCN in ovariectomized rats, the ovarian estrogen would appear to play a facilitatory role in the maturation of synaptic organization in the ARCN in these animals. Similar precocious synaptogenesis in the ARCN has been reported in immature rats in which precocious ovulation can be induced by the administration of estrogen (Clough and Rodriguez-Sierra, 1983) or by the placement of an electrolytic lesion in the basal hypothalamus (Ruf, 1982).

In adult animals, sex steroids can affect synaptic remodelling in certain neuronal substrates which possess considerable plasticity and where vacated synaptic sites are made available by spontaneous degeneration or surgical deafferentation. Three weeks after complete deafferentation of the medial basal hypothalamus (MBH-island), the number of shaft and spine synapses in the ARCN decreases to two-fifths and three-fifths of the intact controls, respectively, because of transection of the inputs to the MBH. However, estrogen treatment for three weeks produces a marked increase in the number of both shaft and spine synapses in ovariectomized MBH-island females, with the number of shaft synapses being restored to almost 75% of the intact level; the number of spine synapses is significantly greater than in the intact control animals (Matsumoto and Arai, 1981b). These results suggest that estrogen enhances neuronal plasticity of adult ARCN neurons and stimulates axonal sprouting and dendritic spine formation in the intact ARCN neurons in the MBH-island. The stimulation of synaptogenesis by estrogen in adult brain may be dependent on the potentials for axonal or synaptic reformation in the neuropil environment. Since estrogen fails to increase the synaptic number in the ARCN in ovariectomized females without MBH-deafferentation, availability of vacated synaptic sites following the deafferentation or during the process of spontaneous remodelling may enable estrogen to stimulate synaptogenesis in the neuropil environment. Whether the responses of the neuropil matrix to estrogen is activational in nature or one of the organizational events continuing into adult life remains to be investigated in the near future.

ACKNOWLEDGEMENTS

The work from the authors' laboratory was supported by research grants from the Japanese Ministry of Education, Culture and Science; the Ministry of Health and Welfare; and the Sasagawa Foundation for Medical Research.

REFERENCES

Arai, Y., 1981, Synaptic correlates of sexual differentiation, Trend. Neurosci., 4:31.

Arai, Y., and Matsumoto, A., 1978, Synapse formation of the hypothalamic arcuate nucleus during postnatal development in the female rat and its modification by neonatal estrogen treatment, Psychoneuroendocrinology, 3:31.

Arimatsu, Y., Seto, A., and Amano, T., 1981, Sexual dimorphism in α-bungarotoxin binding capacity in the mouse amygdala, Brain Res., 213:432.

Arimatsu, Y., Kondo. S., and Kojima, M., 1985, Enhancement by estrogen treatment of α-bungarotoxin binding in fetal mouse amygdala cell reaggregated in vitro, Neurosci. Res., 2:211.

Clough, R. W., and Rodriguez-Sierra, J. F., 1983, Synaptic changes in the hypothalamus of the prepubertal female rat administered estrogen, Am. J. Anat., 167:205.

Döhler, K. D., Coquelin, A., Davis, F., Hines, M., Shryne, J. E., Gorski, R. A., 1982, Differentiation of the sexually dimorphic nucleus in the preoptic area of the rat brain is determined by the perinatal hormone environment, Neurosci. Lett., 33:295.

Gorski, R. A., Gordon, J. H., Shryne, J. E., and Southam, A. M., 1978, Evidence for a morphological sex difference within the medial preoptic area of the rat brain, Brain Res., 148:333.

Goy, R. W., and McEwen, B. S., 1980, "Sexual Differentiation of the Brain," MIT Press, Cambridge.

Guldner, F. H., 1982, Sexual dimorphism of axo-spine synapses and postsynaptic density material in the suprachiasmatic nucleus of the rat, Neurosci. Lett., 28:145.

Lawrence, J. M., and Raisman, G., 1981, Ontogeny of synapses in a sexually dimorphic part of the preoptic area in the rat, Brain Res., 183:466.

Le Blond, C. B., Morris, S., Karaliulakis, G., Powell, R., and Thomas, P. J., 1982, Development of sexual dimorphism in the suprachiasmatic nucleus of the rat, J. Endocrinol., 95:137.

Matsumoto, A., and Arai, Y., 1976a, Effect of estrogen on early postnatal development of synaptic formation in the hypothalamic arcuate nucleus of female rats, Neurosci. Lett., 2:275.

Matsumoto, A., and Arai, Y., 1976b, Developmental changes in synaptic formation in the hypothalamic arcuate nucleus of female rats, Cell Tiss. Res., 169:143.

Matsumoto, A., and Arai, Y., 1977, Precocious puberty and synaptogenesis in the hypothalamic arcuate nucleus in pregnant mare serum gonadotropin (PMSG) treated immature female rats, Brain Res., 129:275.

Matsumoto, A., and Arai, Y., 1980, Sexual dimorphism in "wiring pattern" in the hypothalamic arcuate nucleus and its modification by neonatal hormone environment, Brain Res., 190:238.

Matsumoto, A., and Arai, Y., 1981a, Effect of androgen on sexual differentiation of synaptic organization in the hypothalamic arcuate nucleus: an ontogenetic study, Neuroendocrinology, 33:166.

Matsumoto, A., and Arai, Y., 1981b, Neuronal plasticity in the deafferented hypothalamic arcuate nucleus of adult female rats and its enhancement by treatment with estrogen, J. Comp. Neurol., 197:197.

Matsumoto, A., and Arai, Y., 1983, Sex difference in volume of the ventromedial nucleus of the hypothalamus in the rat, Endocrinol. Japon, 30:277.

Matsumoto, A., and Arai, Y., 1986, Male-female difference in synaptic organization of the ventromedial nucleus of the hypothalamus in the rat, Neuroendocrinology, 42:232.
MacLusky, N. J., and Naftolin, F., 1981, Sexual differentiation of the central nervous system, Science, 211:1294.
Mizukami, S., Nishizuka, M., and Arai, Y., 1983, Sexual difference in nuclear volume and its ontogeny in the rat amygdala, Exp. Neurol., 79:569.
Nishizuka, M., and Arai, Y., 1981a, Organizational action of estrogen on synaptic pattern in the amygdala: implications for sexual differentiation of the brain, Brain Res., 213:422.
Nishizuka, M., and Arai, Y., 1981b, Sexual dimorphism in synaptic organization in the amygdala and its dependence on neonatal hormone environment, Brain Res., 212:31.
Nishizuka, M., and Arai, Y., 1982, Synapse formation in response to estrogen in the medial amygdala developing in the eye, Proc. Natl. Acad. Sci. USA, 79:7024.
Pfaff, D. W., and Keiner, M., 1973, Atlas of estrogen-concentrating cells in the central nervous system of the female rat, J. Comp. Neurol., 151:121.
Raisman, G., and Field, P. M., 1973, Sexual dimorphism in the neuropil of the preoptic area of the rat and its dependence on neonatal androgen, Brain Res., 54:1.
Reier, P. J., Cuellen, M. J., Froelich, J. S., and Rothchild, I., 1977, The ultrastructure of the developing medial preoptic nucleus in the postnatal rat, Brain Res., 122:415.
Ruf, K. B., 1982, Synaptogenesis and puberty, in: "Advances in Neuroendocrine Physiology," Front. Horm. Res., Vol. 10, K. B. Ruf and G. Tolis, eds., p. 65, Karger AG, Basel.
Toran-Allerand, C. D., 1976, Sex steroids and the development of the newborn mouse hypothalamus and preoptic area in vitro: implications for sexual differentiation, Brain Res., 61:63.
Toran-Allerand, C. D., 1984, On the genesis of sexual differentiation of the central nervous system: morphogenic consequences of steroid exposure and possible role of α-fetoprotein, in: "Sex Difference in the Brain The Relation between Structure and Function," Porg. Brain Res., Vol. 61, G. J. De Vries, J. P. C. Bruin, H. B. M. Uyling, and M. A. Corner, eds., p. 63, Elsevier, Amsterdam.
Toran-Allerand, C. D., Hashimoto, K., William, T., Greenough, W. T., and Saltarelli, M., 1983, Sex steroids and the development of the newborn mouse hypothalamus and preoptic area in vitro: effects of estrogen on dendritic differentiation, Develop. Brain Res., 7:97.

THE ELECTROPHYSIOLOGY OF THE HYPOTHALAMIC GONADOTROPIC HORMONE RELEASING HORMONE (GnRH) PULSE GENERATOR IN THE RHESUS MONKEY

Ernst Knobil

Laboratory for Neuroendocrinology
The University of Texas Medical School at Houston
Houston, Texas 77225

The rhythmic pulsatile nature of gonadotropic hormone section first noted in the ovariectomized rhesus monkey (Dierschke et al., 1970) has since been described in every mammal studied in this regard (Pohl and Knobil, 1982). The original supposition that each gonadotropin pulse is the consequence of a pulse of GnRH released by the hypothalamus into the pituitary portal circulation has been amply confirmed (Carmel et al., 1976; Clarke and Cummins, 1982). These findings have given rise to the concept of a neuronal construct in the central nervous system which is responsible for the rhythmic activation of GnRH cells and the release of the neuropeptide from their terminals. While this system has been variously referred to as an oscillator or pulse generator for descriptive convenience, its cellular nature remains unknown. In the rhesus monkey, using conventional neuroendocrinological techniques, the pulse generator has been localized to the mediobasal hypothalamus (MBH) (Krey et al., 1975), and more specifically, to the area of the arcuate nucleus (Plant et al., 1978). An effort to detect the electrophysical basis of pulse generator activity was initiated in the late 1970's and was based on the supposition that each bolus of GnRH released into the pituitary portal circulation must be the consequence of the synchronous firing of a large number of GnRH cells and that the approach most likely to detect their action potentials was the application of multiunit recording techniques as utilized earlier for the study of the macrocellular system in the rhesus monkey by Hayward and his colleagues (Hayward, 1977). Our initial approach utilizing single tungsten electrodes acutely placed in the region of the arcuate nucleus of anesthetized rhesus monkeys, while largely unsuccessful, occasionally yielded evidence of astonishing increases in multiunit electrical activity (MUA) coupled with pulses of LH as measured in the peripheral circulation (Dufy et al., 1979; Knobil, 1981). We assumed that the rarity of successful electrode placements were attributable to the sparsity of the active units in the hypothalamus and changed our strategy to the stereotaxic bilateral implantation of chronic multiple electrode arrays which blanketed the MBH (Wilson et al., 1984). This new approach yielded predictable results from one or more electrodes in most animals and permitted the systematic study of the pulse generator by direct, electrophysiological observation.

Each LH pulse, as monitored in the peripheral circulation, in both anesthetized animals and awake ovariectomized monkeys restrained in primate chairs, is associated by a striking increase in MUA recorded from a number of sites within the MBH (Wilson et al., 1984). These MUA volleys are

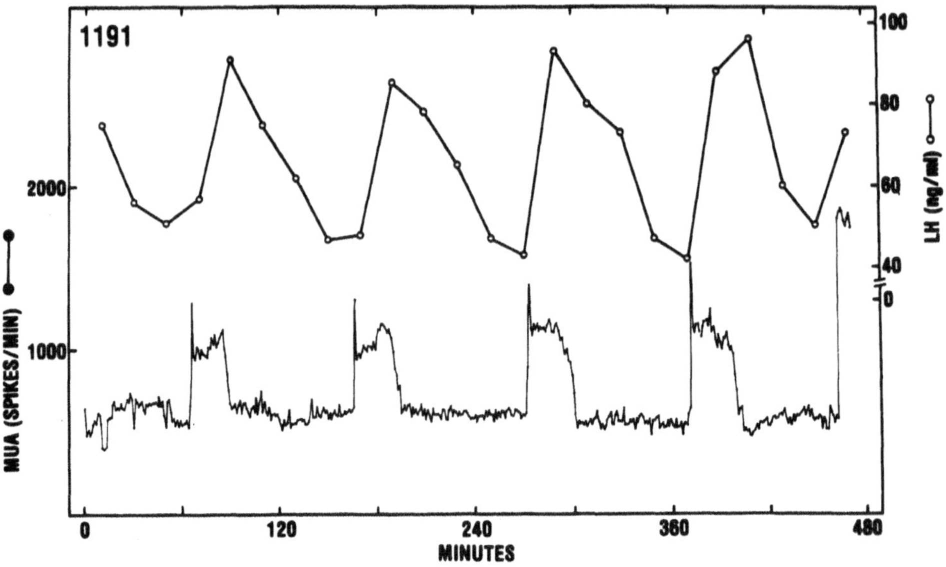

Fig. 1. Temporal relationship between multiunit activity, recorded from the medial basal hypothalamus of a thiopental anesthetized ovariectomized rhesus monkey, and LH pulses in the peripheral circulation. (Modified from Wilson et al., 1984.)

characterized by a sudden and major increase in the frequency of action potentials which remained elevated for varying times following a brief overshoot before abruptly returning to baseline (Fig. 1). An unambiguous unitary relationship between MUA volleys and LH pulses has invariably been observed in all studies to date.

At present, it is not clear whether the electrical activity underlying the operation of the pulse generator is recorded from GnRH cells or fibers or from other neuronal elements which eventually impinge on GnRH cells. This is because the diameter of the electrode tips (50 μm) is too large relative to the structures near which they are found (Silverman et al., in press). In any case, however, the active sites seemingly correspond to the distribution of GnRH neurons and their axons within the MBH (Silverman et al., in press; 1982).

Pentobarbital anesthesia markedly reduces the frequency of the pulse generator as evidence by the underlying volleys of electrical activity and the attendant LH pulses in the peripheral circulation (Wilson et al., 1984). Similarly, GnRH pulse generator activity is inhibited by the α-adrenergic blockers phentolamine, phenoxybenzamine and prazosin as well as by the dopaminergic blocking agent metaclopromide (Kaufman et al., 1985). These drugs either reduce the frequency of the pulse generator or arrest it altogether, depending on the dose, suggesting that central adrenergic and dopaminergic inputs can modulate its activity. Morphine has a similar effect which can be abruptly reversed by the administration of the opiate antagonist naloxone (Kesner et al., 1986a). This is illustrated in Fig. 2. The recent finding by Ferin and his colleagues (Van Vugt et al., 1983) that reduction in LH pulse frequency occasioned by progesterone can also be reversed by the administration of naloxone strongly suggests that the physiological action of this ovarian steroid is mediated by endogenous opiates. The reduction in LH

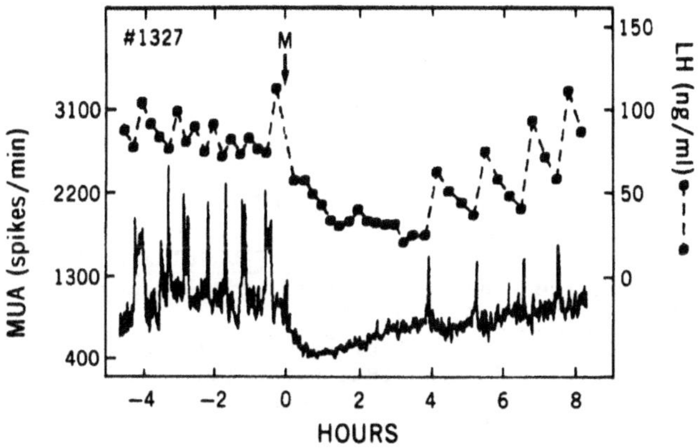

Fig. 2. Effect of a single intravenous injection of morphine (M) on hypothalamic pulse generator activity and LH pulses in an ovariectomized rhesus monkey. Note recovery at a reduced frequency. (Modified from Kesner et al., 1986a.)

pulse frequency effected by testosterone is similarly inhibited by opiate antagonists (Veldhuis et al., 1984). Naloxone has no effect in the absence of the gonads or during the follicular phase of the menstrual cycle indicating that the frequency of the pulse generator in these circumstances may be maximal.

In normal monkeys, estrogens do not appear to have an effect on pulse generator activity, although in ovariectomized animals even subphysiological concentrations of this steroid are markedly inhibitory (Kesner et al., 1986c). The nature and physiological significance of this ovarian influence on the sensitivity of the pulse generator to estrogen is currently under investigation.

All reproductive processes in both sexes are dependent on a cascade of neuroendocrinological events which is initiated by the GnRH pulse generator. Its proper functioning is, therefore, of fundamental importance, yet essentially nothing is known of its structure or manner of operation. All that can be said with certainty is that its rhythmic activation and deactivation does not seem to be the result of a short loop feedback mechanism (Kesner et al., 1986b). Although the nature of the pulse generator and its ultimate cellular origin remain to be elucidated, direct physiological access to the central component of the hypothalamic-hypophysial system which governs ovarian function opens new avenues for the investigation of the neural control of reproductive phenomena such as the initiation of puberty, the normal menstrual cycle and a variety of physiological and pathophysiological amenorrheic states.

REFERENCES

Carmel, P. W., Araki, S., and Ferin, M., 1976, Pituitary stalk portal blood collection in rhesus monkeys: evidence for pulsatile release of GnRH, Endocrinology, 99:243.

Clarke, I. J., and Cummins, J. T., 1982, The temporal relationship between gonadotropic releasing hormone (GnRH) and luteinizing hormone (LH) secretion in ovariectomized ewes, Endocrinology, 11:1737.
Dierschke, D. J., Bhattacharya, A. N., Atkinson, L. E., and Knobil, E., 1970, Circhoral oscillations of plasma LH levels in the ovariectomized rhesus monkey, Endocrinology, 87:850.
Dufy, B., Dufy-Barbe, L., Vincent, J. D., and Knobil, E., 1979, Etude electrophysiologique des neurones hypothalamiques et regulation gonadotrope chez le singe rhesus, Journal de Physiologie, 75:105.
Hayward, J. N., 1977, Functional and morphological aspects of hypothalamic neurons, Physiol. Rev., 57:574.
Kaufman, J. M., Kesner, J. S., Wilson, R. C., and Knobil, E., 1985, Electrophysiologic manifestation of "LHRH pulse generator" activity in the rhesus monkey: influence of α-adrenergic and dopaminergic blocking agents, Endocrinology, 1164:1327.
Kesner, J. S., Kaufman, J. M., Wilson, R. C., Kuroda, G., and Knobil, E., 1986a, The effect of morphine on the electrophysiological activity of the hypothalamic luteinizing hormone releasing hormone pulse generator in the rhesus monkey, Neuroendocrinology, 43:686.
Kesner, J. S., Kaufman, J. M., Wilson, R. C., and Knobil, E., 1986b, On the short-loop feedback regulation of the hypothalamic luteinizing hormone releasing hormone "pulse generator" in the rhesus monkey, Neuroendocrinology, 42:109.
Kesner, J. S., Wilson, R. C., Kaufman, J. M., Hotchkiss, J., and Knobil, E., 1986c, Unphysiological responses of the LHRH pulse generator to physiological estrogen inputs in the absence of the ovary, Proc. Ann. Mtg. Endocrine Soc., Abstract.
Knobil, E., 1981, Patterns of hypophysiotropic signals and gonadotropin secretion in the rhesus monkey, Biol. Reprod., 24:44.
Krey, L. C., Butler, W. R., and Knobil, E., 1975, Surgical disconnection of the medial basal hypothalamus and pituitary function in the rhesus monkey. I. Gonadotropin secretion, Endocrinology, 96:1073.
Plant, T. M., and Krey, L. C., Moossy, J., McCormack, J. T., Hess, D. L., and Knobil, E., 1978, The arcuate nucleus and the control of gonadotropin and prolactin secretion in the female rhesus monkey (Macaca mulatta), Endocrinology, 102:52.
Pohl, C. R., and Knobil, E., 1982, The role of the central nervous system in the control of ovarian function in higher primates, Ann. Rev. Physiol., 44:583.
Silverman, A. J., Antunes, J. L., Abrams, G. M., Nilaver, G., Thau, R., Robinson, J. A., Ferin, M., Krey, L. C., 1982, The luteinizing hormone-releasing hormone pathways in rhesus (Macaca mulatta) and pigtailed (Macaca nemestrina) monkeys: new observations on thick unembedded sections, J. Comp. Neurol., 211:309.
Silverman, A. J., Wilson, R. C., Kesner, J. S., and Knobil, E., Hypothalamic localization of multiunit electrical activity associated with pulsatile LH release in the rhesus monkey, Neuroendocrinology, in press.
Van Vugt, D. A., Bakst, G., Dyrenfurth, I., and Ferin, M., 1983, Naloxone stimulation of luteinizing hormone secretion in the female monkey: influence of endocrine and experimental conditions, Endocrinology, 113:1858.
Veldhuis, J. D., Rogol, A. D., Samojlik, E., and Ertel, N. H., 1984, Role of endogenous opiates in the expression of negative feedback actions of estrogens and androgens on pulsatile properties of luteinizing hormone secretion in man, J. Clin. Invest., 74:47.
Wilson, R. C., Kesner, J. S., Kaufman, J. M., Uemura, T., Akema, T., and Knobil, E., 1984, Central electrophysiologic correlates of luteinizing hormone secretion in the rhesus monkey, Neuroendocrinology, 39:256.

OVARIAN FEEDBACK REGULATION OF GONADOTROPIN

RELEASING HORMONE SECRETION AND ACTION

I. J. Clarke

Medical Research Centre
Prince Henry's Hospital
St. Kilda Road
Melbourne, Australia 3004

INTRODUCTION

Feedback signals from the gonads can regulate gonadotropin secretion by action at the hypothalamic or pituitary level. In the former case, feedback hormones act by regulating the pulsatile secretion of gonadotropin releasing hormone (GnRH). Since the pattern of luteinizing hormone (LH) secretion in the sheep is primarily dependent upon the pulsatile secretion of GnRH (Clarke and Cummins, 1982), the pituitary feedback effects of ovarian hormones are manifest via modulation of GnRH action.

The situation regarding feedback control of follicle stimulating hormone (FSH) in the sheep may be somewhat different to that of LH, in that secretion is not distinctly pulsatile and continues in the absence of GnRH stimulus (Clarke et al., 1986a). With the deprivation of GnRH, the plasma FSH levels fell gradually over a period of weeks to eventually become undetectable (Clarke et al., 1984, 1986a). From these observations we have hypothesized that GnRH pulses stimulate FSH synthesis and that the secretion of this gonadotropin is substantially passive. Thus, after GnRH deprivation, secretion diminishes as pituitary FSH stores are depleted. In terms of feedback effects, long-term alterations (day to day) in FSH secretion might occur as a result of altering GnRH secretion, but short-term (hour to hour) changes would most likely be effected by the pituitary action of ovarian hormones. In this scheme, where GnRH is not obligatory for FSH secretion *per se*, the feedback effects need not be coupled to GnRH action.

This regulatory mechanism for FSH may not hold for species other than the sheep, in which secretion is pulsatile. For example, there is reasonable evidence that LH and FSH are often secreted contemporaneously in the human (Baker et al., 1976), suggesting that, at least in some cases, GnRH actively stimulates secretion of both gonadotropins. The fact that FSH pulses may also occur in the absence of LH pulses has been taken as evidence of a separate FSH releasing factor (Bowers et al., 1973). In the rat, the discrepancy between LH and FSH pulses may be greater than that seen in humans (Lumpkin et al., 1984).

In this paper, I wish to review a series of studies that we have undertaken in sheep to ascertain the relative hypothalamic and pituitary components of the feedback effects of gonadal hormones. These have been made

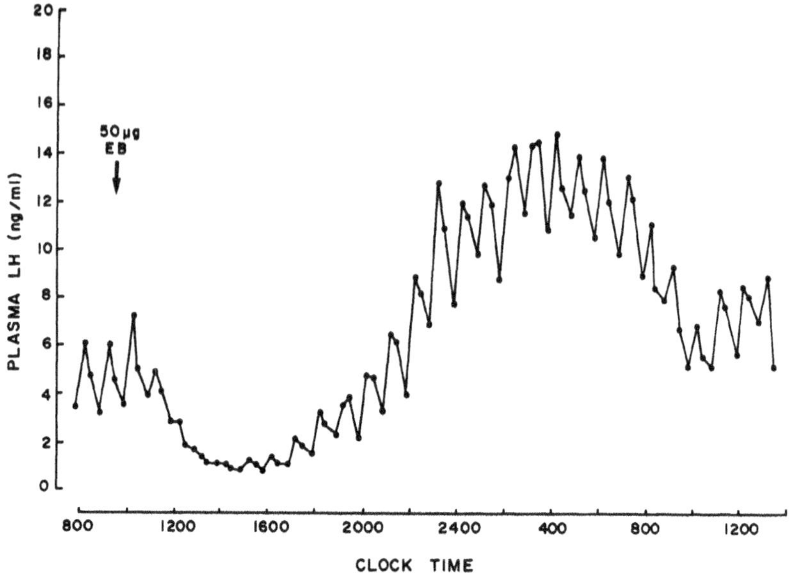

Fig. 1. Plasma LH levels in an OVX-HPD ewe receiving hourly pulses (i.v.) of 500 ng GnRH and a 50 µg i.m. injection of estradiol benzoate (EB). Note the change in pulse shape during the surge period indicated by the arrows. Reproduced from Clarke and Cummins (1984), with permission.

possible by the development of two sheep models. In one model, we are able to measure GnRH in the hypophysial port blood of conscious ewes and thus monitor hypothalamic feedback effects (Clarke and Cummins, 1982). In order to study direct pituitary feedback we have used a surgical method of hypothalamo-pituitary disconnection (HPD) to isolate the pituitary from the brain, without infarction of the gland (Clarke et al., 1983). Gonadotropin secretion can be maintained indefinitely in OVX-HPD ewes with the pulsatile administration of GnRH (Clarké et al., 1984).

TYPES OF FEEDBACK

Our working hypothesis has been that three types of feedback exist. These are best described in terms of the resultant changes in plasma LH levels after estrogen treatment of ovariectomized (OVX) ewes. The immediate effect, upon i.v. administration of a bolus estradiol, is a reduction in LH (Clarke et al., 1982) which we have called short-term negative feedback. Subsequently - some 10-20 h later depending on dose and route of administration - a positive feedback effect is seen in which plasma LH levels rise 10-20 fold in OVX ewes (Radford et al, 1969; Clarke et al., 1982). Continuing estrogen treatment maintains plasma LH levels below those seen in OVX ewes (Diekman and Malven, 1973; Wright et al., 1981) (long-term negative feedback).

Short-term Negative Feedback

In OVX ewes, an i.m. injection of 50 µg estradiol benzoate did not have an immediate effect on GnRH pulsatility (Clarke and Cummins, 1985a). In

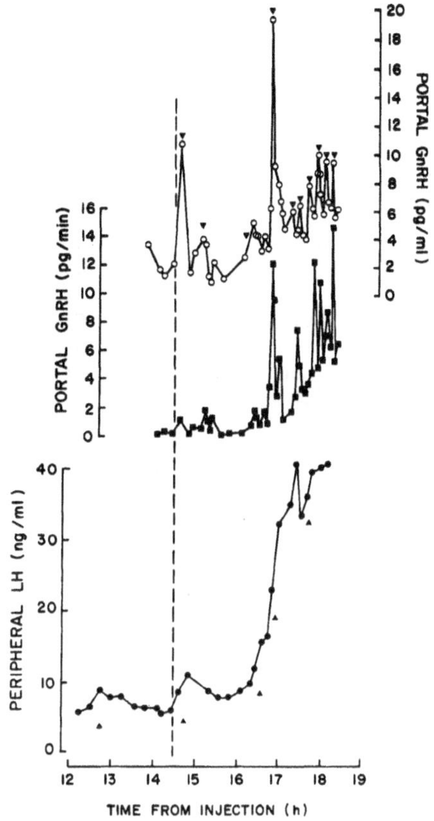

Fig. 2. Portal plasma GnRH (pg/ml o; pg/min ■) concentrations and peripheral plasma LH (●) concentrations in an OVX ewe at the onset of an estrogen-induced LH surge. The dotted vertical line shows the start of the LH surge and indicates GnRH and LH pulses, respectively. Reproduced from Clarke and Cummins (1985a), with permission.

contrast, OVX-HPD ewes given hourly pulses of 500 ng GnRH and the same estrogen treatment showed an immediate reduction in LH pulse amplitude (Fig. 1) (Clarke and Cummins, 1984). This suggested that the short-term negative feedback effect was the result of feedback to the pituitary gland, not the hypothalamus.

Positive Feedback

During the estrogen-induced LH surge in OVX ewes, we have observed an increase in GnRH pulse frequency from approximately one pulse per h to two pulses per h (Clarke and Cummins, 1985a). A spectacular example of increased pulse frequency during the onset of the LH surge is seen in Fig. 2. This indicates the hypothalamic component of the positive feedback response. Using 'push-pull' perfusion of the median eminence and sampling OVX-estrogen treated ewes, Schillo et al. (1985) have found variable patterns of GnRH secretion associated with the LH surge. Some animals showed no substantial rise in GnRH secretion whilst others showed a large 'pulse' at the start of the surge and then a sustained rise; others showed a gradual and sustained rise in GnRH secretion. In OVX monkeys, the positive feedback effect was also seen to be associated with variable patterns of GnRH secretion; increases in frequency and/or amplitude of GnRH pulses were observed during the LH surge (Levine et al., 1985). Most recently, a study of GnRH secretion throughout the ovine estrous cycle has identified three patterns of GnRH secretion associated with the cyclic LH surge (Clarke et al., 1986b), being similar to those described by Schillo et al. (1985). It seems that more than one pattern of GnRH secretion is able to produce an LH surge.

Thus, although there is good evidence from humans (Miyake et al., 1980; Elkind-Hirsch et al., 1982) and rats (Sarkar et al., 1976) that the cyclic LH surge is associated with a rise in GnRH secretion, the definition of a particular pattern of secretion has proved difficult. Studies in a variety of species (sheep, Karsch et al., 1983; cow, Rahe et al., 1980; monkey, Marut et al., 1981; rat, Gallo, 1981; man, Djahanbakhch et al., 1984) have shown that LH secretion is pulsatile during the LH surge and it has been thought that this may reflect GnRH pulsatility. On the other hand, continuous GnRH treatment of gonadotrophs perfused in vitro can yield pulsatile secretory profiles (Magness et al., 1981), and it is possible that an LH surge comcomprised of pulses could be procured with a nonpulsatile pattern of GnRH stimulus. Nevertheless, our studies of cyclic ewes have revealed a pulsatile pattern of GnRH secretion during the LH surge in most (5/6) animals. Interestingly, at this time we recorded approximately 2 pulses per h; similar to the frequency that we found in estrogen-treated, OVX ewes (see above).

In a number of species, progesterone treatment over an estrogen background will result in a positive feedback event (human, Nillius and Wide, 1971; rat, Caligaris et al., 1971; monkey, Clifton et al., 1975) and this is thought to be due to a rise in GnRH secretion (Levine and Ramirez, 1980). Such an effect is of relevance in these species since plasma progesterone levels rise prior to the cyclic LH surge, but a similar phenomenon is not seen in sheep (Baird and Scaramuzzi, 1976). In spite of this, we found that a progesterone background enhanced the short-term negative feedback, and the positive feedback effects of estrogen in GnRH pulsed, OVX-HPD ewes (Clarke and Cummins, 1984). This pituitary effect may account, in part, for the fact that progesterone amplified the positive feedback effects of estrogen on LH secretion in hypogonadal women (Liu and Yen, 1983). Persistent high levels of progesterone will in fact block the estrogen-induced LH surge in hypothalamus-pituitary intact OVX ewes (Scaramuzzi et al., 1971), an effect that is most likely due to the prevention of GnRH secretion.

At the pituitary level, it has long been recognized that responsiveness to GnRH is enhanced in the preovulatory period (sheep, Reeves et al., 1971; women, Yen and Lein, 1976; rats, Cooper et al., 1973). This can also be seen in estrogen-treated OVX ewes at the time of the induced LH surge (Coppings and Malven, 1976). To examine the extent of the phenomenon as an exclusively pituitary event, we studied OVX-HPD ewes receiving hourly GnRH pulses and a single i.m. injection of estradiol benzoate. The plasma LH levels were elevated 2-3 fold during the positive feedback phase (Fig. 1). This contrasts with the 10-20 fold rise seen in hypothalamo-pituitary intact, OVX ewes (Radford et al., 1969; Scaramuzzi et al., 1971). It therefore appears that the positive feedback event in the sheep requires both hypothalamic and pituitary components. One feature of the pattern of LH response seen during the positive feedback period is that pulse shape changes. This is evidenced in Fig. 1 where the LH pulses were more prolonged during the surge phase than prior to estrogen treatment. It is not yet clear as to whether this effect is obtained by prolonged secretion in response to a given GnRH pulse or altered clearance as a result of a change in the molecular character of the secreted LH.

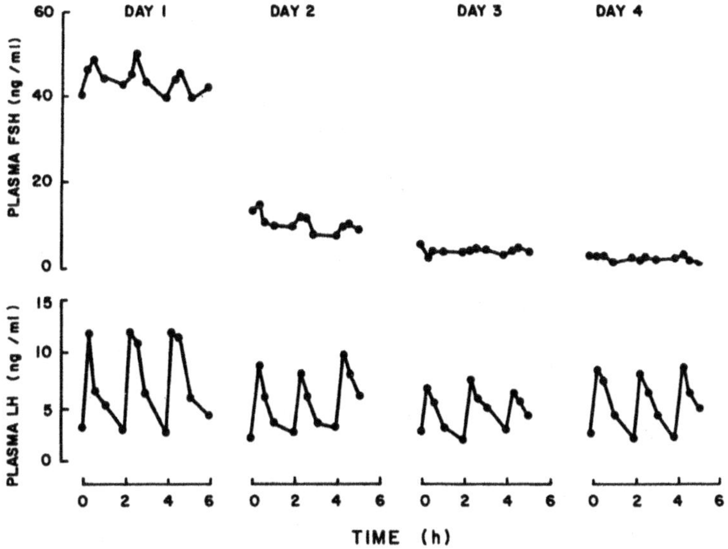

Fig. 3. Effects of subcutaneous injections of charcoal-stripped ovine follicular fluid (2 ml, b.i.d.) on plasma LH and FSH levels in an OVX-HPD ewe receiving 250 ng GnRH (i.v.) each 2 h. Reproduced from Clarke et al. (1986c), with permission.

Long-term Negative Feedback

In Suffolk sheep treated during the breeding season and immediately after ovariectomy, estrogen was found to reduce LH pulse amplitude but not frequency; progesterone treatment reduced pulse frequency but not amplitude (Goodman and Karsch, 1980). In a further study, progesterone at a lower dosage had no effect, but a combination of estrogen and progesterone reduced pulse frequency (Goodman et al., 1981). Again using lower doses than Goodman and Karsch (1980), Martin et al. (1983) found long-term estrogen or progesterone treatment to be without effect in OVX Merino ewes whereas a combined treatment was able to reduce pulse frequency. During a visit to our laboratory by Dr. Fred Karsch, we studied the effects of long-term (1 wk) estrogen and progesterone treatments on GnRH secretion in Corriedale ewes. For this purpose we used the same doses as Goodman and Karsch (1980). In this breed, either steroid completely eliminated GnRH secretion (Clarke et al., 1985; Karsch et al., 1987). Although this prevented any analysis of frequency and amplitude, it demonstrated the powerful long-term negative feedback effects of these steroids at the hypothalamic level. Furthermore, it emphasized the species differences in responsiveness to steroids since the doses of estrogen and progesterone that permitted LH secretion in Suffolks (Goodman and Karsch, 1980) did not do so in Corriedales.

At the level of the pituitary gland we have again used the GnRH pulsed OVX-HPD ewe to define long-term feedback effects. Estrogen alone (same dose as used by Goodman and Karsch, 1980) caused a 20-40% reduction in LH pulse amplitude and a 40-50% fall in plasma FSH (Clarke and Cummins, 1986); progesterone alone (~10 ng/ml) had no effect (Clarke and Cummins, 1984). When progesterone treatment was superimposed on estrogen, after 1 week of estrogen alone, there was a further significant ($p < 0.05$) reduction in plasma LH

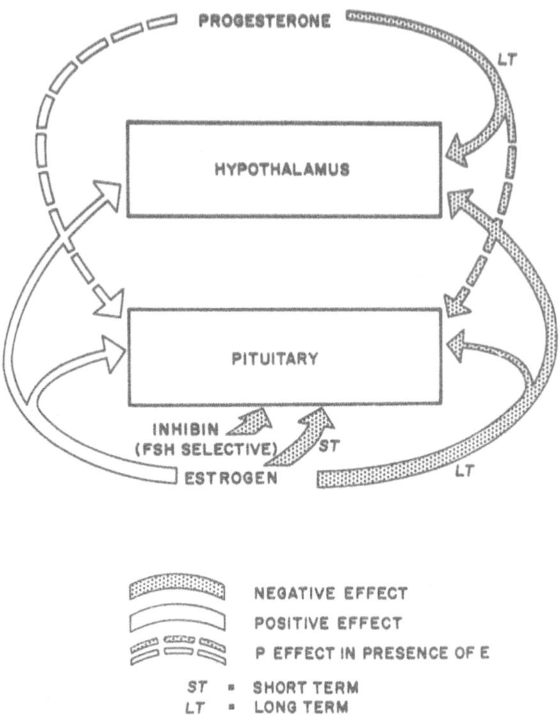

Fig. 4. Schematic representation of the hypothalamic and pituitary feedback effects of estrogen, progesterone and inhibin. The relative importance of the various pathways is indicated by the thickness of the arrows.

levels by 35% (unpublished data). This was not seen when progesterone and estrogen treatments were started together and perhaps indicates that estrogen induces the progesterone receptors in the pituitary gland.

These studies of chronic steroidal treatments suggest that the long-term negative feedback effects are more powerfully expressed at the hypothalamic level than at the pituitary gland.

Inhibin Feedback

Recently, the techniques of molecular biology have permitted the identification of gene sequences for inhibin in porcine (Mason et al., 1985) and bovine ovaries (Robertson et al., 1986). Inhibin is thought to be a selective regulator of FSH secretion and we have studied its hypothalamic and pituitary feedback effects using a crude preparation of charcoal stripped ovine follicular fluid (oFF). In OVX ewes, s.c. injections of 2 ml oFF (b.i.d.) caused a 95% fall in plasma FSH, no change in the LH pulse frequency and a small (30%), but significant ($p < 0.05$), reduction in LH pulse amplitude (Clarke et al., 1986c). These effects were substantially the same when we treated GnRH-pulsed OVX-HPD ewes (Fig. 3) (Clarke et al., 1986c) and we therefore believe that the inhibin effect is manifest at the pituitary level. The small effect on LH pulse amplitude may be a function of the dosage used

(20-30 K units/day); more definitive studies with purified material and physiological levels of treatment will be required to establish this point.

FREQUENCY AND AMPLITUDE MODULATION

Studies in monkeys (Wildt et al., 1981) and in sheep (Clarke et al., 1984) have shown that LH pulse amplitude is inversely related to pulse frequency. When the LH amplitude was manipulated in OVX-HPD ewes by changing GnRH pulse frequency, there was a very high correlation between the releasable pool of LH in the pituitary gland and the LH pulse amplitude (Clarke and Cummins, 1985b). In these circumstances changes in GnRH pulse frequency do not alter GnRH receptor levels (Crowder et al., 1986). Thus, it appears that LH pulse amplitude is determined by the level of pituitary LH stores which, in turn, is determined by GnRH pulse frequency. If this is the case, then changes in plasma LH pulse patterns that are seen during the estrous cycle would reflect changes in GnRH pulse frequency, not GnRH pulse amplitude. The low frequency, high amplitude LH pulses seen during the luteal phase (Baird and Scaramuzzi, 1976; Goodman et al., 1980) would simply reflect a low frequency of GnRH pulses and the high frequency, low amplitude LH pulses seen during the follicular phase (Baird, 1978; Karsch et al., 1983) would reflect a high frequency of pulses.

In essence, the feedback effects that regulate LH secretion throughout the majority of the cycle most likely involve frequency modulation with no requirement for changes in GnRH pulse amplitude. This may explain why long-term tonic negative feedback effects are more powerful at the hypothalamic level - where frequency is determined - than at the pituitary level, where pulse amplitude may be modified. This system breaks down at the time of the preovulatory LH surge when there are substantial increases in GnRH pulse frequency and amplitude (Clarke et al., 1986b) and pituitary responsiveness to GnRH (Reeves et al., 1971).

CONCLUSION

Using two *in vivo* models, we have defined the effects of estrogen, progesterone and inhibin at the hypothalamic and pituitary levels. A summary of our findings is presented in schematic form in Fig. 4. Having identified different types and sites of feedback, we are now in a position to study the cellular mechanisms that are involved in generating these effects.

REFERENCES

Baird, D. T., 1978, Pulsatile secretion of LH and ovarian estradiol during the follicular phase of the sheep estrous cycle, *Biol. Reprod.*, 18:359.

Baker, H. W. G., Santen, R. J., Burger, H. G., de Kretser, D. M., Hudson, B., Pepperell, R. J., and Bardin, C. W., 1975, Rhythms in the secretion of gonadotropins and gonadal steroids, *J. Steroid Biochem.*, 6:793.

Bowers, C. Y., Currie, B. L., Johansson, K. N. G., and Folkers, K., 1973, Biological evidence that separate hypothalamic hormones release the follicle stimulating and luteinizing hormones, *Biochem. Biophys. Res. Commun.*, 50:20.

Caligaris, L., Astrada, J. J., and Taleisnik, 1971b, Biphasic effect of progesterone on the release of gonadotrophin in rats, *Endocrinology*, 89:331.

Clarke, I. J., Burman, K. R., Doughton, B. W., and Cummins, J. T., 1986a, Effects of constant infusion of gonadotrophin-releasing hormone (GnRH) in ovariectomized ewes with hypothalamo-pituitary disconnection: further evidence for differential control of LH and FSH secretion and the lack of a priming effect, J. Endocrinol., in press.

Clarke, I. J., and Cummins, J. T., 1982, The temporal relationship between gonadotropin releasing hormone (GnRH) and luteinizing hormone (LH) secretion in ovariectomized ewes, Endocrinology, 111:1737.

Clarke, I. J., and Cummins, J. T., 1984, Direct pituitary effects of estrogen and progesterone on gonadotropin secretion in the ovariectomized ewe, Neuroendocrinology, 39:267.

Clarke, I. J., and Cummins, J. T., 1985a, Increased GnRH pulse frequency associated with estrogen-induced LH surges in ovariectomized ewes, Endocrinology, 116:2376.

Clarke, I. J., and Cummins, J. T., 1985b, GnRH pulse frequency determines LH pulse amplitude by altering the amount of releasable LH in the pituitary gland, J. Reprod. Fertil., 73:425.

Clarke, I. J., and Cummins, J. T., 1986, Long-term (chronic) negative feedback effects of oestrogen and progesterone at the pituitary level in ovariectomized ewes, Proc. Aust. Soc. Endocr., Brisbane.

Clarke, I. J., Cummins, J. T., and de Kretser, D. M., 1983, Pituitary gland function after disconnection from direct hypothalamic influences in the sheep, Neuroendocrinology, 36:376.

Clarke, I. J., Cummins, J. T., Findlay, J. K., Burman, K. J., and Doughton, B. W., 1984, Effects on plasma luteinizing hormone and follicle stimulating hormone of varying the frequency and amplitude of gonadotropin-releasing hormone pulses in ovariectomized ewes with hypothalamo-pituitary disconnection, Neuroendocrinology, in press.

Clarke, I. J., Funder, J. W., and Findlay, J. K., 1982, Relationship between pituitary nuclear estrogen receptors and the release of LH, FSH and prolactin in the ewe, J. Reprod. Fertil., 64:355.

Clarke, I. J., Findlay, J. K., Cummins, J. T., and Ewens, W. J., 1986c, Effects of ovine follicular fluid on plasma LH and FSH secretion in ovariectomized ewes to indicate the site of action of inhibin, J. Reprod. Fertil., 77, in press.

Clarke, I. J., Karsch, F. J., and Cummins, J. T., 1985, Oestrogen and progesterone stop GnRH pulses in ovariectomized anoestrous ewes, Proc. Aust. Soc. Reprod. Biol. 17th Ann. Conference, Abstract No. 28.

Clarke, I. J., Thomas, G. B., Doughton, B. W., Burman, K. J., Yao, B., and Cummins, J. T., 1986b, Gonadotropin-releasing hormone secretion during the ovine estrous cycle, Proc. 68th Meeting of the Endocrine Soc., Anaheim, California.

Clifton, D. K., Steiner, R. A., Resko, J. A., and Spies, H. G., 1975, Estrogen-induced gonadotropin release in ovariectomized rhesus monkeys and its advancement by progesterone, Biol. Reprod., 13:190.

Cooper, K. J., Fawcett, C. P., and McCann, S. M., 1973, Variation in pituitary responsiveness to luteinizing hormone releasing factor during the rat estrous cycle, J. Endocrinol., 57:187.

Coppings, R. J., and Malven, P. V., 1976, Biphasic effect of estradiol on mechanisms regulating LH release in ovariectomized sheep, Neuroendocrinology, 21:146.

Crowder, M. E., Clarke, I. J., and Nett, T. M., 1986, Effect of GnRH receptors, Proc. Soc. Study of Reprod. Cornell, Abstract.

Diekman, M. A., and Malven, P. V., 1973, Effect of ovariectomy and estradiol on LH patterns in ewes, J. Anim. Sci., 37:562.

Djahanbakhch, O., Warner, P., McNeilly, A. S., and Baird, D. T., 1984, Pulsatile release of LH and estradiol during the periovulatory period in women, Clin. Endocrinol., 20:579.

Elkind-Hirsch, K., Ravnikar, U., Schiff, I., Tulchinsky, D., and Ryan, K. J., 1982, Determinations of endogenous immunoreactive luteinizing hormone-releasing hormone in human plasma, J. Clin. Endocrinol. Metab., 54:602.

Forage, R. G., Ring, J. M., Brown, R. W., McIneney, B. V., Cobon, G. S., Gregson, R. P., Robertson, D. M., Morgan, F. J., Hearn, M. T. W., Findlay, J. K., Wettenhall, E. H., Burger, H. G., and de Kretser, D. M., 1986, Cloning and sequence analysis of cDNA species coding for the two subunits of inhibin from bovine follicular fluid, Proc. Natl. Acad. Sci. USA, 83, in press.

Gallo, R. V., 1981, Pulsatile LH release during the ovulatory LH surge on proestrus in the rat, Biol. Reprod., 24:100.

Goodman, R. L., Legan, S. J., Ryan, K. D., Foster, D. L., and Karsch, F. J., 1980, Importance of variations in behavioural and feedback actions of oestradiol to the control of seasonal breeding in the ewe, J. Endocrinol., 23:404.

Goodman, R. L., Bittman, E. L., Foster, D. L., and Karsch, F. J., 1981, The endocrine basis of the synergistic suppression of luteinizing hormone by estradiol and progesterone, Endocrinology, 109:1414.

Goodman, R. L., and Karsch, R. J., 1980, Pulsatile secretion of luteinizing hormone: differential suppression by ovarian steroids, Endocrinology, 107:1286.

Karsch, F. J., Foster, D. L., Bittman, E. L., and Goodman, R. L., 1983, A role for estradiol in enhancing luteinizing hormone pulse frequency during the follicular phase of the estrous cycle of sheep, Endocrinology, 113:1333.

Karsch, F. J., Cummins, J. T., Thomas, G. B., and Clarke, I. J., 1987, Steroid feedback inhibition of the pulsatile secretion of gonadotropin-releasing hormone in OVX ewes, Biol. Reprod., submitted.

Levine, J. E., Norman, R. L., Gliessman, P. M., Oyama, T. T., Bangsberg, D. R., and Spies, H. G., 1985, In vivo gonadotropin-releasing hormone release and serum luteinizing hormone measurements in ovariectomized, estrogen-treated rhesus monkeys, Endocrinology, 117:711.

Levine, J. E., and Ramirez, V. D., 1980, In vivo release of luteinizing hormone-releasing hormone estimated with push-pull cannulae from the mediobasal hypothalamus of ovariectomized, steroid-primed rats, Endocrinology, 107:1782.

Lumpkin, M. D., De Paolo, L. V., and Negro-Vilar, A., 1984, Pulsatile release of follicle-stimulating hormone in ovariectomized rats is inhibited by porcine follicular fluid (inhibin), Endocrinology, 114:201.

Magness, V. J., Millar, R. P., and Michie, J., 1981, Luteinizing hormone-releasing hormone, calcium and cyclic nucleotide interactions in luteinizing hormone release from ovine pituitary cells in culture, Neuropeptides, 21:190.

Martin, G. B., Scaramuzzi, R. J., and Henstridge, J. D., 1983, Effects of oestradiol, progesterone and androstenedione on the pulsatile secretion of luteinizing hormone in ovariectomized ewes during spring and autumn, J. Endocrinol., 96:181.

Marut, E. L., Williams, R. F., Cowan, B. D., Lynch, A., Lerner, S. P., and Hodgen, G. D., 1981, Pulsatile pituitary gonadotropin secretion during maturation of the dominant follicle in monkeys: estrogen positive feedback enhances the biological activity of LH, Endocrinology, 109:2270.

Mason, A. J., Hayflick, J. S., Ling, N., Esch, F., Ueno, N., Ying, S-Y., Guilleman, R., Niall, H., and Seeburg, P. H., 1985, Complementary DNA sequences of ovarian follicular fluid inhibin show precursor structure and homology with transforming growth factor, Nature (Lond.), 318:659.

Miyake, A., Kawamura, Y., Aono, T., and Kurachi, K., 1980, Changes in plasma LRH during the normal menstrual cycle in women, Acta Endocrinologica, 93:257.

Nillius, S. J., and Wide, L., 1971, Effect of progesterone on the serum levels of FSH and LH in postmenopausal women treated with oestrogen, Acta Endocrinol. Copen., 67:362.

Radford, H. M., Wheatley, I. S., and Wallace, A. L. C., 1969, The effects of oestradiol benzoate and progesterone on the secretion of luteinizing hormone in the ovariectomized ewe, J. Endocrinol., 44:135.

Rahe, C. H., Owens, R. E., Fleeger, J. L., Newton, J. T., and Harms, P. G., 1980, Pattern of plasma luteinizing hormone in the cyclic cow: dependence upon the period of the cycle, Endocrinology, 107:498.
Reeves, J. J., Arimura, A., and Schally, A. V., 1971, Pituitary responsiveness to purified luteinizing hormone-releasing hormone (LHRH) at various stages of the estrous cycle in sheep, J. Anim. Sci., 32:123.
Sarkar, D. K., Chiappa, S. A., Fink, G., and Sherwood, N. M., 1976, Gonadotrophin-releasing hormone surge in proestrous rats, Nature (Lond.), 264:461.
Scaramuzzi, R. J., Tillson, S. A., Thorneycroft, I. H., and Caldwell, B. V., 1971, Action of exogenous progesterone and estrogen on behavioural estrus and luteinizing hormone levels in the ovariectomized ewe, Endocrinology, 88:1184.
Schillo, K. K., Leshin, L. S., Kuehl. D., and Jackson, G. L., 1985, Simultaneous measurement of luteinizing hormone-releasing hormone and luteinizing hormone during estradiol-induced luteinizing hormone surges in the ovariectomized ewe, Biol. Reprod., 33:644.
Wildt, L., Hausler, A., Marshall, G., Hutchison, J. S., Plant, T. M., Belchetz, P. E., and Knobil, E., 1981, Frequency and amplitude of gonadotropin-releasing hormone stimulation and gonadotropin secretion in the rhesus monkey, Endocrinology, 109:376.
Wright, P. J., Geytenbeek, P. E., Clarke, I. J., and Findlay, J. K., 1981, Evidence for a change in estradiol negative feedback and LH pulse frequency in postpartum ewes, J. Reprod. Fertil., 61:97.
Yen, S. S. C., and Lein, A., 1976, The apparent paradox of the negative and positive feedback control system on gonadotropin secretion, Am. J. Obstet. Gynecol., 126:942.

SHORT AND ULTRASHORT FEEDBACK CONTROL OF GONADOTROPIN SECRETION

Béla Flerkó, István Merchenthaler
and György Sétáló

Department of Anatomy
University Medical School
Pécs, Hungary

VASCULAR BASIS FOR THE SHORT FEEDBACK CONTROL MECHANISM

It is a kind of paradox that in the first description of the hypophysial portal system (Popa and Fielding, 1930) a reverse direction of blood flow, i.e. from the pituitary gland towards the hypothalamus, was also mentioned. Their main anatomical findings were soon confirmed by Wislocki and King (1936). However, on histological evidence, Wislocki (1937, 1938) concluded that blood flow in the hypophysial portal vessels was directed downwards, i.e. towards the pituitary gland. Further *in vivo* and *in vitro* studies by Green and Harris (1947, 1949) produced more data supporting this theory, to the extent that finally the original version of Popa and Fielding (1930) regarding the upward blood flow was considered to be a complete misinterpretation.

In our department, Török (1954) studied the vascular supply and blood flow of the dog and cat hypophysis and described small side branches of the median eminence (ME) capillary loops, which could be traced upward into the fine capillary network situated immediately below the ependymal surface of the third ventricle (Fig. 1). Later, his *in vivo* experiments confirmed these observations (Török, 1962), i.e. he observed some upward blood flow towards the hypothalamus in vessels of the subependymal layer of the ME that have connections with the portal vascular system. In Fig. 1, arrows indicate the direction of the blood flow in these subependymal blood vessels.

In the living dog, Török (1962) opened the bulbous upper enlargement of the infundibular recess and brought the opposite ventricular surface of the ME into the visual field of a dissecting microscope. The sequence of events in his observations is indicated diagrammatically in Fig. 2. After intracarotid injection of Indian ink, the dye was first seen in the primary vascular plexus of the pars tuberalis covering the outer surface of the ME, as it became filled up in a downward direction (Fig. 2A). At the same time, or somewhat later, the tops of the capillary loops became visible as darker dots also in a descending sequence (Fig. 2B). Finally, the dye filled up the interior tangential vascular plexus of the ME and the infundibular stem (Fig. 2C). It could clearly be observed that the blood flow in this interior plexus was directed upwards, towards the medial subependymal capillary network of the hypothalamus (Fig. 2C). The same interior tangential plexus of the

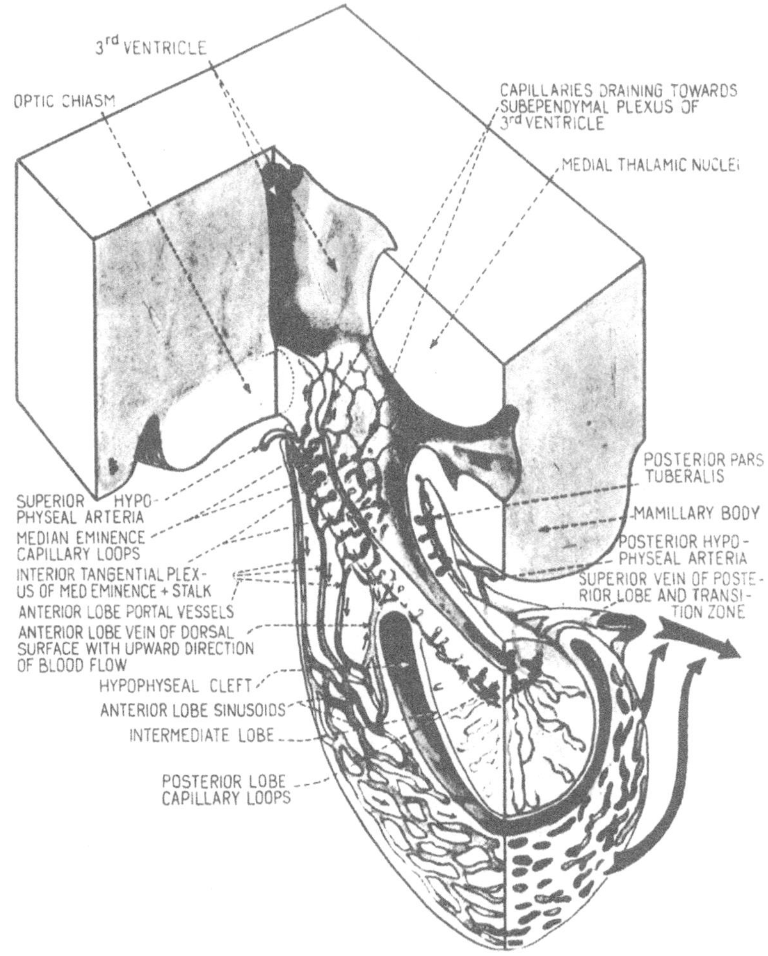

Fig. 1. Diagram illustrating schematically the pituitary circulation in the dog and the cat as revealed by the investigations of Török (1960). Short arrows indicate direction of blood flow. From Szentágothai et al. (1962) by courtesy of the Publishing House of the Hungarian Academy of Sciences, Budapest.

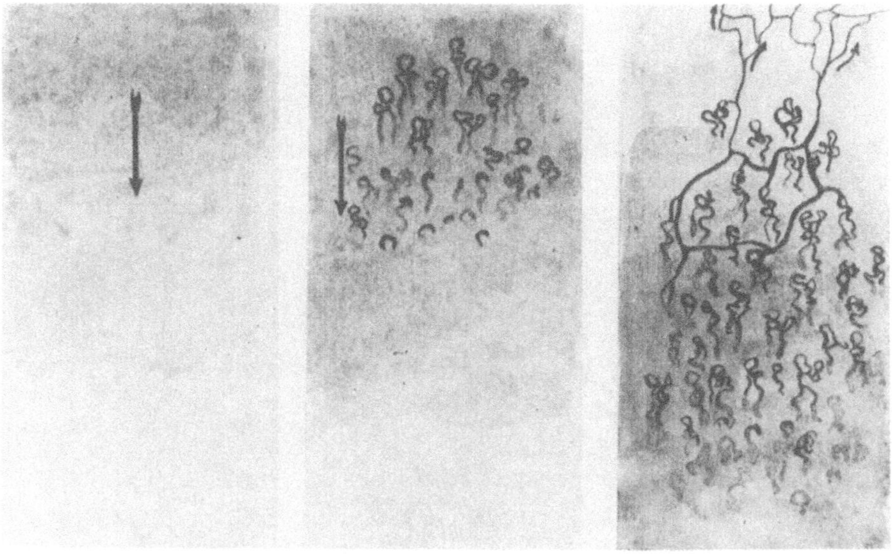

Fig. 2. Sequence of events observed by Török (1960) during intracarotid injection of dyes in the living dog under observation with stereoscopic microscope from the ventricular surface of the bulbous enlargement of the infundibular recess. (a) Initial diffuse descending darkening caused by the downward progress of the dye in the vascular plexus (Mantelplexus) of the pars tuberalis. (b) Appearance of the dye in capillary loops in downward sequence. (c) Filling of interior tangential plexus with dyed blood and escape of this blood towards the hypothalamus. The dye was carried of course by the natural circulation of the living animal. From Szentágothai et al. (1962) by courtesy of the Publishing House of the Hungarian Academy of Sciences, Budapest.

infundibular stem could easily be observed also in the rat (Figs. 3A, C, and D) with similar connections, as seen in the dog, i.e. (i) with the primary plexus of the pars tuberalis, and (ii) with the subependymal vascular network which originates mainly from the capillary loops of the ME, and irrigates a narrow sagittal layer of the hypothalamus on both sides of the third ventricle (Fig. 3C, D). Subsequent observations of Oliver et al. (1977) furnished further evidence for retrograde blood flow in the pituitary stalk.

More recent findings showed that pituitary hormones (Assies et al., 1978) as well as exogenously applied neuropeptides (Mezey et al., 1978) reached the central nervous system from the pituitary within a short period of time and even at a fairly high concentration. Since, following stalk section, there was a marked decrease in this uptake (Mezey et al., 1978), it seems likely that the stalk has a significant role in this transport (Mezey et al., 1979). The solid morphological proof as a basis to explain these findings has been furnished by Ambach et al. (1976), while Porter et al. (1978) have added much to our knowledge concerning the functional role of the retrograde blood flow in the pituitary stalk, especially with respect to the release of hypothalamic dopamine and pituitary LH and prolactin.

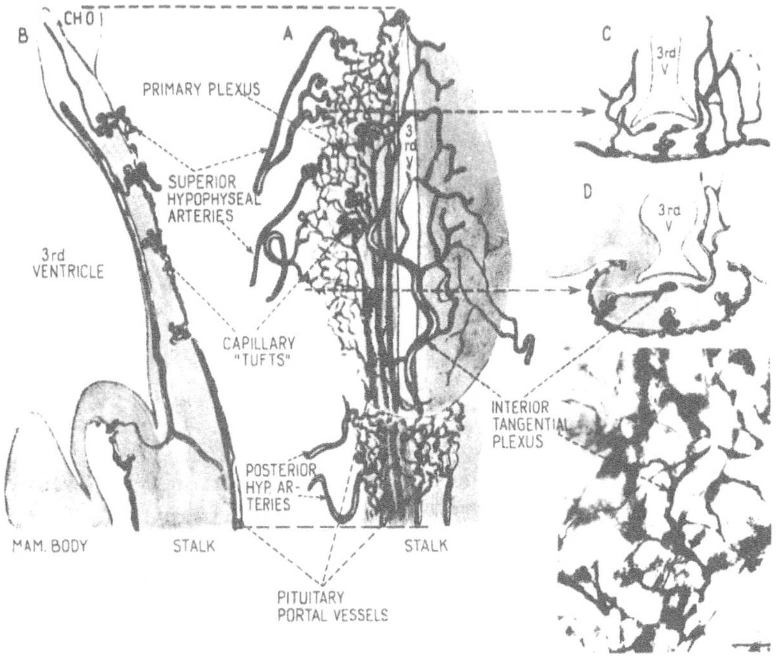

Fig. 3. Semidiagrammatic illustration of median eminence circulation in the rat. Drawing A shows the median eminence as seen from above. On its left side almost all of the brain tissue has been removed, so that the primary vascular plexus on the outer surface of the median eminence is seen, which is fed by the superior hypophysial arteries. From this plexus capillary loops or tufts penetrate into the median eminence. These tufts have two outlets; the main outlet towards the pituitary portal vessels, and smaller connections upwards to the interior tangential plexus of the median eminence. On the right side, a somewhat higher level is shown demonstrating the connections of this interior tangential plexus with the hypothalamic capillary network. B shows a median sagittal section, C and D an anterior and a posterior transection in the levels indicated by dashed lines. In the lower right inset photograph, the internal tangential plexus of a specimen injected with Indian ink is brought into focus. From Szentágothai et al. (1962) by courtesy of the Publishing House of the Hungarian Academy of Sciences, Budapest.

INTERNAL OR SHORT-LOOP FEEDBACK MECHANISM

The circulatory pattern described by Török (1954, 1962) raised the possibility that anterior pituitary hormones might act directly on the hypothalamus. In order to prove this hypothesis, Halász and Szentágothai (1958) implanted anterior pituitary tissue into the hypothalamus of nonhypophysectomized rats. They assumed that the grafted pituitary might release some ACTH. If this ACTH had a direct action on the hypothalamus, it would lead to the inhibition of ACTH secretion by the animal's own hypophysis. In their experiments a rather crude and indirect method was used for evaluation of this proposal. They determined the mean size of the nuclei of the zona fasciculata cells considered to be indicative of the activity of adrenocortical activity. They found that if the pituitary implant was in the medial basal hypothalamus the size of the nuclei of the zona fasciculata cells decreased (Halász and Szentágothai, 1958, 1960), indicating a reduced secretory activity in the ACTH-adrenal axis. This has been the first indirect experimental evidence for the so-called "internal" (according to our nomenclature) or "short-loop" (following the nomenclature of Motta et al., 1969) feedback.

An internal or short-loop feedback action of LH on the hypothalamus has been suggested by the experimental work of Kawakami and Sawyer (1959). They found postcoital changes in the EEG pattern of the rabbit that occurred after reflexogenous stimulation of the pituitary. They concluded that this might possibly not be a correlate of stimulating release of the hormone but might represent feedback action of the hormone on the nervous system. They were able, with large intraperitoneal doses of LH, to evoke this peculiar pattern, which involves both slow wave and paradoxical sleep sequences, but they were not able to induce the pattern with large doses of FSH, TSH, ACTH, or growth hormone (GH). On the basis of these experimental findings, they suggested that such a short feedback mechanism might serve to inhibit the release of LH. The possibility of a short feedback control of gonadotropin secretion has also been suggested by the experimental findings of Szontágh and Uhlarik (1964). Further evidence of an LH feedback action on the nervous system is provided by experiments in which LH was directly applied to the hypothalamus. Corbin and Cohen (1966) and David et al. (1966) have reported that LH implanted into the basal hypothalamus lowered the pituitary and serum LH content.

David et al. (1966) assumed that the same treatments might have lowered even the content of luteinizing hormone-releasing hormone (LHRH) of the ME. Ramirez et al. (1967) have demonstrated that without inducing any particular changes in the EEG, LH altered the pattern of unit firing in the hypothalamus. For several minutes the unit is quiescent following LH injection and its recovery is independent of EEG changes. In the experiment of Terasawa et al. (1969), injections of LH induced an elevation of the multiple-unit activity in the arcuate nucleus-ME region and a simultaneous depression of activity in the preoptic-anterior hypothalamic region. Posterior deafferentation of the hypothalamus did not affect the response to LH, but total, or anterior deafferentation blocked the elevation of activity in the ME. The depression response to LH in the preoptic region survived the hypothalamic deafferentation procedure, suggesting that LH "receptors" lie outside of the deafferented zone. The results are consistent with the hypothesis that LH exerts a short-loop or internal feedback action on brain-pituitary function, and evidence has been adduced that this initial effect is a positive or stimulant feedback action. The experimental findings of Makino et al. (1972) suggest that while large amounts of exogenous LH implanted directly into the ME exhibits a negative effect on pituitary content of LH, endogenous LH appears to exert a positive short feedback on the hypothalamo-hypophyseal-gonadotroph axis.

The experimental results of Corbin and co-workers have first suggested the existence of an internal or short-loop feedback mechanism in the control of FSH secretion (Corbin and Story, 1967; Corbin et al., 1970). Also in the experiment of Arai and Gorski (1968), implantation of FSH in the region of the ME but not in the adenohypophysis, amygdala and cerebral cortex, markedly inhibited ovarian compensatory hypertrophy suggesting that FSH may participate in the regulation of pituitary activity by an internal or short-loop feedback upon the hypothalamus. Ramirez et al. (1971) found that a minute amount of FSH implanted in the medial basal hypothalamus of 24-hr ovariectomized rats is capable of releasing pituitary FSH into the systemic blood as measured by RIA. This positive feedback action of exogenous FSH to increase the release of pituitary FSH in ovariectomized adult rats is quite specific since simultaneous determination of LH in these same animals did not reveal any alteration in immunoreactive plasma LH. Also, the results of Ojeda and Ramirez (1969, 1970, 1972) support the concept that FSH has a postive internal feedback action in the female. Furthermore, the results of Ojeda and Ramirez (1972) indicate the existence of a sexual difference in the participation of this short-loop feedback in the control of gonadotropin secretion: FSH might act as a positive feedback signal in the female but as a negative feedback signal in the male.

Experimental findings of Desjardins (1969) show that exogenous LH suppressed hypophysial LH release independently of FSH and that exogenous FSH suppressed hypophysial FSH release independently of LH. Similarly, Laborde et al. (1981) showed in castrated adult female rabbits that ovine FSH administered intravenously suppressed endogenous FSH secretion but failed to suppress endogenous LH. In contrast, Hirono et al. (1971) concluded that, in the rat, when LH exerts a direct action on the ME, the release of both FSH and LH is inhibited. On the other hand, when LH exerts a direct action on the anterior pituitary, only the release of LH is inhibited. Very recently, Kesner et al. (1986) presented data denying that elevated endogenous plasma LH concentrations had any effect on the LHRH pulse generator activity in the rhesus monkey.

ULTRASHORT FEEDBACK MECHANISM

The possible existence of an ultrashort feedback loop regulating FSH-releasing hormone (FSHRH) release was first suggested by Hyyppa et al. (1971). In their experiments, the subcutaneous administration of a crude FSHRH extract significantly decreased the level of hypothalamic FSHRH in castrated-hypophysectomized rats. Thus, they assumed that the brain contains structures sensitive to changing levels of this releasing hormone. The first clear evidence for ultrashort feedback action of a hypothalamic hormone on its own release was presented by McCann (1980) and Lumpkin et al. (1981). Following the intraventricular injection of somatostatin, they detected an unexpected increase of plasma GH concentration. This increase was observed within 30 min following the injection of either 1 or 5 μg somatostatin; however, the response to the smaller dose was more dramatic but shorter in duration. For explanation of the detected doses difference they assumed that from the higher dose, some somatostatin could reach the pituitary gland and suppressed GH release. As far as the effect of the smaller dose - e.g. increased GH output - is concerned, they supposed that the exogenous somatostatin suppressed the release of the endogenous somatostatin by an ultrashort feedback action. Increased concentration of somatostatin in the immediate vicinity of somatostatinergic neurons could exert inhibitory effect on the release of the endogenous hormone. An alternative explanation is that growth hormone-releasing hormone (GHRH) synthesizing perikarya present in the arcuate nuclei (Merchenthaler et al., 1984a) could be stimulated either directly by the injected somatostatin, or indirectly through the inhibition of some inhibitory neuronal elements synapsing upon GHRH-synthesizing

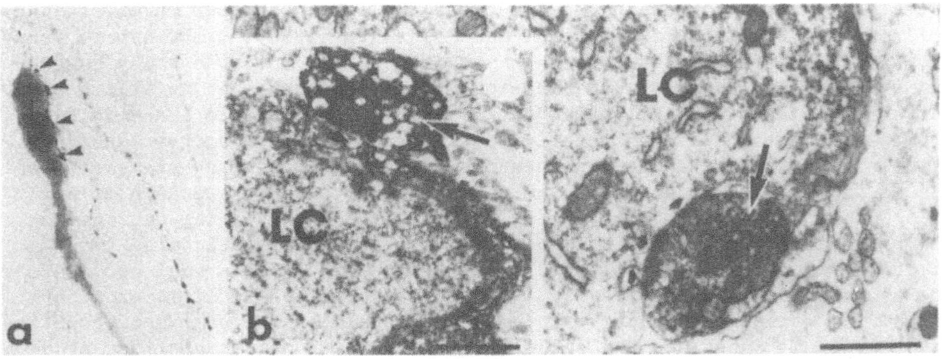

Fig. 4. (a) LHRH cell body in the medial septum. Close contact between axon (arrowheads) of another LHRH neuron and the cell body. Magnification X650. From Merchenthaler et al. (1984), by courtesy of Cell and Tissue Research. (b) Electron micrographs of rat medial preoptic area immunostained for LHRH synaptic connections between labeled axons (arrows) and labeled cell bodies (LC). The synaptic membrane specializations (indicated by the small arrowheads) appear to be of the symmetric Gray II type. All of the labeled axons (arrows) contain pleimorphic-clear vesicles and a few dense-core vesicles. Bar scale = 1 μm. From Leránth et al. (1985), by courtesy of Brain Research.

neurons. Increased GHRH release is then followed by increased GH release from the pituitary gland leading to elevated GH plasma level. Recently, McCann et al. (1984) demonstrated that GHRH administered intraventricularly, also exhibited negative ultrashort feedback effect upon its own hypothalamic release. Corticotropin-releasing factor (CRF), on the other hand, injected intraventricularly, exerted positive ultrashort feedback action upon endogenous CRF release, leading to increased stress induced ACTH release. According to McCann et al. (1984), ultrashort feedback action of brain peptides seems to be "the rule rather than the exception". Keeping in mind that the postulated ultrashort feedback regulation of the hypophysiotropic hormones needs further substantiation, the question has to be raised: Through what routes can hypophysiotropic hormones act on the secretory activity of their own neuronal source? Theoretically, an ultrashort feedback loop can either be (1) neuronal, or (2) humoral. The latter can either be conducted via the cerebrospinal fluid (CSF), or can be bloodborne. In either case, the feedback can act directly upon the peptidergic neurons in question, or can be mediated through interneurons.

Neuronal Ultrashort Feedback Loop

We consider the close apposition of neuronal processes and somata containing the same hypophysiotropic hormone as the anatomical indication of a possible neuronal ultrashort feedback mechanism (Fig. 4a). Apposition of LHRH fibers and LHRH-synthesizing perikarya in the medial preoptic-septal regions may serve as an example of a possible ultrashort feedback circuit whereby LHRH axons might directly influence the activity of LHRH neurons. The existence of synaptic contacts between LHRH terminals and LHRH perikarya was first suggested by Marschall and Goldsmith (1980) and Merchenthaler et al. (1984b) on the basis of light microscopic observations. Marschall and Goldsmith (1980) reported possible synaptic contacts between LHRH-containing axons and neurons. Using a more sensitive technique (silver intensification of the DAB polymer formed in the PAP technique), Merhenthaler et al. (1984b) confirmed the findings of Marschall and Goldsmith (1980). Furthermore, they described LHRH positive structures in their light microscopic study suggesting synaptic contacts between LHRH-containing perikarya and their own recurrent axon collaterals. Silverman et al. (1984) demonstrated LHRH-containing nerve terminals on nonimmunoreactive perikarya by immunoelectronmicroscopy. It was suggested that short interneurons - lacking LHRH - may transfer its signals from one LHRH neuron to the other. Very recently, Léránth et al. (1985) presented evidence for the existence of LHRH-containing boutons in synaptic contact with LHRH containing neurons in the preoptic area (Fig. 4b). All of the connections established by LHRH-immunoreactive axonal terminals were symmetrical (Gray's type II). The existence of peptidergic axon terminals having synaptic contacts with neurons containing the same neuropeptides is not unique. Substance P (Tsuruo et al., 1984), somatostatin (Jew et al., 1984), ACTH (Kiss et al., 1983), cholecystokinin (Takagi et al., 1984), CRF (Liposits et al., 1985) containing terminals on cell bodies containing the same peptide have been reported.

Electrophysiological observations (Renaud et al., 1975) and the fact that synapses between pre- and post-synaptic LHRH components are symmetrical (Leranth et al., 1985) suggest that signals carried by the neuropeptide LHRH may be inhibitory in nature. These morphological observations support the concept of an ultrashort feedback loop. Many subsequent experiments have suggested that through an ultrashort feedback mechanism LHRH may control own production and/or release and as a consequence, the release of gonadotropins from the anterior pituitary.

Humoral Ultrashort Feedback Loop

In the last decade, new studies on the portal circulation have revealed important anatomical details which support our earlier supposition that this circulatory system may participate in ultrashort feedback control of anterior pituitary hormone release (Sétáló et al., 1976a). The most important data from these studies concerning the vascular basis of ultrashort feedback are the following:

(a) The hypophysial arteries entering the median eminence (ME) form the primary capillary plexus and give rise to special capillary loops, which in turn have egress both in the portal supratuberal vessels toward the adenohypophysis, and in the subependymal capillary plexus. The subependymal capillaries have multiple capillary connections to the retrochiasmatic and arcuate nucleus regions through blood vessels curving around the lateral edges of the infundibular recess (Fig. 5) (Török, 1954, 1962; Ambach et al., 1976; Page et al., 1978). It is important to keep in mind that these capillaries have fenestrated endothelial lining (Page and Dovey-Hartman, 1984).

(b) Venous blood from the medial basal hypothalamus (MBH, including the arcuate nuclei) is drained off by the tuberal veins (Ambach et al., 1976). Thus, a relatively high concentration of all bioactive substances characteristic of pituitary portal blood can only be provided for the MBH, but not for the rest of the diencephalon.

Coupling the results of studies on pituitary circulation with those indicating the ultrashort feedback effect of experimentally administered hypophysiotropic substances, the question seems to be close at hand: Are hypophysiotropic hormones transported to the MBH under physiological conditions? In the absence of direct data about the concentrations of the hypophysiotropic hormones present in the capillary blood irrigating the MBH, we have no clearcut answer for this question. We know, however, from pituitary transplantation studies that although more or less release of different anterior pituitary hormones can be detected from all surviving pituitaries, the

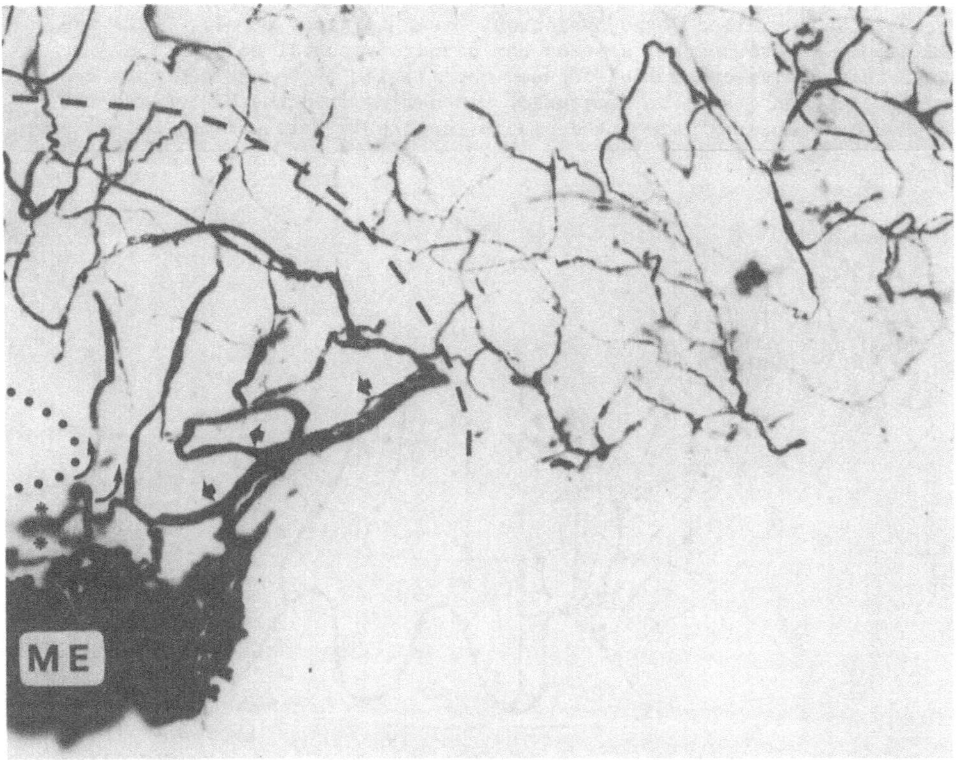

Fig. 5. Blood vessels of the hypothalamus filled with Indian ink. Dashed line indicates the lateral boundary of the MBH where the capillary beds originating from the ME and from other hypothalamic arteries meet. Curved arrows indicate direction of blood flow in the capillaries originating from the subependymal plexus (asterisks) of the ME. Short arrows indicate branches of anterior hypophysial artery. Dotted line shows the ependymal wall of the third ventricle. Magnification X200.

best functional grafts are those located in the MBH (Halász et al., 1962, 1965; Flament-Durand, 1964, 1965; Knigge, 1962). This region of the hypothalamus was termed by Halász et al. (1962, 1965) the "hypophysiotropic area" (HTA). Maintained secretory activity of actopic pituitary grafts is presumed to be due to their access to the relevant hypophysiotropic hormones. The question is how do the hypophysiotropic hormones reach the grafts which are located in the HTA?

Theoretically, these hormones can reach the graft via the cerebrospinal fluid (CSF) with, or without, the active participation of the ependymal cells lining the infundibular recess of the third ventricle. The fact, however, that the grafts implanted into the HTA and being in direct contact with the neuropile were more active than those located in the third ventricle (Halász et al., 1962; Sétáló et al., 1977) strongly suggested that hypophysiotropic hormones reached the grafts through the special blood circulation of the MBH. Consequently, we have to suppose that the HTA is hypophysiotropic because it is supplied with bloodborne hypophysiotropic hormones from the portal circulation. It was in 1975 that the mediation of the hypophysiotropic effect of the so-called "hypophysiotropic area" (Halász et al., 1962) was explained by Sétáló on the basis of the pituitary portal circulation. He assumed that the vascular link between capillaries of the ME and the "hypophysiotropic area", and the retrograde blood flow from the ME towards the "hypophysiotropic area" are the decisive factors in this effect (Sétáló et

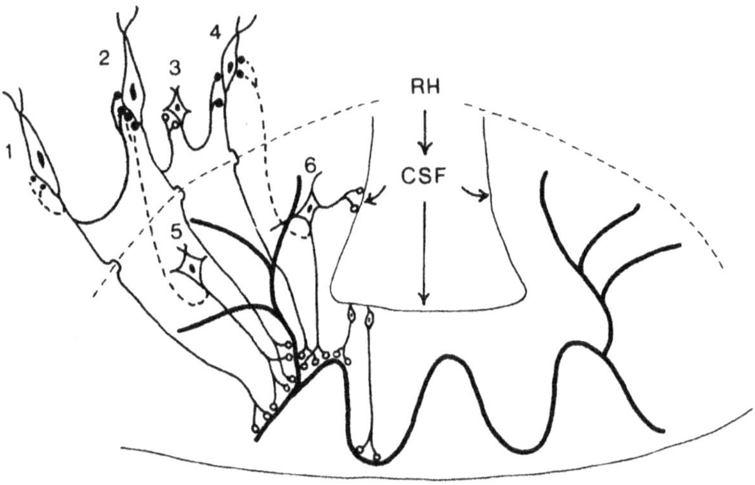

Fig. 6. Schematic illustration of the theoretical possibilities of ultrashort feedback regulation of the synthesis and/or release of releasing hormones (RH). Humoral ultrashort feedback: (a) RH-synthesizing cells (1,2,4,5) release their hormones into the capillaries (heavy lines) of the median eminence (ME). Portal blood is partially drained into the medial basal hypothalamus (delineated by dashed lines), where some neurons (5,6) may have receptors for RH's and can influence the synthesis and/or release of RH's through their axon collaterals abutting upon RH-synthesizing neurons (2,4), or through their axon terminals ending in the ME. (b) Rh's are released into the cerebrospinal fluid (CSF) at the periventricular ependymal organs and can reach the portal blood aided by tanycytes. Actual levels of RH's in CSF can be detected by supposed receptor endings situating amongst the ependymal cells of the third ventricle and axon collaterals of the same cells can reach RH-synthesizing cells at remote places. Neuronal ultrashort feedback: Axon collaterals of RH-synthesizing neurons may synapse (a) upon their own perikaryon (1); (b) on other RH-synthesizing cells (1-2); or on (c) interneurons (2-3) which in turn can end on other RH-synthesizing neurons (3-4). Such connections between RH-neurons can be either inhibitory in nature, or have synchronizing effect.

al., 1976a, 1976b). This opinion was further supported by the findings obtained from animals in which the ME was surgically removed on one side (Sétáló, 1980) and a pituitary graft was implanted into the HTA. The ME is the place from where the MBH receives its blood supply. On the side of the ablated ME, the grafts exhibited the histological signs of an inactive pituitary, while on the other side, where the ME was intact, the grafts seemed to be morphologically active (Sétáló and Liposits, 1986). Depending on the amounts of the active pituitary tissue, some of these animals (which were, of course, hypophysectomized) even showed vaginal estrous cycles indicating the functional value of such grafts (Sétáló et al., 1977). Because these grafts had access to the CSF and to the neuropile of the MGH the same way on both sides, there seems to remain only one explanation of the morphological differences existing between the two sides of the grafts: In the course of revascularization of the implanted pituitary, capillaries grew into the graft from all directions. However, the capillaries on the side of the intact ME are in connection with the portal circulation, while on the other side, where the source of portal blood had previously been removed, they are connected to the general hypothalamic circulation. These experimental findings indirectly indicate that the HTA in the rat is supplied with portal blood, and give morphological basis for the postulation of a humoral ultrashort feedback controlling the release of the hypophysiotropic hormones. One can speculate about neurons located in the HTA which could monitor the actual levels of hypophysiotropic hormones in the portal circulation and influence them either locally or with remote axon collaterals, according to actual needs. The existence of such neurons, however, remains to be verified. The theoretical possibilities of ultrashort feedback regulation of the release and/or synthesis of releasing and inhibiting hormones is schematically represented in Fig. 6.

REFERENCES

Ambach, G., Palkovits, M., and Szentágothai, J., 1976, Blood supply of the rat hypothalamus. IV. Retrochiasmatic area, median eminence, arcuate nucleus, Acta Morph. Acad. Sci. Hung., 24:93.

Arai, Y., and Gorski, R. A., 1968, Inhibition of ovarian compensatory hypertrophy by hypothalamic implantation of gonadotrophin in androgen-sterilized rats: evidence for "internal" feedback, Endocrinology, 82:871.

Assies, J., Schellekens, A. P., and Touber, J. L., 1978, Protein hormones in cerebrospinal fluid. Evidence for retrograde transport of prolactin from the pituitary to the brain in man, Clin. Endocrinol., 8:487.

Corbin, A., and Story, J. C., 1967, "Internal" feedback mechanism: response of pituitary FSH and of stalk-median eminence follicle stimulating hormone-releasing factor to median eminence implants of FSH, Endocrinology, 80:1006.

Corbin, A., and Cohen, A. I., 1966, Effect of median eminence implants of LH on pituitary LH of female rats, Endocrinology, 78:41.

Corbin, A., Daniels, E. L., and Milmore, J. E., 1970, An "internal" feedback mechanism controlling follicle stimulating hormone releasing factor, Endocrinology, 86:735.

David, M. A., Fraschini, F., and Martini, L., 1966, Control of LH secretion: role of a "short" feedback mechanism, Endocrinology, 78:55.

Desjardins, C., 1969, Alteration of hypophysial LH and FSH release by exogenous LH and FSH in orchidectomized mice, Proc. Soc. Exp. Biol. Med., 130:535.

Flament-Durand, J., 1964, Morphologie et fonction de transplants hypophysaires dans l'hypothalamus chez le rat, C. R. Acad. Sci. (Paris), 259:4376.

Flament-Durand, J., 1965, Observations on pituitary transplants into the hypothalamus of the rat, Endocrinology, 77:446.

Green, J. D., and Harris, G. W., 1947, The neurovascular link between the neurohypophysis and adenohypophysis, J. Endocrinol., 5:136.

Green, J. D., and Harris, G. W., 1949, Observation of the hypophysioportal vessels of the living rat, J. Physiol., 108:359.

Halász, B., Pupp, L., and Uhlarik, S., 1962, Hypophysiotrophic area in the hypothalamus, J. Endocrinol., 25:147.

Halász, B., and Szentágothai, J., 1958, Über die unmittelbare Rückwirkung einer vom Hypophysenvorderlappen erzeugten Substanz auf den Hypothalamus, Acta Physiol. Acad. Sci. Hung., 14:Suppl. 6.
Halász, B., and Szentágothai, J., 1960, Control of adrenocorticotrophic function by direct influence of pituitary substance on the hypothalamus, Acta Morph. Acad. Sci. Hung., 9:251.
Halász, B., Pupp, L., Uhlarik, S., Tima, L., 1965, Further studies on the hormone secretion of the anterior pituitary transplanted into the hypophysiotrophic area of the rat hypothalamus, Endocrinology, 77:343.
Hirono, M., Igarashi, M., and Matsumoto, S., 1971, Short- and auto-feedback mechanism of LH, Endocr. Jap., 18:175.
Hyyppa, M., Motta, M., and Martini, L., 1971, "Ultrashort" feedback control of follicle stimulating hormone-releasing factor secretion, Neuroendocrinology, 7:227.
Jew, J., Leránth, C., Arimura, A., and Palkovits, M., 1984, Preoptic LHRH and somatostatin in the rat median eminence: an experimental light and electron microscopic immunocytochemical study, Neuroendocrinology, 38:169.
Kawakami, M., and Sawyer, C. H., 1959, Induction of behavioral and electroencephalographic changes in the rabbit by hormone administration or brain stimulation, Endocrinology, 65:631.
Kesner, J. S., Kaufman, J. M., Wilson, R. C., Kuroda, G., and Knobil, E., 1986, On the short-loop feedback regulation of the hypothalamic luteinizing hormone-releasing hormone "pulse generator" in the rhesus monkey, Neuroendocrinology, 42:109.
Kiss, J. Z., and Williams, T. H., 1983, ACTH-immunoreactive boutons form synaptic contacts in the hypothalamic arcuate nucleus of rat: evidence for local opiocortin connections, Brain Res., 263:142.
Knigge, K. M., 1962, Gonadotrophic activity of neonatal pituitary glands implanted in the rat brain, Am. J. Physiol., 202:387.
Laborde, N. P., Wolfsen, A. R., and Odell, W. D., 1981, Short loop feedback system for the control of follicle-stimulating hormone in the rabbit, Endocrinology, 108:72.
Leránth, C., Segura, L. M. G., Palkovits, M., MacLusky, N. J., Shanabrough, M., and Naftolin, F., 1985, The LHRH-containing neuronal network in the preoptic area of the rat: demonstration of LHRH-containing nerve terminals in symaptic contact with LHRH neurons, Brain Res., 345:332.
Liposits, Z., Paull, W. K., Sétáló, G., and Vigh, S., 1985, Evidence for local corticotropin releasing factor (CRF) immunoreactive neuronal circuits in the paraventricular neucleus of the rat hypothalamus, Histochemistry, 83:5.
Lumpkin, M. D., Negro-Vilar, A., and McCann, S. M., 1981, Paradoxical elevation of growth hormone by intraventricular somatostatin: possible ultrashort-loop feedback, Science, 211:1072.
Makino, T., Fang, V. S., and MacDonald, G. J., 1971, Effect of implantation of anti-LH serum into median eminence on rat pituitary and serum LH, Fed. Proc., 140:703.
Marschall. P. E., and Goldsmith, P. C., 1980, Neuroregulatory and neuroendocrine GnRH pathways in the hypothalamus and forebrain of the baboon, Brain Res., 193:353.
McCann, S. M., 1980, Control of anterior pituitary hormone release by brain peptides, Neuroendocrinology, 31:355.
McCann, S. M., Lumpkin, M. D., Ono, N., Khorram, O., Ottlecz, A., Koenig, J., Bedran De Castro, J., Krulich, L., and Samson, W. K., 1984, Interactions of brain peptides within the hypothalamus to alter anterior pituitary hormone secretion, in: "Endocrinology," F. Labrie and L. Proulx, eds., pp. 185, Elsevier Science Publishers, New York.
Merchenthaler, I., Vigh, S., Schally, A. V., and Petrusz, P., 1984a, Immunocytochemical localization of growth hormone releasing factor in the rat hypothalamus, Endocrinology, 115:1082.

Merchenthaler, I., Görcs, T., Sétáló, G., Petrusz, P., and Flerkó, B., 1984b, Gonadotropin-releasing hormone (GnRH) neurons and pathways in the rat brain, Cell Tiss. Res., 237:15.
Mezey, E., Kivovics, P., and Palkovits, M., 1979, Pituitary-brain retrograde transport, Trends Neurosci., 2:57.
Mezey, E., Palkovits, M., De Kloet, E. R., Verhoef, J., and de Wied, D., 1978, Evidence for pituitary-brain transport of a behaviorally potent ACTH analog, Life Sci., 22:831.
Motta, M., Fraschini, F., and Martini, Z., 1969, "Short" feedback mechanism in the control of anterior pituitary function, in: "Frontiers in Neuro-endocrinology," F. Ganong and L. Martini, eds., p. 241, Oxford University Press, London.
Ojeda, S. R., and Ramirez, V. D., 1969, Automatic control of LH and FSH secretion by short feedback circuits in immature rats, Endocrinology, 84:786.
Ojeda, S. R., and Ramirez, V. D., 1970, Failure of estrogen to block compensatory ovarian hypertrophy in prepubertal rats bearing medial basal hypothalamic FSH implants, Endocrinology, 86:50.
Ojeda, S. R., and Ramirez, V. D., 1972, Different pituitary-gonadal response to hemicastration in female and in male rats bearing intrahypothalamic FSH implants, Neuronendocrinology, 10:161.
Oliver, C., Mical, R. S., and Porter, J. C., 1977, Hypothalamic-pituitary vasculature. Evidence for retrograde blood flow in the pituitary stalk, Endocrinology, 101:598.
Page, R. B., Leure-DuPree, A. E., and Bergland, R. M., 1978, The neurohypophyseal capillary bed. II. Specializations within median eminence, Am. J. Anat., 153:33.
Page, R. B., and Dovey-Hartman, B. J., 1984, Neurohemal contact in the internal zone of the rabbit median eminence, J. Comp. Neurol., 226:274.
Porter, J. C., Barnea, A., Cramer, O. M., and Parker, C. R., 1978, Hypothalamic peptide and catecholamine secretion. Roles for portal and retrograde blood flow in the pituitary stalk in the release of hypothalamic dopamine and pituitary prolactin and LH, Clin. Obstet. Gynecol., 5:271.
Popa, G. T., and Fielding, W., 1930, A portal circulation from the pituitary to the hypothalamic region, J. Anat., 65:88.
Ramirez, V. D., Komisaruk, B. R., Whitmoyer, D. I., and Sawyer, C. H., 1967, Effects of hormones and vaginal stimulation on the EEG and hypothalamic units in rats, Am. J. Physiol., 212:1376.
Ramirez, V. D., Ojeda, S. R., and Alvarez, E. O., 1971, Hypothalamic receptors for FSH and estrogen. Proceedings of the Seventh Pan-American Congress of Endocrinology, Sao Paulo, Brazil, August 16-21, 1970, Excerpta Medica, Amsterdam.
Renaud, L. P., Martin, J. B., and Brazeau, P., 1975, Depressant action of TRH, LHRH and somatostatin on activity of central neurons, Nature (Lond.), 255:233.
Sétáló, G., 1980, Vascular communications between the brain and the anterior pituitary gland, in: "Advances in Physiological Science, Vol. 15, Reproduction and Development," B. Flerkó, G. Sétáló, L. Tima, eds., pp. 55, Pergamon Press, Oxford, Akademiai Kiado Budapest.
Sétáló, G., Vigh, S., Schally, A. V., Arimura, A., and Flerkó, B., 1975, Immunohistological investigations of the LHRH-synthesizing neuron system of the rat, Symp. Int. Soc. Psychoneuroendocrinology, Visegrád, p. 77.
Sétáló, G., Vigh, S., Hagino, N., and Flerkó, B., 1976a, Immunihistological observations on the "hypophysiotrophic" area of the hypothalamus, Acta Morph. Acad. Sci. Hung., 24:79.
Sétáló, G., Vigh, S., Schally, A. V., Arimura, A., and Flerkó, B., 1976b, Immunohistological investigations on the LHRH-synthesizing neuron system of the rat, in: "Cellular and Molecular Bases of Neuroendocrine Processes," Symp. Int. Soc. Psychoneuroendocrinology, Visegrád 1975, E. Endroczi, ed., p. 77, Akadémiai Kiadó, Budapest.

Sétáló, G., Horváth, J., Schally, A. V., Arimura, A., and Flerkó, B., 1977, Effect of the isolated removal of the median eminence (ME) and pituitary stalk (PS) on the immunohistology and hormone release of the anterior pituitary gland grafted into the hypophysiotrophic area (HTA) and/or of the in situ pituitary gland, Acta Biol. Acad. Sci. Hung., 28:333.

Sétáló, G., and Liposits, Z., 1986, The possible role of the portal circulation in ultrashort feedback control of the anterior pituitary gland, in: "Pars Distalis of the Pituitary Gland. Structure, Function and Regulation," F. Yoshimura and A. Gorbman, eds., Excerpta Medica, pp. 315.

Silverman, A. J., 1984, Luteinizing hormone-releasing hormone containing synapses in the diagonal band and preoptic area of the guinea pig, J. Comp. Neurol., 227:452.

Szentágothai, J., Flerkó, B., and Halász, B., 1962, "Hypothalamic Control of the Anterior Pituitary," Publishing House of the Hungarian Academy of Sciences, Budapest.

Szontágh, F. E., and Uhlarik, S., 1964, The possibility of a direct "internal" feedback in the control of pituitary gonadotrophin secretion, J. Endocrinol., 29:203.

Takagi, H., Kubota, Y., Mori, S., Tateishi, K., Hamaoka, T., and Tokyama, M., 1984, Fine structural studies of cholecystokinin-8-like immunoreactive neurons and axon terminals in the nucleus of tractus solitarius of the rat, J. Comp. Neurol., 227:369.

Terasawa, E., Whitmoyer, D. I., and Sawyer, C. H., 1969, Effect of luteinizing hormone on multiple-unit activity in the rat hypothalamus, Am. J. Physiol., 217:1119.

Török, B., 1954, Lebendbeobachtung des Hypophysen Kreislaufes an Hunden, Acta Morph. Acad. Sci. Hung., 4:83.

Török, B., 1962, Neue Angaben zum Blutkreislauf der Hypophyse. Verh. 1. Eur. Anatomen-Kongr., Strasbourg 1960, Anat. Anz., 109:622.

Tsuruo, Y., Hisano, S., and Daikoku, S., 1984, Morphological evidence for synaptic junctions between substance P-containing neurons in the arcuate nucleus of the rat, Neurosci. Lett., 46:65.

Wislocki, G. B., 1937, The vascular supply of the hypophysis cerebri of the cat, Anat. Rec., 69:361.

Wislocki, G. B., 1938, The vascular supply of the hypophysis cerebri of the rhesus monkey and man, Res. Publ. Ass. Nerv. Ment. Dis., 17:48.

Wislocki, G. B., and King, L. S., 1936, The permeability of the hypophysis and the hypothalamus to vital dyes, with a study of the hypophysial vascular supply, Am. J. Anat., 58:421.

THE HYPOTHALAMO-PITUITARY-GONADAL SYSTEM:

ROLE OF PEPTIDES AND SEX STEROIDS

M. Motta, D. Dondi, R. Maggi, E. Messi,
S. Zoppi, M. Zanisi, and F. Piva

Department of Endocrinology
University of Milano
21, Viz Andrea del Sarto
Milano, Italy 20129

INTRODUCTION

The activity of the hypothalamo-pituitary unit is modified by a variety of modulatory influences either originating in the central nervous system or coming directly from the periphery. One type of input is represented by nervous signals which are transferred to the hypothalamic hormone producing neurons through the release, at neuronal junctions, of "classical" neurotransmitters as well as of different classes of peptides. Although the available evidence cannot be reviewed in the present report, it has been clearly established that catecholamines, acetylcholine, serotonin, prostaglandins, histamine, gamma-aminobutyric acid, substance P, neurotensin, peptides common to the brain and to the gastrointestinal tract, opioid peptides are all involved in the regulation of the hypothalamo-pituitary function (see Müller et al., 1977; McCann, 1980; Morley, 1981, for references). It is interesting that some of the peptides present in the mammalian brain have also been identified in the amphibian skin (TRH, neurotensin, bombesin, etc.). Recently two new peptides, dermorphin and sauvagine, have been isolated from the skin of the South American frog, *Philomedusa sauvagei*; they have been subsequently chemically identified and obtained by synthesis (Montecucchi et al., 1980; 1981). Evidence is accumulating which suggests that also these peptides may play a physiological role as modulators of the activity of the hypothalamo-pituitary unit (see Motta, 1985, for references). The other type of influence is hormonal in nature and is provided by the hormones produced in the peripheral target glands, by the pituitary hormones and possibly by the hypothalamic hormones themselves, which all act through different classes of feedback mechanisms in regulating the activity of the hypothalamo-pituitary system (see Piva et al., 1979, for references).

This paper reports the results obtained in the authors' laboratory on the effects exerted by some families of peptides in the regulation of the activity of the hypothalamo-pituitary-gonadal axis. Attention will be particularly given to the following classes of peptides: opioids, neurotensin, and sauvagine. In addition, recent evidence supporting the presence of an ultrashort loop feedback effect of LHRH will be presented.

ROLE OF OPIOID AGONISTS AND ANTAGONISTS IN THE CONTROL OF GONADOTROPIN SECRETION

The presence of elevated concentrations of endogenous opioid-like peptides in the hypothalamus and other brain structures involved in the control of neuroendocrine processes has suggested that these principles might participate in the regulation of anterior pituitary function (Cuello, 1983). In both experimental animals and humans, the administration of the naturally occurring peptides (beta-endorphin, met-enkephalin, leu-enkephalin, dynorphin, etc.) or of their synthetic analogs stimulates PRL release and, in general, inhibits LH and FSH secretion (Morley, 1981; Clement-Jones and Rees, 1982). This observation receives support also from the data obtained in the authors' laboratory while testing a new potent morphine-like peptide, dermophin (see Giudici et al., 1984, for references). This peptide, when given either subcutaneously to normal male rats or intraventricularly (i.v.t.) to long-term castrated male rats, is able to decrease serum LH levels, while it is unable to modify serum FSH titers (Motta et al., 1984; Motta, 1985).

However, a few recent reports suggest that, in particular experimental conditions, brain opioids and their synthetic analogs may stimulate rather than inhibit gonadotropin release. Motta and Martini (1982) have demonstrated that the i.v.t. injection of either methionine-enkephalin or one of its synthetic analogs (D-Ala^2)-methionine-enkephalinamide induces a significant increase of serum LH levels in long-term castrated female rats, while is unable to modify serum levels of FSH. A stimulatory effect of the opioids on LH secretion has also been obtained following the i.v.t. infusion of beta-endorphin and leu-enkephalin (Takahara et al., 1978; Leadem and Kalra, 1985), and the systemic administration of morphine (Pang et al., 1977; Cicero et al., 1980).

Recently, it has been suggested that tht effects of the opioids and of their antagonists on gonadotropin secretion largely depend on the steroid "milieu" present in the rat at time of drug administration (Sylvester et al., 1982; Bhanot and Wilkinson, 1983; Piva et al., 1984, 1985, 1986). The influence of the steroid "milieu" on the hypothalamo-pituitary responses to the administration of opioid antagonists has been established also in subhuman primates (Gasselin et al., 1983), and in women (Reid et al., 1983). To examine this issue closely, two sets of experiments have been performed respectively in adult male and female rats.

Experiments in Male Animals

For these experiments, morphine (an opioid agonist) and naloxone (an opioid antagonist) have been injected i.v.t. into normal or long-term castrated male rats. The animals were killed by decapitation at different time intervals after treatment; serum LH and FSH were measured by specific radioimmunoassays. When morphine (200 μg/rat) and naloxone (7.5 and 15 μg/rat) are injected i.v.t. into normal male rats, a significant increase of serum levels of LH is observed 10 and 20 min after treatment. Conversely, when the same dose of morphine is injected i.v.t. to castrated male rats, it significantly decreases serum LH levels beginning 40 min after treatment and lasting up to 180 min. None of these treatments exert any consistent effect on serum FSH levels either in normal or castrated male rats.

These findings allow some considerations: the stimulatory effect exhibited by naloxone on LH secretion in normal male rats confirms that endogenous opioids exert mainly an inhibitory influence on this gonadotropin; the stimulatory effect exerted by morphine on LH secretion in normal male rats suggests that opioid pathways (or receptors), influencing in a stimulatory way the mechanisms controlling LH secretion, might also exist. In addition, the opposite effects (stimulatory and inhibitory) exerted by morphine

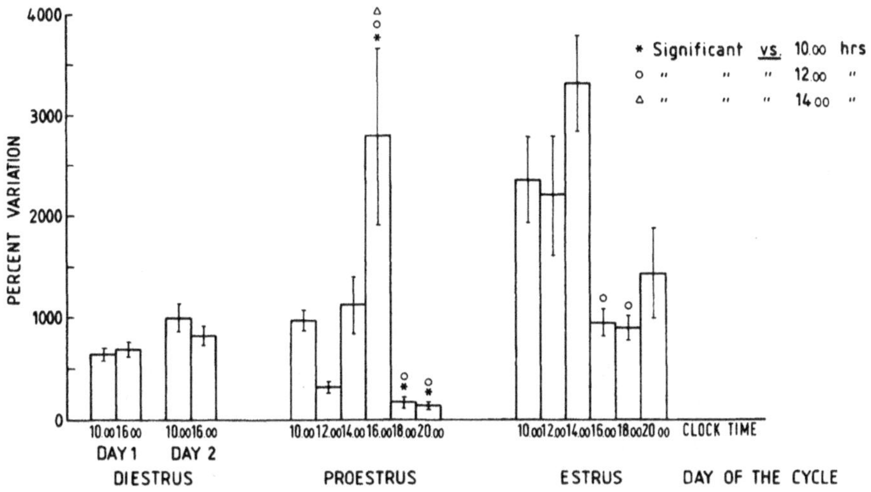

Fig. 1. Percent variations vs. controls of serum LH levels in adult cycling female rats treated s.c. with naloxone (2.5 mg/kg) at different hours of the day throughout the estrous cycle and killed 20 min after treatment.

respectively in normal and castrated male rats indicate that the endocrine "milieu" present in the animal at the time of the experiment is crucial in directing the effects exerted by the opioids on LH secretion. Finally, the lack of effect of morphine and naloxone on the release of FSH suggests that brain opioids do not participate in the regulation of the release of this gonadotropin, indicating that the brain mechanisms controlling LH and FSH release are distinct (Motta and Martini, 1982; Piva et al., 1984, 1985, 1986).

Experiments in Female Animals

To further elucidate the possibility of the involvement of the hormonal environment on the endocrine effects of opioids, experiments have been performed in adult female rats with a regular four-day cycle, characterized by spontaneous changes in the levels of estrogens and progesterone during the different phases of the estrous cycle (Neill and Smith, 1974).

Effect of naloxone on gonadotropin release. The rats have been injected subcutaneously with naloxone (2.5 mg/kg) at different hours of the day, during the various phases of the estrous cycle and sacrificed 20 min after treatment. Serum LH and FSH were assayed by specific radioimmunoassays. The subcutaneous injections of naloxone stimulate LH release in every phase of the estrous cycle; however the magnitude of the responses appears to be highly variable (Fig. 1). Increases of the order of 700-1000% (expressed as percent variations over saline injected controls) are observed during the two days of diestrus, at 1000 and 1400 h of the day of proestrus, and at 1600, 1800 and 2000 h of the day of estrus. Much higher responses (of the order of 2700-3300%) are recorded at 1600 h of the day of proestrus and at 1000, 1200 and 1400 h of the day of estrus. The LH response to naloxone appears to be obliterated at 1800 and 2000 h of the day of proestrus (at the time of the preovulatory surge in the authors' laboratory conditions). Serum levels of FSH are not affected by the treatment at any of the time intervals considered (results not shown).

These data show that in normally cycling female rats naloxone is effective in stimulating LH secretion during each day of the estrous cycle. This finding underlines once more that endogenous opioids influence in an inhibitory way LH secretion also in normally cycling female rats. The stimulatory effect of naloxone has been shown to be maximal just before the spontaneous proestrous surge of LH and in the morning of estrus and minimal at the time of the spontaneous proestrous LH surge (Piva et al., 1985). It is possible that the enhanced LH response to naloxone observed immediately before the proestrous LH surge does not reflect changes in the modulatory influences endogenous opioids exert on LH release; this major response might actually be due to the sensitizing effect exerted by estrogens (secreted at this time of the estrous cycle) on the pituitary responses to LHRH (Kalra and Kalra, 1974). On the contrary, the enhanced LH response to naloxone observed in the morning of the day of estrus is probably due mainly to a change of the opiatergic tone controlling LHRH release (Ching, 1983). The present results confirm that the opiatergic tone controlling LH secretion may be influenced by the steroid "milieu" existing in the animal, as shown by the conspicuous and significant variations in the responses of LH after naloxone administration during the different phases of the estrous cycle. Finally, the observation that naloxone does not affect FSH release underlines once more that the central mechanisms controlling LH and FSH secretion are substantially different (Piva et al., 1985).

Brain opioid receptors. It is known that the effects of the naturally occurring opioids (met-enkephalin, leu-enkephalin, endorphin, dynorphin, etc.) are exerted through the interaction with specific binding sites. Different classes of brain opioid receptors (mu, kappa, delta, etc.) have been described (Cuello, 1983; Paterson et al., 1983); among these the mu receptors appear to be particularly relevant for the control of gonadotropin secretion (Pfeiffer et al., 1983).

The possibility that changes of the binding characteristics of brain mu opioid receptors may be linked to variations in the hormonal environment has been investigated. To this purpose, the number and the affinity of mu receptors have been evaluated in the whole brain membrane preparations of regularly cycling female rats killed in the various phases of the estrous cycle. As previously mentioned, conspicuous variations in the levels of estrogen and progesterone spontaneously occur during the different phases of the estrous cycle (Neill and Smith, 1974).

Fig. 2 shows that the number of the mu opioid receptors (expressed as fmoles of ^{3}H-dihydromorphine bound/mg protein) shows remarkable changes in the whole brain of female rats during the different phases of the estrous cycle. In particular, their number is rather low during the second day of diestrus and in the morning of proestrus, while an increase in the number of such receptors is observed at 1200 h of the day of proestrus. After this time, there is a progressive decline in the number of mu receptors which remains low and rather constant throughout the day of estrus. A significant increase occurs again late in this day (1800 h) followed by a decrease which gradually brings the number of receptors, during diestrus 1, to the low levels observed during diestrus 2. No modifications in the affinity constant is every observed.

These data suggest that the estrus-linked modifications in the number of brain mu opioid receptors are induced by changes in the steroid "milieu" occurring during the different phases of the estrous cycle. It is possible that the number of brain mu receptors increases during the day of proestrus under the influence of estrogens secreted during this phase of the estrous cycle (Neill and Smith, 1974) and that their decline during late proestrus and during estrus are the consequence of the concomitant effects of estrogens and progesterone secreted during these phases (Neill and Smith, 1974).

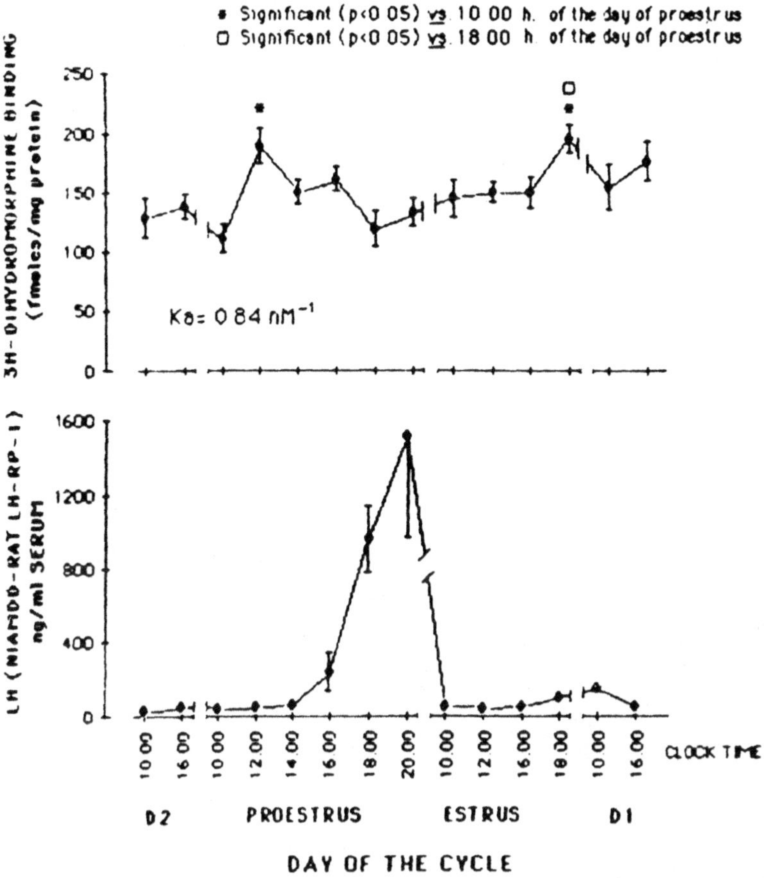

Fig. 2. Number of brain mu opioid receptors during the various phases of the estrous cycle of female rats.

It is difficult to strictly compare the results here obtained with those previously reported on the effects of naloxone on LH release during the different phases of the estrous cycle. As previously shown, naloxone induces a moderate LH response at 1200 h of the day of proestrus and at 1800 h of the day of estrus, when the number of brain mu receptors is elevated. The apparent discrepancy between the two groups of results may be due to several reasons. First of all, the changes of the effects of naloxone on serum LH levels are probably due not only to a stimulation of LHRH secretion, but also to changes in the sensitivity of the anterior pituitary to the liberated LHRH (Kalra and Kalra, 1974). Secondly, even if the majority of the data available clearly indicate that LHRH secretion is due mainly to an interaction of brain opioids with the mu receptors (Pfeiffer et al., 1983), the possibility is not totally excluded that also other types of opioid receptors might participate in such a control.

An additional piece of evidence confirming the existence of an interaction between sex steroids and brain opioids has been obtained recently in the authors' laboratory. The binding characteristics of mu opioid receptors have been evaluated in the hypothalami of rats submitted to different endocrine manipulations at birth. The following groups of animals have been studied at different ages: normal male rats, normal female rats, male rats orchidectomized on the second day of life (deandrogenized males) and female rats treated on the second postnatal day with testosterone (androgenized females).

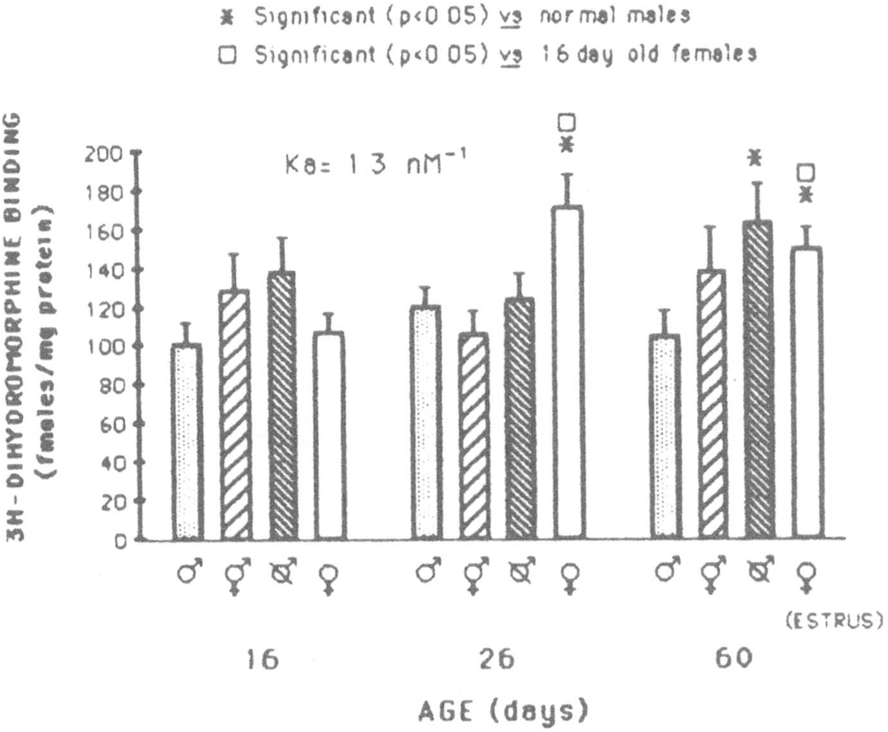

Fig. 3. Effect of neonatal endocrine environment on the number of hypothalamic mu opioid receptors in rats of different ages.

Fig. 3 shows that at 60 days of age the number of mu hypothalamic receptors does not differ in normal males and in androgenized females. In the hypothalami of normal females the number of mu receptors, which is similar to that of normal males and of androgenized females at 16 days, increases significantly at 26 days and remains high at 60 days. In deandrogenized males of 60 days, the number of mu receptors is not different from that found in normal females of the same age. Normal females and deandrogenized males of 60 days have a number of mu receptors significantly higher than that present in normal males.

These findings seem to indicate that a sex difference exists in the maturation of hypothalamic mu opioid receptors and that the differentiation in the number of these receptors towards a male or a female pattern appears to be due to the endocrine environment present at birth. It is then conceivable that the interaction between sex steroids and opioid systems already shown to be influential for the control of gonadotropin secretion (see the preceding sections), develops early during postnatal life.

EFFECTS OF NEUROTENSIN IN THE CONTROL OF GONADOTROPIN SECRETION

Neurotensin is a linear tridecapeptide which was first isolated from hypothalamic extracts (Carraway and Leeman, 1973). This peptide has been shown to possess a large variety of biological effects on the gastrointestinal tract (Carraway and Leeman, 1973; Fernstrom et al., 1980), and on the cardiovascular (Carraway and Leeman, 1973) and endocrine (Brown and Vale, 1976; McCann, 1980) systems.

The effect of neurotensin on the release of LH and FSH was analyzed in long-term castrated female rats. The i.v.t. injection of neurotensin induces a significant decrease of serum LH levels. This decrease was not accompanied by any change in FSH secretion (Motta and Martini, 1981). The present results on the effects of neurotensin on LH and FSH secretion agree with those of Vijayan and McCann (1978) obtained in ovariectomized freely moving rats, but are not consistent with those of Makino et al. (1973) who found a stimulation of the release of both gonadotropins following the intravenous injection of neurotensin into normal rats.

These results suggest that neurotensin may participate in the control of LH release; as in the case of the opioids, this effect seems to be specific, since the administration of neurotensin is not associated with any modification in FSH secretion. Hence the present observations provide additional data on the separate control of the release of the two gonadotropins (Motta and Martini, 1981; Piva et al., 1984).

EFFECT OF SAUVAGINE IN THE CONTROL OF GONADOTROPIN SECRETION

As mentioned in the Introduction, sauvagine is a 40 amino acid peptide originally isolated from the amphibian skin and presently obtained by synthesis (Montecucchi et al., 1980). Preliminary data seem to indicate that sauvagine may also be present in the mammalian brain. This peptide exhibits biological activities on the cardiovascular and renal systems (Erspamer et al., 1980).

Recent evidence suggests that sauvagine might intervene also in the control of anterior pituitary function. This peptide causes a significant increase in serum levels of ACTH and of beta-endorphin-like immunoreactivity in the rat. The stimulatory effect of sauvagine on the release of ACTH and beta-endorphin was confirmed *in vitro* using the isolated and dispersed rat pituitary cell system (Falaschi et al., 1984). These findings are not surprising, since sauvagine shows striking similarities (number of amino acids, several homologous amino acid sequences) with the corticotropin-releasing hormone (CRF) isolated by Vale et al. (1981). In addition, sauvagine induces a significant reduction in serum prolactin levels both in normal male and lactating rats (Falaschi et al., 1982). The potent and long-lasting prolactin inhibitory effect of this peptide is also present *in vitro* using the isolated and dispersed rat pituitary cell system (Falaschi et al., 1982; see also Motta, 1985, for references).

In the light of these results, it was of interest to investigate the possible effects of sauvagine on the release of LH and FSH using *in vivo* as well as *in vitro* methods. It has been found that sauvagine, given i.v.t. to castrated male rats does not exert any effect on serum LH and FSH levels. Similar negative results have been obtained, when this peptide was assayed in an *in vitro* system (Motta, 1985).

It emerges from these results that sauvagine does not appear to be involved in the control of the release of the two gonadotropins. Thus, a dissimilarity of effect between sauvagine and the other peptides so far considered (opioids, neurotensin, etc.) seems to exist: in analogy with them, sauvagine does not show any influence on FSH release, but, unlike them, it exhibits a lack of effect also on LH secretion.

ULTRASHORT LOOP FEEDBACK EFFECT OF LHRH

Different levels of feedback mechanisms are involved in the regulation of the activity of the hypothalamo-pituitary complex. In addition to long feedback mechanisms in which the signal is offered by the peripheral steroid hormones, the existence of short loop feedback mechanisms in which the regulatory signal is represented by the pituitary hormones has been shown for ACTH, LH, FSH, TSH and prolactin (Motta et al., 1965; David et al., 1966; Fraschini et al., 1968; Kakita et al., 1984; Suda et al., 1986).

The possibility of a third type of modulatory feedback mechanism in which the hypothalamic hormones themselves play an autoregulatory role in the control of their own secretion has been put forward. Hyyppa et al. (1971) have been the first to propose that the synthesis, the storage and/or the release of the hypothalamic releasing hormones might be influenced by changes of their own titers in the general circulation. They have shown that the treatment of castrated-hypophysectomized male rats with a hypothalamic extract containing FSH-RF is able to bring back to normal the high levels of FSH-RF usually found in the hypothalami of these animals. These findings have been interpreted to mean that the hypothalamus probably contains elements that are sensitive to changing levels of FSH-RF and which participate in the control of FSH secretion.

The hypothesis that other hypothalamic hormones might directly control their own secretion, and consequently influence the release of their corresponding pituitary hormones has also been recently proposed for the control of GH. Peterfreund and Vale (1984) and Richardson and Twente (1986), using dispersed fetal and adult rat hypothalamic cell systems, respectively, have shown that both somatostatin and its synthetic analogs inhibit the release of endogenous somatostatin from the cultured hypothalamic cells. Indirect evidence, which is consistent with this observation, has been obtained *in vivo* (Lumpkin et al., 1981). These authors have demonstrated that the i.v.t. injection of somatostatin induces an increase rather than a decrease in the release of GH. More recently, the suggestion that also GHRH release might be under the control of an ultrashort loop feedback mechanism has been postulated. Lumpkin et al. (1985) have reported that low doses of GHRH, given i.v.t. to adult male rats, are able to decrease the release of GH.

In a preliminary *in vivo* experiment using adult male rats submitted to hypophysectomy and castration, chronic treatment with an LHRH analog has been shown to decrease the hypothalamic stores of LHRH to levels lower than those found in untreated hypophysectomized-castrated animals (Zanisi, Messi, Martini and Motta, unpublished observations). In the light of these findings the possible existence of an ultrashort feedback mechanism for the control of LHRH secretion might be put forward. Support to this hypothesis comes from the following *in vitro* experiment.

A perfusion system on single rat hypothalami developed in the authors' laboratory (Zanisi et al., 1986) has been used. This *in vitro* technique

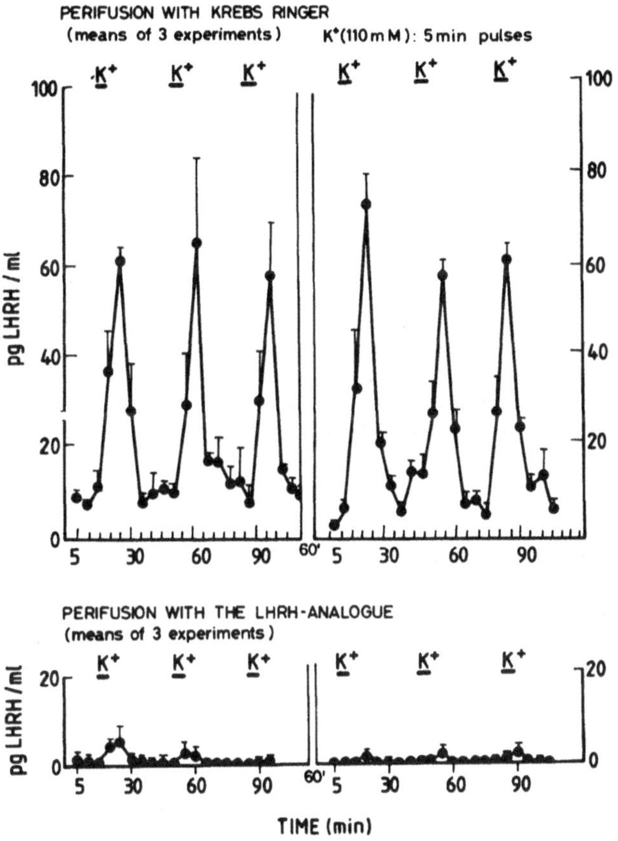

Fig. 4. Effect of continuous perfusion with an LHRH-analog (buserelin 5×10^{-6} M) on K^+-induced LHRH release from single hypothalami of normal male rats.

provides a useful tool for evaluating the direct effect of LHRH on its own release, since the interferences exerted by gonadal and pituitary hormones are eliminated. The endogenous LHRH release from adult male rat hypothalami has been stimulated by pulses of K^+ both in the presence or in the absence of D-Ser-(tBU)^6-desGly^{10}-LHRH-N-ethylamide (buserelin). Since the final parameter of such experiments is represented by the quantification of radioimmunoassayable LHRH in the perfusion effluent, it has been ascertained in preliminary studies that the analog selected does not crossreact with the antiserum used to measure LHRH by radioimmunoassay.

Fig. 4 shows the results of an experiment in which one single hypothalamus has been perfused continuously either with control medium or with medium containing the LHRH analog. K^+ pulses of 5-min duration were applied every 30 min. When buserelin is added to the perfusion medium, the analog is able to inhibit the K^+-induced release of endogenous LHRH. The prolonged exposure to the analog completely abolishes the response to K^+-depolarization; this, on the contrary, persists in the absence of the analog for the whole time of perfusion (90 min). The ability of the hypothalamus to release LHRH in response to K^+-stimulation reappears after removal of the analog from the medium (data not shown). The failure of TRH, when added to the perfusion medium, to alter the LHRH release induced by K^+ stimulation, suggests that the effect of LHRH analog on LHRH output is specific (data not shown).

The present findings clearly demonstrate that the perfusion with an LHRH analog inhibits the release of endogenous LHRH from the perfused hypothalamus of adult male rats, when this is stimulated by K^+ pulses. The data are consistent with the possibility that LHRH acting at the level of the hypothalamus participates in the physiological control of its own release.

ACKNOWLEDGEMENTS

The experiments performed in the authors' laboratory and here reported were supported by grants from the Consiglio Nazionale delle Ricerche, Roma, Italy (through the Special Projects "Preventive and Rehabilitative Medicine," contracts no. 85.00715.56 and 85.00679.56 and "Endocrinology", contract no. 85.00786.04) and by grants of the Ministero della Pubblica Istruzione. Such support is gratefully acknowledged. Thanks are also due to Mrs. Paola Assi, Giovanna Miccichè and Mr. Enrico Croce for their skillful technical assistance.

REFERENCES

Bhanot, R., and Wilkinson, M., 1983, Opiatergic control of LH secretion is eliminated by gonadectomy, Endocrinology, 112:399.

Brown, M., and Vale, W., 1976, Effects of neurotensin and substance P on plasma insulin, glucagon, and glucose levels, Endocrinology, 98:819.

Carraway, R., and Leeman, S. E., 1983, The isolation of a new hypotensive peptide, neurotensin, from bovine hypothalami, J. Biol. Chem., 248:6854.

Ching, M., 1983, Morphine suppresses the proestrus surge of GnRH in the pituitary portal plasma of rats, Endocrinology, 112:2209.

Cicero, T. J., Meyer, E. R., Gabriel, S. M., Bell, R. D., and Wilcox, C. E., 1980, Morphine exerts testosterone-like effects in the hypothalamus of the castrated male rat, Brain Res., 202:151.

Clement-Jones, V., and Rees, L. H., 1982, Neuroendocrine correlates of the endorphins and enkephalins, in: "Clinical Neuroendocrinology," G. M. Besser and L. Martini, eds., p. 139, Academic Press, New York.

Cuello, A. C., 1983, Central distribution of opioid peptides, Br. Med. Bull., 39:11.

David, M. A., Fraschini, F., and Martini, L., 1966, Control of LH secretion: role of a "short" feedback mechanism, Endocrinology, 78:55.
Erspamer, V., Falconieri Erspamer, B., Improta, G., Negri, L., and de Castiglione, R., 1980, Sauvagine, a new polypeptide from Phyllomedusa sauvagei skin, Naunyn Schm. Arch. Pharmacol., 312:265.
Falaschi, P., D'Urso, R., Negri, L., Rocco, A., Montecucchi, P. C., Henshen, A., Melchiorri, P., and Erspamer, V., 1982, Potent in vivo and in vitro prolactin inhibiting activity of sauvagine, a frog skin peptide, Endocrinology, 111:693.
Falaschi, P., Melchiorri, P., D'Urso, R., Negri, L., Rocco, M., Motta, M., and Erspamer, V., 1984, Sauvagine and ovine CRF: a frog skin peptide as anticipator of mammalian brain hormones, in: "Pituitary Hyperfunction," E. E. Müller, G. M. Molinatti and F. Camanni, eds., p. 41, Raven Press, New York.
Fernstrom, M. H., Carraway, R. E., and Leeman, S. E., 1980, Neurotensin, in: "Frontiers in Neuroendocrinology," L. Martini and W. F. Ganong, eds., vol. 6, p. 103, Raven Press, New York.
Fraschini, F., Motta, M., and Martini, L., 1968, A "short" feedback mechanism controlling FSH secretion, Experientia, 24:270.
Giudici, D., D'Urso, R., Falaschi, P., Negri, L., Melchiorri, P., and Motta, M., 1984, Dermorphin stimulates prolactin secretion in the rat, Neuroendocrinology, 39:236.
Gosselin, R. E., Blankstein, J., Dent, D. W., Hobson, W. C., Fuller, G. B., Reyes, F. I., Winter, J. S. D., and Faiman, C., 1983, Effects of naloxone and an enkephalin analog on serum prolactin, cortisol, and gonadotropins in the chimpanzee, Endocrinology, 112:2168.
Hyyppa, M., Motta, M., and Martini, L., 1971, "Ultrashort" feedback control of follicle-stimulating hormone-releasing factor secretion, Neuroendocrinology, 7:227.
Kakita, T., Laborde, N. P., and Odell, W. D., 1984, Autoregulatory control of thyrotropin in rabbits, Endocrinology, 114:2301.
Kalra, S. P., and Kalra, P. S., 1974, Effects of circulating estradiol during rat estrous cycle on LH release following electrochemical stimulation of preoptic brain or administration of synthetic LRF, Endocrinology, 94:845.
Leadem, C. A., and Kalra, S. P., 1985, Effects of endogenous opioid peptides and opiates on luteinizing hormone and prolactin secretion in ovariectomized rats, Neuroendocrinology, 41:342.
Lumpkin, M. D., Negro-Vilar, A., and McCann, S. M., 1981, Paradoxical elevation of growth hormone by intraventricular somatostatin: possible ultrashort loop feedback, Science, 211:1972.
Lumpkin, M. D., Samson, W. K., and McCann, S. M., 1985, Effects of intraventricular growth hormone-releasing factor on growth hormone release: further evidence for ultrashort loop feedback, Endocrinology, 116:2070.
Makino, R., Carraway, R., Leeman, S. E., and Greep, R. O., 1973, In vitro and in vivo effects of a newly purified hypothalamic tridecapeptide on rat LH and FSH release, Progr. 6th Mtg. Group Study Reprod., p. 26.
McCann, S. M., 1980, Control of anterior pituitary hormone release by brain peptides, Neuroendocrinology, 31:355.
Montecucchi, P. C., Anastasi, A., de Castiglione, R., and Erspamer, V., 1980, Isolation and amino acid composition of sauvagine, an active polypeptide from methanol extracts of the skin of the South American frog, Phyllomedusa sauvagei, Int. J. Peptide Prot. Res., 16:191.
Montecucchi, P. C., de Castiglione, R., Piani, S., Gozzini, L., and Erspamer, V., 1981, Amino acid composition and sequence of dermorphin, a novel opiate-like peptide from skin of Phillomedusa sauvagei, Int. J. Peptide Prot. Res., 17:275.
Morley, J. E., 1981, The endocrinology of the opiates and opioid peptides, Metabolism, 30:195.
Motta, M., 1985, Neuroendocrine effects of some amphibian peptides, Peptides, 6:131.

Motta, M., and Martini, L., 1981, Neurotensin inhibits LH release, Proc. Soc. Exp. Biol. Med., 168:62.

Motta, M., and Martini, L., 1982, Effect of opioid peptides on gonadotropin secretion, Acta Endocr., 99:321.

Motta, M., Mangili, G., and Martini, L., 1965, A "short" feedback loop in the control of ACTH secretion, Endocrinology, 77:392.

Motta, M., Falaschi, P., Zoppi, S., and Martini, L., 1984, Role of peptides in the regulation of gonadotropin and prolactin secretion, in: "Hormonal Control of the Hypothalamo-Pituitary-Gonadal Axis," K. W. McKerns and Z. Laron, eds., p. 39, Plenum Press, New York.

Müller, E. E., Nistico, G., and Scapagnini, U., 1977, "Neurotransmitters and Anterior Pituitary Function," Academic Press, New York.

Neill, J. D., and Smith, M. S., 1974, Pituitary-ovarian interrelationships in the rat, in: "Current Topics in Experimental Endocrinology," V. H. T. James and L. Martini, eds., vol. 2, p. 73, Academic Press, New York.

Pang, C. N., Zimmermann, E., and Sawyer, C. H., 1977, Morphine inhibition of the preovulatory surges of plasma luteinizing hormone and follicle stimulating hormone in the rat, Endocrinology, 101:1726.

Paterson, S. J., Robson, L. E., and Kosterlitz, H. W., 1983, Classification of opioid receptors, Br. Med. Bull., 39:31.

Pfeiffer, D. G., Pfeiffer, A., Shimohigashi, Y., Merriam, G. R., and Loriaux, D. L., 1983, Predominant involvement of mu rather than delta or kappa opiate receptors in LH secretion, Peptides, 4:647.

Peterfreund, R. A., and Vale, W. W., 1984, Somatostatin analogs inhibit somatostatin secretion from cultured hypothalamus cells, Neuroendocrinology, 39:397.

Piva, F., Motta, M., and Martini, L., 1979, Regulation of hypothalamic and pituitary function: long, short, and ultrashort feedback loops, in: "Endocrinology," L. J. De Groot, ed., p. 21, Grune and Stratton, New York.

Piva, F., Limonta, P., Maggi, R., and Martini, L., 1984, Dual effects of the opioids in the control of gonadotropin secretion, in: "Opioid Modulation of Endocrine Function," G. Delitala, M. Motta and M. Serio, eds., p. 155, Raven Press, New York.

Piva, F., Maggi, R., Limonta, P., Motta, M., and Martini, L., 1985, Effect of naloxone on luteinizing hormone, follicle-stimulating hormone, and prolactin secretion in the different phases of the estrous cycle, Endocrinology, 117:766.

Piva, F., Limonta, P., Maggi, R., and Martini, L., 1986, Stimulatory and inhibitory effects of the opioids on gonadotropin secretion, Neuroendocrinology, 42:504.

Reid, R. L., Quigley, M. E., and Yen, S. S.C., 1983, The disappearance of opioidergic regulation of gonadotropin secretion in postmenopausal women, J. Clin. Endocr. Met., 57:1107.

Richardson, S. B., and Twente, S., 1986, Inhibition of rat hypothalamic somatostatin release by somatostatin: evidence for somatostatin ultrashort loop feedback, Endocrinology, 118:2076.

Suda, T., Yajima, F., Tomori, N., Sumitomo, T., Nakagami, Y., Ushiyama, T., Demura, H., and Shizume, K., 1986, Inhibitory effect of adrenocrticotropin on corticotropin-releasing factor release from rat hypothalamus in vitro, Endocrinology, 118:459.

Sylvester, P. W., Van Vugt, D. A., Aylsworth, C. A., Hanson, E. A., and Meites, J., 1982, Effects of morphine and nalozone on inhibition by ovarian hormones of pulsatile release of LH in ovariectomized rats, Neuroendocrinology, 34:269.

Takahara, J., Kageyama, J., Yunoki, S., Yakushiji, W., Yamauchi, J., Kageyama, N., and Ofuji, T., 1978, Effects of 2-bromo-alpha-ergocryptine on beta-endorphin-induced growth hormone, prolactin and luteinizing hormone release in urethane anesthetized rats, Life Sci., 22:2205.

Vale, W., Spiess, J., Rivier, C., and Rivier, J., 1981, Characterization of a 41-residue ovine peptide that stimulates secretion of corticotropin and beta-endorphin, Science, 213:1394.

Vijayan, E., and McCann, S. M., 1978, Effects of intraventricular injection of substance P (SP), neurotensin, (NT), and gastrin (G) on pituitary hormone release in conscious ovariectomized (OVX) rats, Endocrinology, 102:217A.

Zanisi, M., Messi, E., and Martini, L., 1986, "In vitro" release of luteinizing hormone-releasing hormone from the hypothalamus of old male rats, Endocrinology, in press.

INVOLVEMENT OF GABA IN THE NEUROENDOCRINOLOGY

OF REPRODUCTION

W. Wuttke, H. Jarry, J. Demling,
R. Wolf, and E. Düker

Division of Clinical Experimental Endocrinology
Department of Obstetrics and Gynecology
Humboldtalle 19
Göttingen 3400, Federal Republic of Germany

INTRODUCTION

Pulsatile LHRH secretion from hypothalamic neurons is a prerequisite for proper pituitary LH and FSH release (Knobil, 1980). As will be shown in other chapters in this volume, pulsatile LHRH secretion is the result of synchronous, phasic activation of LHRH neurons. This pulsatility can be disrupted by noradrenergic antagonists and this mechanism involves most probably α-adrenoceptors (Kaufmann et al., 1985). Furthermore, the amino acid neurotransmitter γ-aminobutyric acid (GABA) can also disrupt pulsatile LHRH secretion (Lamberts et al., 1982; Fuchs et al., 1984). GABA was recently shown to be the neurotransmitter of many estrogen-receptive preoptic/anterior hypothalamic (MPO/AH) neurons (Flügge et al., 1986). The majority of LHRH neurons in the rat are also located in this area. In previous publications, it was shown that both positive as well as negative feedback effects of estrogens on LH secretion involve among other structures also the MPO/AH area (Mansky et al., 1982; Fuchs et al., 1984).

Utilizing two different methods, we were able to demonstrate that GABA turnover rates in the MPO/AH or *in vivo* GABA release rates are low in ovariectomized (ovx) rats and high in ovx rats in which the negative feedback of estrogens is exerted (Mansky et al., 1982; Demling et al., 1985). Fig. 1 shows *in vivo* GABA release rates in the MPO/AH of ovx rats, which have high LH levels. These values compare to significantly higher GABA release rates ($p < 0.01$) in ovx estrogen treated animals in which the steroid treatment results in a negative feedback action on blood LH levels. As mentioned above, a large number (approx. 60%) of MPO/AH estrogen-receptive neurons utilize GABA as neurotransmitter. Pulsatile LH secretion can be modulated by estrogens and this effect may involve estrogen-receptive GABAergic neurons. Therefore, we were interested in studying GABA release rates and to determine whether these estrogen-receptive neurons contribute to pulsatile LHRH secretion.

GABA AND PULSATILE LHRH RELEASE

Ovx rats were implanted with push-pull cannulae into the MPO/AH area prior to experimentation. For withdrawal of blood samples, a jugular vein catheter was also implanted. On the experimental day, perfusion of the

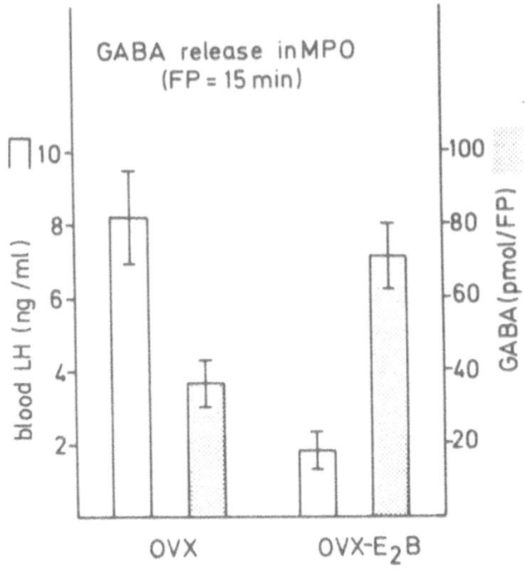

Fig. 1. GABA release rates as measured in push-pull perfusates of the preoptic area (MPO) and blood LH levels in ovariectomized (ovx) and in ovx-estrogen primed (ovx-E_2B) rats under negative feedback conditions. Note high blood LH levels and low preoptic GABA release rates in ovx rats; lower blood LH levels are present in ovx-E_2B rats which have high preoptic GABA release rates. FP = fraction period; SEM on top of each bar.

MPO/AH started at 1100 h and fractions of perfusates were collected every 5 min. GABA was measured in the perfusates as described earlier. Fig. 2 details pulsatile pituitary LH secretion, indicating intact, synchronous phasic activation of hypothalamic LHRH neurons. GABA release rates in the MPO/AH are also shown in this figure. Each LH episode is preceded by a drop of GABA release. GABA release increases as blood LH levels decrease or are low. This inverse release pattern was seen in many animals and suggests an interaction of GABAergic neurons with LHRH neurons. In previous publications we discussed the possibility that GABA may be presynaptically inhibitory to those noradrenergic terminals innervating LHRH neurons (Fuchs et al., 1984; Demling et al., 1985). As mentioned above, α-adrenoreceptive noradrenergic mechanisms are stimulatory to LHRH pulse generating mechanisms. From previous push-pull cannula experiments, it is known that norepinephrine (NE) is being released in a pulsatile fashion in the MPO/AH, but the occurrence of these pulses appeared to be erratic and did not correlate with the occurrence of blood LH pulses (Demling et al., 1985). The question arises therefore whether a direct correlation between preoptic NE release and LHRH neuronal activity needs to exist. It is possible that the occasional presence of NE sets the membrane of LHRH neurons such that the neurons are able to exhibit bursting activity. It has indeed been shown that specific α-adrenergic stimulating drugs cause bursting of hypothalamic LHRH neurons (Kelly et al., 1986). We may therefore explain pulsatile LHRH release on the basis of these two observations: The occasional presence of NE may result in membrane effects such that each LHRH neuron is ready for phasic activation. The

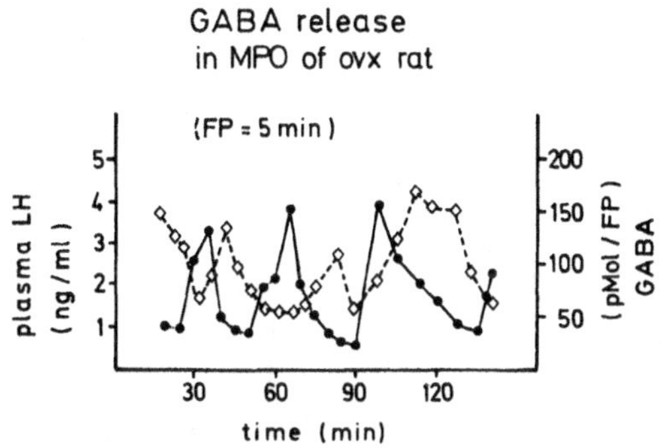

Fig. 2. Blood LH levels and preoptic GABA release rates in an individual ovariectomized rat. Note the inverse pattern of these two parameters.

activation of LHRH neurons however would occur in a random fashion, i.e. would be asynchronous. The decreased GABA release prior to each LHRH episode suggests that this relieves inhibition of LHRH neurons. As the LHRH neurons were exposed to enough NE and are therefore entrained for phasic activation, this relief of GABAergic inhibition results now in synchronous phasic activation of LHRH neurons.

EFFECTS OF ESTROGEN ON GABA RELEASE

In the following, we will describe estrogen-dependent effects on GABA release in the mediobasal hypothalamus (MBH) of which we do not know the physiological importance. These results stem also from experiments using push-pull cannulae (Jarry et al., 1987). The tips of the cannulae were implanted into the arcuate nucleus/median eminence complex; the other experimental details were the same as described above. Fig. 3 shows GABA release rates in the MBH of an ovx rat as measured in 5 min fractions. It is obvious that GABA release is of pulsatile nature with pulses exceeding baseline GABA release rates often by a factor of 5 to 10. Pulse frequency analysis (Clifton and Steiner, 1983) of these GABA pulses yielded the information that the pulses occur every 36 ± 7 min (mean ± SEM). From the GABA release pattern observed in ovx-estrogen treated animals (Fig. 4), it is evident that the occurrence of GABA pulses is much less frequent than in ovx non-estrogen treated rats. In the estrogen-treated group of animals, GABA pulses were observed only every 117 ± 15 min ($p < 0.001$ vs. ovx rats). It is interesting that mean GABA release rates (calculated from all fractions and from all animals) did not differ significantly; mean hypothalamic GABA release in ovx rats was 282 ± 95 pmol/fraction period. In the estrogen-treated animals, GABA release rates were 217 ± 35 pmol/fraction period. Due to the more frequently occurring pulses in ovx animals, the variances of the means were significantly larger than in the estrogen-treated group.

This allows the conclusion that it is not the absolute GABA release within a given time period which is of importance for the mediobasal hypothalamic function(s) influenced by GABA, but that it is the pulse frequency

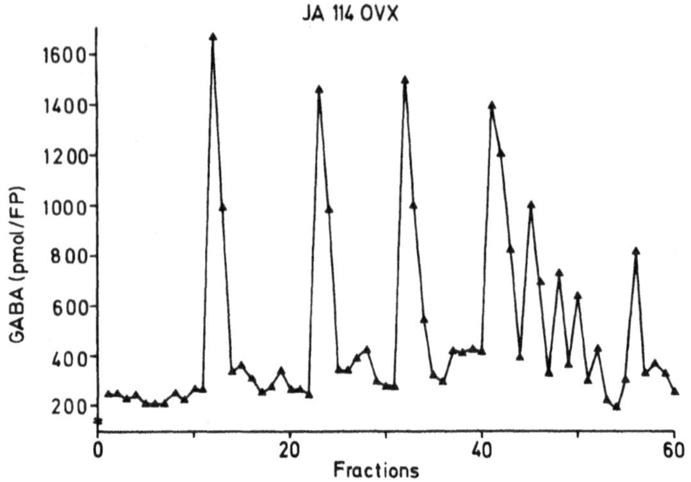

Fig. 3. GABA release rates in the mediobasal hypothalamus of an ovariectomized rat. Note the frequent occurrence of GABA release pulses.

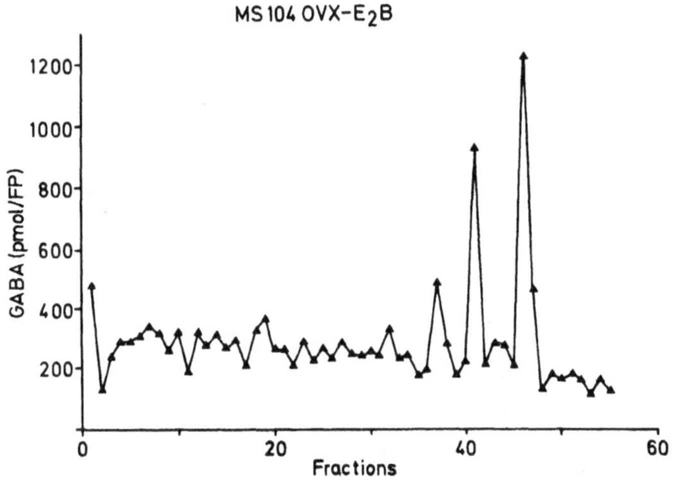

Fig. 4. GABA release rates in the mediobasal hypothalamus of an ovariectomized rat under estrogen influence. Note the infrequent occurrence of GABA release pulses.

that bears the signal character for these functions. Since the pulse frequency of MBH GABA neurons is affected by estrogen, it is feasible to assume that the functions affected by those neurons are also affected by estrogens. This is, to our knowledge, the first time that pulsatile release of a neurotransmitter within the CNS has been demonstrated. We favor the idea that this hypothalamic mechanism is involved in the control of pituitary prolactin release (Jarry et al., 1987).

In summary, we have demonstrated an inverse relationship between preoptic GABA and pituitary LH release which may involve estrogen-receptive GABAergic neurons. Mediobasal hypothalamic GABA release occurs in a pulsatile nature. Pulse frequency but not mean release rate is affected by estrogens, which indicates that pulsatility and not the amount of GABA bears the signal character.

REFERENCES

Clifton, D. K., Steiner, R. A., 1983, Cycle detection: a technique for estimating the frequency and amplitude of episodic fluctuations in blood hormone and substrate concentrations, Endocrinology, 112:1057

Demling, J., Fuchs, E., Baumert, M., and Wuttke, W., 1985, Preoptic catecholamines, GABA and glutamate release in ovariectomized and ovariectomized, estrogen-primed rats utilizing the push-pull cannula technique, Neuroendocrinology, 41:212.

Flügge, G., Oertel, W. H., and Wuttke, W., 1986, Evidence for estrogen-receptive GABAergic neurons in the preoptic-anterior hypothalamic area of the rat, Neuroendocrinology, 43:1.

Fuchs, E., Mansky, T., Stock, K., Vijayan, E., and Wuttke, W., 1984, Involvement of catecholamines and glutamate in GABAergic mechanisms regulatory to LH and prolactin secretion, Neuroendocrinology, 38:484.

Jarry, H., Sprenger, M., and Wuttke, W., 1987, Rates of release of GABA and catecholamines in the mediobasal hypothalamus of ovariectomized and ovariectomized, estrogen-treated rats: correlation with blood prolatin levels, Neuroendocrinology, in press.

Kaufmann, J. M., Kerner, J. S., Wilson, R., and Knobil, E., 1985, Electrophysiological manifestation of luteinizing hormone-releasing hormone pulse generator activity in the Rhesus monkey: influence of α-adrenergic and dopaminergic blocing agents, Endocrinology, 116:1327.

Kelly, M. J., Condon, T. P., and Ronnekleiv, O. K., 1986, Estrogen modulation of the α_1-adrenergic excitation of hypothalamic parvocellular neurons, 1st International Congress on Neuroendocrinology, Abstract No. 24.

Knobil, E., 1980, The neuroendocrine control of the menstrual cycle, Rec. Prog. Horm. Res., 36:53.

Lamberts, R., Vijayan, E., Graf, M., Mansky, T., and Wuttke, W., 1983, Involvement of preoptic-anterior hypothalamic GABA neurons in the regulation of pituitary LH and prolactin release, Exp. Brain Res., 52:356.

Mansky, T., Mestres-Ventura, P., and Wuttke, W., 1982, Involvement of GABA in the feedback of estradiol on gonadotropin and prolactin release: hypothalamic GABA and catecholamine turnover rates, Brain Res., 231:352.

HYPOTHALAMIC BIOGENIC AMINES AND THE REGULATION OF LUTEINIZING HORMONE RELEASE IN THE RAT

Clive W. Coen

Department of Anatomy and Human Biology
King's College
London, England

INTRODUCTION

The possibility that biogenic amines play a physiological role in the spontaneous regulation of luteinizing hormone (LH) release was first indicated by Sawyer, Everett and Markee in 1949. They demonstrated that rats treated with dibenamine, an α-adrenergic receptor blocking drug, failed to ovulate (Sawyer et al., 1949; Everett et al., 1949; Everett and Sawyer, 1949). A thorough development of their original research became possible only with the introduction of the radioimmunoassay for LH twenty years later. By 1976 it had been established that an increased discharge of LH releasing hormone (LHRH) into pituitary portal blood on the day of proestrus causes a marked surge in the systemic concentration of LH and results in ovulation (Sarkar et al., 1976). This surge is entrained to the light-dark cycle and its occurrence (peaking at approximately 1800 h with lighting from 0600 to 2000 h) depends upon an elevated level of serum estrogen and a central circadian "clock" which is probably located in the suprachiasmatic nuclei of the hypothalamus (Coen and MacKinnon, 1976, 1977, 1980; Raisman and Brown-Grant, 1977). In 1976 there was also the first report clearly indicating that the majority of the LHRH-containing perikarya which project to the median eminence in the rat are located in the medial preoptic area near the organum vasculosum of the lamina terminalis (Sétáló et al., 1976).

The first studies which specifically investigated the effects of α-adrenergic antagonists on serum LH concentrations demonstrated that the estrogen- and progesterone-induced LH surges in ovariectomized rats are blocked by phenoxybenzamine (Kalra and McCann, 1973) and that the pulsatile mode of secretion following ovariectomy is suppressed by the same drug or by phentolamine (Gnodde and Schuiling, 1976; Weick, 1978). More recently it has been demonstrated that prazosin, an α_1-adrenergic receptor antagonist, can block the proestrous surge (Coen and Coombs, 1983) and the estrogen-induced surge (Drouva et al., 1982). Since the α-adrenergic antagonists will interrupt neurotransmission involving either noradrenaline or adrenaline, the relative significance of these and other amines must be assessed (1) by the effects of alternative pharmacological treatments, (2) by the estimation of rates of turnover of these and related "transmitters" at particular times and in relevant hypothalamic sites, and (3) by neuroanatomical studies involving immunohistochemical identification of the neurons presumed to be involved in

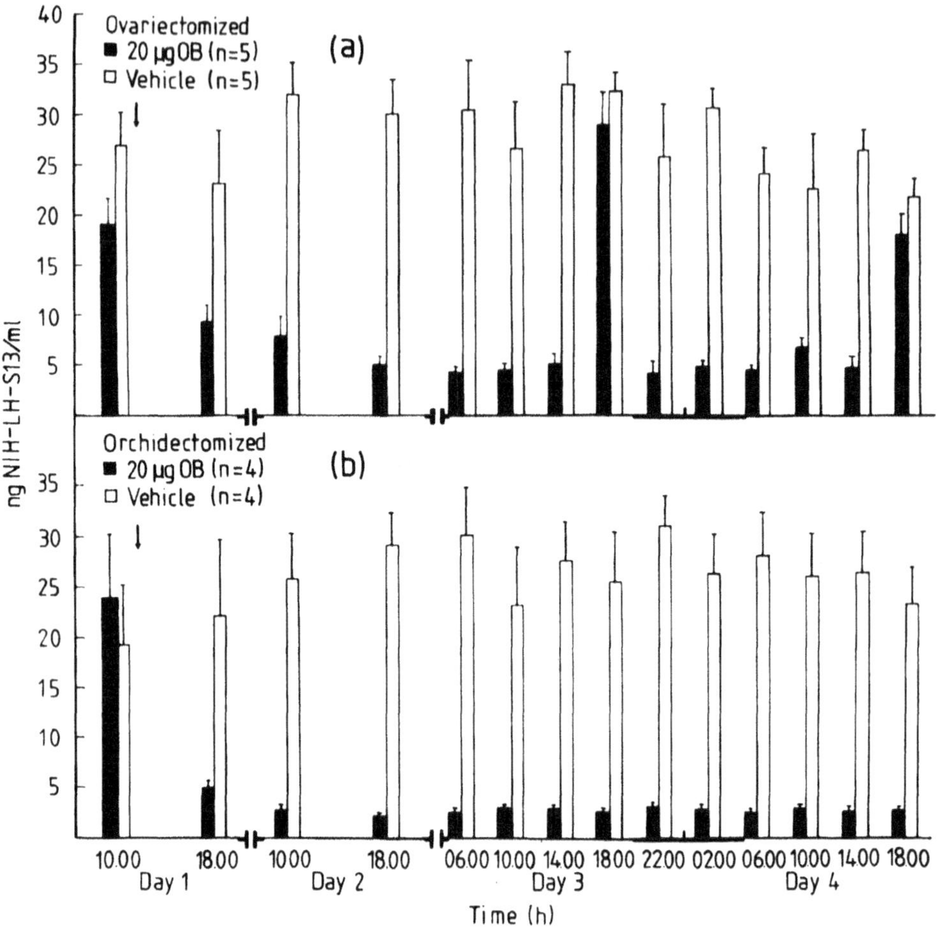

Fig. 1. Serum LH concentration (± SEM), at times indicated, expressed as ng NIH-LH-S13/ml, in (a) chronically ovariectomized (n = 5), or (b) orchidectomized (n = 4) rats, treated with 20 µg estradiol benzoate (OB) (solid bars) or vehicle (open bars) at 1200 h on day 1. The black bars on the abscissa indicate hours of darkness (2000-0600 h).

LH control. It should be emphasized that none of these approaches is capable of generating unequivocal evidence; provisional conclusions can be based only on the "weight" of the evidence.

EFFECTS OF PHARMACOLOGICAL INTERVENTION

The immense range of research which had its inception with the early discoveries of Sawyer et al. cannot be adequately reviewed here. Only in vivo studies will be considered and these will be discussed in the context of four modes of spontaneous LH release: (1) the preovulatory LH surge; (2) the analogous surge which is generated daily in ovariectomized rats treated with a long-acting estrogen (Fig. 1); (3) the comparable surge which occurs without subsequent repetition when the ovariectomized rats are given

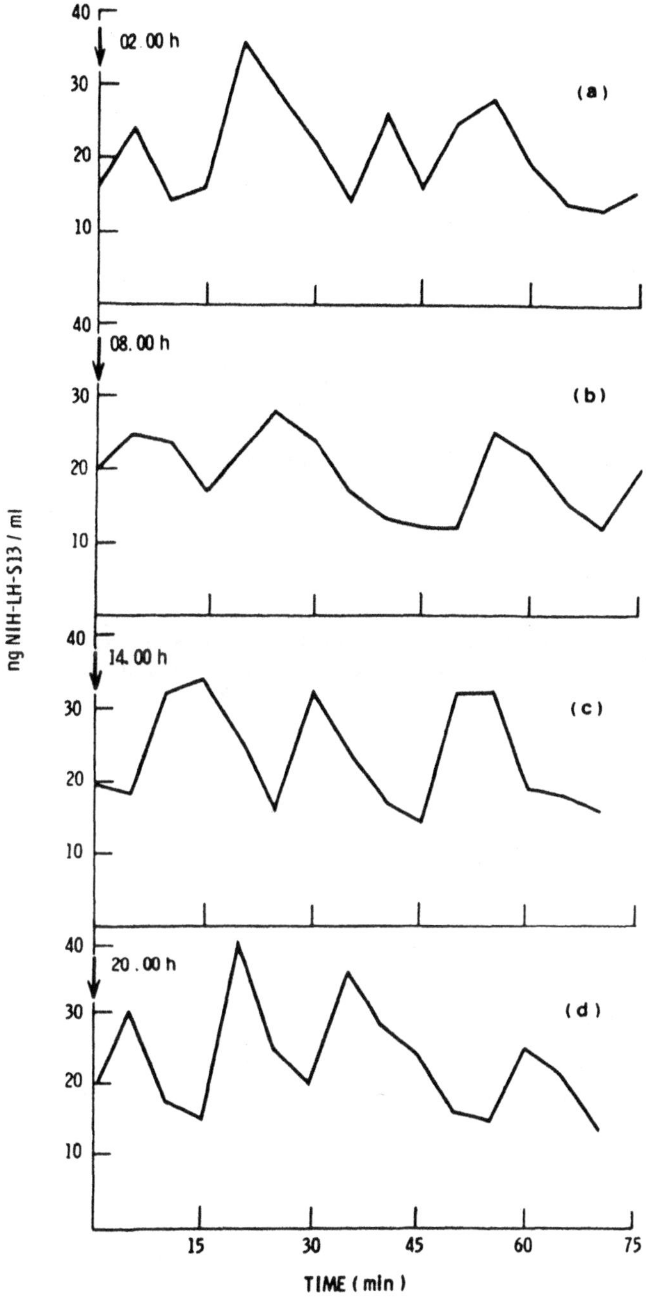

Fig. 2. Serum LH concentration expressed as ng NIH-LH-S13/ml in a chronically ovariectomized rat bled through an atrial cannula every 5 min over a period of 70 or 75 min beginning at (a) 0200, (b) 0800, (c) 1400, and (d) 2000 h.

progesterone after estrogen priming (Caligaris et al., 1971); and (4) the LH pulses which are readily observed when samples are taken with sufficient frequency either following ovariectomy, when they lack any circadian modulation (Fig. 2), or during the estrous cycle and the preovulatory surge (Gallo, 1981a, 1981b).

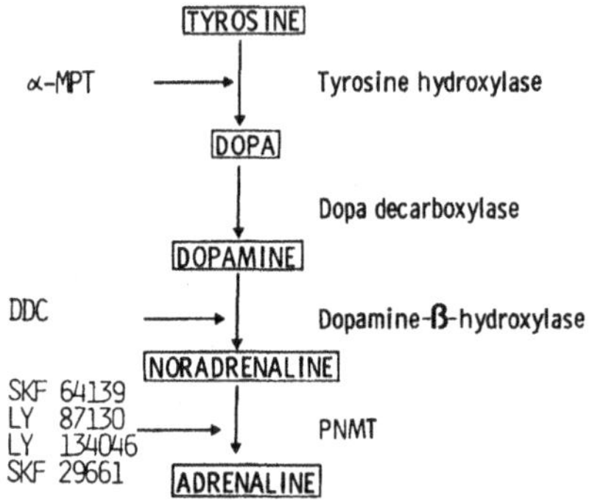

Fig. 3. The biosynthetic pathway for catecholamines together with inhibitors of enzymes involved in that pathway: α-methyl-p-tyrosine (αMPT); diethyldithiocarbamate (DDC); compounds donated by Smith Kline and French (SKF) and Eli Lilly and Co. (LY).

Catecholamine Synthesis Inhibitors and the Proestrous LH Surge

Our studies on the role of catecholamines initially concentrated on the effects of various inhibitors of their synthesis (Fig. 3) when administered on the day of proestrus (Coen and Coombs, 1983). Tyrosine hydroxylase inhibition by means of α-methyl-p-tyrosine (αMPT) fails to affect the occurrence of the LH surge but 41% of the animals producing an apparently normal surge fail to ovulate (Table 1), suggesting a direct effect of the drug on the ovaries. In contrast, the inhibition of dopamine β-hydroxylase (DBH) or phenylethanolamine N-methyltransferase (PNMT) with, respectively, diethyldithiocarbamate (DDC) or SKF64139 leads, at relatively low doses, to the occurrence of ovulation without the 1800 h LH surge (Table 1); when samples are taken at additional times, it is apparent that all cases of ovulation without a surge at 1800 h are associated with the occurrence of a surge at 2400 h (Table 2). Relatively high doses of DDC and SKF64139 result in a loss of the surge at 1800 and 2400 h and a consequent failure of ovulation (Table 2). Remarkably similar results (Coen et al., 1985) were achieved with a range of doses of LY87130, another PNMT inhibitor (Table 2); treatment with SKF29661, a PNMT inhibitor which fails to cross the blood-brain barrier, is, however, without effect on the normal occurrence of the surge (Table 2) (Coen et al., 1986).

Table 1. Incidence of Ovulation and of the Surge in Serum Luteinizing Hormone (LH) Concentration at 1800 h on Proestrus following Treatment with Vehicle, α-methyl-p-tyrosine (αMPT), Diethyldithiocarbamate (DDC) or SKF64139 at Doses, Times and Frequency as indicated.

CATECHOLAMINE SYNTHESIS INHIBITORS

	OVULATING		NOT OVULATING	
TREATMENT PRO-OESTRUS 10.00–17.00h (mg/kg x Frequency)	LH SURGE 18.00h %	BASAL LH 18.00h %	LH SURGE 18.00h %	BASAL LH 18.00h %
VEHICLE (x 1 or 2; n = 45)	91	9		
α MPT (300 to 400 x 1 or 2; n = 76)	50	1	41	8
DDC (600 or 650 x 1; n = 46)	9	59	2	30
DDC (600 or 650 x 2; n = 32)		25		75
SKF64139 (50 x 1; n = 20)	75	20		5
SKF64139 (100) (50 x 2 or 100 x 1; n = 38)		21		79

Table 2. Incidence of Ovulation and of the Surge in Serum LH at Either 1800 or 2400 h on the Day of Proestrus following Treatment with Vehicle, Diethyldithiocarbamate (DDC), SKF64139, LY87130 or SKF29661 at Doses, Times and Frequency as indicated.

CATECHOLAMINE SYNTHESIS INHIBITORS

TREATMENT PRO-OESTRUS 12.00-19.00h (mg/kg x Frequency)	OVULATING 18.00h SURGE %	OVULATING 24.00h SURGE %	NOT OVULATING 24.00h SURGE %	NOT OVULATING NO SURGE %
VEHICLE (x 1, 2 or 3; n = 20)	95	5		
DDC (600 or 650 x 1 or 2; n = 20)		30		70
SKF64139 (50 x 2 or 100 x 1; n = 19)		21		79
LY87130 (20 or 40 x 1; n = 11)		82		18
LY87130 (20 x 2 or 3; n = 12)		8	17	75
SKF29661 (50 x 2; n = 4)	100			

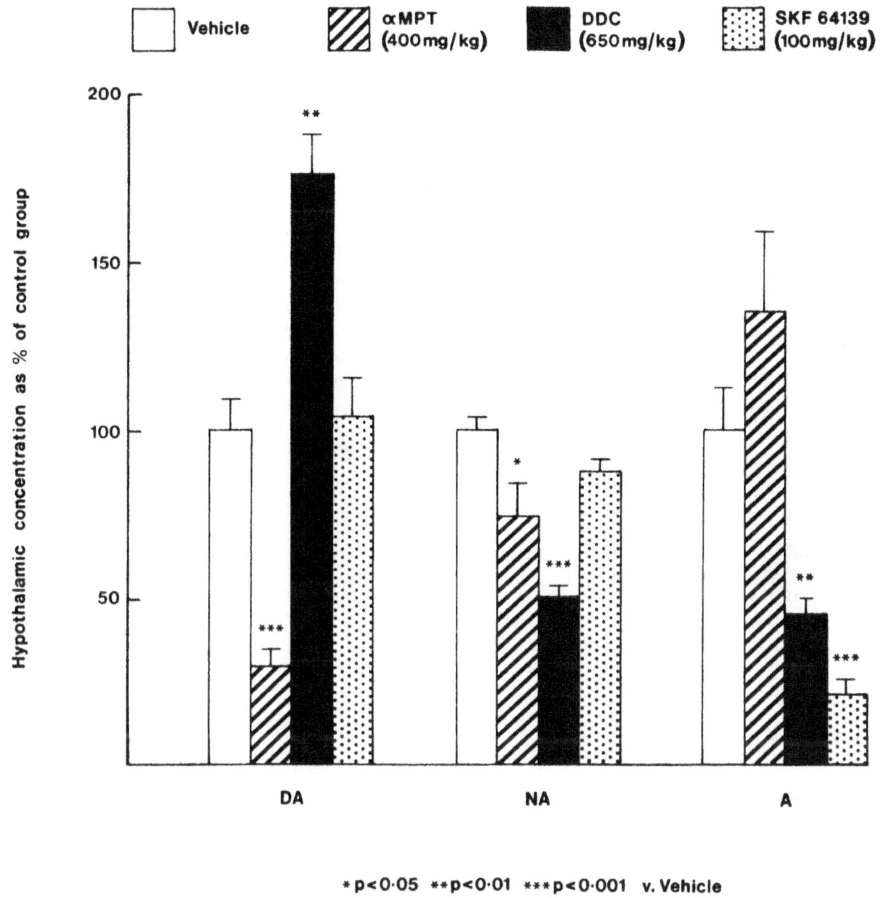

Fig. 4. Mean (+ SEM) concentration (pg/μg protein expressed as percentage of control value) of dopamine (DA), noradrenaline (NA) and adrenaline (A) in hypothalami of proestrous rats treated with vehicle, α-methyl-p-tyrosine (αMPT), diethyldithiocarbamate (DDC) or SKF64139 at 1200 and killed at 1800 h.

In order to elucidate the basis for the lack of effect of αMPT and for the comparable blockade induced by DDC and SKF64139, the effect of these drugs on the hypothalamic concentration of catecholamines was measured (Coen and Coombs, 1983). At this level of analysis, the only common effect of DDC and SKF64139 at doses which block the LH surge was a severe depletion of adrenaline (Fig. 4); αMPT failed to produce this effect on adrenaline or to block the surge despite inducing a marked depletion of dopamine and a moderate loss of noradrenaline. The basis for the apparent resistance of hypothalamic adrenaline to the depleting effects of tyrosine hydroxylase inhibition remains to be established (Coen and Coombs, 1983). Since neither DDC nor SKF64139 affects pituitary responsiveness to LHRH (Coen and Coombs, 1983), it seems possible that the significant action of these drugs in blocking the surge is associated with a depletion of the neuronal pools of adrenaline within the brain. The occurrence of a 'midnight' surge after treatment with relatively low doses of the drugs may reflect a sufficient recovery of adrenaline-mediated neurotransmission to fulfil the requirements of the 'clock' that governs the timing of this event. It is unclear whether the unconventional surges are the consequence of a simple shift in a timing mechanism or a change of state in which an underlying capacity to produce a surge in the middle of the dark phase is given expression. The latter possibility is suggested by the observation of a similarly timed second surge of LHRH in hypophysial portal blood, which, in the event of a normal LH surge at 1800 h, fails to produce a corresponding discharge of LH (Sarkar et al., 1976).

Table 3. Incidence of Ovulation and of the Surge in Serum LH at either 1800 or 2400 h on the Day of Proestrus following Treatment with Prazosin, Piperoxane or Clonidine at Doses, Times, and Frequency as indicated.

ADRENERGIC ANTAGONISTS OR AGONISTS

TREATMENT PRO-OESTRUS 12.00-19.00h (mg/kg x Frequency)	OVULATING 18.00h SURGE %	24.00h SURGE %	NOT OVULATING 24.00h SURGE %	NO SURGE %
PRAZOSIN (5 x 2; n = 5)				100
PIPEROXANE (50 x 2; n = 8)	100			
CLONIDINE (0.1 x 1; n = 12)	25	67		8
CLONIDINE + PIPEROXANE (0.1 x 1 and 50 x 1; n = 16)	63		31	6
CLONIDINE 0.1 x 2; n = 8)		38		62

Adrenergic Antagonists and Agonists and the Proestrous LH Surge

The failure of the proestrous surge with prazosin, an α_1-adrenergic receptor antagonist (Table 3), suggests that the contribution of adrenaline to the production of the surge may be mediated by postsynaptic α_1-receptors (Coen and Coombs, 1983). It remains to be established whether these receptors are directly associated with the LHRH-containing neurons of the preoptic area. The PNMT inhibitors SKF64139 and LY87130 are also known to have α_2-adrenergic receptor blocking properties (Toomey et al., 1981); nevertheless, the possibility that such actions might be significant in the effect of these drugs on the surge seems to be denied by the absence of effect on the preovulatory surge of the α_2-receptor antagonist piperoxane (Table 3) (Coen and Coombs, 1983). The normal proestrous surge may, however, be prevented by treatment with clonidine (Table 3), a drug which reduces the turnover of hypothalamic adrenaline by a putatively presynaptic action on α_2-receptors (Coen and Coombs, 1983); this effect may be counteracted by pretreatment with piperoxane (Table 3).

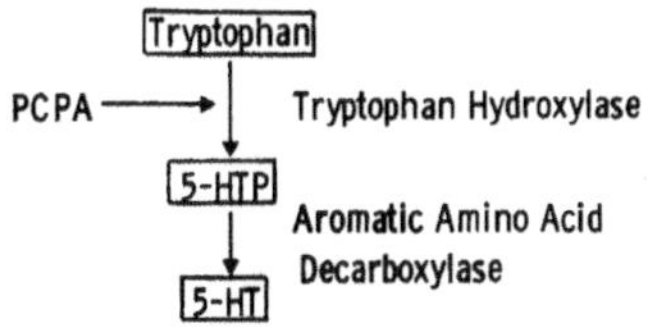

Fig. 5. The biosynthetic pathway for 5-hydroxytryptamine together with parachlorophenylalanine (PCPA), an inhibitor of its synthesis at the level of tryptophan hydroxylase.

5-Hydroxytryptamine and the Proestrous or Steroid-induced LH Surge

In addition to the studies involving manipulation of catecholaminergic systems, we have undertaken comparable research on the role of 5-hydroxytryptamine (5-HT; serotonin) in the regulation of the LH surge. Parachlorophenylalanine (PCPA), a drug which includes among its actions the inhibition of tryptophan hydroxylase and thus the synthesis of 5-HT (Fig. 5), prevents the preovulatory LH surge; restoration of 5-HT synthesis by treatment with its immediate precursor, 5-hydroxytryptophan (5-HTP) reinstates the surge (Coen et al., 1980). These procedures were originally elaborated in ovariectomized estrogen-treated rats which generate a surge of LH daily at 1800 h (Fig. 6) (Coen and MacKinnon, 1979; 1980); it was established that restoration of the 1800 h surge is dependent (1) upon the time of 5-HTP treatment, 1000 h being optimal, and (2) upon the integrity of the suprachiasmatic nuclei. The eight-hour delay between treatment and restoration (Fig. 7) led us to conclude that the role of 5-HT was no more than permissive. Because of the evidence indicating that 5-HT is capable of inhibiting neural mechanisms involved in LH release, we suggested that restoration of the surge by 5-HTP might involve a preparatory inhibition of LHRH release followed by a period of disinhibition (Coen and MacKinnon, 1979). Nevertheless, a physiological role for 5-HT in the production of the LH surge seems improbable since treatment with the 5-HT neurotoxin 5,7-dihydroxytryptamine is compatible with a normal occurrence of the surge despite a depletion of more than 95% of hypothalamic 5-HT (Coen, unpublished). It is therefore probable that the role of the suprachiasmatic nuclei in the 5-HTP-induced restoration of the surge (Coen and MacKinnon, 1980) is specifically associated with the status of these nuclei as a circadian 'clock' and does not involve their dense concentration of 5-HT terminals. Although 5-HT may not actively participate in the normal occurrence of the surge, it does appear to be involved in the 5-HTP-induced restoration of the surge after PCPA since previous treatment with the neurotoxin 5,7-DHT prevents this effect (Fig. 8).

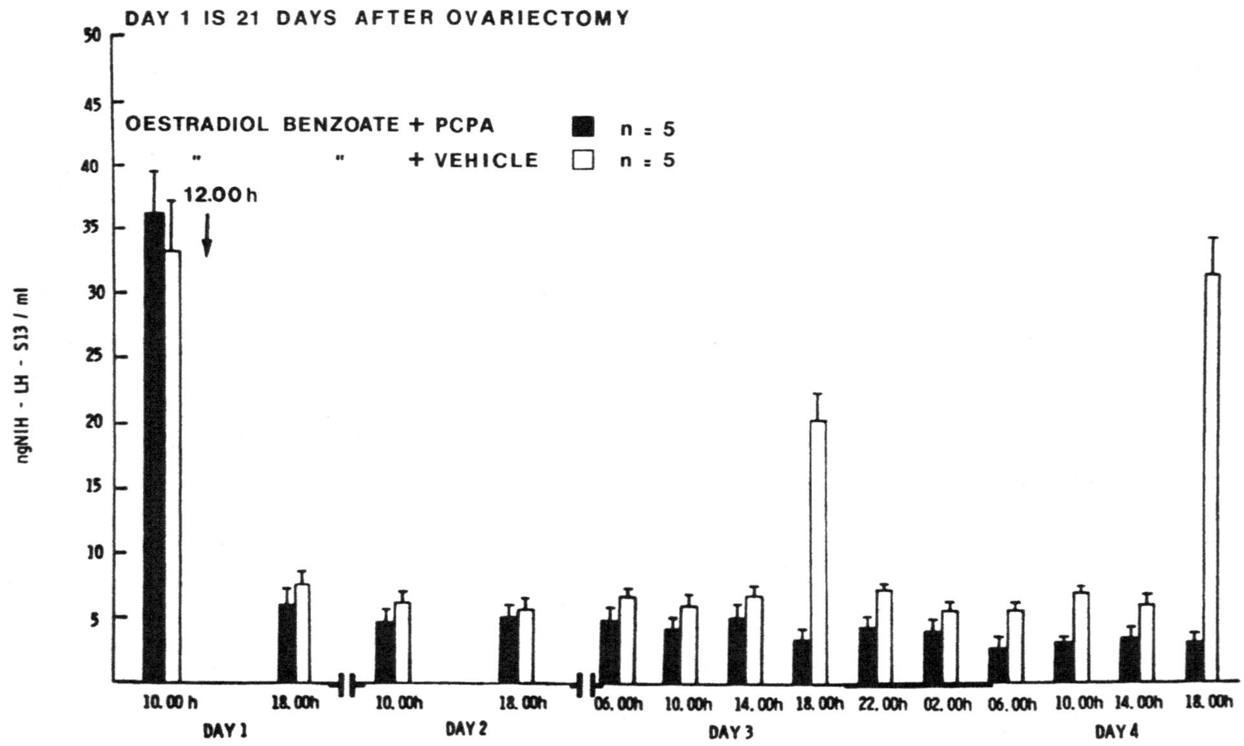

Fig. 6. Mean (+ SEM) serum LH concentration, at times indicated, expressed as ng NIH-LH-S13/ml in chronically ovariectomized rats treated with 50 μg estradiol benzoate at 1200 h on day 1 and bled through an atrial cannula. Animals were given 400 mg p-chlorophenylalanine (PCPA)/kg (n = 5, solid bars) or vehicle alone (n = 5, open bars) at 1200 h on day 1. The black bars on the abscissa indicate hours of darkness (2000-0600 h).

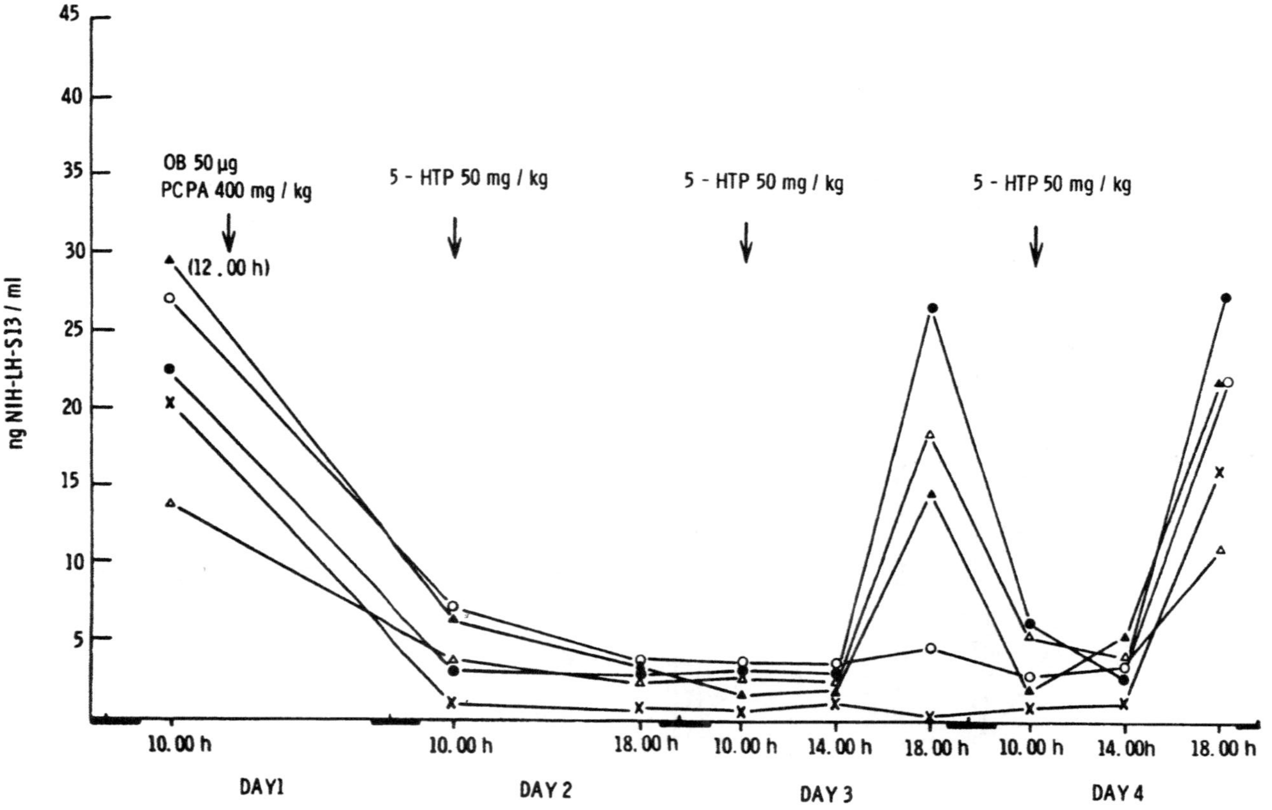

Fig. 7. Serum LH concentrations, at times indicated, expressed as ng NIH-LH-S13/ml, in five chronically ovariectomized rats treated with 50 µg estradiol benzoate and 400 mg p-chlorophenylalanine (PCPA)/kg at 1200 h on day 1. At 1000 h on days 2-4, the animals were given 50 mg 5-hydroxytryptophan/kg (arrows). The black bars on the abscissa indicate hours of darkness (2000-0600 h).

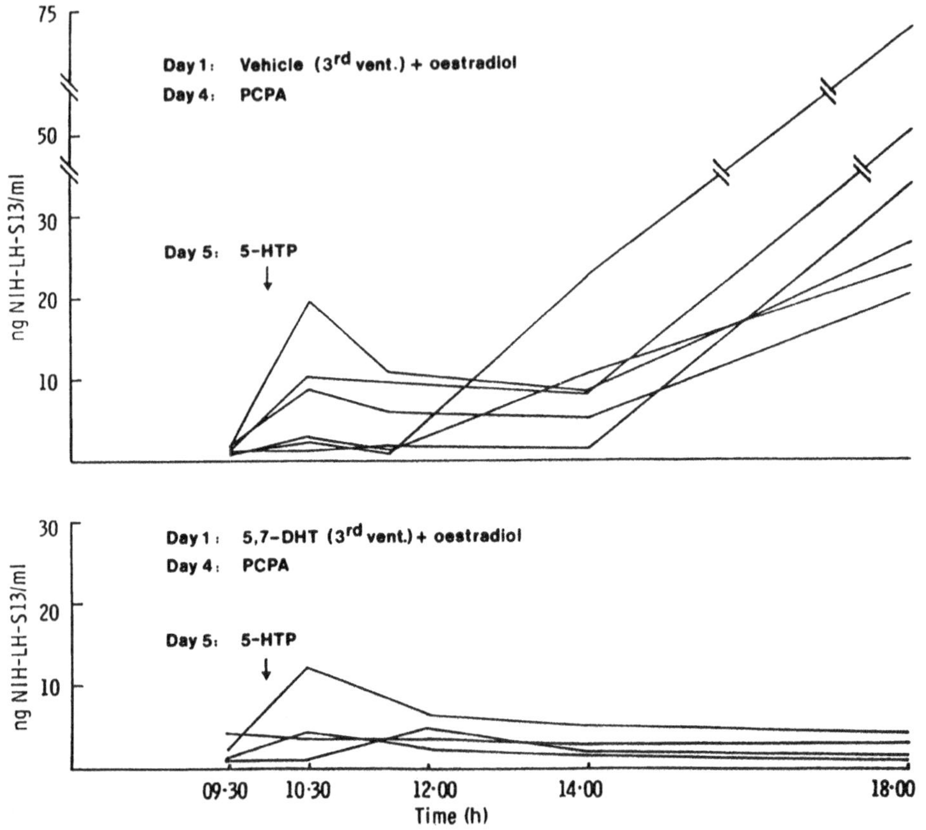

Fig. 8. Serum LH concentration, at times indicated, expresses as ng NIH-LH-S13/ml, in chronically ovariectomized rats treated on day 1 with 20 μg estradiol benzoate, 25 mg DMI/kg and a third ventricular infusion of either vehicle or 150 μg 5,7-dihydroxytryptamine (5,7-DHT); on day 4 at 1200 h they were given 250 mg p-chlorophenylalanine/kg and at 1000 h on day 5 they received 50 mg 5-HTP/kg.

Nevertheless, an involvement of 5-HT in the control of a particular function is usually considered to be established when the effect of PCPA on that function can be overcome by 5-HTP. Since the LH surge is left unaffected by the 5-HT neurotoxin it seems possible that the assumptions underlying the use of PCPA and 5-HTP are in error and that these compounds have reciprocal effects on putative neurotransmitter systems in addition to that of 5-HT. In order to clarify the neurochemical effects of treatment with PCPA and 5-HTP, we examined the effects of these drugs on hypothalamic monoamines and found that they respectively suppress and elevate the concentration of adrenaline in addition to that of 5-HT (Fig. 9) and, furthermore, that PNMT is inhibited in vivo by PCPA and in vitro by its metabolite, parachlorophenylethylamine (Fig. 10) (Coen et al., 1983). Consequently, the studies with PCPA and 5-HTP provide further circumstantial evidence for the involvement of hypothalamic adrenaline in the production of the surge. As a possible resolution of the paradoxes concerning PCPA and 5-HTP, we have argued that the initial rise in the hypothalamic content of adrenaline after 5-HTP in PCPA-treated animals (Fig. 9) is the consequence of a brief, 5-HT-mediated inhibition of the turnover rate of adrenaline and that this

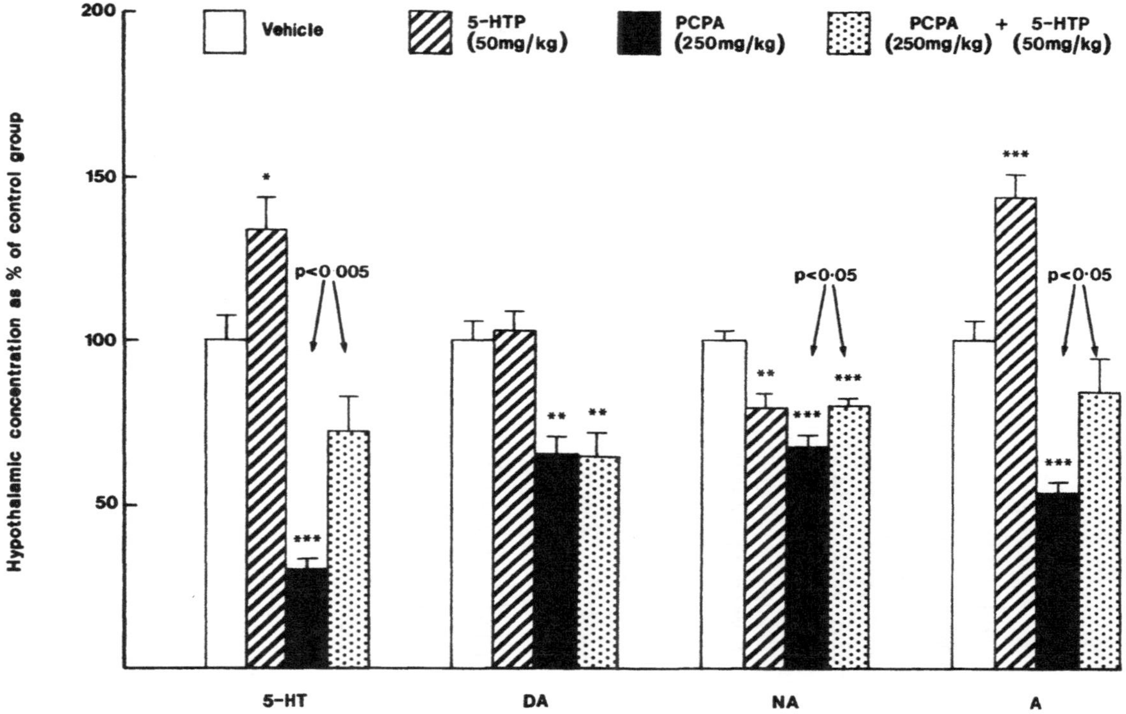

Fig. 9. Mean (+ SEM) concentration (pg/μg protein expressed as percentage of control value) of 5-hydroxytryptamine (5-HT), dopamine (DA), noradrenaline (NA) and adrenaline (A) in hypothalami of ovariectomized rats given 20 μg estradiol benzoate (OB) at 1200 h on day 1 and treated at 1200 h on day 4 and 1000 h on day 5 with, respectively, (1) vehicle and vehicle (open bars); (2) vehicle and 50 mg 5-hydroxytryptophan (5-HTP)/kg (hatched bars); (3) 250 mg p-chlorophenylalanine (PCPA)/kg and vehicle (solid bars); or (4) 250 mg PCPA/kg and 50 mg 5-HTP/kg (stippled bars) and killed on day 5 at 1130 h. *$p < 0.05$; **$p < 0.01$; ***$p < 0.001$; v. Vehicle.

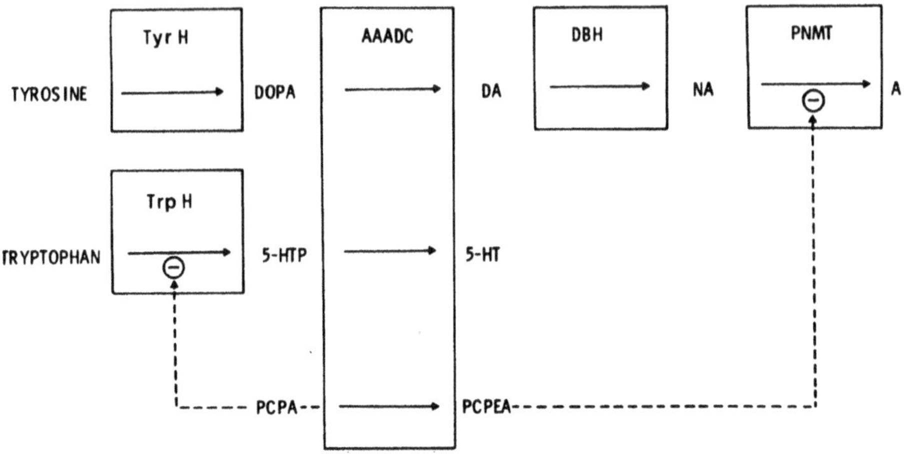

Fig. 10. Some pathways of amine synthesis involving adrenaline (A), aromatic amino acid decarboxylase (AAADC), dihydroxyphenylalanine (DOPA), dopamine (DA), dopamine β-hydroxylase (DBH), 5-hydroxytryptamine (5-HT), 5-hydroxytryptophan (5-HTP), noradrenaline (NA), p-chlorophenylalanine (PCPA), p-chlorophenylethylamine (PCPEA), phenylethanolamine N-methyltransferase (PNMT), tryptophan, tryptophan hydroxylase (Trp H), tyrosine and tyrosine hydroxylase (Tyr H). Steps at which p-chlorinated compounds inhibit enzymes are indicated by dashed arrows.

replenishment of the neuronal pools of adrenaline is sufficient to fulfil subsequent demands from the circadian oscillator which governs the timing of the surge (Coen et al., 1983). Whether the capacity of 5-HT to inhibit adrenaline release may underlie the proposed inhibitory role for 5-HT in the control of LH secretion remains to be established.

Catecholamines and the Steroid-induced LH Surge

Direct evidence indicating that hypothalamic adrenaline is involved in the LH surge which occurs daily at 1800 h in ovariectomized estrogen-treated rats is provided by the failure of this surge following PNMT inhibition with SKF64139 or LY134046 and by its normal occurrence in the presence of SKF29661, an inhibitor of peripheral PNMT (Table 4) (Coen et al., 1986); the LH surge induced by progesterone following estrogen priming is affected in precisely the same way by these drugs (Table 4) (Coen et al., 1986). The effects of LY134046 are particularly significant since this drug has been reported to have negligible α_1- or α_2-adrenergic antagonist properties (Fuller et al., 1981).

Catecholamines and Pulsatile LH Release

Many studies have implicated catecholamines in the regulation of pulsatile LH release in rats. The pulses are suppressed following treatment with antagonists of α-adrenergic receptors (Gnodde and Schuiling, 1976; Weick, 1978; Gallo and Kalra, 1983) or inhibitors of DBH (Drouva and Gallo, 1976; Gnodde and Schuiling, 1976; Gallo and Kalra, 1983) but unaffected by β-adrenergic or dopamine receptor antagonists (Drouva and Gallo, 1976; Weick, 1978; Gallo and Kalra, 1983). Tyrosine hydroxylase inhibition by means of αMPT does not, however, affect the pulses (Carr et al., 1975; Drouva and

Table 4. Incidence of the Surge in Serum LH at either 1800 or 2400 h on the Fourth or Fifth Day after 50 μg Estradiol Benzoate following Treatment with SKF64139, LY134046 or SKF29661 at Doses, Times and Frequency as indicated.

STEROID-INDUCED LH SURGES

TREATMENT 13.00-16.00h (mg/kg x Frequency)	N.	OESTRADIOL BENZOATE 18.00h SURGE %	OESTRADIOL BENZOATE 24.00h SURGE %	OESTRADIOL BENZOATE NO SURGE %	N.	OESTRADIOL BENZOATE + PROGESTERONE 18.00h SURGE %	OESTRADIOL BENZOATE + PROGESTERONE 24.00h SURGE %	OESTRADIOL BENZOATE + PROGESTERONE NO SURGE %
VEHICLE (x 1 or 2)	63	89		11	8	100		
SKF64139 (50 x 2)	8			100	8			100
LY134046 (20 or 40 x 2)	35	17	3	80	10	10		90
SKF29661 (50 x 2)	6	100			5	100		

Gallo, 1976; Gnodde and Schuiling, 1976); this result, like the failure of αMPT to affect the proestrous surge, may be related to the lack of effect on the concentration of hypothalamic adrenaline (Fig. 4).

The first study to implicate adrenaline in the maintenance of pulsatile LH release (Coen et al., 1985) demonstrated a marked suppression of the pulses following treatment with the PNMT inhibitor LY87130 at a dose which blocks the proestrous surge (Fig. 11). We have subsequently found (Coen and Gallo, 1986) that SKF64139 and LY134046 also suppress the pulses when administered in doses which result in a depletion of at least 70% of hypothalamic adrenaline but no change in the concentration of dopamine or noradrenaline; the peripheral PNMT inhibitor SKF29661 does not, however, inhibit the pulses.

LY134046 and LH Release

Since it has been claimed that LY134046 is a centrally acting PNMT inhibitor with negligible α_1- or α_2-antagonist properties (Fuller et al., 1981), the most convincing pharmacological results implicating adrenaline in the regulation of LH release are the suppression of LH pulses (Coen and Gallo, 1986) and the blockade of the steroid-induced surges following treatment with this drug (Table 4). With the other PNMT inhibitors, SKF64139 and LY87130, we have invariably found that the dose that blocks pulsatile release also blocks the proestrous surge. The only major anomaly within the range of in vivo studies focussing on adrenaline concerns the relative ineffectiveness of LY134046 in preventing the occurrence of ovulation (Sheaves et al., 1985); when serum LH levels are measured on proestrus following a dose of LY134046

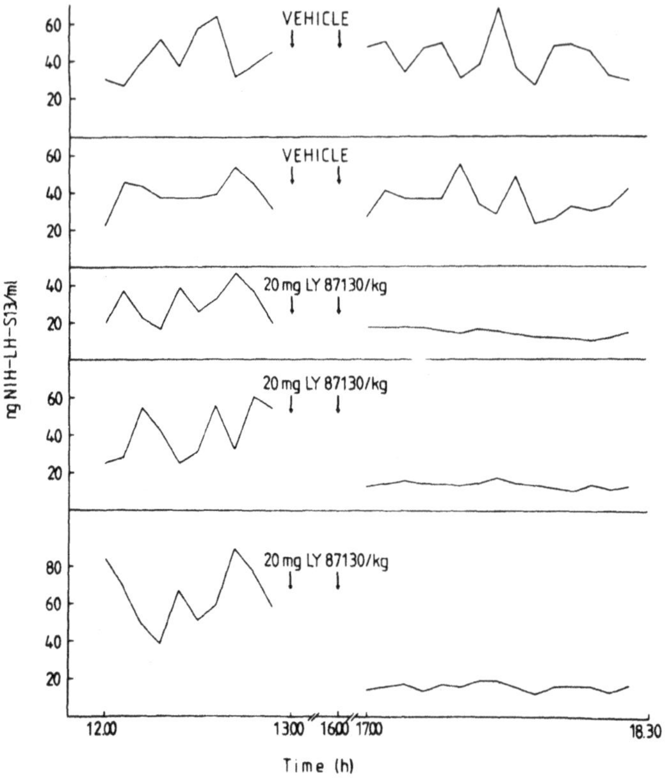

Fig. 11. Serum LH concentrations in 5 representative ovariectomized rats. Blood samples were taken every 6 min between 1200 and 1300 h and between 1700 and 1830 h. The animals were given the vehicle or 20 mg LY87130/kg at 1300 and 1600 h.

which suppresses pulsatile release in ovariectomized rats, 50% of the animals still produce a normal LH surge and the others surge at 2400 h (Table 5; Coen et al., 1986). It is certainly possible that the proestrous rat is less susceptible to the depleting effects of a pure PNMT inhibitor, perhaps due to a reduced rate of turnover, than rats in the other endocrine conditions and that the presence of the α_2-antagonist properties in SKF64139 and LY87130 (Toomey et al., 1981) actually potentiates adrenaline depletion because of the enhanced turnover rate of hypothalamic adrenaline associated with α_2 blockade (Fuller et al., 1979). Although these speculations have not yet been fully tested, we have established (Coen et al., 1986) that supplementing LY134046 with the α_2-antagonists piperoxane or idazoxane, at doses

Table 5. Incidence of Ovulation and of the Surge in Serum LH at either 1800 or 2400 h on the Day of Proestrus following Treatment with Vehicle, LY134046, Piperoxane or Idazoxane at Doses and Times as indicated.

PNMT INHIBITION AND α_2-ANTAGONISM

TREATMENT (mg/kg) PRO-OESTRUS 13.00 & 16.00h	OVULATING 18.00h SURGE %	OVULATING 24.00h SURGE %	OVULATING NO SURGE %	NOT OVULATING 18.00h SURGE %	NOT OVULATING 24.00h SURGE %	NOT OVULATING NO SURGE %
VEHICLE (n = 10)	100					
LY134046 (20) (n = 6)	83			17		
LY134046 (40) (n = 10)	40	50		10		
PIPEROXANE (50) (n = 6)	100					
LY134046 (20) + PIPEROXANE (50) (n = 10)		50	20			30
LY134046 (40) + PIPEROXANE (50) (n = 7)			14			86
IDAZOXANE (2) (n = 4)	100					
LY134046 (40) + IDAZOXANE (2) (n = 6)		33	17			50

which do not by themselves affect the surge, does indeed block the surge at 1800 h and, in most cases, at 2400 (Table 5); furthermore, we have recently found that treatment with piperoxane or idazoxane in conjunction with the peripheral PNMT inhibitor SKF29661 has no deleterious effect on the surge.

TRANSMITTER TURNOVER STUDIES

Our studies involving drugs which manipulate aminergic systems indicate that a disruption of adrenaline-mediated neurotransmission is frequently the common factor associated with suppressed LH release. The results suggest that the contribution of noradrenaline to the preovulatory LH surge may be limited to its role in the biosynthetic pathway for adrenaline. It is therefore important to examine the case for a direct noradrenergic involvement in this process as presented in several studies reporting an increased rate of turnover of hypothalamic noradrenaline on the day of proestrus (Löfström, 1977; Honma and Wuttke, 1980; Demarest et al., 1981; Rance et al., 1981; Rance and Barraclough, 1981); various sites have been reported to show this increased noradrenergic activity including the medial preoptic area, the suprachiasmatic nuclei, the arcuate nuclei and the median eminence. The turnover rate of hypothalamic noradrenaline has also been examined in association with the LH surge induced in ovariectomized rats by estrogen alone (Wise et al., 1981; Rance et al., 1981) or by progesterone following a priming dose of estrogen (Crowley et al., 1978; Simpkins et al., 1979; Honma and Wuttke, 1980; Wise et al., 1981; Crowley, 1982); the results implicated the same wide range of sites mentioned in connection with the proestrous surge.

In these studies the rates of turnover of dopamine and noradrenaline were estimated according to the degree of depletion of the amine following inhibition of tyrosine hydroxylase (Brodie et al., 1966); it should therefore be noted that this procedure inhibits the synthesis of dopamine more effectively than that of noradrenaline (Thierry et al., 1971; Widerlöv and Lewander, 1978), and also has a marked stimulatory effect on the release of prolactin, adrenocorticotropic hormone, corticosterone and progesterone (Coen and Coombs, 1983). Furthermore, since studies of neurotransmitter turnover rates provide only correlative information, any of the spontaneous changes in the endocrine system occurring in association with the LH surge may be pertinent; these include the surge of prolactin, follicle stimulating hormone and progesterone, an increase in the amplitude of the circadian release of adrenocorticotropic hormone and corticosterone, a biphasic pattern in the secretion of thyroid stimulating hormone and thyroxine and a fall in the serum concentration of estradiol either before or during the LH surge (see Coen and Coombs, 1983).

In several of the reports concerning noradrenergic activity in association with the LH surge, comparisons have been made merely between the rate of turnover at a particular time on the day of the presumptive surge and that observed at the same time on a day without such an occurrence (Löfström, 1977; Crowley et al., 1978; Honma and Wuttke, 1980; Crowley, 1982). These studies consequently fail to identify a change in activity during the day of the surge and leave open the possibility that significant differences may reflect less abrupt endocrine changes such as those occurring in, for example the serum concentration of estrogen. Furthermore, many of the studies have failed to consider the concentration of the amine at the time at which the synthesis inhibitor is administered as the initial value from which the decline is calculated (Crowley et al., 1978; Simpkins et al., 1979; Honma and Wuttke, 1980; Demarest et al., 1981; Crowley, 1982) and have assumed that the concentration observed in a vehicle-treated group killed at the same time as the drug-treated group is an adequate substitute. This practice contravenes the principle of estimating the rate of turnover as a function of the degree

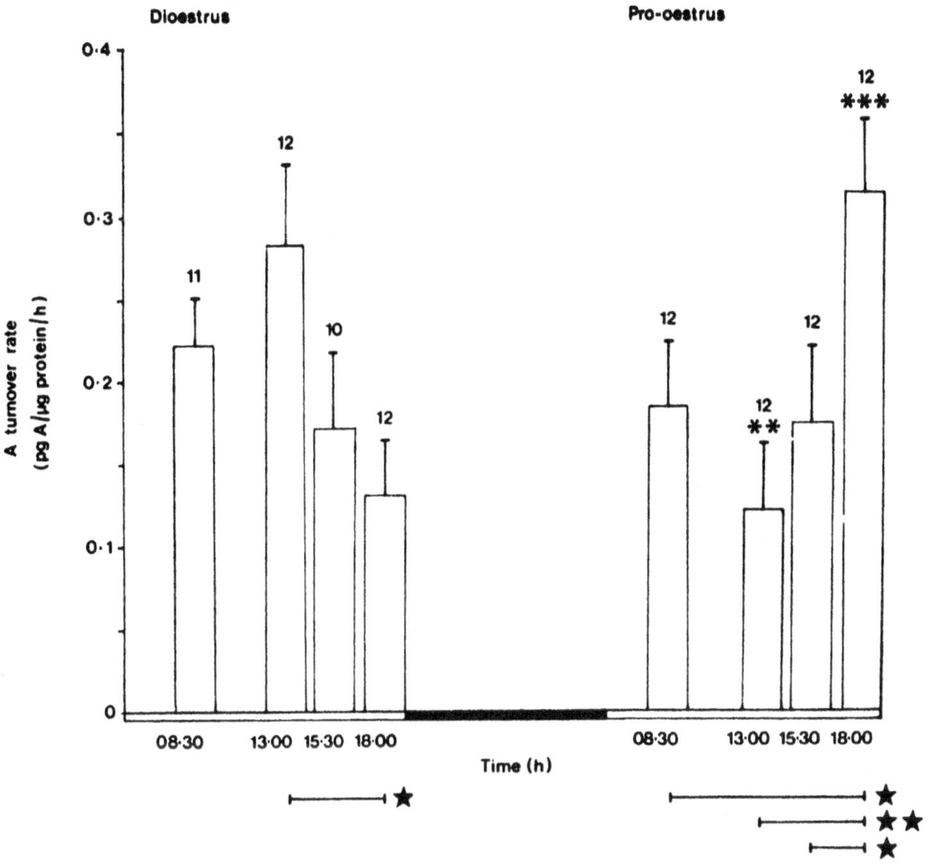

Fig. 12. Mean (+ SEM) turnover rate (pg/µg protein/h) of adrenaline (A) in the medial preoptic area during the 2-h period following 0830, 1300, 1530 and 1800 h on the days of diestrus and proestrus. The number of animals in each group is given above each bar. The hours of darkness are represented by a black bar on the abscissa. Significant differences between results are shown by stars (★ $p < 0.05$, ★★ $p < 0.02$) for those at different times within the same day (comparisons indicated by lines), and by asterisks (**, $p < 0.01$, ***, $p < 0.005$) for those at the same time on the different days.

of depletion following a cessation of synthesis (Brodie et al., 1966); it may also lead to inappropriately high "initial" levels of catecholamines given that tyrosine hydroxylase activity has been reported to rise in conjunction with the LH surge (Carr and Voogt, 1980).

Nevertheless, Barraclough and his associates have presented data on the rate of noradrenaline turnover at various times on the day of a presumptive LH surge using the true initial concentration in their calculations (Rance and Barraclough, 1981; Rance et al., 1981a; Wise et al., 1981; Rance et al., 1981b). Of their two studies concerned with the day of proestrus, one (Rance and Barraclough, 1981) indicates a marked elevation in noradrenergic activity in the medial preoptic area and suprachiasmatic nuclei at the time of the LH surge. This level of activity in the medial preoptic area is, however, present even when the processes underlying the LH surge have been blocked by phenobarbital (Rance and Barraclough, 1981). Furthermore, although their other paper on proestrous rats (Rance et al., 1981a) includes the same data for the two times of day prior to the LH surge, the different results presented in association with the surge indicate a level of noradrenergic activity in the medial preoptic area and suprachiasmatic nuclei which is not significantly higher than than observed three hours earlier, thus suggesting that the changes in these parameters during proestrus are only gradual and not comparable to the abrupt discharge of LH. Similarly, the increase in the rate of noradrenaline turnover in the median eminence is found three hours before the LH surge (Rance and Barraclough, 1981; Rance et al., 1981a). An apparently abrupt increase in noradrenergic activity at the time of the surge is, however, observed in the arcuate nuclei (Rance et al., 1981a), but the importance of this enhanced activity is unclear since it is corroborated in only one (Wise et al., 1981) of their two studies concerned with the LH surge in estrogen-treated ovariectomized rats (Wise et al., 1981; Rance et al., 1981b).

In order to complement our functional studies on the role of hypothalamic adrenaline, we assessed the turnover rate of adrenaline in discrete regions of the brain by measuring the decline in the concentration of adrenaline following inhibition of PNMT (Coombs and Coen, 1983). This enzyme, together with adrenaline, has been demonstrated in the medial preoptic area and mediobasal hypothalamus (Saavedra et al., 1974; Van der Gugten et al., 1976), areas which, respectively, include the LHRH-containing cell bodies and terminals of the preoptico-infundibular tract (Réthelyi et al., 1981). Turnover rates of adrenaline were examined in relation to the serum concentration of LH during proestrus and the preceding day of diestrus (Coombs and Coen, 1983).

We found that the concentration and turnover rate of adrenaline are increased in the medial preoptic area coinciding with the peak of the preovulatory LH surge at 1800 h; moreover, the turnover rate of adrenaline in the medial preoptic area at that time is higher than at the corresponding time on diestrus (Fig. 12). No variation was observed in the turnover rate of adrenaline in the mediobasal hypothalamus within either of the days. This increase in activity of the adrenaline-containing neurons at the time of the surge may be dependent upon the prior rise in the secretion of estrogen; it may be relevant that enhanced hypothalamic PNMT activity has been found following administration of estradiol benzoate to ovariectomized rats (Coen et al., 1983) (Fig. 13).

The possibility that the rise in the turnover rate of adrenaline at 1800 h on proestrus is related to physiological events other than the LH surge should not be ignored. Nevertheless, the results indicate that the adrenaline-containing projections to the preoptic area may be actively involved in the production of the spontaneous preovulatory surge of LH in rats.

Comparable studies were also undertaken on the rate of turnover of 5-HT in the medial preoptic area, suprachiasmatic nuclei and arcuate nuclei (Coombs and Coen, in preparation). Consistent with the conclusions reached in our functional studies, the onset of the preovulatory surge was not associated with a significant change in the rate of 5-HT turnover in any of the areas examined.

IMMUNOHISTOCHEMICAL STUDIES

Since the medial preoptic area contains the perikarya of the LHRH-containing neurons which project to the median eminence (Réthelyi et al., 1981) and represents the site of increased adrenaline turnover at the time of the preovulatory LH surge (Fig. 12), it is possible that these perikarya receive, either directly or indirectly, a stimulatory input from adrenaline-containing neurons. The preoptic site of the LHRH perikarya is certainly characterized by a remarkably dense concentration of PNMT-containing processes (Coen, 1984) (Fig. 14).

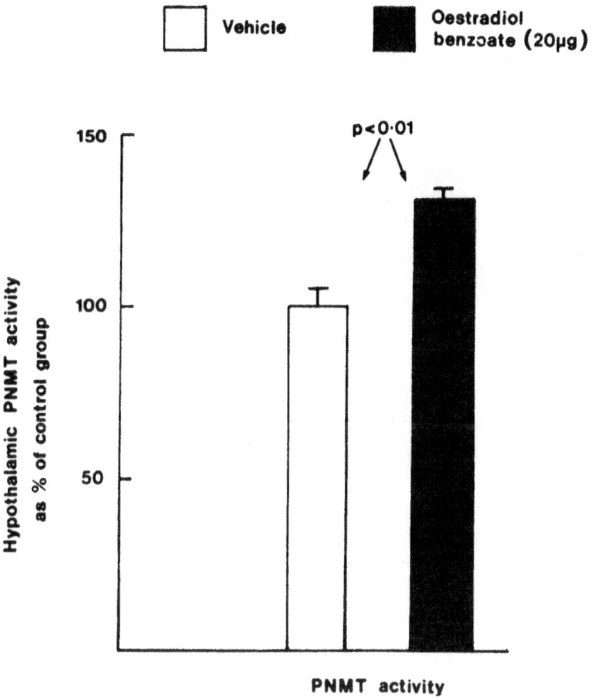

Fig. 13. Mean (+ SEM) phenylethanolamine N-methyltransferase (PNMT) activity expressed as pmol N-methylphenylethanolamine (N-Me-PEA) formed per mg protein per hour, in hypothalami of ovariectomized rats given 20 µg estradiol benzoate (OB) or vehicle at 1200 h on day 1 (n = 4 in both groups). Animals were killed on day 5 at 1130 h.

Until recently the parent cells for immunoreactive PNMT in the forebrain were believed to be exclusively within the C1-C3 groups in the medulla: C1 ventrolaterally, C2 dorsomedially (Hökfelt et al., 1974) and C3 within the medial longitudinal fasciculus (Howe et al., 1980). It is now clear, however, that PNMT is present in cell bodies not only in the medulla but also in the hypothalamus (Coen, 1984; Ross et al., 1984). We have recently established that horseradish peroxidase conjugated to wheatgerm agglutinin is transported retrogradely from injection sites in the vicinity of the LHRH

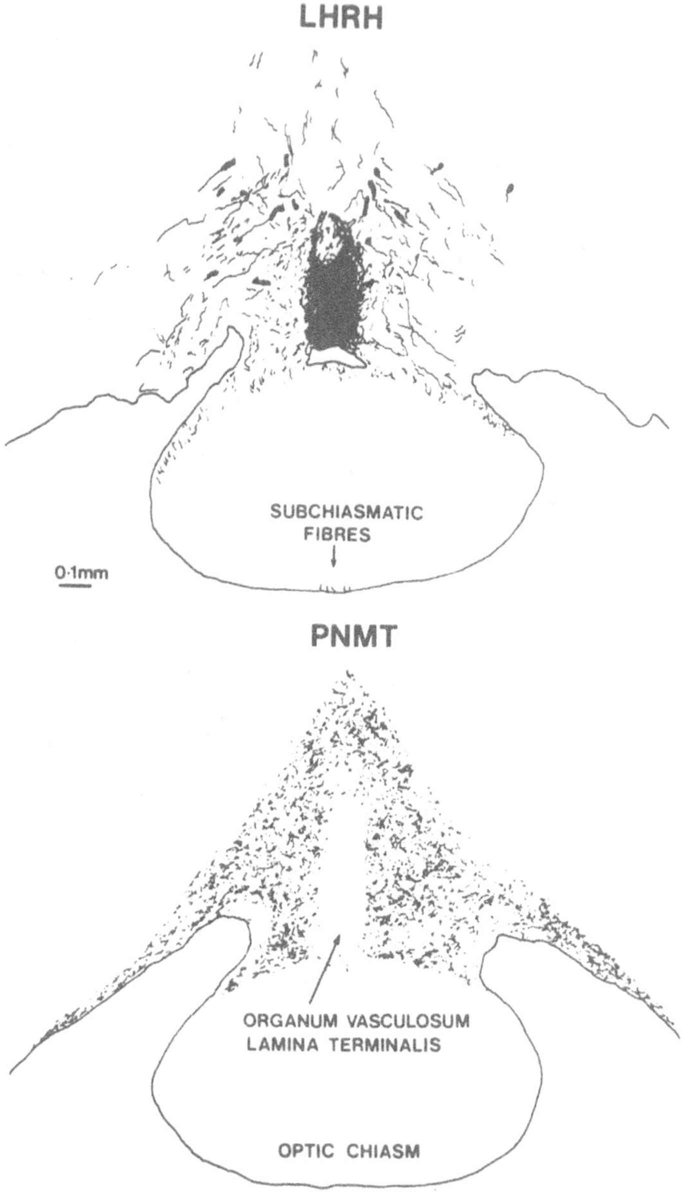

Fig. 14. Camera lucida drawings of coronal sections of rat brain at the level of the organum vasculosum of the lamina terminalis; immunoreactivity for LHRH-containing perikarya and fibers (above) and immunoreactivity for PNMT-containing processes (below).

cell bodies to PNMT-immunoreactive cells within the C2 group (Taylor and Coen, unpublished). Whether the more rostral PNMT cell groups contribute to the innervation of the medial preoptic area remains to be determined. Nevertheless, using a double-immunoperoxidase method involving silver-intensification of one of the diaminobenzidine reaction products (Liposits et al., 1986), it has been possible to identify at the light microscope level PNMT-immunoreactive fibers in apposition to LHRH-containing structures in the preoptic area (Liposits and Coen, unpublished); analysis of these presumptive interactions at the electron microscope level is currently in progress.

CONCLUSIONS

An involvement of adrenergic processes in the control of LH release was first proposed nearly forty years ago. It was subsequently demonstrated that intraventricular adrenaline can restore ovulation in rats when the otherwise spontaneous preovulatory surge of LH is blocked by barbiturate-treatment (Rubinstein and Sawyer, 1970); furthermore, intraventricular adrenaline was found to be more potent than noradrenaline in restoring barbiturate-blocked ovulation (Rubinstein and Sawyer, 1970) and in stimulating LH release in steroid-primed ovariectomized rats (Vijayan and McCann, 1978). Evidence indicating that this process may be specifically mediated by adrenaline rather than noradrenaline was obtained in 1981 with the discovery that SKF64139, an inhibitor of PNMT, the enzyme which catalyzes the conversion of noradrenaline to adrenaline, blocks the proestrous surge (Coen and MacKinnon, 1981) and the surge induced by progesterone in estrogen-primed ovariectomized rats (Crowley and Terry, 1981).

While recognizing that considerable caution is required in interpreting studies involving either pharmacological manipulation, estimation of neurotransmitter "turnover" or immunohistochemical analysis of neuronal connectivity, we propose as a working hypothesis: (1) that the pulsatile release of LH is a necessary condition for the preovulatory surge (i.e. the surge is a special instance of pulsatile release); (2) that adrenaline is specifically involved in producing the pulsatile activity within the LH releasing system upon which the successful triggering of the surge depends; and (3) that the increase in the turnover rate of preoptic area adrenaline at the time of the surge (Coombs and Coen, 1983) is associated with the increase in the frequency and amplitude of the pulses underlying the surge (Gallo, 1981a). The mechanisms underlying the transformation of the normal pulses into the surge (Fig. 15) involve a circadian clock probably located in the suprachiasmatic nuclei (receiving direct neural input from the retina) and an active role for estrogen which may include, among other actions, an increase in PNMT activity (Coen et al., 1983).

Many of the drug treatments which have been shown to inhibit various modes of LH release include among their known actions a deleterious effect on adrenaline-mediated neurotransmission. The discovery of the "midnight window" at which the LH surge may be triggered if the animal has been given a relatively low dose of the drug emphasizes the importance of taking sequential samples from the same animal. Nevertheless, it is clear that many substances other than biogenic amines participate in the regulation of LHRH release. Given the vast range of research on the other candidate substances, it may be particularly useful to note (1) the coexistence of PNMT and neuropeptide Y (NPY) in brainstem perikarya (Everitt et al., 1984); (2) the striking similarity between immunoreactivity for NPY and PNMT in the medial preoptic area (Coen and Taylor, 1986); and (3) the recent evidence indicating that NPY can exert powerful actions on LH release when administered centrally (McDonald et al., 1985). Other research based on drug administration has demonstrated that the inhibitory effects of β-endorphin on LHRH secretion may be mediated by a preterminal inhibition of adrenaline release (Kalra and

Crowley, 1982). It is now essential to establish whether the anatomical basis for such an effect and for the proposed presynaptic inhibitory action of 5-HT (Fig. 15) can be identified.

The LHRH neurons are few in number and their activity cannot as yet be monitored "on-line" with confidence. Nevertheless, it is clear that the testing of hypotheses concerning the processes which underlie the preovulatory LH surge will continue to provide an excellent opportunity for multidisciplinary research.

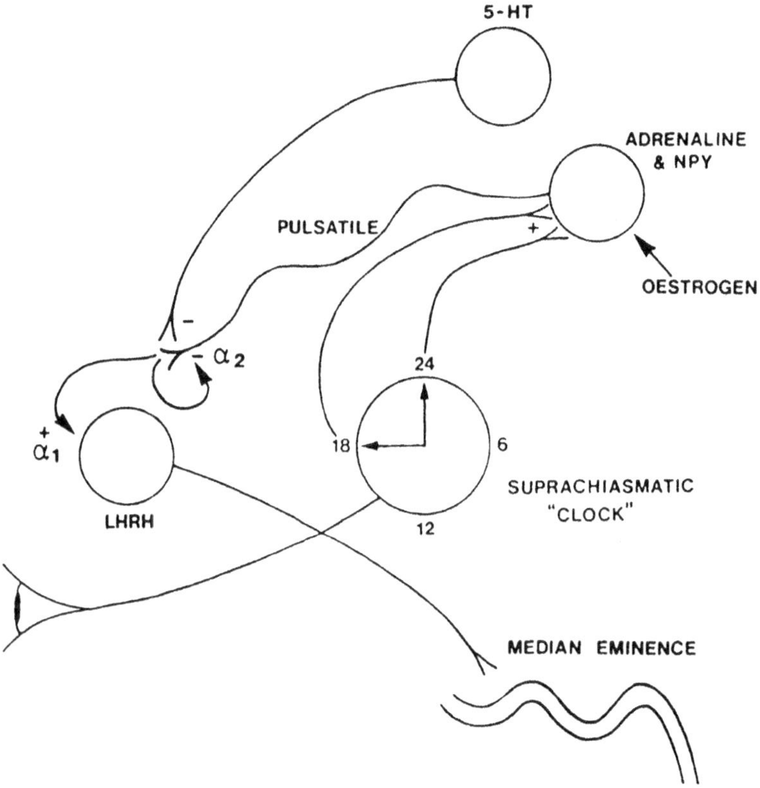

Fig. 15. Schematic representation of the proposed neural systems underlying the control of LH secretion.

REFERENCES

Brodie, B. B., Costa, E., Dlabac, A., Neff, N. H., and Smookler, H. H., 1966, Application of steady state kinetics to the estimation of synthesis rate and turnover time of tissue catecholamines, J. Pharmac. Exp. Ther., 35:161.

Caligaris, L., Astrada, J. J., and Taleisnik, S., 1971, Biphasic effect of progesterone on the release of gonadotropin in rats, Endocrinology, 89:331.
Carr, L. A., Conway, P. M., and Voogt, J. L., 1975, Inhibition of brain catecholamine synthesis and release of prolactin and luteinizing hormone in the ovariectomized rat, J. Pharmacol. Exp. Therap., 192:15.
Coen, C. W., 1984, Immunohistochemical localization of phenylethanolamine N-methyltransferase in the rat brain, Physiologist, 27:223.
Coen, C. W., and Coombs, M. C., 1983, Effects of manipulating catecholamines on the incidence of the preovulatory surge of luteinizing hormone and ovulation in the rat: evidence for a necessary involvement of hypothalamic adrenaline in the normal or "midnight" surge, Neuroscience, 10:187.
Coen, C. W., Coombs, M. C., Wilson, P. M. J., Clement, E. M., and MacKinnon, P. C. B., 1983, Possible resolutionof a paradox concerning the use of p-chlorophenylalanine and 5-hydroxytryptophan: evidence for a mode of action involving adrenaline in manipulating the surge of luteinizing hormone in rats, Neuroscience, 8:583.
Coen, C. W., Franklin, M., Laynes, R. W., and MacKinnon, P. C. B., 1980, Effects of manipulating serotonin on the incidence of ovulation in the rat, J. Endocrinol., 87:195.
Coen, C. W., and Gallo, R. V., 1986, Effects of various inhibitors of phenylethanolamine N-methyltransferase on pulsatile release of luteinizing hormone in ovariectomized rats, J. Endocrinol., 111:51.
Coen, C. W., and MacKinnon, P. C. B., 1976, Serotonin involvement in oestrogen-induced luteinizing hormone release in ovariectomized rats, J. Endocrinol., 71:49P.
Coen, C. W., and MacKinnon, P. C. B., 1977, Evidence for a critical period and a possible site of action for serotonin involvement in oestrogen-induced luteinizing hormone release in ovariectomized rats, J. Endocrinol., 75:43P.
Coen, C. W., and MacKinnon, P. C. B., 1979, Serotonin involvement in the control of phasic luteinizing hormone release in the rat: evidence for a critical period, J. Endocrinol., 82:105.
Coen, C. W., and MacKinnon, P. C. B., 1980, Lesions of the suprachiasmatic nuclei and the serotonin-dependent release of luteinizing hormone in the rat: effects on drinking rhythmicity and on the consequences of preoptic area stimulation, J. Endocrinol., 84:231.
Coen, C. W., and MacKinnon, P. C. B., 1981, The effects of blocking catecholamine synthesis on the incidence of the LH surge and ovulation in the rat, Neuroscience Lett. Suppl., 7:252.
Coen, C. W., Simonyi, A., and Fekete, M. I. K., 1985, Effects of phenylethanolamine N-methyltransferase inhibition on serum concentrations of luteinizing hormone, follicle stimulating hormone and prolactin in pro-oestrous and ovariectomized rats, Brain Res., 343:383.
Coen, C. W., Tacconelli, F., and MacKenzie, F. J., 1986, Further studies on the effects of inhibitors of phenylethanolamine N-methyltransferase on luteinizing hormone release, in preparation.
Coen, C. W., and Taylor, A. M., 1986, Immunohistochemical localization of neurotransmitter-related antigens in the division of the preoptic area containing luteinizing hormone releasing hormone in the rat, Soc. Neuroscience Abstracts, 12:297.
Coombs, M. C., and Coen, C. W., 1983, Adrenaline turnover rates in the medial preoptic area and mediobasal hypothalamus in relation to the release of luteinizing hormone in female rats, Neuroscience, 10:207.
Crowley, W. R., and Terry, L. C., 1981, Effects of an epinephrine synthesis inhibitor, SKF64139, on the secretion of luteinizing hormone in ovariectomized female rats, Brain Res., 204:231.
Crowley, W. R., 1982, Effects of ovarian hormones on norepinephrine and dopamine turnover in individual hypothalamic and extrahypothalamic nuclei, Neuroendocrinology, 34:381.

Crowley, W. R., O'Donohue, T. L., Wachslicht, H., and Jacobowitz, D. M., 1978, Effects of estrogen and progesterone on plasma gonadotropins and catecholamine levels and turnover in discrete brain regions of ovariectomized rats, Brain Res., 154:345.
Crowley, W. R., Terry, L. C., and Johnson, M. D., 1982, Evidence for the involvement of central epinephrine systems in the regulation of luteinizing hormone, prolactin, and growth hormone release in female rats, Endocrinology, 110:1102.
Demarest, K. T., Johnston, C. A., and Moore, K. E., 1981, Biochemical indices of catecholamine neuronal activity in the median eminence during the estrous cycle of the rat, Neuroendocrinology, 32:24.
Drouva, S. V., and Gallo, R. V., 1976, Catecholamine involvement in episodic luteinizing hormone release in adult ovariectomized rats, Endocrinology, 99:651.
Drouva, S. V., Laplante, E., and Kordon, C., 1982, α_1-Adrenergic receptor involvement in the LH surge in ovariectomized estrogen-primed rats, Eur. J. Pharmacol., 81:341.
Everett, J. W., and Sawyer, C. H., 1949, A neural timing factor in the mechanism by which progesterone advances ovulation inthe cyclic rat, Endocrinology, 45:581.
Everett, J. W., Sawyer, C. H., and Markee, J. E., 1949, A neurogenic timing factor in control of the ovulatory discharge of luteinizing hormone in the cyclic rat, Endocrinology, 44:234.
Everitt, B. J., Hökfelt, T., Terenius, L., Tatemoto, K., Mutt, V., and Goldstein, M., 1984, Differential co-existence of neuropeptide Y (NPY-)-like immunoreactivity with catecholamines in the central nervous system of the rat, Neuroscience, 11:443.
Fuller, R. W., Hemrick-Luecke, S., Toomey, R. E., Horng, J.-S., Ruffolo, R. R. Jr., and Molloy, B. B., 1981, Properties of 8,9-dichloro-2,3,4,5-tetrahydro-1H-2-benzazepine, an inhibitor of norepinephrine N-methyltransferase, Biochem. Pharmacol., 30:1345.
Fuller, R. W., Perry, K. W., Hemrick, S. E., and Molloy, B. B., 1979, Lowering of brain epinephrine by inhibition of norepinephrine N-methyltransferase in rats, in: "Catecholamines: Basic and Clinical Frontiers," E. Usdin, I. J. Kopin and J. Barchas, eds., pp. 186-188, Pergamon Press, New York.
Gallo, R. V., 1981a, Pulsatile LH release during the ovulatory LH surge on proestrus in the rat, Biol. Reprod., 24:100.
Gallo, R. V., 1981b, PUlsatile release during periods of low level LH secretion in the rat estrous cycle, Biol. Reprod., 24:771.
Gallo, R. V., and Kalra, P. S., 1983, Pulsatile LH release on diestrus 1 in the rat estrous cycle: relation to brain catecholamines and ovarian steroid secretion, Neuroendocrinology, 37:91.
Gnodde, H. P., and Schuiling, G. A., 1976, Involvement of catecholaminergic and cholinergic mechanisms in the pulsatile release of LH in the long-term ovariectomized rat, Neuroendocrinology, 20:212.
Hökfelt, T., Fuxe, K., Goldstein, M., and Johansson, O., 1974, Immunohistochemical evidence for the existence of adrenaline neurons on the rat brain, Brain Res., 66:239.
Honma, K., and Wuttke, W., 1980, Norepinephrine and dopamine turnover rates in the medial preoptic area and the mediobasal hypothalamus of the rat brain after various endocrinological manipulations, Endocrinology, 106:1848.
Howe, P. R. C., Costa, M., Furness, J. B., and Chalmers, J. P., 1980, Simultaneous demonstration of phenylethanolamine-N-methyltransferase immunofluorescent and catecholamine fluorescent nerve cell bodies in the rat medulla oblongata, Neuroscience, 5:2229.
Kalra, S. P., and Crowley, W. R., 1983, Epinephrine synthesis inhibitors block naloxone-induced LH release, Endocrinology, 82:1403.
Kalra, P. S., and McCann, S. M., 1973, Involvement of catecholamines in feedback mechanisms, Prog. Brain Res., 39:185.

Liposits, Zs., Phelix, C., and Paull, W. K., 1986, Adrenergic innervation of corticotropin releasing factor (CRF)-synthesizing neurons in the hypothalamic paraventricular nucleus of the rat, Histochem., 84:201.

Löfström, A., 1977, Catecholamine turnover alterations in discrete areas of the median eminence of the 4- and 5-day cycling rat, Brain Res., 120:131.

Pendleton, R. G., Gessner, G., Weiner, G., Jenkins, B., Sawyer, J., Bondinell, W., and Intoccia, A., 1979, Studies on SK&F29661, an organ-specific inhibitor of phenylethanolamine N-methyltransferase, J. Pharmacol. Exp. Therap., 208:24.

Raisman, B., and Brown-Grant, K. I., 1977, The suprachiasmatic syndrome: endocrine and behavioral abnormalities following lesions of the suprachiasmatic nuclei in the female rat, Proc. Royal Soc. London, Series B, 198:297.

Rance, N., and Barraclough, C. A., 1981, Effects of phenobarbital on hypothalamic LHRH and catecholamine turnover rates in proestrous rats, Proc. Soc. Exp. Biol. Med., 166:425.

Rance, N., Wise, P. M., and Barraclough, C. A., 1981, Negative feedback effects of progesterone correlated with changes in hypothalamic norepinephrine and dopamine turnover rates, median eminence luteinizing hormone-releasing hormone, and peripheral plasma gonadotropins, Endocrinology, 1098:2194.

Rance, N., Wise, P. M., Selmanoff, M. K., and Barraclough, C. A., 1981, Catecholamine turnover rates in discrete hypothalamic areas and associated changes in median eminence luteinizing hormone-releasing hormone and serum gonadotropins on proestrus and diestrous day 1, Endocrinology, 108:1795.

Réthelyi, M., Vigh, S., Sétáló, G., Merchenthaler, I., Flerkó, B., and Petrusz, P., 1981, The luteinizing hormone releasing hormone-containing pathways and their cotermination with tanycyte processes in and around the median eminence and in the pituitary stalk of the rat, Acta Morphologica Acad. Sci. Hung., 29:259.

Ross, C. A., Ruggiero, D. A., Meeley, M. P., Park, D. H., Joh, T. H., and Reis, D. J., 1984, A new group of neurons in hypothalamus containing phenylethanolamine N-methyltransferase (PNMT) but not tyrosine hydroxylase, Brain Res., 306:349.

Rubinstein, L., and Sawyer, C. H., 1970, Role of catecholamines in stimulating the release of pituitary ovulatory hormone(s) in rats, Endocrinology, 86:988.

Saavedra, J. M., Palkovits, M., Brownstein, M. J., and Axelrod, J., 1974, Localization of phenylethanolamine N-methyltransferase in rat brain nuclei, Nature (Lond)., 248:695.

Sarkar, D. K., Chiappa, S. A., Fink, G., and Sherwood, N. M., 1976, Gonadotropin-releasing hormone surge in pro-oestrous rats, Nature (Lond)., 264:461.

Sawyer, C. H., Everett, J. W., and Markee, J. E., 1949, A neural factor in the mechanism by which estrogen induces the release of luteinizing hormone in the rat, Endocrinology, 44:218.

Sétáló, G., Vigh, S., Schally, A. V., Arimura, A., and Flerkó, B., 1976, Immunohistological study of the origin of LHRH-containing nerve fibers of the rat hypothalamus, Brain Res., 103:597.

Simpkins, J. W., Huang, H. H., Advis, J. P., and Meites, J., 1979, Changes in hypothalamic NE and DA turnover resulting from steroid-induced LH and prolactin surges in ovariectomized rats, Biol. Reprod., 20:625.

Thierry, A. M., Blanc, G., and Glowinski, J., 1971, Dopamine-norepinephrine: another regulatory step of norepinephrine synthesis in central noradrenergic neurons, Eur. J. Pharmacol., 14:303.

Toomey, R. E., Horng, J. S., Hemrick-Luecke, S. K., and Fuller, R. W., 1981, α_2-Adrenoreceptor affinity of some inhibitors of norepinephrine N-methyltransferase, Life Sci., 29:2467.

Van der Gugten, J., Palkovits, M., Wijnen, H. L. J. M., and Versteeg,

D. H. G., 1976, Regional distribution of adrenaline in the brain, Brain Res., 107:171.
Vijayan, E., and McCann, S. M., 1978, Re-evaluation of the role of catecholamines in control of gonadotropin and prolactin release, Neuroendocrinology, 25:150.
Voogt, J. L., and Carr, L. A., 1981, Inhibition of LH and prolactin release in the cycling rat following inhibition of dopamine-β-hydroxylase, Brain Res., 209:411.
Weick, R. F., 1978, Acute effects of adrenergic receptor blocking drugs and neuroleptic agents on pulsatile discharges of luteinizing hormone in the ovariectomized rat, Neuroendocrinology, 26:108.
Wise, P. M., Rance, N., and Barraclough, C. A., 1981, Effects of estradiol and progesterone on catecholamine turnover rates in discrete hypothalamic regions in ovariectomized rats, Endocrinology, 108:2186.

DUAL ACTION OF NOREPINEPHRINE IN THE CONTROL OF GONADOTROPIN RELEASE

Hugo Bergen and Peter C.K. Leung

Departments of Ob/Gyn and Physiology
University of British Columbia
Vancouver, British Columbia, Canada

INTRODUCTION

A role for catecholamines in the regulation of ovulation was first implicated when blockade of α-adrenergic receptors prevented ovulation in rabbits while infusion of norepinephrine (NE) into the third ventricle was able to induce ovulation (Sawyer et al., 1947; Sawyer, 1952). Likewise, in the rat, administration of α-adrenergic antagonists blocked ovulation suggesting that on the afternoon of proestrus a critical period exists during which time the hypothalamus becomes "activated" to secrete some gonadotropin-releasing factor (Everett et al., 1949; Everett and Sawyer, 1950). Since then a great deal of work has been done supporting an excitatory role for NE in the regulation of luteinizing hormone (LH) release and/or luteinizing hormone-releasing hormone((LHRH) (Barraclough and Wise, 1982; Ramirez et al., 1984). More recently, the possibility of an inhibitory role for NE In the control of LH release has been raised when it was found that infusion of NE into the third ventricle of ovariectomized (ovx) rats inhibited LH release (Gallo and Drouva, 1979). This chapter will briefly review the evidence of the dual role of NE in the regulation of LH release.

THE LHRH PULSE GENERATOR

The release of LH from the pituitary appears to be primarily a function of LHRH release from the LHRH nerve terminals in the median eminence into the hypothalamo-hypophyseal portal blood system. In the rat, these LHRH neurons have their cell bodies located diffusely in the medial preoptic area, lateral anterior hypothalamic areas (King et al., 1982; Hoffman and Gibbs, 1982), and possibly also the medial basal hypothalamus (Kelly et al., 1982). LH release has been shown to be pulsatile throughout the estrous cycle as well as in the ovx rat (Gallo, 1981a, 1981b; Gay and Sheth, 1972). This pulsatile release is thought to be due to pulsatile LHRH release from the median eminence (Clarke and Cummins, 1982; Levine et al., 1982).

The effects of NE on LH release could be due to its actions directly or indirectly (via interneurons) on LHRH neurons either at the level of the LHRH cell body or terminal. In the hypothalamus, NE is restricted to nerve terminals and axons whose cell bodies are located in the NE cell groups of the brainstem (Moore and Bloom, 1979). NE fibers originating in the brainstem pass through the midbrain via two major pathways, the ventral noradrenergic tract (VNT) and the dorsal noradrenergic tract (DNT) (Lindvall and Bjorklund, 1978). Although the relative roles of these two pathways in gonadotropin release are not fully understood, it is thought that impulses travelling via these pathways are able to modulate the frequency and/or the amplitude of the LHRH pulses, thus affecting LH release. Moreover, the effects of NE on the LHRH neuron also appear to depend on a variety of factors such as the steroidal environment and the activity of other inputs to the LHRH neurons.

STIMULATORY ACTION OF NE ON LH RELEASE

As mentioned previously, an excitatory role for NE in the control of gonadotropin release was first suggested by Sawyer and co-workers (1947). It was later demonstrated that NE, when infused into the third ventricle, could induce ovulation in pentobarbital-treated proestrous rats as well as rats made anovulatory by anterior hypothalamic lesions or continuous illumination (Rubinstein and Sawyer, 1970; Tima and Flerko, 1974). Ventricular infusion of NE was also effective in eliciting LH release in estrogen-progesterone primed ovx rats (Krieg and Sawyer, 1976; Gallo and Drouva, 1979). It has also been shown that drugs which deplete brain monamines or inhibit catecholamine synthesis also inhibit LH release and block ovulation (Barraclough and Sawyer, 1957; Kalra and McCann, 1974). Similarly, depletion of hypothalamic NE with 6-hydroxy-dopamine decreased preovulatory and steroid-induced LH surge release (Hancke and Wuttke, 1979; Simpkins et al., 1979). Sarkar and Fink (1981) also reported that catecholamine synthesis inhibitors reduced the height of the LHRH and LH surge. The stimulatory effect of NE on the release is probably mediated via α-adrenergic receptors since phenoxybenzamine (an α-adrenergic antagonist) reduced both the LHRH and LH surge whereas propranolol and yohimbine (a β- and α_2-antagonist, respectively) had no effect on the estrogen-induced LH surge (Sarkar and Fink, 1981; Drouva et al., 1982).

A stimulatory role for NE on LHRH release (mediated by α-adrenergic receptors) is also suggested by a number of <u>in vitro</u> studies. It has been shown that NE is able to induce LHRH release in median eminence fragments in a dose-dependent manner and these effects are blocked by α-adrenergic antagonists but not β-adrenergic antagonists (Negro-Vilar et al., 1979; Ojeda et al., 1982). Using a similar median eminence incubation system, Heaulme and Dray (1984) have recently demonstrated that phenylephrine (an α_1-adrenergic agonist) is effective in stimulating LHRH release whereas clonidine (an α_2-adrenergic agonist) and isoproterenol (a β-adrenergic agonist) are ineffective. As well, they also found that an α_1-adrenergic antagonist, but not an α_2- or β-antagonist, was able to block the stimulatory effects of NE on LHRH release. NE has also been shown to elicit LHRH release from superfused hypothalami (Nowak and Swerdloff, 1985). As with the median eminence fragments, the effects of NE were dose dependent and blocked by an α-adrenergic antagonist.

An excitatory role for NE in LH release is also supported by studies which have examined catecholamine turnover rates in specific hypothalamic regions. Changes in turnover rates have been correlated with alterations in plasma LH levels in intact cycling rats, ovx rats, ovx plus estrogen, or estrogen- and progesterone-primed rats (Honma and Wuttke, 1980; Rance et al., 1981a; Rance et al., 1981b; Crowley, 1982). In proestrous rats, NE turnover increases initially in the median eminence prior to initiation of the LH surge; once the surge has started NE turnover rates are also increased in the medial preoptic, suprachiasmatic, and arcuate nuclei (Rance et al., 1981b). Honma and Wuttke (1980) also reported an increase in NE turnover rates on the afternoon of proestrus in the medial preoptic area and medial basal hypothalamus; the increase in NE turnover in the median eminence and medial basal hypothalamus correlates well with increased LH levels in ovx rats primed either with estrogen alone, or with estrogen and progesterone. Rance et al., (1981a) found that in ovx rats primed with estrogen, increases in NE turnover occurred in the median eminence, medial preoptic area, and suprachiasmatic nuclei at the time of increased LH secretion. In rats treated with estrogen and progesterone, these increases in NE turnover are advanced in time as is the LH surge (Wise et al., 1981). Taken together, these studies suggest that increased NE activity is involved in triggering and maintaining the LH surge, and the feedback effects of estrogen and progesterone on LH release may be mediated by their effects on hypothalamic NE activity which presumably alters LHRH secretion.

As well as being implicated in the onset of the proestrous and steroid-induced LH surge, NE is also thought to be involved in maintaining the pulsatile release of LH observed in ovx rats, as well as in intact rats on diestrus 1. Systemic injection of phentolamine or phenoxybenzamine (α-adrenergic antagonists) into ovx rats results in marked suppression of pulsatile LH release (Gnodde and Schuiling, 1976; Weick, 1978). Similarly, inhibition of NE synthesis in ovx rats which results in decreased hypothalamic NE concentration also results in marked suppression of pulsatile LH release (Gnodde and Schuiling, 1976; Drouva and Gallo, 1976; Negro-Vilar et al., 1982). Gallo and Kalra (1983) found that, as in ovx rats, inhibition of NE synthesis in intact rats in diestrus 1 also blocks pulsatile LH release. Similar to ovx rats, α-adrenergic but not β-adrenergic antagonists were able to reduce mean blood LH levels by decreasing both LH pulse amplitude and pulse frequency in intact rats on diestrus 1. Taken together, these results suggest that NE acting via α-adrenergic receptors is important in maintaining the pulsatile pattern of LH release in ovx and intact rats.

INHIBITORY ACTION OF NE ON LH RELEASE

It is only rather recently that an inhibitory action of NE on LH release has been described. Gallo and Drouva (1979) first suggested the possibility that activation of noradrenergic receptors could result in inhibition of LH release when it was demonstrated that NE infusion into the third ventricle inhibited pulsatile LH release in ovx rats. This inhibitory effect is probably via α-adrenergic receptors since α-adrenergic agonists such as clonidine and phenylephrine were more effective than isoproterenol, a β-agonist, in suppressing LH release (Leung et al., 1982a). The proposal that NE is acting via α-adrenergic receptors to inhibit LH release is also supported by our recent report that pretreatment with α-adrenergic antagonists, but not a β-adrenergic antagonist, prevented the NE-induced suppression of pulsatile LH release in ovx rats (Bergen and Leung, 1986a). Activation of adrenergic receptors inhibitory to pulsatile LH release in ovx rats resulted in decreased LH pulse frequency while pulse amplitude remained unchanged, suggesting that it is the frequency of the hypothetical LHRH pulse generator that is affected and not the amount of LHRH released per pulse into the portal veins (Gallo, 1984). Leipheimer and Gallo (1985) found that the medial preoptic area is one site where NE could act to suppress the putative LHRH pulse generator and subsequent LH release in ovx rats. Using a push-pull perfusion cannula they found that perfusion of the medial preoptic area with NE resulted in suppression of LH pulse frequency while pulse amplitude remained unchanged, similar to infusion of NE into the third ventricle (Leipheimer and Gallo, 1985). It has also been demonstrated that electrical stimulation of ascending noradrenergic fibers at the level of the mesencephalon results in inhibition of pulsatile LH release similar to intraventricular infusion of NE (Leung et al., 1981; Bergen and Leung, 1986b). More recently, it was shown in the free-moving ovx rats that electrical stimulation in the region of the dorsal noradrenergic tract inhibited pulsatile LH release (Fig. 1) while electrical stimulation in the region of the ventral noradrenergic tract had no effect on pulsatile LH release (Bergen and Leung,

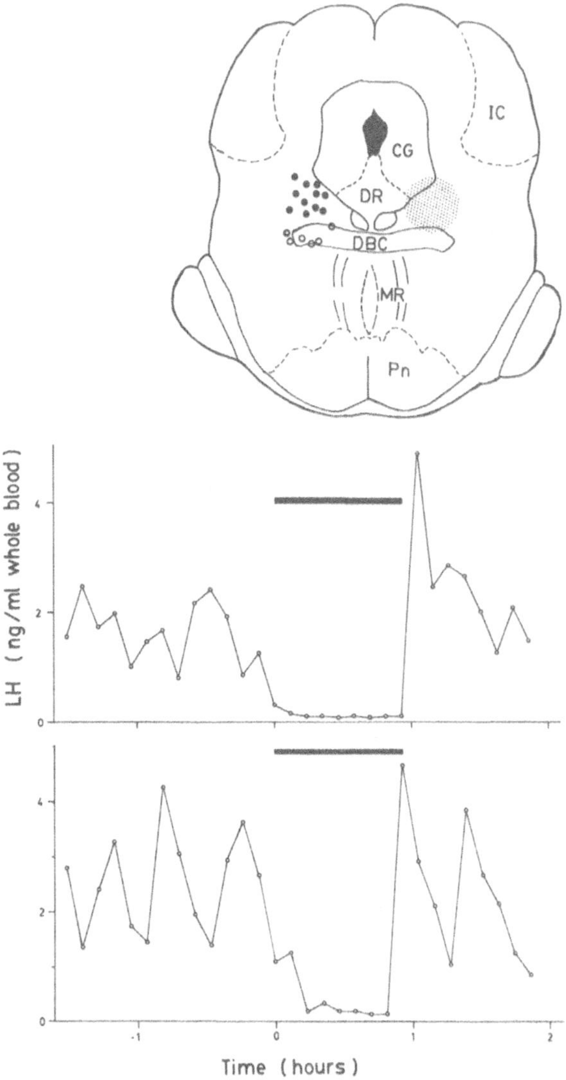

Fig. 1. Effects of electrical stimulation of the dorsal tegmentum region of the midbrain on pulsatile LH release in ovariectomized rats. Two representative animals are shown with the period of electrical stimulation indicated by the shaded bars. The locations of electrode tips are denoted on the frontal section of the brain (● = inside the dorsal noradrenergic tract, o = outside).

1986b). It is not known where the effect of activation of these fibers is mediated but one possibility is the medial preoptic area since the suppression of LH release is similar to that seen with NE perfusion of this area, i.e. a decrease in pulse frequency (Leipheimer and Gallo, 1985). Whether these inhibitory effects are mediated by an inhibitory interneuron between the ascending NE fibers and the LHRH neuron is not known.

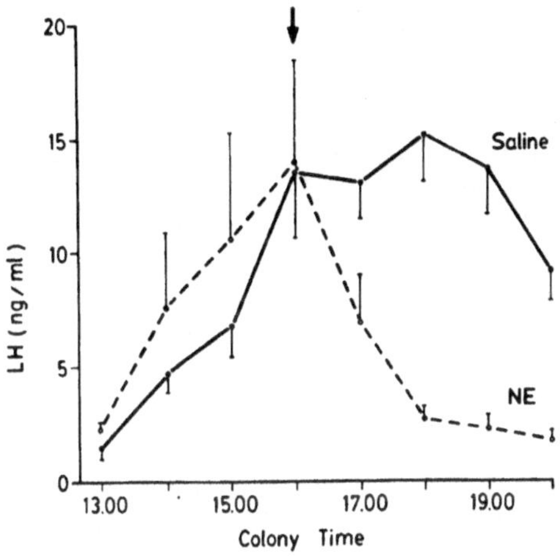

Fig. 2. Effects of intraventricular administration of norepinephrine (NE) on progesterone-induced LH release in estrogen-primed ovariectomized rats.

An inhibitory role for NE in regulating LH release has also been proposed by Taleisnik and co-workers (Caceres and Taleisnik, 1980b; Dotti and Taleisnik, 1982, 1984). It was demonstrated that electrochemical stimulation of frontal lobe cortex on the day of proestrus blocked the preovulatory LH surge and ovulation (Caceres and Taleisnik, 1980a). Stimulation of the anterior cingulate cortex also blocked the steroid-induced surge in ovx rats and LH release induced by electrical stimulation of the medial preoptic area. This inhibition was found to be mediated via β-adrenergic receptors since propranolol, a β-adrenergic antagonist, prevented the inhibitory effects of cortical stimulation (Caceres and Taleisnik, 1980b). It was also shown that the release of LH induced by intraventricular infusion of NE was enhanced if the β-adrenergic receptor had been blocked, and intraventricular infusion of isoproterenol (a β-adrenergic agonist) inhibited LH release induced by medial preoptic area stimulation (Caceres and Taleisnik, 1980b; 1982). Interestingly, activation of NE cell groups in the brainstem (locus coeruleus, A5, A1, cell groups) inhibited ovulation and LH release (Dotti and Taleisnik, 1982). The inhibitory NE fibers from these cell groups ascend through the mesencephalon in the dorsal noradrenergic tract and eventually end in the premammillary nucleus, where NE activates β-adrenergic receptors located on interneurons which are thought to either directly or indirectly inhibit LHRH release (Dotti and Taleisnik, 1982; 1984). Beltramino and Taleisnik (1984) found that premammillary cuts and lesions of the posterior hypothalamus which presumably interrupt these ascending inhibitory fibers, resulted in an earlier proestrous LH surge while stimulation of the premammillary or posterior hypothalamic nuclei results in blockade of the LH surge and ovulation. It was suggested that these inhibitory inputs acting via β-adrenergic receptors may be involved in setting the timing and height of the LH surge. A role for β-receptors in mediating the inhibitory effect of NE on LH release has also been proposed when it was demonstrated that intraventricular infusion of a β-adrenergic agonist resulted in suppression of a steroid-induced LH surge (Leung et al., 1982b). Interestingly, in addition to β-agonists, it has been shown recently (Bergen and Leung, 1986c) that NE and α-adrenergic agonists (phenylephrine and methoxamine) can also inhibit an ongoing LH surge (Figs. 2 and 3). These rather unexpected results suggest the possibility that once an LH surge has been initiated, the elevated LH levels can become suppressed with activation of α- and/or β-adrenergic receptors. In this context, it is of interest to note that the increase in LH release induced by electrochemical stimulation of the medial preoptic area could also be partially suppressed by perfusion of the third ventricle with high concentrations of NE and epinephrine (Cramer and Barraclough, 1978).

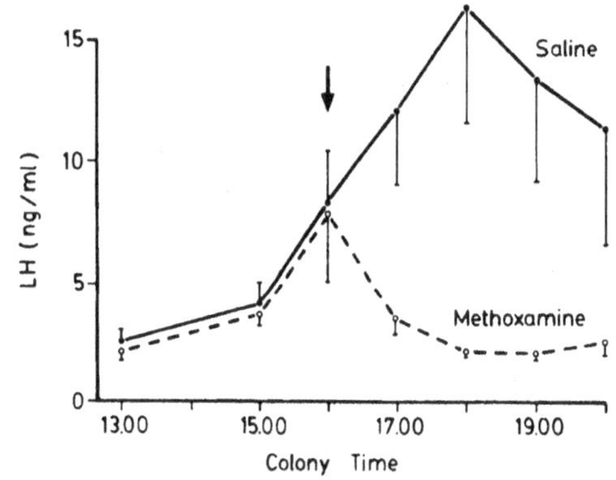

Fig. 3. Effects of intraventricular administration of methoxamine on progesterone-induced LH release in estrogen-primed ovariectomized rats.

The mechanism and physiological significance of NE inhibition of LH release is not known. One possibility is that the LHRH neurons during the ascending phase of LH secretion are under increased excitatory influence from NE inputs and these LHRH neurons may become desensitized to additional exogenous noradrenergic input, resulting in reduced LHRH output and subsequently lower LH release. Gallo (1982) showed that continuous ventricular infusion of NE in ovx steroid-primed rats initially increased LH release but LH levels returned to preinfusion levels within 90 min thus demonstrating that the LHRH neurons can become desensitized to continuous NE input. On the other hand, bolus intraventricular infusions of NE in similarly high concentrations results in elevations of LH levels in estrogen-progesterone primed rats (Krieg and Sawyer, 1976; Vijayan and McCann, 1978; Caceres and Taleisnik, 1980b). Caceres and Taleisnik (1980b) also found that intraventricular infusion of NE had no effect on LH release induced by medial preoptic area stimulation and NE was more than 2.5 times more effective in releasing LH when β-adrenergic receptors were blocked. These results suggest that desensitization of the NE receptors does not readily occur.

Another possibility is that once an LH surge has been triggered, inhibitory components could become activated to suppress the amplitude and duration of the LH surge. If the changes in LH release are seen as alterations in the tone of excitatory and inhibitory inputs, as has recently been suggested by Beltramino and Taleisnik (1984), initiation of the LH surge could result from a relative increase in excitatory tone whereas an increase in inhibitory tone could act to restrain the height and/or duration of the surge. Interestingly, while desensitization to the stimulatory effects of NE occurs within 90 min, the receptors activated by NE that are inhibitory to pulsatile LH release maintain their responsiveness to continuous NE infusion even after 20 h and therefore are not easily desensitized (Gallo, 1982). It has been demonstrated that some inhibitory tone is present even during an LH surge and the amount of LH released can be enhanced with elimination of this inhibitory tone. Blockade of opioid receptors with naloxone enhances LH release in proestrous rats (Kubo et al., 1983; Ieiri et al., 1980) and the estrogen-induced LH surge in ovx rats (Sylvester et al., 1980). The proestrous LH surge was enhanced by transections of fibers in the ventrolateral region of the midbrain (Kawakami and Arita, 1980). These data suggest the presence of fibers inhibitory to LH release even though the nature of these fibers is not known.

ROLE OF BRAIN NE PATHWAYS

One approach to better understanding the role of NE in the regulation of LH release has involved lesioning or stimulating the ascending NE pathways and monitoring ovulation and/or LH release. Noradrenergic nerve terminals in the hypothalamus have their cell bodies in the brainstem which send their axons rostrally via the DNT and VNT (Lindvall and Bjorklund, 1978; Moore and Bloom, 1979). The DNT contains primarily axons originating in the locus coeruleus and A5 cell groups and the VNT is made up of fibers originating from the lateral tegmentum.

The inhibitory effects of NE on LH release are probably mediated, at least in part, by NE release from nerve fibers ascending via the DNT. Electrical stimulation of the DNT in ovx rats inhibits pulsatile LH release (Leung et al., 1981; Bergen and Leung, 1986b). Although the precise mechanism is still unclear, this inhibition is probably mediated by the activation of α-adrenergic receptors since (1) the inhibition is similar to that observed with intraventricular infusion of NE in ovx rats, i.e. a decrease in pulse frequency, (2) α-adrenergic agonists are more effective than β-agonists in suppressing pulsatile release, and (3) α-adrenergic antagonists but not β-antagonists are able to block the NE-induced inhibition of pulsatile release. Also, the location of the inhibitory α-adrenergic receptor(s) is not known, but the medial preoptic area is a possible site since perfusion of this area with NE is effective in inhibiting pulsatile LH release (Leipheimer and Gallo, 1985).

The effect of locus coeruleus activation in suppressing ovulation and LH release is also mediated by fibers in DNT since the inhibition was blocked by lesions of the DNT but not of the VNT (Dotti and Taleisnik, 1982). Interestingly, this inhibition is mediated by activation of a β-adrenergic receptor situated in the premammillary nucleus (Dotti and Taleisnik, 1984). It is not known whether the inhibitory effects of β-agonists on the LH surge is via this same receptor or another population of β-receptors possibly located elsewhere. Dotti and Taleisnik (1984) suggest that β-receptors inhibitory to LH release are not situated anterior to the premammillary nucleus. While transections of the DNT may disrupt ovarian cyclicity (Clifton and Sawyer, 1979), Kawakami and Arita (1980) reported that transections of the DNT on the morning of proestrus did not affect the LH surge or ovulation. Also, electrical stimulation of the DNT in ovx, steroid-primed rats had no effect on LH release (Leung et al., 1981). Taken together, these results suggest that the DNT contains fibers which are primarily, although not necessarily exclusively, inhibitory to LH release and that the inhibition can occur via activation of either α- or β-adrenergic receptors.

Although studies which have examined the possible role of the VNT in gonadotropin release are not in total agreement, it has been suggested that the fibers in this tract are primarily excitatory to LH release (Ramirez et al., 1984). Martinovic and McCann (1977) reported that chemical lesioning of the VNT with 6-hydroxy-dopamine, which selectively destroys catecholaminergic neurons, blocked the proestrous and steroid-induced gonadotropin surges. On the other hand, others (Nicholson et al., 1978; Hancke and Wuttke, 1979) showed that regular estrous cycles occurred even after 6-hydroxy-dopamine was injected into the VNT, suggesting that NE may not be mandatory for ovulation but the VNT is involved in facilitating LH release (Hancke and Wuttke, 1979). Kawakami and Arita (1980) reported that transections of the VNT on proestrus blocked the LH surge and ovulation. They also demonstrated that if NE was infused into the third ventricle at the time of transection, ovulation occurred (Kawakami and Ando, 1981). It is not reported whether normal estrous cycles were able to resume after VNT transections. These results suggest that the VNT is probably facilitatory to LHRH release, but whether it is indispensable for ovulation and regular estrous cyclicity is less certain. As discussed earlier, excitation of the LHRH neuron is probably by activation of α-adrenergic receptors (either directly or indirectly) as suggested by studies on LHRH release _in vitro_ (Heaulme and Dray, 1984), and LH release _in vivo_ (Drouva et al., 1982; Coen and Coombs, 1983).

SUMMARY

The precise nature of noradrenergic inputs to LHRH neurons is unknown nor is their role in the regulation of LH understood. Since NE can be both excitatory and inhibitory to LH release, it seems unlikely that the same receptor populations could mediate both of these effects. This prompted us to propose that NE exerts its differential effects via at least two separate and anatomically distinct receptor populations (Leung et al., 1982; Bergen and Leung, 1986c). As hypothesized in Fig. 4, one group of NE receptors (α-receptors) presumably located directly on LHRH neurons which would be stimulatory to LH release; tonic activation of these neurons would maintain pulsatile LH release. Increased activation of these receptors might also mediate facilitation of the LH surge. Another population of α- and β-adrenergic receptors may be located on inhibitory interneurons which could act to suppress LH release. It has been suggested that these inhibitory inputs may be involved in offsetting the excitatory inputs to the LHRH neurons. These inputs may prevent inappropriate excitatory stimulation of the LHRH neurons, as well as providing information to the LHRH neurons which allows for the proper timing and optimum release of LHRH and subsequent LH secretion.

The role of NE in controlling LH release at any point in time is probably a function of ovarian steroid levels and previous LH levels. In situations of low steroid levels, such as in ovx or intact rats on diestrus,

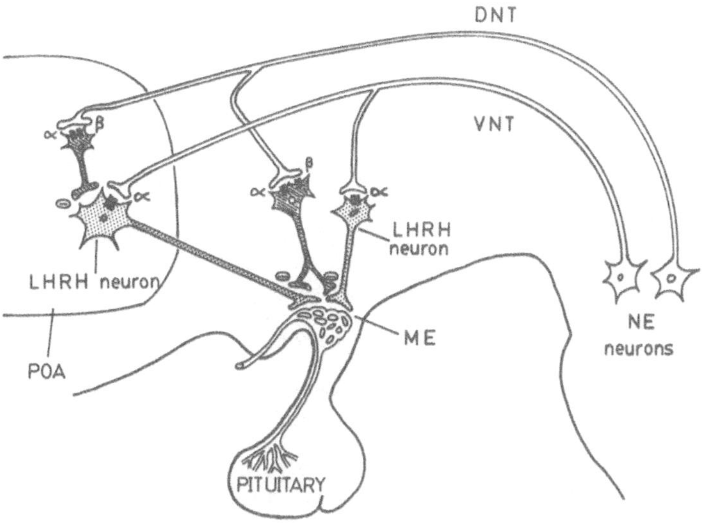

Fig. 4. Proposed mechanism for the dual action of norepinephrine (NE) on the regulation of hypothalamic LHRH release in the rat. Stimulatory effects of NE are exerted primarily via α-receptors on LHRH neurons. On the other hand, inhibitory effects could be mediated by α- and β-receptors on putative inhibitory interneurons (-).

a low level of NE release is present which maintains LH pulses since it is well established that NE synthesis inhibitors or α-adrenergic antagonists can suppress pulsatile release. However, further increases in NE levels in this same low steroidal background would inhibit LH release, presumably via inhibitory interneurons. Under conditions of high steroid levels such as in steroid-primed ovx rats or at proestrus, there is increased activation of adrenergic receptors which results in the initiation of the LH surge. It is interesting to note that the neural mechanisms by which ovarian steroids modulate the noradrenergic regulation of LH release in female rats is also present in the male rat (Condon et al., 1986). Nevertheless, our recent data suggest that continued elevation of NE levels once again results in a suppression of the elevated LH levels even in the presence of a high steroid background. This could occur through inactivation of the excitatory receptor and/or activation of inhibitory NE inputs to the LHRH neurons. This dynamic interplay of excitatory versus inhibitory influences of NE on the neurons may be important in the fine tuning of LHRH and gonadotropin release at any given time.

REFERENCES

Barraclough, C. A., and Sawyer, C. H., 1957, Blockade of the release of pituitary ovulation hormone in the rat by chlorpromazine and reserpine: possible mechanisms of action, Endocrinology, 61:341.

Barraclough, C. A., and Wise, P. M., 1982, The role of catecholamines in the regulation of pituitary luteinizing hormones and follicle-stimulating hormone secretion, Endocrine Rev., 3:91.
Beltramino, C., and Taleisnik, S., 1984, Inhibitory influence of the nuclei of the posterior hypothalamus on the proestrous surge of LH, Acta Endo., 105:433.
Bergen, H., and Leung, P. C. K., 1986a, Norepinephrine inhibition of pulsatile LH release: receptor specificity, Am. J. Physiol. 250 (Endocrin. Metab. 13), E205.
Bergen, H., and Leung, P. C. K., 1986b, Electrical stimulation of ventral versus dorsal mesencephalic tegmental areas in the conscious rat: effects on luteinizing hormone release, Neuroendocrinology, in press.
Bergen, H., and Leung, P. C. K., 1986c, Suppression of progesterone-induced gonadotropin surge by adrenergic agonists in estrogen-primed ovariectomized rats, Neuroendocrinology, 43:397.
Caceres, A., and Taleisnik, S., 1982, Dissociation of the inhibitory and facilitatory effects of norepinephrine on the release of LH by premammillary lesions, Neuroendocrinology, 35:98.
Caceres, A., and Taleisnik, S., 1980a, Blockade of ovulation and release of LH in the rat by electrochemical stimulation of the frontal lobe cortex, Brain Res., 188:411.
Caceres, A., and Taleisnik, S., 1980b, Inhibition of secretion of luteinizing hormone induced by electrochemical stimulation of the anterior cingulate cortex mediated by an α-adrenergic mechanism, J. Endocrinol., 87:419.
Clarke, I. J., and Cummins, J. T., 1982, The temporal relationship between gonadotropin releasing hormone and luteinizing hormone secretion in ovariectomized ewes, Endocrinology, 111:1737.
Clifton, D. K., and Sawyer, C. H., 1979, LH release and ovulation in the rat following depletion of hypothalamic norepinephrine: chronic vs. acute effects, Neuroendocrinology, 28:442.
Coen, C. W., and Coombs, M. C., 1983, Effects of manipulating catecholamines on the incidence of the preovulatory surge of luteinizing hormone and ovulation in the rat: evidence for a necessary involvement of hypothalamic adrenaline in the normal or "midnight" surge, Neuroscience, 10:187.
Condon, T. P., Handa, R. J., Gorski, R. A., Sawyer, C. H., Whitmoyer, D. I., 1986, Ovarian steroid modulation of norepinephrine action on luteinizing hormone release: analogous effects in male and female rats, Neuroendocrinology, 43:550.
Cramer, O. M., and Barraclough, C. A., 1978, The actions of serotonin, norepinephrine, and epinephrine on hypothalamic processes leading to adenohypophyseal luteinizing hormone release, Endocrinology, 103:694.
Crowley, W. R., 1982, Effects of ovarian hormones on norepinephrine and dopamine turnover in hypothalamic and extrahypothalamic nuclei, Neuroendocrinology, 34:381.
Dotti, C., and Taleisnik, S., 1982, Inhibition of the release of LH and ovulation by activation of the noradrenergic system. Effect of interrupting the ascending pathways, Brain Res., 249:281.
Dotti, C., and Taleisnik, S., 1984, Beta-adrenergic receptors in the premammillary nucleus mediate the inhibition of LH release evoked by locus ceruleus stimulation, Neuroendocrinology, 38:6.
Drouva, S. V., and Gallo, R. V., 1976, Catecholamine involvement in episodic luteinizing hormone release in adult ovariectomized rats, Endocrinology, 99:651.
Drouva, S. V., Laplante, E., and Kordon, C., 1982, α_1-adrenergic receptor involvement in the LH surge in ovariectomized estrogen-primed rats, Eur. J. Pharmcol., 81:341.
Everett, J. W., and Sawyer, C. H., 1950, A 24-hour periodicity in the "LH-release apparatus" of female rats, disclosed by barbiturate sedation, Endocrinology, 47:198.

Everett, J. W., Sawyer, C. H., and Markee, J. E., 1949, A neurogenic timing factor in control of the ovulatory discharge of luteinizing hormone in the cyclic rat, Endocrinology, 44:234.
Gallo, R. V., 1982, Luteinizing hormone secretion during continuous or pulsatile infusion of norepinephrine: central nervous system desensitization to constant norepinephrine input, Neuroendocrinology, 35:380.
Gallo, R. V., 1984, Further studies on norepinephrine-induced suppression of pulsatile luteinizing hormone release in ovariectomized rats, Neuroendocrinology, 39:120.
Gallo, R. V., 1981a, Pulsatile LH release during the ovulatory LH surge on proestrus in the rat, Biol. Reprod., 24:100.
Gallo, R. V., 1981b, Pulsatile LH release during low level LH secretion in the rat estrous cycle, Biol. Reprod., 24:771.
Gallo, R. V., and Drouva, S. V., 1979, The effect of intraventricular infusion of catecholamines on luteinizing hormone release in ovariectomized and ovariectomized, steroid-primed rats, Neuroendocrinology, 29:149.
Gallo, R. V., and Kalra, P. S., 1983, Pulsatile LH release on diestrus 1 in the rat estrous cycle: relation to brain catecholamines and ovarian steroid secretion, Neuroendocrinology, 37:91.
Gay, V. L., and Sheth, N. A., 1972, Evidence for a periodic release of LH in castrated male and female rats, Endocrinology, 90:158.
Gnodde, H. P., and Schuiling, G. A., 1976, Involvement of catecholaminergic and cholinergic mechanisms in the pulsatile release of luteinizing hormone in the long-term ovariectomized rat, Neuroendocrinology, 20:212.
Hancke, J. L., and Wuttke, W., 1979, Effects of chemical lesion of the ventral noradrenergic bundle or of the medial preoptic area on preovulatory LH release in rats, Exp. Brain Res., 35:127.
Heaulme, M., and Dray, F., 1984, Noradrenaline and prostaglandin E_2 stimulate LHRH release from rat median eminence through distinct 1-alpha-adrenergic and PGE_2 receptors, Neuroendocrinology, 49:403.
Hoffman, G. E., ad Gibbs, F. P., 1982, LHRH pathways in rat brain: "deafferentation" spares a subchiasmatic LHRH projection to the median eminence, Neuroscience, 7:1979.
Honma, K., and Wuttke, W., 1980, Norepinephrine and dopamine turnover rates in the medial preoptic area and the mediobasal hypothalamus of the rat brain after various endocrinological manipulations, Endocrinology, 106:1848.
Ieiri, T., Chen, H. T., Campbell, G. A., and Meites, J., 1980, Effects of naloxone and morphine on the proestrous surge of prolactin and gonadotropins in the rat, Endocrinology, 106:1568.
Kalra, S. P., and McCann, S. M., 1974, Effects of drugs modifying catecholamine synthesis on plasma LH and ovulation in the rat, Neuroendocrinology, 15:79.
Kawakami, M., and Ando, S., 1981, Lateral hypothalamic mediation of midbrain catecholaminergic influences on preovulatory surges of serum gonadotropin and prolactin in female rats, Endocrinology, 108:66.
Kawakami, M., and Arita, J., 1980, Involvement of the ventromedial part of the midbrain in the control of the proestrous surge of gonadotropins and prolactin in the rat, Neuroendocrinology, 30:337.
Kelly, M. J., Ronnekleiv, O. K., and Eskay, R. L., 1982, Immunocytochemical localization of luteinizing hormone-releasing hormone in neurons in the medial basal hypothalamus of the female rat, Exp. Brain Res., 48:97.
King, J. C., Tobet, S. A., Snavely, F. L., and Arimura, A. A., 1982, LHRH immunopositive cells and their projections to the median eminence and organum vasculosum of the lamina terminalis, J. Comp. Neurol., 209:287.
Krieg, R. J., and Sawyer, C. H., 1976, Effects of intraventricular catecholamines on luteinizing hormone release in ovariectomized-steroid-primed rats, Endocrinology, 99:411.

Kubo, K., Kiyota, Y., and Fukunga, S., 1983, Effects of third ventricular injection of β-endorphin on luteinizing hormone surges in female rat: sites and mechanisms of opioid actions in the brain, Endocrinol. Japon, 30:419.

Leipheimer, R. E., and Gallo, R. V., 1985, Medial preoptic area involvement in norepinephrine-induced suppression of pulsatile luteinizing hormone release in ovariectomized rats, Neuroendocrinology, 40:345.

Leung, P. C. K., Arendash, G. W., Whitmoyer, D. I., Gorski, R. A., and Sawyer, C. H., 1981, Electrical stimulation of mesencephalic noradrenergic pathway: effects on luteinizing hormone levels in blood of ovariectomized and ovariectomized, steroid-primed rats, Endocrinology, 109:720.

Leung, P. C. K., Arendash, G. W., Whitmoyer, D. I., Gorski, R. A., and Sawyer, C. H., 1982a, Differential effects of central adrenoceptor agonists on luteinizing hormone release, Neuroendocrinology, 34:207.

Leung, P. C. K., Whitmoyer, D. I., Garland, K. E., and Sawyer, C. H., 1982b, Beta-adrenergic suppression of progesterone-induced luteinizing hormone surge in ovariectomized, steroid-primed rats, Proc. Soc. Exp. Biol. Med., 169:161.

Levine, J. E., Pau, K.-Y. F., Ramirez, V. D., and Jackson, G. L., 1982, Simultaneous measurement of luteinizing hormone-releasing hormone and luteinizing hormone release in unanesthetized, ovariectomized sheep, Endocrinology, 111:1449.

Lindvall, O., and Bjorklund, A., 1978, Organization of catecholamine neurons in the rat central nervous system, in: "Handbook of Psychopharmacology," I. L. Iversen, S. O. Iverson, S. H. Snyder, eds., p. 139, Plenum Press, New York.

Martinovic, J. V., and McCann, S. M., 1977, Effect of lesions in the ventral noradrenergic tract produced by microinjection of 6-hydroxydopamine on gonadotropin release in the rat, Endocrinology, 100:1206.

Moore, R. Y., and Bloom, F. E., 1979, Central catecholamine neuron systems: anatomy and physiology of the norepinephrine and epinephrine systems, Ann. Rev. Neurosci., 2:113.

Negro-Vilar, A., Advis, J. P., Ojeda, S. R., and McCann, S. M., 1982, Pulsatile luteinizing hormone (LH) patterns in ovariectomized rats: involvement of norepinephrine and dopamine in the release of LH releasing hormone and LH, Endocrinology, 111:932.

Negro-Vilar, A., Ojeda, S. R., and McCann, S. M., 1979, Catecholaminergic modulation of luteinizing hormone-releasing hormone release by median eminence terminals in vitro, Endocrinology, 104:1749.

Nicholson, G., Greeley, G., Humm, J., Youngblood, W., and Kizer, J. S., 1978, Lack of effect of noradrenergic denervation of the hypothalamus and medial preoptic area on the feedback regulation of gonadotropin secretion and the estrous cycle of the rat, Endocrinology, 103:559.

Nowak, F. V., and Swerdloff, R. S., 1985, Gonadotropin-releasing hormone release by superfused hypothalami in response to norepinephrine, Biol. Reprod., 33:790.

Ojeda, S. R., Negro-Vilar, A., and McCann, S. M., 1982, Evidence for involvement of alpha-adrenergic receptors in norepinephrine-induced prostaglandin E_2 and luteinizing hormone-releasing hormone release from the median eminence, Endocrinology, 110:409.

Ramirez, V. D., Feder, H. H., Sawyer, C. H., 1984, The role of brain catecholamines in the regulation of LH secretion: a critical inquiry, in: "Frontiers in Neuroendocrinology," L. Martini and W. F. Ganong, eds., p. 27-84, Raven Press, New York.

Rance, N., Wise, P. M., and Barraclough, C. A., 1981a, Negative feedback effects of progesterone correlated with changes in hypothalamic norepinephrine and dopamine turnover rates, median eminence luteinizing hormone-releasing hormone, and peripheral plasma gonadotropins, Endocrinology, 108:2194.

Rance, N., Wise, P. M., Selmanoff, M. K., and Barraclough, C. A., 1981b, Catecholamine turnover rates in discrete hypothalamic areas and associated changes in median eminence luteinizing hormone-releasing hormone and serum gonadotropins on proestrus and diestrus day 1, Endocrinology, 108:1795.

Rubinstein, L., and Sawyer, C. H., 1970, Role of catecholamines in stimulating the release of pituitary hormone(s) in rats, Endocrinology, 86:988.

Sarkar, D. K., and Fink, G., 1981, Gonadotropin-releasing hormone surge: possible modulation through postsynaptic α-adrenoceptors and two pharmacologically distinct dopamine receptors, Endocrinology, 108:862.

Sawyer, C. H., 1952, Stimulation of ovulation in the rabbit by the intraventricular injection of epinephrine or norepinephrine, Anat. Rec., 112:385.

Sawyer, C. H., Markee, J. E., and Hollinshead, W. H., 1947, Inhibition of ovulation in the rabbit by the adrenergic blocking agent dibenamine, Endocrinology, 41:395.

Simpkins, J. W., Advis, J. P., Hodson, C. A., and Meites, J., 1979, Blockade of steroid-induced luteinizing hormone release by selective depletion of anterior hypothalamic norepinephrine activity, Endocrinology, 104:506.

Sylvester, P. W., Chen, H. T., and Meites, J., 1980, Effects of morphine and naloxone on phasic release of luteinizing hormone and follicle-stimulating hormone, Proc. Soc. Exp. Biol. Med., 164:207.

Tima, L., and Flerko, B., 1974, Ovulation induced by norepinephrine in rats made anovulatory by various experimental procedures, Neuroendocrinology, 15:346.

Vijayan, E., and McCann, S. M., Re-evaluation of the role of catecholamines in control of gonadotropin and prolactin release, Neuroendocrinology, 25:150.

Weick, R. F., 1978, Acute effects of adrenergic receptor blocking drugs and neuroleptic agents on pulsatile discharges of luteinizing hormone in the ovariectomized rat, Neuroendocrinology, 26:108.

Wise, P. M., Rance, R., and Barraclough, C. A., 1981, Effects of estradiol and progesterone on catecholamine turnover rates in discrete hypothalamic regions in ovariectomized rats, Endocrinology, 108:2186.

PHYSIOLOGICAL AND BIOCHEMICAL DISSECTION OF MECHANISMS UNDERLYING PUBERTY

S. R. Ojeda, H. F. Urbanski, C. E. Ahmed,
L. Rogers, and D. Gonzalez

Department of Physiology
University of Texas Health Science Center at Dallas
5323 Harry Hines Boulevard
Dallas, Texas 75235-0940

INTRODUCTION

The last few years have witnessed the emergence of a more comprehensive understanding of the mechanisms underlying the onset of mammalian puberty. At present, several groups utilizing different animal models are actively engaged in studying various aspects of the pubertal process, more noticeably the luteinizing hormone-releasing hormone (LHRH)-gonadotropin component of the reproductive axis.

Comprehensive reviews dealing with the neuroendocrinology of sexual development have recently appeared (Plant, 1983; Terasawa et al., 1983; Kelch et al., 1988; Ojeda et al., 1986) and the interested reader is referred to them for further consultation. The present article rather than attempting to again review this area presents an account of some of the current efforts made in our laboratory to elucidate the intimate mechanisms responsible for the acquisition of female reproductive competence. In pursuing this goal we have made extensive use of a variety of physiological and biochemical techniques; more recently, we have begun to explore the approaches of cell biology and molecular biology to obtain a deeper understanding of these developmental events. The animal model we have employed is the female rat.

The initiation, progression and completion of puberty is a remarkably complex process. In the most general terms, however, the process may be viewed as being primarily dependent on two components: an "initiator" which resides within the central nervous system (CNS) and a "chronometer" which is represented by the ovary. While the initiator operates through the LHRH-secreting system, the effector of the chronometer is estradiol (E_2). Before the positive feedback of E_2 on gonadotropin release can be expressed as the first preovulatory gonadotropin surge, the ovary must first acquire the capacity to produce E_2 in sufficiently large amounts and for a sufficiently long period of time.

This large discharge of gonadotropins, which occurs in the afternoon of the first proestrus, can be considered as the neuroendocrine culmination of puberty. Its immediate and most direct consequence, the first ovulation, can be viewed (in the most general sense) as the culmination of the development of female reproductive function.

Intrinsic to the development of the hypothalamic-pituitary-ovarian axis is the establishment of negative and positive steroid feedback mechanisms. Little is known at present regarding the maturation of ovarian peptidergic feedback systems controlling gonadotropin release, but it is obvious that the recent isolation and characterization of inhibin and its associated FSH-releasing factor protein (FRP) from follicular fluid (Vale et al., 1986; Ling et al., 1986) has paved the way for further studies along these lines.

THE DEVELOPMENT OF ESTRADIOL (E_2) NEGATIVE FEEDBACK

For several years the prevailing explanation for the onset of puberty has been the "gonadostat" hypothesis (Ramirez and McCann, 1963) which states that as the animal matures the sensitivity of the hypothalamic-pituitary system to steroid negative feedback gradually declines, thereby allowing gonadotropin release to increase and to further stimulate the secretion of gonadal steroids.

While there is no doubt that pubertal and mature animals are less sensitive than immature individuals to the inhibitory effect of E_2 (Steele and Weisz, 1974; Eldridge et al., 1977), we have presented evidence that in the rat such a change in sensitivity does not actually occur before the first preovulatory surge of gonadotropins, but rather becomes fully expressed after the first ovulation (Andrews et al., 1981). It is our view that the gonadostat resetting is a consequence rather than the cause of puberty. It is possible, however, that part of the change may occur earlier; recent experiments by Docke et al. (1984) have shown that a reduction in gonadotropin responsiveness to E_2 negative feedback already occurs on the day of first proestrus. The reasons for the difference between these experiments and our own are not clear. Nevertheless, even if part of the resetting occurs on the day preceding the first ovulation, the initial diurnal change in LH release which represents the onset of puberty (vide infra) occurs earlier, clearly indicating that a change in steroid negative feedback effectiveness is not the primary event responsible for the initiation of puberty. Indeed several investigators working with monkeys and rats have recently arrived at similar conclusions (Plant, 1983, Terasawa et al., 1984; Matsumoto et al., 1986; Raum and Swerdioff, 1986).

Juvenile rats are exquisitively sensitive to E_2 negative feedback (Steele and Weisz, 1974; Eldridge et al., 1977; Andrews et al., 1981; Andrews and Ojeda, 1981). In contrast, infantile animals are much less sensitive (Andrews and Ojeda, 1977; Meijs-Roelofs et al., 1979), mainly because of the presence of high serum levels of alpha fetoprotein which binds estrogen avidly and thus diminishes the amount of free E_2 that may reach the gonadotropin-releasing system. As alpha fetoprotein decreases from the circulation, more E_2 becomes available to the tissues (Germain et al., 1978) and its effectiveness as a negative feedback agent increases (Andrews and Ojeda, 1981; Meijs-Roelofs et al., 1979). As previously indicated, this level of sensitivity remains elevated even until the day when the first preovulatory surge of gonadotropins occurs.

THE DEVELOPMENT OF ESTRADIOL (E_2) POSITIVE FEEDBACK

For many years it has been known that E_2 positive feedback can be activated long before puberty (McCormack and Meyer, 1964; Caligaris et al., 1972). Indeed, exogenously administered E_2 (via silastic capsules) can induce LH release as early as postnatal day 15 (Kronibius and Wuttke, 1977), provided that adequate levels of serum E_2 are maintained for at least 48 h (Andrews et al., 1981a). In our studies, these levels appear to be twice as high as those seen on the day of first proestrus. If even larger E_2 values are produced in infantile 12-day-old rats, no LH release is observed (Andrews et al., 1981a). In contrast, proestrous levels of E_2 are sufficient to evoke an LH discharge in juvenile (22 - 32-day-old) rats. It is evident, therefore, that had the juvenile ovaries the capability of producing proestrous levels of E_2 then a discharge of LH release would have occurred well ahead of the normal time of puberty. The ovary, however, acquires this capability only after its follicles have matured to a preovulatory condition. Physiologically, this process lasts about 19 days (Hage et al., 1978), and occurs under strict CNS control. As we will discuss later, an emerging body of evidence suggests that in addition to controlling ovarian function through the secretions of the anterior pituitary gland, the CNS sends direct information to the gonads via the ovarian nerves.

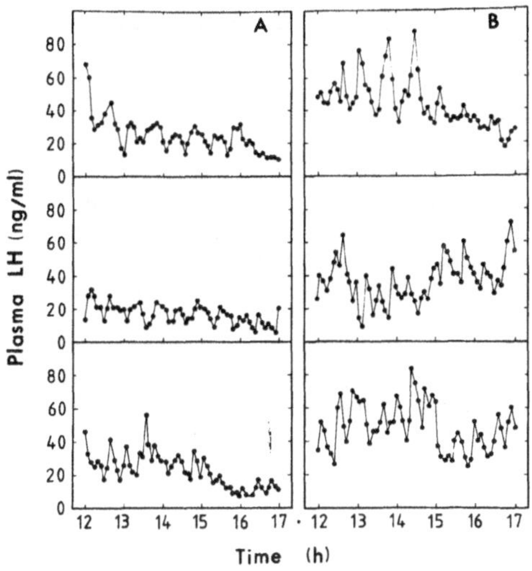

Fig. 1. Representative afternoon plasma LH profiles from juvenile (A; 27-29 days old) and peripubertal (B; 30-38 days old) female rats bled continuously for 5 h. Six individual profiles from a total of 16 are depicted. (From Urbanski and Ojeda, 1985, with permission.)

THE MATURATION OF GONADOTROPIN RELEASE

In previous reviews we have discussed extensively the factors that may control gonadotropin release during the infantile phase of development (Ojeda et al., 1984; Ojeda et al., 1986).

During juvenile days (day 21 to about day 30-32) plasma FSH levels are declining and plasma LH levels, though low, are pulsatile (Kimura and Kawakami, 1981; Andrews and Ojeda, 1981a; Urbanski and Ojeda, 1985). Pulsatile LH profiles are very similar in the morning and in the afternoon, but by the end of juvenile development a diurnal pattern, characterized by a higher baseline and amplitude of the LH pulses in the afternoon than in the morning, becomes established (Urbanski and Ojeda, 1985; Fig. 1).

This is, in our view, the first direct neuroendocrine manifestation of the onset of puberty. In addition to the increase in LH pulse amplitude, some animals show a more sustained mid-afternoon episode of LH release which we have termed "LH minisurge." This minisurge of LH secretion has a magnitude of about 1/10 of the proper preovulatory surge and appears to be elicited by small increases in the secretion of ovarian E_2 since it fails to occur in long- or short-term ovariectomized rats but can be readily evoked in such animals by increasing their serum E_2 to a level that is only 30% higher than basal values observed in intact rats (Urbanski and Ojeda, 1986). In contrast to these minisurges of LH, the afternoon changes in pulse amplitude and LH baseline appear to be originated by a central, ovarian-independent mechanism (Urbanski and Ojeda, submitted). This conclusion is based on the findings that short-term (2 or 4 days) ovariectomized peripubertal rats exhibit more elevated plasma LH levels in the afternoon than in the morning. Early juvenile rats, on the other hand, fail to show such a difference. Detailed analysis of the LH secretory profiles has revealed that the afternoon change in LH release is mostly due to an increase in trough values and and overall greater secretory output.

We have also examined the effect that this peripubertal change in LH release may have on ovarian function. Ovaries from peripubertal rats were perfused with LH following a pattern that closely simulated either the morning or afternoon pulsatile release of LH seen in vivo (Urbanski and Ojeda, 1985a). The results indicate that the output of E_2 increases significantly when the ovaries are exposed to "afternoon" LH pulses, but not when they are presented with "morning" LH pulses. These data indicte that the establishment of a diurnal change in LH release is an important factor involved in stimulating the immature ovary to produce E_2. By inference it can be suggested that such a change in LH release plays a definitive role in the control of ovarian maturation.

THE DEVELOPMENT OF THE LHRH-SECRETING SYSTEM

What are the central mechanisms by which the diurnal change in LH release becomes established? A clear answer has not yet been forwarded. It appears, however, that the intracellular machinery involved in transducing the stimulatory signals to LHRH neurons is operative long before puberty. For instance, exposure of median eminence (ME) terminals from juvenile rats to prostaglandin E_2 (PGE_2) - an intracellular mediator implicated in the process of norepinephrine (NE)-induced LHRH release - results in LHRH release. This effect can be demonstrated early in juvenile development (postnatal day 22, the earliest age studied) (Ojeda et al., 1986a). Recent evidence implicates the involvement of protein kinase C (Ojeda et al., 1986b). Median eminence nerve terminals from juvenile rats release LHRH in response to substances such as diacylglycerol and phorbol esters which activate PKC indicating that both the PGE_2 and the PKC pathways are functional during prepubertal development. Remarkably, not only these intracellular mechanisms appear to be established long before puberty, but also the capacity of the preoptic area-medial basal hypothalamus (POA-MBH) to release LHRH in a rhythmic, pulsatile fashion can be detected as early as the first week of postnatal life (Urbanski and Ojeda, in preparation).

These observations suggest that the activation of the LHRH secreting system at puberty depends upon the amplification of signals which originate outside the LHRH neuron. Whether these signals are only excitatory in nature or they, in addition, involve the removal of an inhibitory component is nuclear at present. It does appear, however, that for the activity of LHRH neurons to reach a pubertal level the neuronal circuitry that normally regulates their secretory function needs to become established. Indirect evidence for this view comes from the recent observation made by Wray and Hoffman (1986) that LHRH neurons exist in two forms: a "smooth" type, the soma of which has an even membrane surface and a "spiny" type which is characterized by an irregular surface. Most remarkably, the "smooth" neurons are more abundant during infantile days, but during the days preceding puberty the "spiny" neurons become predominant. These authors have interpreted their findings as indicative of the establishment of synaptic connections between LHRH neurons and other neural cells during the juvenile and peripubertal phases of development. This may, in fact, be the case at it is well established that a marked increase in hypothalamic synaptic contacts occurs during prepubertal development (Matsumoto and Arai, 1976; Ruf, 1982; Clough and Rodriguez-Sierra, 1983).

In recent experiments we have begun to examine the hypothesis that the molecular mechanisms underlying these changes in synaptic circuitry involve changes in the gene expression of cytoskeleton proteins. Indeed, initial experiments have shown that two mRNAs (a 1.8 and a 2.9 kilobase species) encoding for neural cell specific β-tubulins (Bond et al., 1984) are strikingly more abundant in the juvenile hypothalamus than in the cerebral cortex or cerebellum of the same animals (Rogers and Ojeda, unpublished).

Intrinsic to the question of the mechanisms responsible for the pubertal activation of LHRH neurons is the issue of the factors involved in controlling the development and differentiation of LHRH neurons. Very recent experiments in our laboratory (Gonzalez, Kozlowski, Costa and Ojeda, unpublished) have demonstrated that fetal LHRH neurons can be cultured in serum-free medium. Although attachment of the cells to he culture well occurs readily in the presence of a polyornithine-laminin matrix the cells differentiate little and after 4-6 days begin to die. Addition of E_2 at physiological concentrations (10^{-11}-10^{-9} M), however, results in an increased amount of LHRH immunoreactivity within the cell and an increase in soma volume. More LHRH-positive cells were also found, suggesting an improved survival rate.

It thus appears that the maturation of both LHRH neurons themselves and the neuronal circuitry associated with them may play a fundamental role in the acquisition of female reproductive competence.

As indicated before, the CNS not only regulates ovarian development through its control over anterior pituitary function, but also directly through peripheral nerves.

THE DEVELOPMENT OF THE OVARY

In previous reviews we have discussed several aspects concerning the regulation of ovarian development (Ojeda et al., 1984; Ojeda and Urbanski, in press). In this context it is important to mention the fact that, in addition to the well-established role of LH and FSH, two other pitutiary hormones prolactin (Prl) and growth hormone (GH) appear to play an important supportive role in ovarian maturation. While hyperprolactinemia advances the onset of puberty and sensitizes the ovary to the stimulatory effect of gonadotropins (Advis and Ojeda, 1978), hypoprolactinemia produces the opposite effects (Advis et al., 1981). On the other hand, GH deficiency delays puberty and results in a decreased number of ovarian hCG-LH receptors (Advis et al., 1981a), an effect that appears to be, to a significant extent, a direct consequence of withdrawal of GH support to the ovary. Very recent reports have revealed that GH enhances FSH-stimulated differentiation of granulosa cells (Jia et al., 1986), and that this effect may be exerted by increasing the ovarian production of somatomedin C immunoreactive material (Davoren and Hsueh, 1986).

It is evident, therefore, that several anterior pituitary hormones work in concert to promote follicular maturation and hence the acquisition of ovulatory competence at puberty.

For many years it has been known that the ovary receives an adrenergic innervation (for a review see Burden, 1985). The adrenergic nerves, which essentially contain only NE reach the ovary via two routes: the superior ovarian nerve (SON) and the plexus nerve (PN) (Lawrence and Burden, 1980). Experiments conducted in our laboratory have revealed that section of SON leads to a rapid drop (within 4 min) in E_2 and progesterone (P) secretion when the section is performed on the day of proestrus, but not on the day of estrus (Aguado and Ojeda, 1984a). On a longer term basis (7 days), a compensatory increase in β-adrenergic receptor number is observed, a phenomenon that is associated with hypersensitivity by denervation (Aguado and Ojeda, 1984). This hypersensitivity, detected using cultured granulosa cells, is manifested as an increase in the capacity of the cells to produce P in response to β_2-adrenergic stimulation (Aguado and Ojeda, 1984). It is well established that the predominant population of adrenergic receptors in the rat ovary is of the β_2-subtype (for a review see Ojeda and Aguado, 1985) and that responsiveness of granulosa cells to β-adrenergic stimulation is dependent on prior exposure to FSH (Adashi and Hsueh, 1981).

Circulating epinephrine of adrenomedullary origin also appears to affect ovarian function since selective removal of the adrenal medulla depresses circulating levels of epinephrine, blunts ovarian responsiveness to gonadotropins and delays the onset of puberty (Aguado and Ojeda, 1984b).

Very recent experiments have revealed that in addition to noradrenergic nerves the immature rat ovary receives peptidergic innervation. Such an innervation includes fibers containing substance P (SP), vasoactive intestinal peptide (VIP) and neuropeptide Y (NPY) (Dees et al., 1985; Ojeda et al., 1985; Ahmed et al., 1986; McDonald et al., submitted).

While it appears that neither SP nor NPY can affect ovarian steroidogenesis (in granulosa cells) it is now clear that VIP is an effective stimulus for E_2, P and androgen secretion from the ovary (Davoren and Hseuh, 1985; Ahmed et al., 1986). This effect of VIP Is not reproduced by any of the members of the VIP family (Fig. 2) with the exception of peptide histidine isoleucine (PHI), which has 50% sequence homology to VIP.

VIP appears to act on a population of granulosa cells which is different from that stimulated by FSH (Kasson et al., 1985). This observation and our own findings that VIPergic fibers can be detected around small, preantral follicles (Ahmed et al., 1986) and that VIP can induce aromatase activity long before the ovary acquires responsiveness to FSH (i.e. during late fetal-early perinatal life) (George and Ojeda, unpublished) strongly suggest that VIP may act on a more immature population of ovarian (granulosa) cells than FSH, and thus may represent one of the first stimulatory inputs to the developing ovary.

That VIP may indeed play a hitherto unsuspected role in ovarian maturation is suggested by the recent demonstration that the peptide acts on granulosa cells to stimulate the synthesis of the cholesterol side chain cleavage enzyme complex (SCC), the rate limiting step in the biosynthesis of P (Trezeciak et al., in press). VIP can induce the synthesis of the three components of SCC (cytochrome P_{450}, iron sulphur protein and NADPH-adrenodoxin reductase). This effect of VIP appears to be exerted at the gene

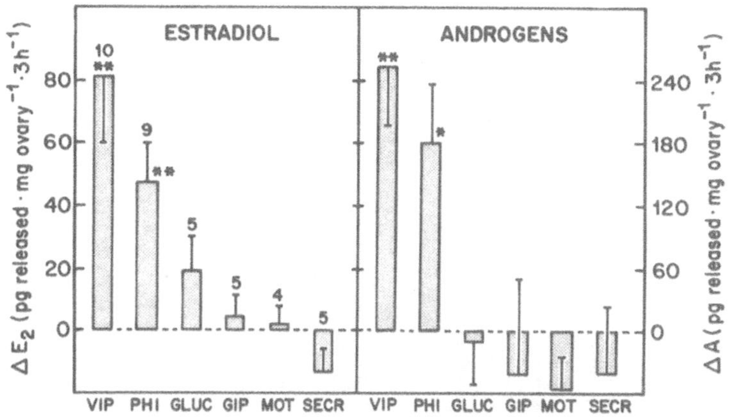

Fig. 2. Specificity of the stimulatory effect of VIP on ovarian steroidogenesis. Ovaries from rats injected with 2 IU PMSG at 29 days of age were obtained 24 h after the injection. Ovary halves were incubated as described (Ahmed et al., 1986), with one pair of halves receiving no hormone (basal release) and the remaining contralateral halves receiving VIP, PHI, glucagon (GLUC), gastric inhibitory polypeptide (GIP), motilin (MOT) or secretin (SECR) at 1 μM each. The mean ± SEM are shown. The number of ovaries per group is indicated above each bar. *, $p < 0.05$; **, $p < 0.01$ (vs. basal values). Δ, change from basal values. (From Ahmed et al., 1986, with permission.)

expression level as in very recent experiments we have been able to show that VIP induces an increase in both translatable mRNA and the steady state level of the mRNA encoding for the cytochrome P_{450} component of SCC (Trzeciak et al., unpublished). In regard to SP and NPY, we suspect that these peptides may be involved in regulating ovarian blood flow. Other functions, however, may be suspected.

The aforementioned observations strongly suggest that the capacity of the ovary to produce preovulatory amounts of E_2 is acquired under the concerted influence of circulating hormones and the direct influence of noradrenergic and peptidergic nerves.

THE ACTIVATION OF THE CENTRAL COMPONENT OF ESTRADIOL (E_2) POSITIVE FEEDBACK AT PUBERTY

Development of large, estrogen producing preovulatory follicles results in an increase in circulating levels of E_2 to levels 6-8 times greater than in juvenile rats (Meijs-Roelofs et al., 1975; Parker and Mahesh, 1976). Estradiol then acts at both the hypothalamic and pituitary levels to induce a discharge of LHRH, on the one hand, and to sensitize the pituitary gonadotrophs to the neuropeptide on the other.

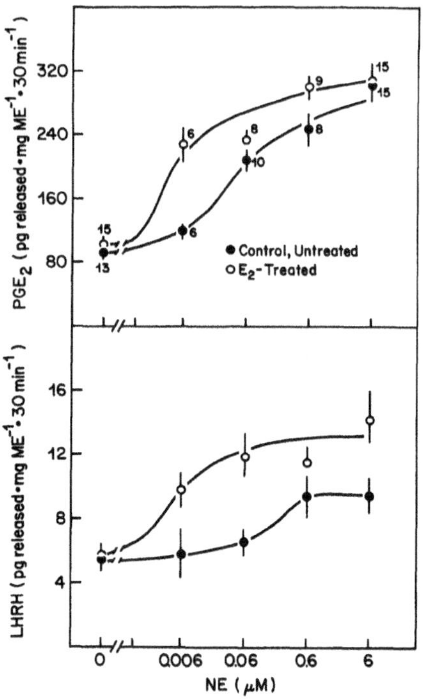

Fig. 3. Effect of in vivo simulation of proestrous serum E_2 levels on the in vitro release of PGE_2 (upper panel) and LHRH (lower panel) from the ME of juvenile 28-day-old female rats in response to NE. E_2 was provided in s.c. silastic capsules at a concentration (400 μg/ml corn oil) that reproduces serum E_2 levels on the day of first proestrous. The E_2-containing capsules were implanted 48 h before the experiment. (From Ojeda et al., 1986a, with permission.)

At the hypothalamic level, E_2 is known to increase the turnover of NE (for a review see Ramirez et al., 1984) and to enhance the capacity of the tissue to synthesize PGE_2 from (exogenous) arachidonic acid (Ojeda and Campbell, 1982). More recently, we have found that preovulatory levels of serum E_2 increase the capacity of ME nerve terminals to produce PGE_2 in response to NE and to release LHRH in response to PGE_2 (Ojeda et al., 1986a) (Fig. 3). Not surprisingly, E_2 also enhances the LHRH response to NE.

As mentioned earlier, the PGE_2-dependent pathway is not the only biochemical route involved in transducing the effect of excitatory signals impinging on LHRH neurons. When ME terminals were exposed to phospholipase C (PLC), the enzyme that catalyzes the hydrolysis of polyphosphoinositides into inositol triphosphate and diacylglycerol, a marked increase in PGE_2 production and LHRH release was observed (Ojeda et al., 1986b). Suppression of prostaglandin synthesis with indomethacin completely obliterated the PGE_2 response, but reduced the LHRH response by only about 50%. This and other observations described in the same report (Ojeda et al., 1986b) have led us to conclude that the E_2-dependent LHRH discharge which occurs in the after-

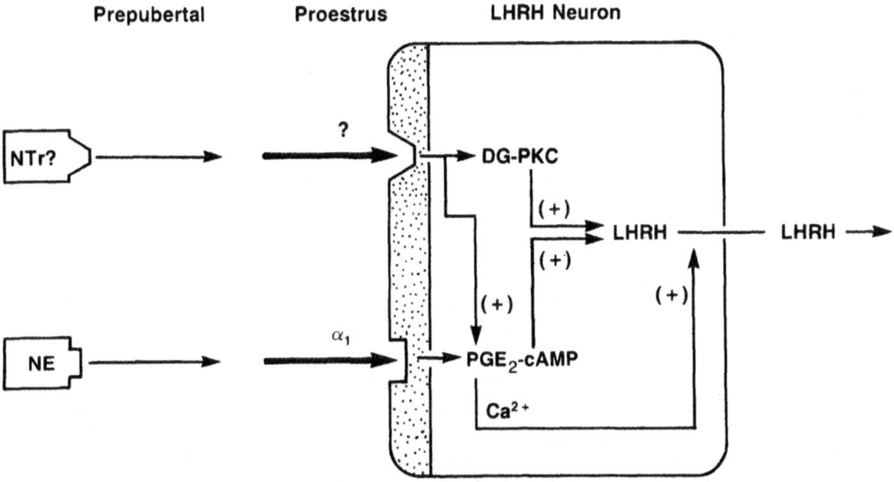

Fig. 4. Postulated intracellular pathways involved in the transduction of neurotransmitter signals leading to LHRH release during puberty in the female rat. During prepubertal days both pathways may be operative but the intensity of the extracellular inputs activating them is low (thin arrows). In the afternoon of proestrus, an increased activity (thick arrows) of NE neurons and of another neurotransmitter system (that presumably activates the PKC pathway) induces the preovulatory discharge of LHRH. NE = norepinephrine; PKC = protein kinase C; Ntr = unidentified neurotransmitter activating PKC; PGE_2 = prostaglandin E_2; cAMP = cyclic AMP; Ca^{2+} = calcium. (From Ojeda and Urbanski, in press.)

noon of first proestrus may depend on the simultaneous activation of two independent, but complementary mechanisms, i.e. a PGE_2-cyclic AMP and a PKC mediated pathway (Fig. 4).

While much more work is necessary to firmly substantiate this view, it may be suggested that the expression of two key events determines the initiation of neuroendocrine reproductive competence in the female rat: an ovarian ovarian-independent diurnal change in the pattern of LH release which initiates puberty, and an E_2-induced surge of LHRH release that triggers the first preovulatory surge of gonadotropins and hence represents the neuroendocrine culmination of the pubertal process.

ACKNOWLEDGEMENTS

We thank Ms. Judy Scott for typing the manuscript. This investigation was supported by NIH grant HD-09988, Project IV and by NSF grant BNS-8318017.

REFERENCES

Adashi, E. Y., and Hsueh, A. J. W., 1981, Stimulation of β_2-adrenergic responsiveness by follicle-stimulating hormone in rat granulosa cells in vitro and in vivo, Endocrinology, 108:2170.

Advis, J. P., and Ojeda, S. R., 1978, Hyperprolactinemia induced precocious puberty in the female rat: ovarian site of actions, Endocrinology, 103:924.

Advis, J. P., Smith-White, S. S., and Ojeda, S. R., 1981, Delayed puberty induced by chronic suppression of prolactin release in the female rat, Endocrinology, 109:1321.

Advis, J. P., Smith-White, S. S., and Ojeda, S. R., 1981a, Activation of GH short loop negative feedback delays puberty in the female rat, Endocrinology, 108:1343.

Aguado, L. I., and Ojeda, S. R., 1984, Prepubertal ovarian function is finely regulated by direct adrenergic influences. Role of the adrenergic innervation, Endocrinology, 114:1845.

Aguado, L. I., and Ojeda, S. R., 1984a, Ovarian adrenergic nerves play a role in maintaining preovulatory steroid production, Endocrinology, 114:1944.

Aguado, L. I., and Ojeda, S. R., 1984b, Effect of selective removal of the adrenal medulla on female sexual development, Biol. Reprod., 31:605.

Ahmed, C. E., Dees, W. L., and Ojeda, S. R., 1986, The immature rat ovary is innervated by vasoactive intestinal peptide (VIP)-containing fibers and responds to VIP with steroid secretion, Endocrinology, 118:1682.

Andrews, W. W., and Ojeda, S. R., 1977, On the feedback actions of estrogen on gonadotropins and prolactin release in infantile female rats, Endocrinology, 101:1517.

Andrews, W. W., and Ojeda, S. R., 1981, A quantitative analysis of the maturation of steroid negative feedbacks controlling gonadotropin release in the female rat: transition from an androgenic to a predominantly estrogenic control, Endocrinology, 108:1313.

Andrews, W. W., and Ojeda, S. R., 1981a, A detailed analysis of the serum LH secretory profiles of conscious, free-moving female rats during the time of puberty, Endocrinology, 109:2032.

Andrews, W. W., Advis, J. P., and Ojeda, S. R., 1981, The maturation of estradiol negative feedback in fmeale rats: evidence that the resetting of the hypothalamic "gonadostat" does not precede the first preovulatory surge of gonadotropins, Endocrinology, 109:2022.

Andrews, W. W., Mizejewski, G. J., and Ojeda, S. R., 1981a, Development of estradiol positive feedback on LH release in the female rat: a quantitative study, Endocrinology, 109:1404.

Bond, J. F., Robinson, G. S., and Farmer, S. R., 1984, Differential expression of two neural cell-specific β-tubulin mRNAs during rat brain development, Mol. Cell. Biol., 4:1313.

Burden, H. W., 1985, The adrenergic innervation of mammalian ovaries, in: "Catecholamines as Hormone Regulators," N. Ben-Jonathan, J. M. Bahr, and R. I. Weiner, eds., p. 261, Raven Press, New York.

Caligaris, L., Astrada, J. J., and Taleisnik, S., 1972, Influence of age on the release of luteinizing hormone induced by estrogen and progesterone in immature rats, J. Endocrinol., 55:97.

Clough, R. W., and Rodriguez-Sierra, J. F., 1983, Synaptic changes in the hypothalamus of the prepubertal female rat administered estrogen, Am. J. Anat., 167:205.

Davoren, J. B., and Hsueh, A. J. W., 1984, Vasoactive intestinal peptide: a novel stimulator of steroidogenesis by cultured rat granulosa cells, Biol. Reprod., 33:37.

Davoren, J. B., and Hsueh, A. J. W., 1986, Growth hormone increases ovarian levels of immunoreactive somatomedin C/insulin-like growth factor I in vivo, Endocrinology, 118:888.

Dees. W. L., Kozlowski, G. P., Dey, R., and Ojeda, S. R., 1985, Evidence for

the existence of substance P in the prepubertal rat ovary. II. Immunocytochemical localization, Biol. Reprod., 33:471.
Docke, F., Rohde, W., Gerber, P., and Dorner, G., 1984, Evidence that desensitization to the negative estrogen feedback is a prepubertal and not a postpubertal event in female rats, Exp. Clin. Endocrinol., 84:1.
Eldridge, J. C., McPherson III, J. C., and Mahesh, V. D., 1977, Maturation of the negative feedback control of gonadotropin secretion in the female rat, Endocrinology, 94:1536.
Germain, B. L., Campbell, P. S., and Anderson, J. N., 1978, Role of the serum estrogen-binding protein in the control of tissue estradiol levels during postnatal development of the female rat, Endocrinology, 103:1401.
Hage, A. J., Groen-Klevant, A. C., and Weischen, R., 1978, Follicle growth in the immature rat ovary, Acta Endocrinol. (Copenh.), 88:375.
Jia, X-C., Kalmijn, J., and Hsueh, A. J. W., 1986, Growth hormone enhances follicle-stimulating hormone induced differentiation of cultured rat granulosa cells, Endocrinology, 118:1401.
Kasson, B. G., Meidan, R., Davoren, J. B., and Hsueh, A. J. W., 1985, Identification of subpopulations of rat granulosa cells: sedimentation properties and hormonal responsiveness, Endocrinology, 117:1027.
Kelch, R. P., Marshall, J. C., Saunder, S. E., Hopwood, N. J., and Reame, N. E., 1983, Gonadotropin regulation during human puberty, in: "Neuroendocrine Aspects of Reproduction," R. L. Norman, ed., p. 229, Academic Press, New York.
Kimura, F., and Kawakami, M., 1982, Episodic LH secretion in the immature male and female rat as assessed by sequential blood sampling, Neuroendocrinology, 35:128.
Kronibius, J., and Wuttke, W., 1977, Positive feedback action of estradiol on gonadotropin release in 15 day old rats, Acta Endocrinol., 86:263.
Lawrence, I. E., and Burden, H. W., 1980, The origin of the extrinsic adrenergic innervation to the rat ovary, Anat. Rec., 196:51.
Ling, N. Ying, S-Y., Ueno, N., Shimasaki, S., Esch, F., Hotta, M., and Guillemin, R., 1986, Pituitary FSH is released by a hetero dimer of the β-subunits from the two forms of inhibin, Nature (Lond.), 321:779.
Matsumoto, A., and Arai, Y., 1976, Developmental changes in synaptic formation in the hypothalamic arcuate nucleus of female rats, Cell Tiss. Res., 169:143.
Matsumoto, A. M., Karpas, A. E., Southworth, M. B., Dorsa, D. M., and Bremmer, W. J., 1986, Evidence for activation of the central nervous system-pituitary mechanism for gonadotropin secretion at the time of puberty in the male rat, Endocrinology, 119:362.
McCormack, C. E., and Meyer, R. K., 1964, Minimal age for induction of ovulation with progesterone in rats: evidence for neural control, Endocrinology, 74:793.
McDonald, J. K., Dees, W. L., Ahmed, C. E., Noe, B. D., and Ojeda, S. R., in preparation, Biochemical and immunocytochemical characterization of Neuropeptide Y in the immature rat ovary.
Meijs-Roelofs, H. M. A., Uilenbroek, J. Th. J., de Greef, W. J., de Jong, F. H., and Kramer, P., 1975, Gonadotropin and steroid levels around the time of first ovulation in the rat, J. Endocrinol., 67:275.
Meijs-Roelofs, H. M. A., and Kramer, P., 1979, Maturation of the inhibitory feedback action of estrogen on follicle-stimulating hormone secretion in the immature female rat: a role for alpha fetoprotein, J. Endocrinol., 81:199.
Ojeda, S. R., and Campbell, W. B., 1982, An increase in hypothalamic capacity to synthesize PGE_2 precedes the first preovulatory surge of gonadotropins, Endocrinology, 111:1031.
Ojeda, S. R., and Aguado, L. I., 1985, Adrenergic control of the prepubertal ovary: involvement of local innervation and circulating catecholamines, in: "Catecholamines as Hormone Regulators," N. Ben-Jonathan, J. M. Bahr, and R. I. Weiner, eds., p. 293, Raven Press, New York.

Ojeda, S. R., Costa, M. E., Katz, K. H., and Hersh, L. B., 1985, Evidence for the existence of substance P in the prepubertal rat ovary. I. Biochemical and physiological studies, Biol. Reprod., 33:471.
Ojeda, S. R., Smith (White), S. S., Urbanski, H. F., and Aguado, L. I., 1984, The onset of female puberty: underlying neuroendocrine mechanisms, in: "Neuroendocrine Perspectives," vol. 3, p. 225, Elsevier, Amsterdam.
Ojeda, S. R., Urbanski, H. F., and Ahmed, C. E., 1986, The onset of female puberty: studies in the rat, in: "Recent Progress in Hormone Research," R. O. Greep, ed., vol. 42, p. 385, Academic Press, New York.
Ojeda, S. R., Urbanski, H. F., Katz, K. H., and Costa, M. E., 1986a, Activation of estradiol-positive feedback at puberty: estradiol sensitizes the LHRH-releasing system at two different biochemical steps, Neuroendocrinology, 43:259.
Ojeda, S. R., Urbanski, H. F., Katz, K. H., Costa, M. E., and Conn, P. M., 1986b, Activation of two different but complementary biochemical pathways induces release of hypothalamic luteinizing hormone-releasing hormone, Proc. Natl. Acad. Sci., 83:4932.
Ojeda, S. R., and Urbanski, H. R., in press, Puberty in the rat, in: "Physiology of Reproduction," E. Knobil and J. D. Neill, eds, Raven Press, New York.
Parker, C. R. Jr., and Mahesh, V. B., 1976, Hormonal events surrounding the natural onset of puberty in female rats, Biol. Reprod., 14:347.
Plant, T. M., 1983, Ontogeny of gonadotropin secretion in the Rhesus macaque (Macaca mulatta), in: "Neuroendocrine Aspects of Reproduction," R. L. Norman, ed., p. 133, Academic Press, New York.
Ramirez, V. D., and McCann, S. M., 1963, Comparison of the regulation of luteinizing hormone (LH) secretion in immature and adult rats, Endocrinology, 72:452.
Ramirez, V. D., Feder, H. H., and Sawyer, C. H., 1984, The role of brain catecholamines in the regulation of LH secretion: a critical inquiry, in: "Frontiers in Neuroendocrinology," L. Martini and W. F. Ganong, eds., vol. 8, p. 27, Raven Press, New York.
Raum, W. J., and Swerdloff, R. S., 1986, The effect of hypothalamic catecholamine synthesis inhibition by α-methyltyrosine on gonadotropin secretion during sexual maturation in the male Wistar rat, Endocrinology, 119:168.
Ruf, K. B., 1982, Synaptogenesis and Puberty, Front. Horm. Res., 10:65.
Steele, R. E., and Weisz, J., 1974, Changes in sensitivity of the estradiol-LH feedback system with puberty in the female rat, Endocrinology, 95:513.
Terasawa, E., Nass, T. E., Yeoman, R. R., Loose, M. D., and Schultz, N. J., 1983, Hypothalamic control of puberty in the female Rhesus macaque, in: "Neuroendocrine Aspects of Reproduction," R. L. Norman, ed., p. 149, Academic Press, New York.
Terasawa, E., Bridson, W. E., Nass, T. E., Noonan, J. J., and Dierschke, D. J., 1986, Developmental changes in the luteinizing hormone secretory pattern in peripubertal female Rhesus monkeys: comparisons between gonadally intact and ovariectomzied animals, Endocrinology, 115:2233.
Trzeciak, W. H., Ahmed, C. E., Simpson, E., and Ojeda, S. R., in press, Vasoactive intestinal peptide induces the synthesis of the cholesterol side chain cleavage enzyme complex in cultured rat ovarian granulosa cells, Proc. Natl. Acad. Sci.
Urbanski, H. F., and Ojeda, S. R., 1985, The juvenile-peripubertal transition period in the female rat: establishment of a diurnal pattern of pulsatile luteinizing hormone secretion, Endocrinology, 117:644.
Urbanski, H. F., and Ojeda, S. R., 1985a, In vitro stimulation of prepubertal changes in pulsatile luteinizing hormone release enhances progesterone and estradiol 17β-secretion from immature rat ovaries, Endocrinology, 117:638.
Urbanski, H. F., and Ojeda, S. R., 1986, Development of afternoon minisurges of LH secretion is ovarian-dependent, Endocrinology, 118:1187.

Urbanski, H. F., and Ojeda, S. R., submitted, Gonadal-independent activation of enhanced afternoon luteinizing hormone release during pubertal development in the female rat.
Vale, W. , Rivier, J., Vaugham, J., McClintock, R., Corrigan, A, Woo, W., Carr, D., and Spiess, J., 1986, Purification and characterization of an FSH-releasing protein from porcine ovarian follicular fluid, Nature (Lond.), 321:776.
Wray, S., and Hoffman, G., 1986, Postnatal morphological changes in rat LHRH neurons correlated with sexual maturation, Neuroendocrinology, 43:93.

BIOLOGICAL ACTIVITY OF NON-GnRH SYNTHETIC

PEPTIDE SEQUENCES OF THE GnRH PRECURSOR

R. P. Millar, P. J. Wormald, M. J. Abrahamson,
R. C. deL. Milton and K. Waligora

MRC Regulatory Peptides Research Unit
Department of Chemical Pathology, and
Endocrine Laboratory, Department of Medicine
University of Cape Town Medical School
and Groote Schuur Hospital
Observatory 7925, South Africa

Introduction

The demonstration of GnRH biosynthesis by tryptic and carboxypeptidase B cleavage of a higher molecular precursor in sheep, pig and rat hypothalamus, and the postulate that $pGlu^1$ is derived from cyclization of Gln while the amide of Gly^{10} is generated from an additional Gly followed by a pair of basic amino acid residues (Millar et al., 1977, 1978, 1981), was recently definitively demonstrated by nucleotide sequencing of human placental cDNA (Seeburg and Adelman, 1984) and human and rat hypothalamic cDNA (Adelman et al., 1986).

This breakthrough has laid the foundations for a host of new studies on the regulation of the reproductive system. Amongst these is the possibility that non-GnRH peptide sequences of the GnRH precursor may be processed to biologically active peptides which play a role in the pituitary, gonads, placenta and other tissues in which GnRH-like peptides are present, and in which regulatory effects of GnRH have been reported (see review Millar and King, 1984).

Seeburg and colleagues reported that the entire COOH-terminal extension peptide sequence of GnRH in the human precursor (sequence 14-69 in Fig. 1) stimulates gonadotropin and inhibits prolactin release from rat pituitary cells in culture (Nikolics et al., 1985). In independent studies, our laboratory found that the sequence 14-26 stimulates gonadotropin release from human pituitary cells in culture (Wormald et al., 1985; Millar et al., 1986). In this chapter, we review our studies on the processing and biological activity of a series of synthetic peptides corresponding to inter-basic amino acid sequences of the GnRH precursor. We also report on the activity of another series of peptides designed to more clearly delineate the minimal sequence required to elicit gonadotropin secretion.

Biological Activity of GnRH Precursor Peptides

The peptides indicated in Fig. 1 were synthesized by solid phase methodology and purified by preparative HPLC as previously described (Millar

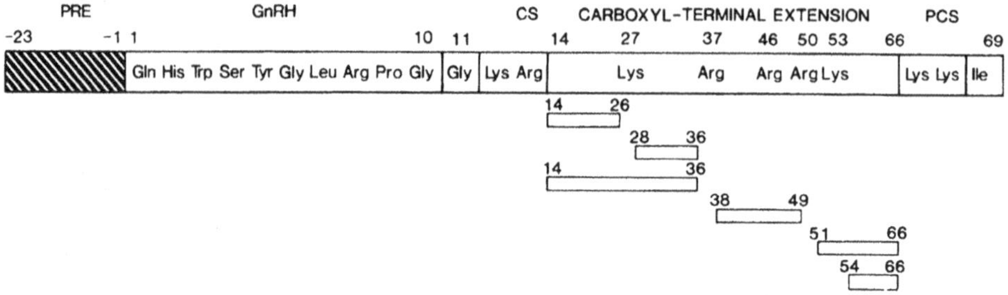

Fig. 1. Diagrammatic representation of human pre-Pro GnRH. The signal sequence (PRE) is followed by the GnRH sequence and the Lys-Arg cleavage site (CS). A second potential cleavage site (PCS) precedes the carboxyl-terminal isoleucine residue. Single basic amino acid residues which may be additional post-translational processing sites are also indicated. Gln^1 undergoes cyclization to pGlu at the amino terminus of GnRH and Gly^{11} is the donor for amidation of Gly^{10}.

et al., 1986). The amino acid composition and sequence of the peptides was verified by amino acid analysis and by gas phase sequencing.

The biological activity associated with the synthetic peptides was assessed by their ability to alter the secretion of trophic hormones from human, baboon, sheep, rat and chicken anterior pituitary cells (Millar and King, 1983). Initially the peptides were screened for activity at a final concentration of 10^{-5}M in the tissue culture medium. In human pituitary cell cultures, peptides 14-26 and 14-36 stimulated LH and FSH secretion while the other peptides were ineffective (Table 1). None of the peptides affected the secretion of TSH, GH, PRL and ACTH (data not shown). Peptide 14-36 was more effective than 14-26 as it stimulated LH and FSH secretion at 10^{-6}M in comparison to $5x10^{-6}$M for 14-26. These high doses suggest a relatively low activity of the peptides, and may indicate that longer sequences are required for full activity, especially as Nikolics et al. have found that 14-69 has high potency in stimulating gonadotropin release from rat pituitary cell cultures (Nikolics et al., 1985).

Baboon pituitary cells also responded to 14-26 (Millar et al., 1986) while rat pituitary cells did not release LH in response to this peptide even at high doses (10^{-5}M) but did respond to 14-36 at a dose of 10^{-7}M or higher (Fig. 2, Table 1). The ED_{50} for 14-36 was $2x10^{-6}$M. As with human pituitary cells, prolactin secretion was unaffected by any of the peptide sequences. Sheep and chicken pituitary cells did not release LH in response to even higher doses (10^{-5}M) of 14-26. In view of the lack of activity of this peptide in the rat but distinct gonadotropin releasing activity of longer sequences, our previous interpretation of a species specificity must be reassessed after testing longer sequences in sheep and chicken.

Specificity and Mechanism of 14-26 and 14-36 Stimulation of Gonadotropin Release

In view of the lack of sequence homology between the active pHGnRH peptides and GnRH, it was of interest to determine whether stimulation of gonadotropin release by these different peptides was mediated by a common membrane receptor on pituitary gonadotrophs. The antagonist analogue [Ac-D-Nal(2)^1-Me-4Cl-D-Phe^2-D-Trp^3, D-Arg^6, D-Ala^{10}]GnRH, which competi-

Table 1. Summary of Gonadotropin-releasing Activity of GnRH Precursor C-Peptides in Human and Rat Pituitary Cell Cultures

	HUMAN LH	HUMAN FSH	RAT LH
14–26	↑	↑	0
14–36	↑↑	↑↑	↑↑
28–36	0	0	0
38–49	0	0	0
51–66	0	0	0
54–66	0	0	0

(Double arrows indicate higher activity than single arrows.)

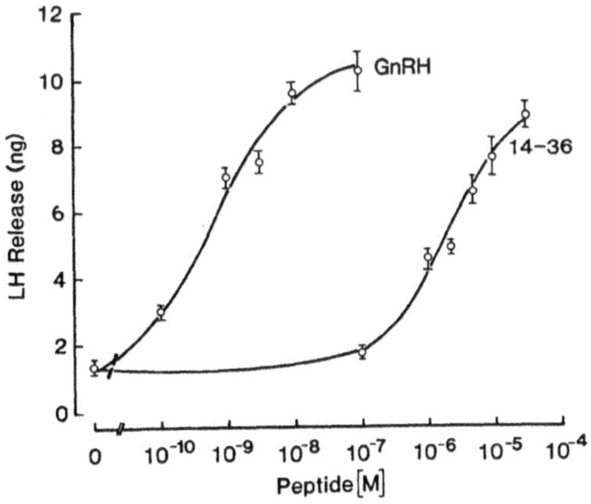

Fig. 2. Stimulation of LH release from cultured rat pituitary cells by GnRH and peptide 14–36.

Table 2. Gonadotropin-releasing Activity of C-Peptides in Human and Rat Pituitary Cell Cultures

	HUMAN		RAT
	LH	FSH	LH
14-26	↑	↑	0
17-26	↑	↑	0
14-36	↑↑	↑↑	↑↑
14-37.NH_2	↑↑	↑↑	↑↑
17-36	↑↑	↑↑	↑↑

(Double arrows indicate higher activity than single arrows.)

tively inhibited GnRH binding to its receptor at an equivalent concentration without stimulating the release of LH and FSH, was unable to block 14-26 stimulation of LH or FSH release from human pituitary cells suggesting that 14-26 does not stimulate the gonadotrophs via the GnRH receptor in human pituitary cells (Millar et al., 1986). This conclusion was supported by the inability of 14-26 to compete with an ^{125}I GnRH-analogue for binding to human anterior pituitary cell membranes.

The specificity of effect is of concern when ascribing biological properties to novel peptides. The effect of 14-26 and 14-36 on gonadotropin appears specific by several criteria. Firstly, like GnRH, their stimulation of gonadotropin secretion is dependent on extracellular Ca^{2+} such that secretion remains at basal levels when the peptides are added to cells in the presence of EGTA. Secondly, secretion of thyrotropin, growth hormone, prolactin and adrenocorticotropin was unaffected. Thirdly, gonadotropin release is unlikely to be due to a general toxic effect of the peptides giving rise to gonadotropin release from the cells as lactate dehydrogenase was not released from the cells, trypan blue staining of cells was not increased, and protein synthesis (which is a good integrated index of normal function in cells) was unaffected.

Delineation of the Sequence with the Gonadotropin Releasing Activity

In order to determine more specifically the peptide sequence responsible for eliciting gonadotropin release, an additional series of peptides was synthesized and compared with 14-26 and 14-36 for gonadotropin release in human and rat pituitary cells. The biological activity of these peptides is summarized in Table 2.

In human pituitary cells 14-26, 14-36, 17-36 and 14-37.NH_2 consistently (seven experiments) significantly stimulated LH and FSH release. The decapeptide sequence (17-26), which incorporates the sequence Asn-Leu-Ile-Asp-Ser-Phe-Gln-Glu-Ile-Val present in all of the active peptides, significantly stimulated LH and FSH secretion from human pituitary cells in two of four experiments, albeit at higher concentrations than the other peptides. This suggests that this peptide contains the essential sequence for activity and approaches the minimal sequence required for gonadotropin release. In rat pituitary cells (Table 2), 14-26 and 17-26 did not stimulate LH release. However, the longer peptides which incorporate carboxyl-terminal extensions, and which were more active in human pituitary cells, stimulated LH release. Incorporation of $Arg^{37}.NH_2$ to stimulate further extension of the molecule ($14\text{-}37.NH_2$) was even more active. Further characterization of the active sequence with respect to the role of specific amino acid residues would probably be better pursued by the synthesis of peptide analogues with amino acid substitutions in the 17-26 sequence or in longer sequences than by further shortening of the minimal active peptide sequence.

Post-translational Processing of the GnRH Precursor

Nucleotide sequencing of rat and human hypothalamic cDNA and human placental cDNA (Seeburg and Adelman, 1984; Adelman et al., 1986) have indicated that the GnRH precursor is of identical length (69 amino acids plus a signal sequence of 23 amino acids) (Fig. 1). Immunoprecipitation of reticulocyte lysate translation products of rat and human hypothalamic mRNA (Curtis and Fink, 1983; Curtis et al., 1983) and radioimmunoassay of gel filtration separated extracts of placenta and hypothalamus (Gautron et al., 1981) revealed GnRH precursors of 26-28K molecular weight. Whether these are artifacts or reflect possible alternative splicing of GnRH mRNA to produce a higher molecular weight mRNA is unknown. Studies in our laboratory on sheep and rat hypothalamic GnRH with an antiserum directed at the middle region of GnRH which recognizes NH_2- and COOH-terminally extended GnRH, detected a GnRH precursor of approximately 16K on Sephadex G-50 under dissociating conditions. This material was not detected by a COOH-terminally directed antiserum and poorly by an NH_2-terminally directed antiserum, suggesting that the GnRH sequence had extensions at both ends. A smaller species of about 8K was also detected by the middle or NH_2-directed antisera but not by the COOH-directed antiserum. This molecule corresponds in size, therefore, to the GnRH precursor derived from the cDNA sequence. Interestingly, on repeated treatment of the larger 16K species with urea or guanidinium hydrochloride and rechromatography on HPLC gel permeation chromatography, the peptide eluted in the position of the 8K species. This suggests that a strong self association of the 8K precursor or association with a binding protein increases the apparent molecular weight and obscures the sequences of GnRH recognized by the NH_2-directed antiserum. Although this may account for previous reports of GnRH precursors of >8K (Gautron et al., 1981; Curtis and Fink, 1983; Curtis et al., 1983), the possibility of alternatively spliced longer GnRH mRNA cannot be ruled out.

In addition to cleavage of GnRH from the precursor and amidation at the sequence Gly^{10}-Gly^{11}-Lys^{12}-Arg^{13}, there are a number of potential sites of processing of the C-terminal 69 amino acid extension. The pair of basic amino acid residues Lys^{67}-Lys^{68} is a characteristic cleavage site. Processing also occurs at single basic amino acid residues with a preference of Arg over Lys (Schwartz, 1986). A cleavage could occur at Lys^{26} (rat and human), Arg^{37} (human), Arg^{46} (rat and human), Arg^{50} (rat and human), Lys^{53} (human), Arg^{53} (rat) and Arg^{58} (rat). Schwartz (1986) has pointed out that Pro-Arg sequences are particularly favored for cleavage so that Pro^{45}-Arg^{46} which is conserved in human and rat must be seriously considered.

Our studies on processing are presently in preliminary stages. An antiserum to 14-26 which is C-terminally directed and does not crossreact significantly with peptides extended at amino acid 26 revealed hypothalamic neuronal axons and termini in immunocytochemical studies on rat hypothalamus (Rubin et al., submitted) but another study on monkey, rat and sheep hypothalamus revealed cell bodies, axons and termini (Silverman et al., submitted). Sephadex G-25 separation of peptides extracted from sheep hypothalami and human placentae revealed a peptide of approximately 1K molecular weight with this antiserum. These findings together with the immunocytochemical observations suggest some processing of 14-26 occurs in the hypothalamus and placenta. Another antiserum to 51-66 revealed high molecular weight peptides (GnRH precursor and entire C-peptide) as well as lower molecular weight material suggestive of cleavage after Arg^{46}, Arg^{50} or Lys^{53}. The availability of the range of antisera to peptide fragments indicated in Fig. 1, together with others specifically recognizing the cleavage site, will allow resolution of the processing of the GnRH precursor.

It is feasible that processing varies between tissues and with physiological status. Indeed, certain peptides which are without activity in one tissue may have unique activity in another. In this regard, the GnRH precursor C-peptides require testing for effects in the gonads and placenta.

Concluding Remarks

Gonadotropin releasing activity of peptides of the C-terminal extension of GnRH in its precursor has been independently reported (Nikolics et al., 1985; Wormald et al., 1985; Millar et al., 1986). While Nikolics et al. (1985) have found that the peptide comprising the entire 59 amino acid extension to GnRH in its precursor (sequence 14-69) stimulates LH and FSH secretion from rat pituitary cells in culture, we have found that a peptide comprising the first 14 amino acids (sequence 14-26) stimulates LH and FSH secretion from human pituitary cells in culture (Wormald et al., 1985; Millar et al., 1986). The studies reported here on a further series of overlapping peptide sequences have localized the activity to the sequence 17-26, which is therefore considered to represent a defined region of the GnRH precursor with intrinsic gonadotropin releasing activity. The increased activity of longer peptides which incorporate carboxyl-terminal extensions of this sequence suggests that a putative physiological gonadotropin releasing peptide is extended to at least the second potential cleavage site at Arg^{37}. The increased activity of $14\text{-}37.NH_2$ and $17\text{-}37.NH_2$ in rat pituitary cells indicates that longer peptides are even more active. Indeed, the entire 59 amino acid peptide tested by Nikolics et al. is considerably more active than the most active of our peptides. Thus, the physiologically relevant peptide may be the entire 59 amino acid sequence or peptides cleaved at Arg^{46} or more C-terminal sites. In view of the favored cleavage at the Pro-Arg sequence present at Arg^{46}, the synthesis and testing of 14-46 is of high priority.

Our data indicating that the active gonadotropin-releasing peptides operate via a mechanism independent of occupancy of the GnRH receptor raise important questions as to a potential ancilliary role for these peptides in regulating gonadotropin secretion. The demonstration that the ED_{50} for FSH release from rat pituitary cells by 14-69 is 0.5 nM compared with 2 nM for LH release, points to a special role of the peptide in FSH secretion. Our studies in human pituitary cells show variability in the relative LH and FSH releasing activity of the C-peptides. In spite of these interesting observations on gonadotropin releasing activity of GnRH precursor C-peptides, the ability of synthetic GnRH to restore complete reproductive function in laboratory animals and humans with hypothalamic lesions, and the marked inhibitory effect of treatment with GnRH antagonist or GnRH antiserum, suggests that these peptides play a relatively minor role in reproduction. Possibly the effects are more important in the tonic regulation of gonadotropin secretion. However, studies on the effects of longer sequences, the effects of *in vivo* neutralization with antisera, treatment of hypothalamically lesioned animals with the peptides, and the insertion of GnRH gene constructs into transgenic hpg mice (A. Mason, personal communication) are required before definite conclusions can be reached.

Nikolics and colleagues have observed that the entire C-peptide (sequence 14-69) is a potent inhibitor of prolactin secretion from rat pituitary cells in culture (Nikolics et al., 1985). On the basis of this and the elevation of prolactin in rabbits actively immunized with synthetic peptides, they have proposed that the peptide is the long sought-after prolactin inhibiting factor (Nikolics et al., 1985). It is of interest, however, that prolactin secretion in vivo in rats and from perifused rat pituitary cells is unaffected by the peptide (Schally et al., 1986). None of the peptides synthesized by us (14-26, 17-26, 17-36, 17-37.NH_2, 14-36, 14-37.NH_2, 28-36, 38-49, 51-66, 54-36) affected prolactin secretion from human or rat pituitary cells. Thus, it may be that the entire 14-69 sequence or other processed sequences such as 14-45 or 47-69 are required to elicit prolactin inhibition.

Extensive studies on the nature of peptides processed from the GnRH precursor in the hypothalamus, placenta and gonads are required to set the stage for further examination of potential physiological roles for GnRH gene products. Although only a single gene has been identified to date, the identification of several forms of GnRH within a single species of several non-mammalian vertebrates and recent observations in our laboratory (King, Hassan and Millar, in preparation) of several immunoreactive GnRHs in sheep and rat hypothalamus points to multiple physiological roles for GnRH and GnRH-related gene products.

Acknowledgements

We wish to acknowledge support from the Medical Research Council, the National Cancer Association, and the University of Cape Town.

References

Adelman, J. P., Mason, A. J., Hayflick, J. S., and Seeburg, P. H., 1986, Isolation of the gene and hypothalamic cDNA for the common precursor of gonadotropin-releasing hormone and prolactin release-inhibiting factor in human and rat, Proc. Natl. Acad. Sci. USA, 83:179.

Curtis, A., and Fink, G., 1983, A high molecular weight precursor of luteinizing hormone-releasing hormone from rat hypothalamus, Endocrinology, 112:390.

Curtis, A., Lyons, V., and Fink, G., 1983, The human hypothalamic LHRH precursor is the same size as that in rat and mouse hypothalamus, Biochem. Biophys. Res. Commun., 117:872.

Gautron, J. P., Pattou, E., and Kordon, C., 1981, Occurrence of higher molecular weight forms of LHRH in fractionated extracts from rat hypothalamus, cortex and placenta, Mol. Cell. Endocrinol., 24:1.

Millar, R. P., Aehnelt, C., and Rossier, G., 1977, Higher molecular weight immunoreactive species of luteinizing hormone-releasing hormone: possible precursors of the hormone, Biochem. Biophys. Res. Commun., 74:720.

Millar, R. P., Denniss, P., Tobler, C., King, J. C., Schally, A. V., and Arimura, A., 1978, Presumptive prohormonal forms of hypothalamic peptide hormones, in: "Cell Biology of Hypothalamic Neurosecretion," J. D. Vincent, and C. Kordon, eds., p. 487, Centre National de la Recherche Scientifique 280, Bordeaux.

Millar, R. P., and King, J. A., 1983, Synthesis, luteinizing hormone-releasing activity, and receptor binding of chicken hypothalamic luteinizing hormone-releasing hormone, Endocrinology, 113:1364.

Millar, R. P., and King, J. A., 1984, Molecular heterogeneity of luteinizing hormone-releasing hormone, in: "Hormonal Control of the Hypothalamo-Pituitary-Gonadal Axis," K. W. McKerns, ed., p. 437, Plenum Publishing Corporation, New York.

Millar, R. P., Wegener, I., and Schally, A. V., 1981, Putative prohormone of luteinizing hormone-releasing hormone, in: "Neuropeptides: Biochemical and Physiological Studies," R. P. Millar, ed., p. 111, Churchill-Livingstone, New York.
Millar, R. P., Wormald, P. J., and Milton, R. C. deL., 1986, Stimulation of gonadotropin release by a non-GnRH peptide sequence of the GnRH precursor, Science, 232:68.
Nikolics, K., Mason, A. J., Szonyi, E., Ramachandran, J., and Seeburg, P. H., 1985, A prolactin-inhibiting factor within the precursor for human gonadotropin-releasing hormone, Nature 316:511.
Schally, A. V., Olsen, D. B., Gulyas, J., Szoke, B., Horvath, J., Karashima, T., Redding, T. W., Nikolics, K., and Seeburg, P. H., 1986, In vitro and in vivo studies with synthetic gonadotropin releasing hormone-associated peptide (GAP), a proposed prolactin release-inhibiting peptide, Proc. 68th Ann. Meeting Endocr. Soc., Anaheim, p. 26.
Schwartz, T. W., 1986, The processing of peptide precursors 'Proline-directed arginyl cleavage' and other monobasic processing mechanisms, FEBS Lett., 200:1.
Seeburg, P. H., and Adelman, J. P., 1984, Characterization of cDNA for precursor of human luteinizing hormone releasing hormone, Nature, 311:666.
Wormald, P., Milton, R. deL., Abrahamson, M., Duflou, J., Roberts, J., Brandt, W., and Millar, R., 1985, A peptide fragment from the GnRH precursor stimulates gonadotropin release from cultured human pituitary cells, S. Afr. J. Sci., 81:579.

MECHANISMS OF GnRH ACTION: INTERACTIONS BETWEEN GnRH-STIMULATED CALCIUM-PHOSPHOLIPID PATHWAYS MEDIATING GONADOTROPIN SECRETION

John P. Chang, Ernest McCoy*,
Reginald O. Morgan, and Kevin J. Catt

Endocrinology and Reproduction Research Branch
National Institute of Child Health
and Human Development
National Institutes of Health
Bethesda, Maryland, USA 20892

INTRODUCTION

Hypothalamic regulation of gonadotropin secretion is exerted through receptor-mediated actions of gonadotropin-releasing hormone (GnRH) on pituitary gonadotropes. In rat pituitary cells, the GnRH receptor consists of two binding components of molecular weight 53,000 and 42,000 as revealed by photoaffinity labelling techniques (Iwashita and Catt, 1985). As with all peptide hormones, GnRH-receptor binding leads to activation of intracellular "second messenger" systems that culminate in the release of gonadotropins. The regulation of luteinizing hormone (LH) secretion by GnRH has been extensively studied in cultured rat anterior pituitary cells and in gonadotrope-enriched fractions prepared by centrifugal elutriation for the identification of specific effects of GnRH on the activation of second messenger systems. Although both cAMP and cGMP levels are increased in pituitary cell cultures following GnRH stimulation (for review see Conn et al., 1981), cAMP (Sen and Menon, 1979) and cGMP (Naor and Catt, 1980) do not appear to mediate acute LH responses to the releasing peptide. GnRH stimulation of LH release is a Ca^{2+}-dependent process (Conn et al., 1981; Conn, 1986). Arachidonic acid (AA) and its lipoxygenase products, as well as the activation of protein kinase C, a Ca^{2+}- and phospholipid-dependent enzyme, have been implicated as mediators of GnRH-induced LH secretion (for reviews see Catt et al., 1984, 1985). More recently, GnRH-stimulated hydrolysis of phosphoinositides has also been shown to participate in the mechanism of GnRH action (Catt et al., 1984, 1985; Schrey, 1985). In this article, results from recent investigations into the mechanisms of action of GnRH, as well as evidence suggesting that the LH response to GnRH requires the coordination of several intracellular messenger systems, will be presented and discussed.

COORDINATION OF PROTEIN KINASE C AND AA IN GnRH ACTION

The presence of protein kinase C activity in the rat pituitary gland has been demonstrated in several recent reports. In the rat pituitary gonadotrope, protein kinase C can be activated by diacylglycerols (DGs) and

*Department of Pediatrics, University of Alberta, Edmonton, Alberta, Canada.

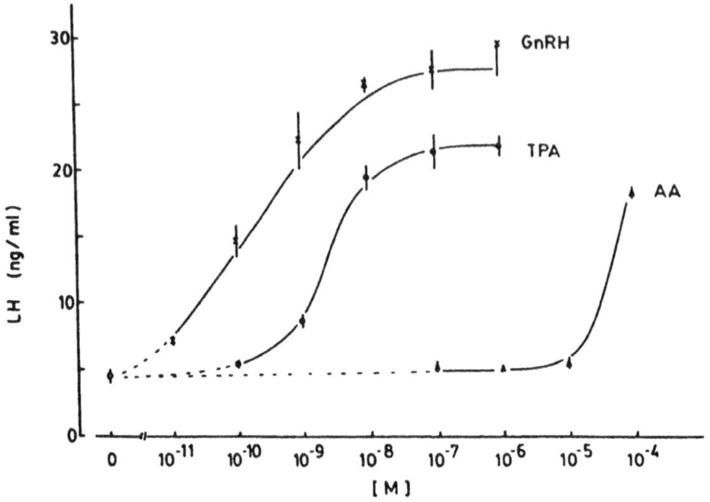

Fig. 1. LH secretory response to increasing doses of GnRH, TPA, and AA in static pituitary cell cultures. Means ± SE of quadruplicate incubations are shown.

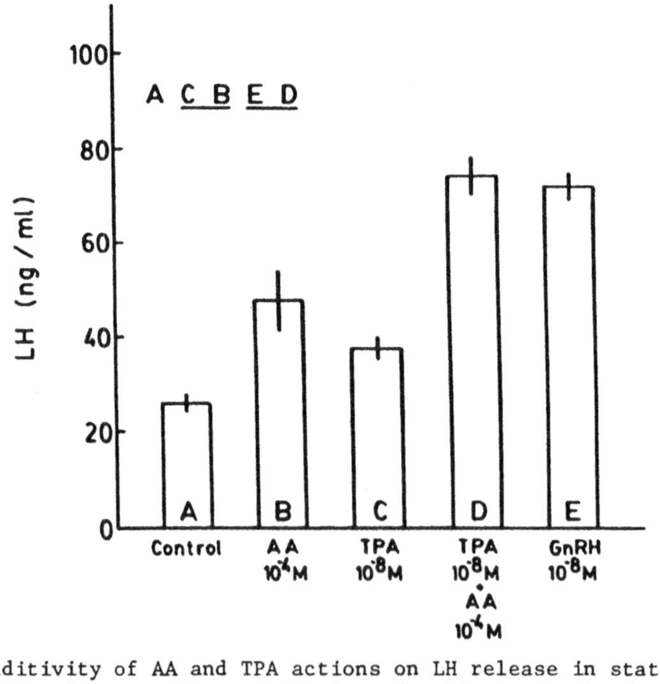

Fig. 2. Additivity of AA and TPA actions on LH release in static pituitary cell cultures. Means ± SE of quadruplicate incubations are shown. Groups joined by the same underline are not different from one another ($p > 0.05$).

phorbol esters in the presence of Ca^{2+} and phosphatidylserine, resulting in the phosphorylation of proteins that includes several kinase C-specific target proteins (Turgeon et al., 1984; Hirota et al., 1985; Naor et al., 1985; Ngo and Catt, unpublished results). The phorbol ester, 12-o-tetradecanoyl phorbol-13-acetate (TPA), and a synthetic DG, dioctanoylglycerol, stimulate LH release (Catt et al., 1985). In cultured rat anterior pituitary cells, the ability of various phorbol ester analogs to stimulate LH release parallels their ability to activate protein kinase C. Both TPA and GnRH cause a redistribution of protein kinase C activity from the cytosol to membrane fractions of rat gonadotropes; the time course and dose-dependence of protein kinase C translocation are similar to those of LH release (Hirota et al., 1985). Translocation of protein kinase C is thought to be a phenomenon associated with enzyme activation, and the redistribution of enzyme activity is maximal within 10 min of GnRH addition. The ability of GnRH to increase protein kinase C activity in fractions of pituitary cells prepared by centrifugal elutriation is correlated with the degree of enrichment with gonadotropes. Retinal, an inhibitor of protein kinase C activation, suppresses the GnRH-induced increase in protein kinase C activity in pituitary cytosol, and also decreased GnRH-stimulated LH release. These results are consistent with the activation of protein kinase C as an early event in the stimulation of LH release by GnRH.

In dispersed rat pituitary cells prelabeled with $[^{14}C]$AA, GnRH stimulation of LH secretion was accompanied by the release of ^{14}C-radioactivity; GnRH stimulation of both LH and ^{14}C-radioactivity were inhibited by the TnRH antagonists, [Ac-D-p-Cl-$Phe^{1,2}$, D-Trp^3, D-Lys^6, D-Ala^6]-GnRH and [D-Phe^2, Pro^3, D-Phe^6]-GnRH. AA and its lipoxygenase metabolite, 5-hydroxy-6,8,11,14-eicosatetraenoic acid, consistently stimulate LH release _in vitro_ Catt et al., 1983) and the LH response to AA can be abolished by the lipoxygenase inhibitor nordihydroguaiaretic acid (NDGA). Similarly, GnRH-induced LH release can be decreased by NDGA and other inhibitors of AA metabolism such as 5,8,11,14-eicosatetraynoic acid (ETYA) and 3-amino-1-[m-(trifluoromethyl)-phenyl]-2-pyrazoline (BW755c). These observations suggest that AA metabolites, especially those of the lipoxygenase pathway, may mediate GnRH-induced LH release. In pituitary cells prelabeled with $[^3H]$AA, $[^3H]$AA metabolites of the lipoxygenase, mono-oxygenase, and cyclo-oxygenase pathways are readily measurable by reverse-phase HPLC following 3-h incubation with GnRH, consistent with the proposal that GnRH enhances AA metabolism (Catt et al., 1984).

Although stimulation of both protein kinase C and AA metabolism appear to mediate GnRH action, the amount of LH release by treatment of cultured dispersed anterior pituitary cells with maximal doses of TPA or AA is less than that obtained with maximal GnRH stimulation (Fig. 1). However, the effect of a combination of maximal doses of AA and TPA is equivalent with that of GnRH (Fig. 2). It seems likely that GnRH stimulation of LH release is mediated by the combined action of at least two major pathways, those of protein kinase C activation and AA metabolism. This hypothesis was further tested by the use of retinal, the protein kinase C inhibitor, and the lipoxygenase inhibitors, ETYA, 2-(12-hydroxydodeca-5,10-diynyl)-3,5,6-trimethyl-1,4-benzoquinone (AA 861), and NDGA. Inhibition of protein kinase C activation by retinal reduced but did not abolish the GnRH-stimulated LH response (Fig. 3). Retinal, at doses that inhibited TPA-induced LH release, also did not alter the LH response to AA treatment (Fig. 4). Blockade of the lipoxygenase pathway by ETYA reduced but did not abolish GnRH-stimulated LH release (Fig. 3). Neither NDGA nor ETYA affected the LH response to TPA treatment. However, addition of retinal together with a lipoxygenase inhibitor, e.g. ETYA (Fig. 3), AA861 (Fig. 5) or NDGA (Chang et al., 1986), completely abolished the LH response to GnRH. Combined treatment with retinal and ETYA also abolished the LH response to AA + TPA treatment (Fig. 3). These findings confirm that in static pituitary cell cultures, GnRH stimulation of

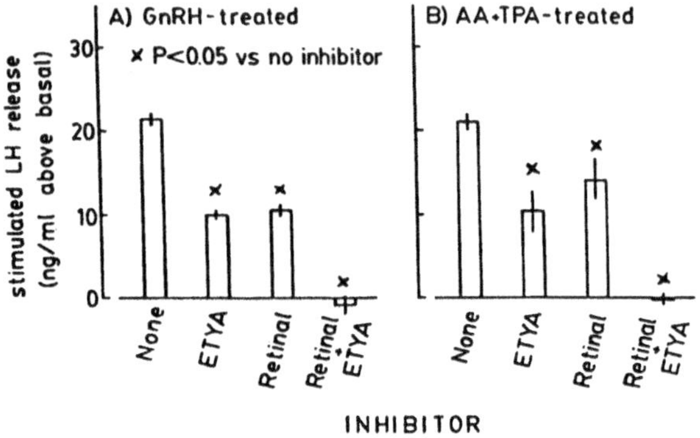

Fig. 3. Combined actions of 10^{-5} M retinal and/or 2×10^{-5} M ETYA on A) 10^{-8} M GnRH-, and B) 5×10^{-5} M AA + 10^{-7} M TPA-stimulated LH release. Means ± SE of quadruplicate incubations are shown.

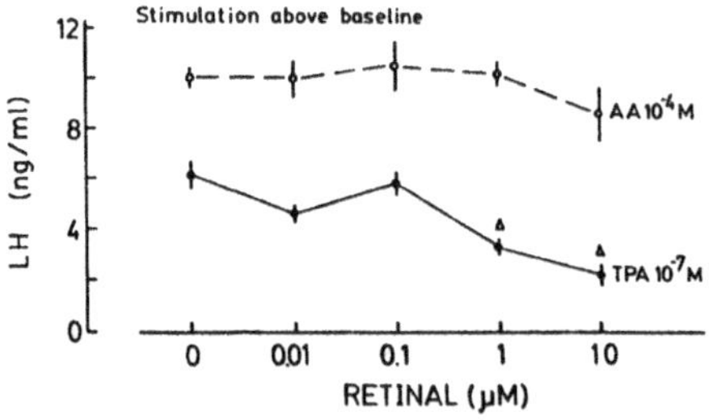

Fig. 4. Effects of increasing concentrations of retinal on 10^{-7} M TPA- and 10^{-4} M AA-stimulated LH release. Means ± SE of quadruplicate incubations are shown. (Δ, $p < 0.05$ vs. zero retinal.)

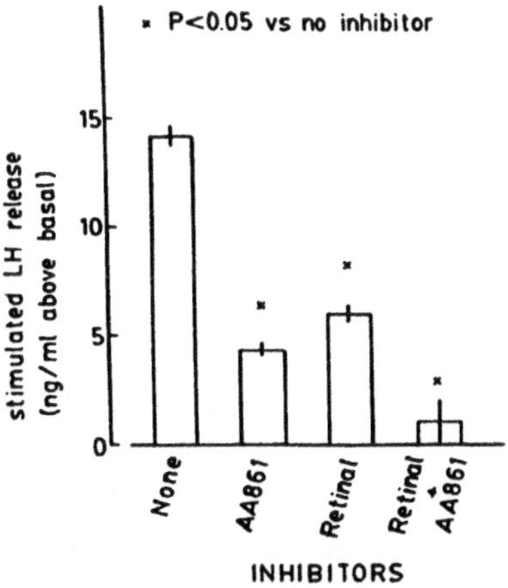

Fig. 5. Combined actions of 10^{-5} M retinal and 10^{-5} M AA861 on 10^{-8} M GnRH-stimulated LH release. Means ± SE of quadruplicate incubations are shown.

LH release includes a protein kinase C as well as an AA component, and suggests that these two independent pathways interact to stimulate LH release.

When dispersed rat anterior pituitary cells are cultured on Cytodex beads and stimulated in a column perifusion system, the LH responses to sequential pulses of GnRH are consistent and reproducible within individual columns. In such a system, the onset of the LH response to 2-min pulses of 10 nM GnRH occurs simultaneously with the arrival of the releasing peptide. This LH response consists of a rapid increase in LH release rate, reaching a maximum at about one minute after the initiation of the LH response; this is followed by a second phase of LH release comprised of a second peak or a plateau in the LH release rate, and a gradual return to the basal level. Treatment with retinal did not alter the initial rapid LH response to GnRH but significantly shortened the total duration of the response, suggesting that blockade of protein kinase C activation preferentially inhibits the second phase of the GnRH-stimulated pulse of LH release. In contrast, although NDGA treatment also did not delay the onset of the LH response, it significantly decreased the height and advanced the position of the first LH peak, suggesting that blockade of AA metabolism had a detrimental effect on the first phase of the LH response. Consistent with the idea that AA and/or its metabolites play a role in the first phase while protein kinase C activation participates in the second phase of the LH response to GnRH, perifusion of dispersed rat anterior pituitary cells with AA results in an LH

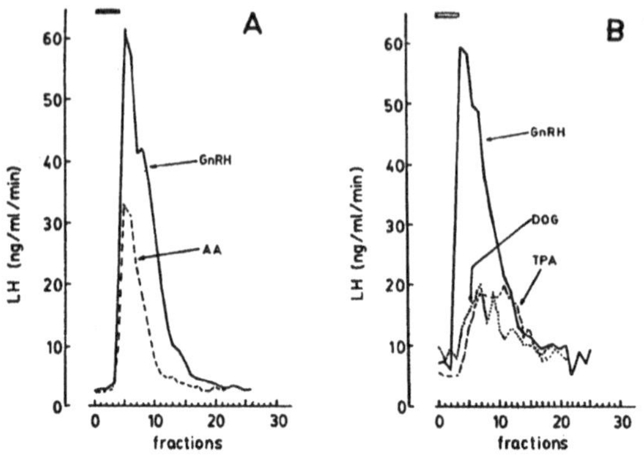

Fig. 6. Comparison of relative potencies of 2-min pulses of A) 10^{-4} M AA and 10^{-8} M GnRH; and B) 10^{-7} M TPA, 10^{-4} M dioctoanoylglycerol (DOG) and 10^{-8} M GnRH in stimulating LH release in a cell column perifusion system. Fractions were collected every 30 sec. The duration of application of the secretagogues is indicated by the bar at the top of each figure. The fraction corresponding to the start of each test pulse was assigned a serial number of zero.

release pattern that corresponds to the first, rapid phase of the LH response to GnRH (Fig. 6A). In contrast, treatment with TPA or dioctanoylglycerol produces LH peaks with delayed onsets and release profiles that coincide only with the second phase of the GnRH-stimulated LH response (Fig. 6B). These results indicate that GnRH-induced LH release is mediated by at least two major pathways, the mobilization and metabolism of AA mediating the initial phase of LH release and the activation of protein kinase C-dependent mechanisms participating in the maintenance and prolongation of the response to GnRH.

INVOLVEMENT OF VOLTAGE-SENSITIVE CALCIUM CHANNELS (VSCC) AND DIFFERENT CALCIUM POOLS IN GnRH-STIMULATED LH RELEASE

The requirement for Ca^{2+} in GnRH-stimulated LH release is well documented, and depletion of Ca^{2+} by EDTA or EGTA abolishes the LH response to GnRH in static culture systems (for review see Conn et al., 1981). The stimulatory actions of GnRH on LH release are suppressed by certain calcium blockers (e.g. verapamil; Naor et al., 1980) and calmodulin antagonists, including W_5, W_7, and pimozide (Catt and Iwashita, unpublished); whereas high concentrations of Ca^{2+} ionophores, A23187 and ionomycin (Kiesel and Catt, 1984; Chang et al., 1986) increased LH secretion. Studies using $^{45}Ca^{2+}$ have suggested that intracellular Ca^{2+} levels increase following GnRH-receptor occupancy (Hopkins and Walker, 1978). Also, based on measurements of $^{45}Ca^{2+}$ fluxes, it has been propsed that GnRH action is exclusively dependent on the entry of external Ca^{2+} (Bates and Conn, 1984).

Recently, we have analyzed the roles of VSCC and Ca^{2+} entry in mediating GnRH-induced LH release using dihydropyridine Ca^{2+} channel agonist [methyl-1,4-dihydro-2,6-methyl-3-nitro-4-(2-phenyl)-pyridine-5-carboxylate; BK 8644) and antagonist analogs (ethyl-1,4-dihydro-2,6-dimethyl-3-carboxymethyl-4-(3-nitro-phenyl)-pyridine-5-carboxylate; nitrendipine; Chang et al., 1986].

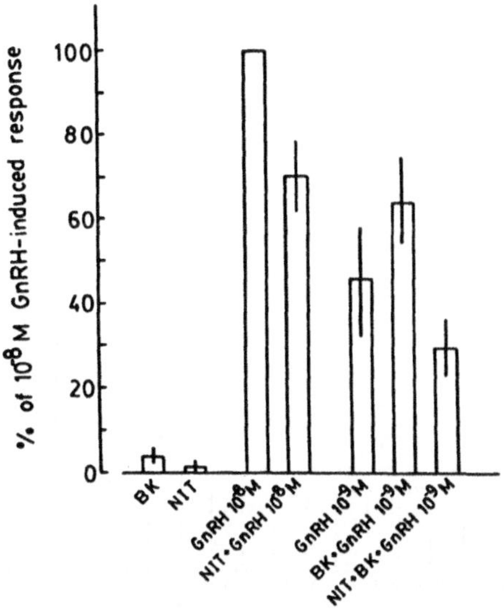

Fig. 7. Inhibitory effects of 10^{-8} M nitrendipine (Nit) on the stimulatory actions of 10^{-8} M GnRH and 10^{-8} M BK 8644 (BK)-potentiated 10^{-9} M GnRH. Data are expressed as % of the 10^{-8} M GnRH-stimulated response and pooled results (mean ± SE) from 4 similar experiments are shown.

These dihydropyridines are among the most potent and specific VSCC blockers or agonists (Triggle, 1984; Towart and Schramm, 1984). The Ca^{2+} channel agonist, BK 8644, elevated basal LH levels when applied at concentrations from 10^{-7} to 10^{-5} M. It has been proposed that BK 8644 does not stimulate VSCC opening but acts as an allosteric effector, exerting its action by shifting a pre-existing equilibrium between open and closed channels in favor of channel opening time (Schramm et al., 1983; Towart and Schramm, 1984). It seems likely that small numbers of VSCC are operational in the gonadotrope under unstimulated conditions. At 10 nM concentration, BK 8644 did not alter basal LH release, but it significantly potentiated the ability of submaximal (0.1 and 1 nM) doses of GnRH (Fig. 7), as well as 20 mM K^+ (Fig. 8) to increase LH release. The ability of BK 8644 to potentiate GnRH- and K^+-induced LH secretion can be blocked by a low concentration (10 nM) of the Ca^{2+} channel antagonist, nitrendipine (Figs. 7 and 8). Furthermore, nitrendipine doses that do not alter basal LH release (10^{-8} to 10^{-5} M), decreased the LH secretory response to maximal (10 nM) doses of GnRH by 30-40%. These results with the Ca^{2+} channel agonist and antagonist suggest that GnRH action on LH release involves the activation of VSCC.

Using Quin-2 fluorescence as an index of cytoplasmic Ca^{2+} concentration, the action of GnRH on increases in intracellular Ca^{2+} levels can be directly estimated. We have found that cultured rat gonadotropes which have been carefully harvested to minimize cell damage have a basal intracellular Ca^{2+} concentration of 100-200 nM. In Quin-2 loaded rat gonadotropes, supramaximal

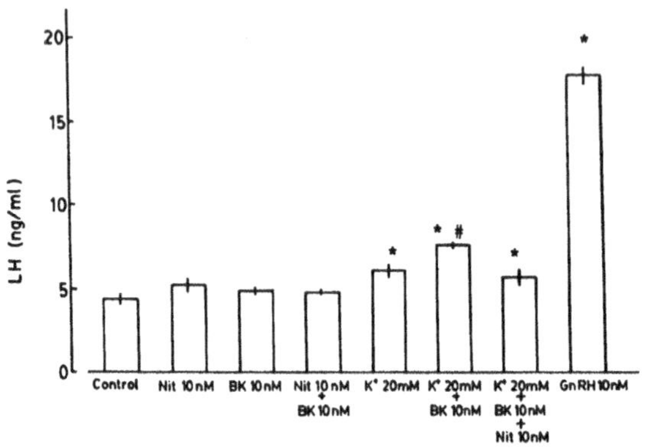

Fig. 8. Inhibition of BK 8644 (BK)-potentiation of 20 mM K^+ action on LH release by nitrendipine (Nit). Means ± SE of quadruplicate incubations are shown. (*, significantly different from control, $p < 0.05$; #, significantly different from 20 mM K^+, $p < 0.05$.)

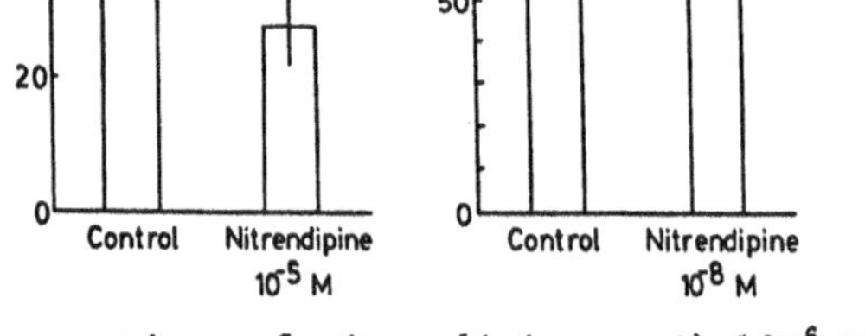

Fig. 9. Inhibitory actions of nitrendipine on A) 10^{-6} M GnRH agonist-stimulated increase in intracellular Ca^{2+} levels; and B) 10^{-8} M GnRH-induced increase in LH release. Means ± SE are shown.

(10^{-6} M) doses of a potent GnRH agonist, [D-Ala^6]-des Gly^{10}-GnRH-N-ethyl amide, rapidly increased cytoplasmic Ca^{2+} levels by 50-100 nM within 15 sec (Chang et al., 1986). Concomitant treatment with nitrendipine decreased the GnRH agonist-induced Ca^{2+} rise by about 40% (Fig. 9). During cell column perifusion studies, in which rapid changes in LH release can be monitored, nitrendipine treatment also decreased the LH release to GnRH stimulation by approximately 40% (Fig. 9). These observations indicate that a correlation exists between increases in cytoplasmic Ca^{2+} and LH-responses, and provide further evidence that GnRH action is mediated, at least in part, by activation of VSCC. However, the failure of high doses of nitrendipine (10^{-5} M) to abolish the LH and Ca^{2+} responses to GnRH indicate that the increase in cytoplasmic Ca^{2+} concentration during GnRH action originates in part from mobilization of internal Ca^{2+} stores.

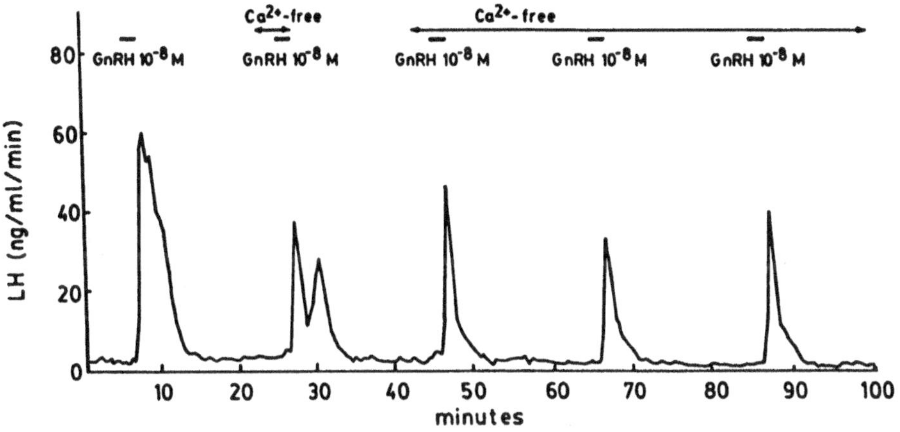

Fig. 10. GnRH-stimulated LH release in a cell column perifusion system before and during Ca^{2+}-free conditions. LH contents in perifusate collected at 30-sec intervals are plotted as ng/ml/min.

The role of external Ca^{2+} in mediating the short-term action of GnRH upon LH release has also been studied in the cell column perifusion system. Continuous perifusion with Ca^{2+}-free medium (Fig. 10) diminished but did not abolish the GnRH-induced LH response. Under these conditions, the duration but not the onset of the LH response was affected. When Ca^{2+}-free treatment was terminated together with the end of the 2-min GnRH pulse, a partial recovery of the second-phase LH response was observed. The total duration of this type of "split-response" was identical to that observed with normal GnRH treatment. Similar effects on the delayed occurrence of the second phase of LH release could also be obtained with concomitant applications of 5mM EGTA or EDTA together with the 2-min GnRH pulse. The total time-lag between the first LH peak and the beginning of the recovery of the second phase in these short-term Ca^{2+}-free conditions is approximately 2 min, the exact time period during which GnRH is active in the absence of extracellular Ca^{2+}. These results indicate that the presence of external Ca^{2+} is not essential for the

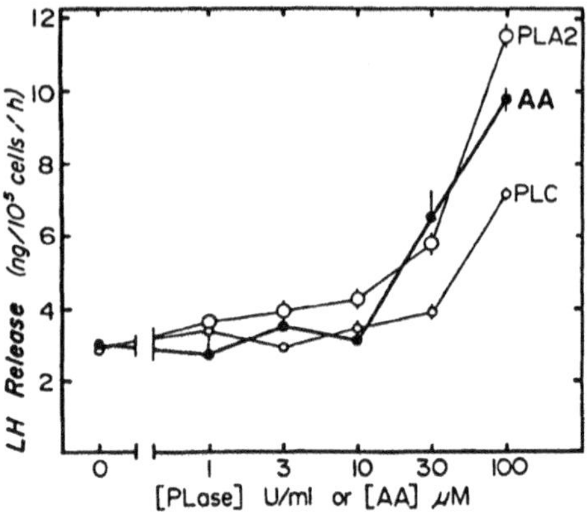

Fig. 11. LH secretory response to increasing concentrations of AA, PLA_2, and PLC in static pituitary cell cultures. Means ± SE of triplicate incubations are shown.

initiation of GnRH action, but is required for the prolongation of the LH response. These observations, together with the inability of nitrendipine to completely abolish the GnRH-induced increase in intracellular Ca^{2+} concentrations and LH release, strongly suggest that the increase in cytoplasmic Ca^{2+} levels during GnRH action must originate in part from mobilization of internal Ca^{2+} stores.

INVOLVEMENT OF PHOSPHOINOSITIDES IN GnRH ACTION

As in other secretory cells in which the mechanism of hormone release involves Ca^{2+}-dependent pathways, GnRH-stimulation of LH secretion has been shown to involve phospholipid metabolism. GnRH stimulates ^{32}P incorporation into phosphatidylinositol (PI), phosphatidylinositol-4-phosphate (PIP), phosphatidylinositol-4,5-bisphosphate (PIP_2), and phosphatidic acid (PA) in pituitary cells (Snyder and Bleasdale, 1982; Kiesel and Catt, 1984; Raymond et al., 1984). GnRH has also been shown to decrease the radioactivity of PI, PIP, and PIP_2 fractions in ^{32}P-prelabeled gonadotropes (Andrews and Conn, 1986). Addition of phospholipase C (PLC) stimulates LH release in pituitary cell cultures (Fig. 11) and the metabolism of phosphoinositides produces inositol phosphates, DG and PA. Among the inositol phosphates, inositol-1,4,5-trisphosphate (I-1,4,5-P_3, IP_3) has been shown to mobilize Ca^{2+} from intracellular non-mitochondrial (probably endoplasmic reticulum) pools (for reviews see Michell, 1983; Berridge and Irvine, 1984; Berridge, 1985). IP_3 probably acts via specific IP_3 receptors to mobilize Ca^{2+} from intracellular storage sites. Specific IP_3 binding sites have been identified in membrane preparations of bovine adrenals (Baukal et al., 1985). As discussed in an earlier section, DG has been shown to activate protein kinase C pathways leading to LH secretion. The ability of PA to stimulate LH secretion has also been demonstrated in previous studies using cultured rat pituitary cells Kiesel and Catt, 1984). Thus, the hydrolysis of phosphoinositides generated several active metabolic products, all of which are potential mediators of GnRH action.

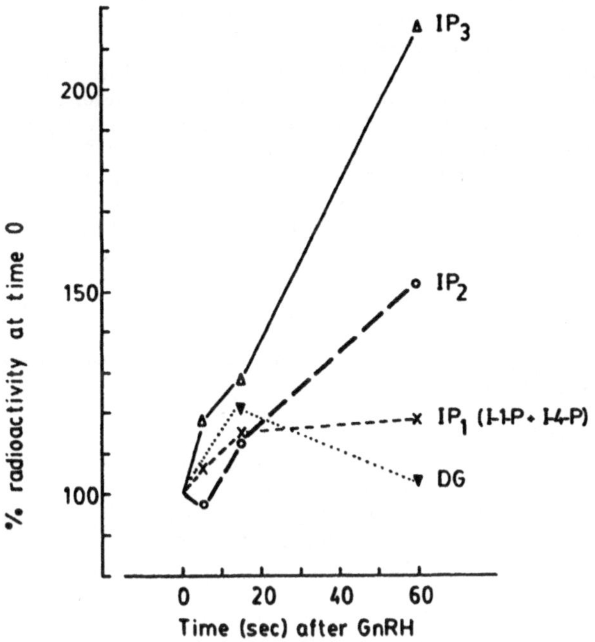

Fig. 12. Changes in radioactivity in IP_3-, IP_2-, IP_1-, and DG-fractions following 10^{-7} M GnRH stimulation of dispersed rat anterior pituitary cells prelabeled to equilibrium with [^{3}H]inositol and [^{14}C]-glycerol. Results are expressed as % of time zero controls. Radioactivity in the DG and inositol phosphate fractions did not change significantly in control treatments during the 60 sec test period.

In view of the possible importance of IP_3 as a mediator of intracellular Ca^{2+} mobilization, the time course of the effect of GnRH on inositol phosphates production was monitored in pituitary cells and gonadotropes. Preliminary results from this laboratory have already shown that in purified rat gonadotropes prelabeled with [^{3}H]inositol, stimulation of LH release by GnRH was associated with prominent increases in [^{3}H]inositol-monophosphate (IP_1) formation, as well as increases in [^{3}H]inositol di- and tri-phosphates, consistent with substantial breakdown of polyphosphoinositides by PLC (Catt et al., 1985). In this study, separation of inositol phosphates was performed by anion exchange chromatography using step-wise elution from Dowex columns. Transient increases in inositol triphosphates were detected as early as 1 min after GnRH application, whereas sustained increases in inositol mono- and di-phosphates were observed from 2-5 min until the end of the 60-min GnRH-treatment period. Early increases in total inositol tri-, di-, and mono-phosphates, as well as dose-related accumulation of these phosphates in rat pituitary cells, were confirmed by Kiesel et al. (1986) also using Dowex columns. However, elution from Dowex columns does not allow for the separation and identification of the different isomers of mono-, di-, and tri-phosphates. The ability to separate various inositol phosphate isomers is critical for the determination of the substrate for GnRH-induced PLC action, and is also important because of the differential activity of the known inositol trisphosphate isomers in Ca^{2+} mobilization. Whereas inositol-1,4,5-trisphosphate, IP_3, is a potent effector of Ca^{2+} mobilization, high concentrations of inositol-2,4,5-trisphosphate are required to increase intracellular Ca^{2+}; also, an inositol-1,3,4-trisphosphate isomer of yet unknown function has recently been identified (for reviews see Berridge and Irvine, 1984; Parthasarathy and Eisenberg, 1986). Recently, the GnRH-induced increases in inositol phosphates have been further characterized by the use of HPLC ion exchange techniques.

In studies using strong anion exchange (SAX) HPLC, the changes in inositol phosphates following GnRH application were confirmed, and increases in IP_3 were detected 5 sec after GnRH addition to [^{3}H]inositol- and [^{14}C]-glycerol-prelabeled pituitary cells (Fig. 12). This rapid increase in IP_3 approximates the time course of the change in intracellular Ca^{2+} concentration detected following GnRH treatment. The increase in IP_3 is followed by the accumulation of inositol-1,4-bisphosphate (IP_2) as well as by inositol monophosphate (Fig. 12). In the same experiment, an increase in ^{14}C-labeled DG was observed at 15 sec; however, the elevation of DG was not sustained (Fig. 12). In subsequent experiments, prolonged (30 min) elevation of IP_3 levels was observed in the continuous presence of GnRH (Morgan and Catt, 1986). Using a gradual elution gradient, all isomers of inositol monophosphates have now been resolved. Accumulation of inositol-4-phosphate (I-4-P), as opposed to inositol-1-phosphate (I-1-P), is preferentially enhanced by GnRH stimulation (Fig. 13). I-4-P has been shown to be a product of IP_3 metabolism and cannot arise directly from PI hydrolysis (Morgan and Catt, 1986). The GnRH-induced changes in I-1-P, I-4-P, IP_2, and IP_3 can be reversed by the addition of a GnRH antagonist (Morgan and Catt, 1986). The continual and specific production of IP_3 and I-4-P indicates that PLC hydrolysis of PIP_2 and/or PIP is sustained during GnRH action. The lack of DG accumulation in the face of continual PIP_2 hydrolysis suggests that DG metabolism is increased by GnRH. Preliminary results already showed that in rat pituitary gonadotropes, DG is a known source of AA through DG lipase action, and DG lipase activity may contribute to AA metabolism. The presence of inositol tetra- and penta-phosphates, as well as phytate (inositol hexaphosphate) in rat gonadotropes have also been demonstrated using SAX-HPLC analysis of aqueous extracts of [^{3}H]inositol- and ^{32}P or ^{33}P-prelabeled cells (Morgan and Catt, 1986). In some experiments, the levels of these higher inositol phosphates was also altered with GnRH treatment. However, the significance and the role of these higher phosphates in GnRH action remains to be established.

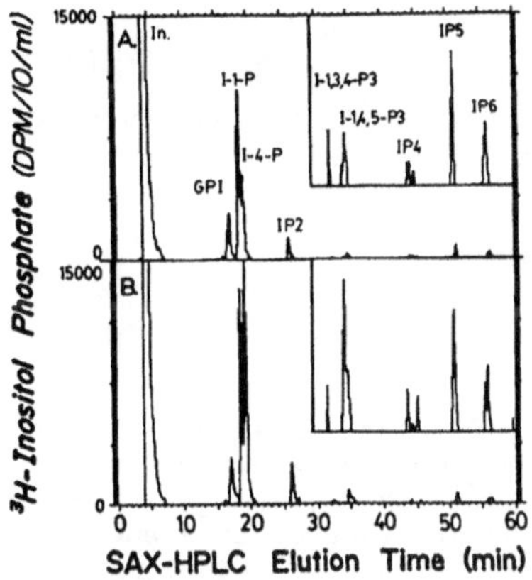

Fig. 13. Paired chromatograms of HPLC analysis of inositol phosphates after 30 min of A) control; and B) 10^{-7} M GnRH treatment. Higher inositol phosphates are shown in insets at 10-fold magnification of scale.

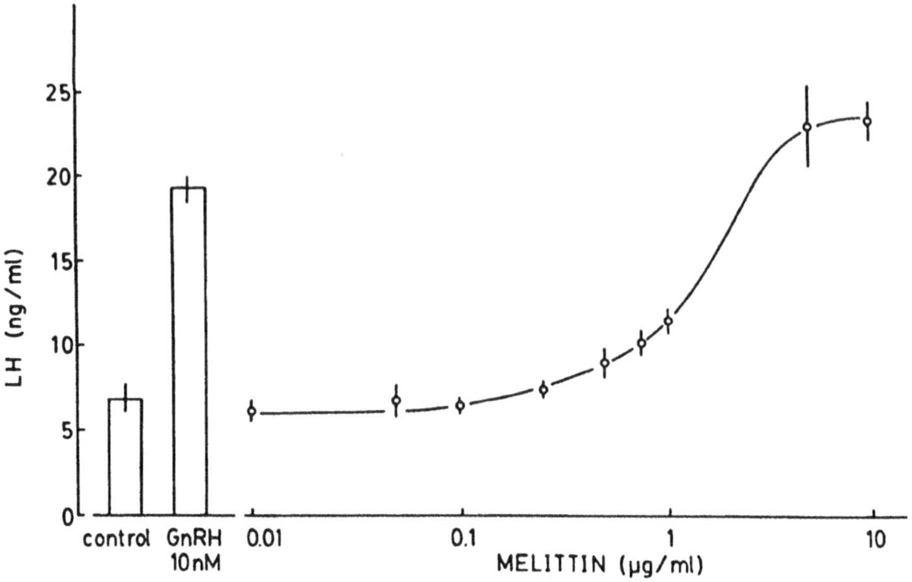

Fig. 14. Dose-related increase of LH release in static pituitary cell cultures with melittin treatment. Means ± SE of quadruplicate incubations are plotted.

These findings suggest that hydrolysis of PIP_2 by PLC may be a primary event in GnRH action. One of the hydrolysis products, IP_3, can mediate increases in intracellular Ca^{2+} levels, and the other metabolite, DG, activates protein kinase C in the presence of Ca^{2+} and is also a potential source of AA. As discussed earlier, AA- and protein kinase C-dependent mechanisms now appear to contribute significantly to the rapid secretory response to GnRH.

PHOSPHOLIPASE A_2-INDUCED AA RELEASE AS A POSSIBLE MECHANISM OF GnRH ACTION

Although AA can be generated from DG via DG lipase activity, the better known route of AA mobilization is by the action of phospholipase A_2 (PLA_2). Addition of PLA_2 and synthetic melittin, an activator of PLA_2, stimulates LH release from cultured dispersed rat pituitary cells (Figs. 11 and 14). GnRH, PLA_2, and melittin treatment all increased ^{14}C[AA] release from [^{14}C]AA-prelabeled gonadotropes. Both GnRH- and melittin-stimulated LH release can be reduced by pretreatment with inhibitors of PLA_2, such as bromophenacyl bromide, chloroquine, and quinacrine (Catt et al., 1983), as well as by pretreatment with the lipoxygenase inhibitor, NDGA. These results are consistent with the participation of PLA_2-induced AA release in mediating GnRH actions.

In Quin-2 loaded cells, treatment with melittin also increased the level of cytoplasmic Ca^{2+}, suggesting that PLA_2 activation and the subsequent release of AA also induces elevation in intracellular Ca^{2+}. Consistent with this hypothesis, both the GnRH and melittin-induced increase in intracellular Ca^{2+} levels can be reduced by pretreatment with the PLA_2 inhibitor,

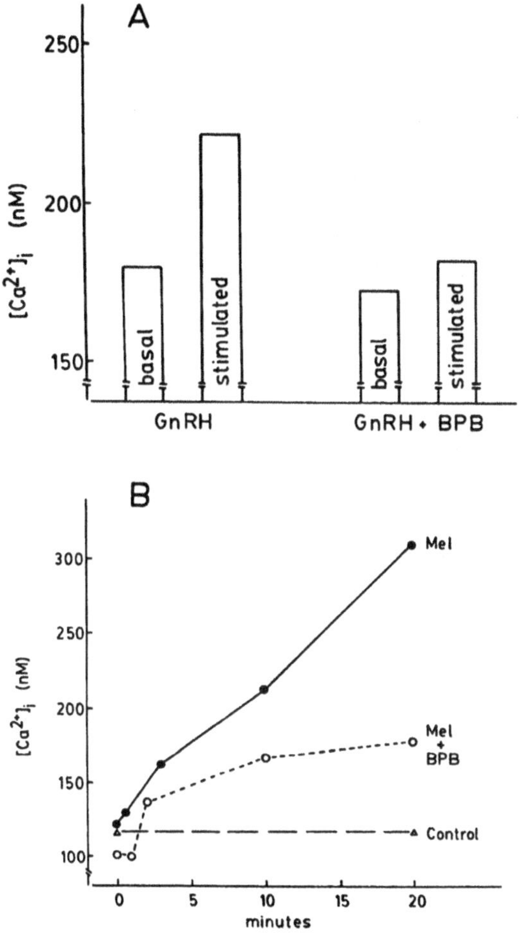

Fig. 15. Inhibition of A) GnRH- and B) melittin (Mel)-stimulated increase in intracellular Ca^{2+} levels by pretreatment with 40 mM bromophenacyl-bromide (BPB).

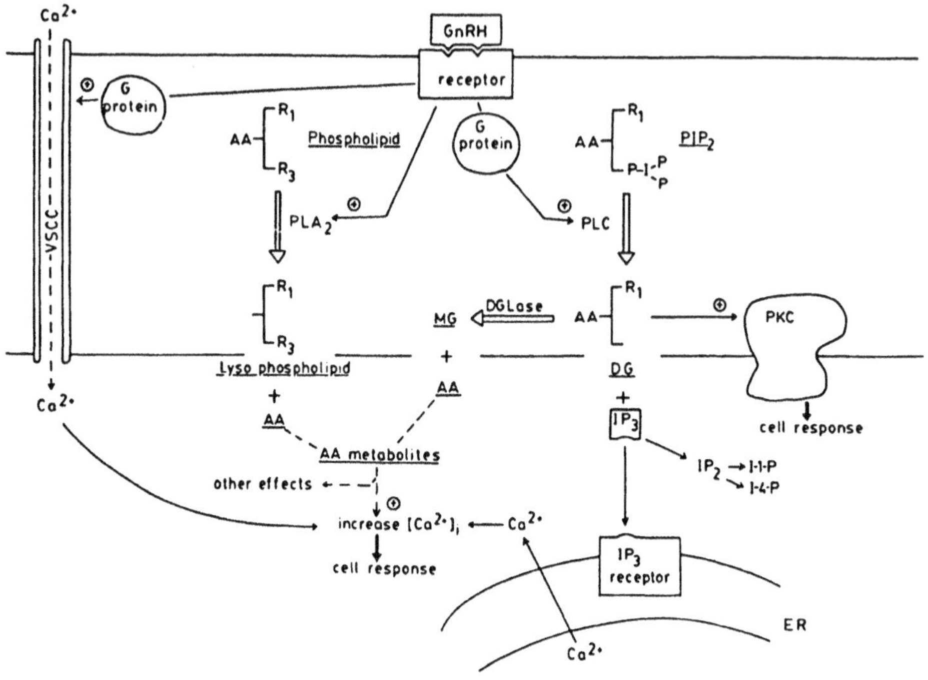

Fig. 16. Schematic representation of intracellular pathways mediating the acute actions of GnRH on LH release.

bromophenacyl bromide (Fig. 15 A and B). However, the PLA_2/AA-induced increase in intracellular Ca^{2+} concentration does not seem to be mediated by nitrendipine-sensitive VSCC. Recently, Wolf et al. (1986) have shown that exogenous AA induces a rapid release of Ca^{2+} from the endoplasmic reticulum of pancreatic islets within 2 min with similar potency as IP_3. It is likely that GnRH-stimulation of PLA_2 activity and the subsequent release of AA also participate in the mobilization of intracellular Ca^{2+}.

SUMMARY

Although there is increasing evidence that cAMP may affect long-term responsiveness to GnRH (Cronin et al., 1984), the short-term action of GnRH on LH release is achieved by the coordination of several other intracellular mechanisms (Fig. 16). AA mobilization and its metabolism, as well as protein kinase C activation, are major pathways mediating the secretory actions of GnRH. AA and its metabolites may mediate the first, rapid phase of the LH response, whereas protein kinase C activation participates in the maintenance and prolongation of the GnRH-stimulated LH release. GnRH action also involves the mobilization of intracellular Ca^{2+} as well as the induction of external Ca^{2+} entry through VSCC. Activation of VSCC may be regulated through a guanine nucleotide-binding protein (G protein; Holtz et al., 1986). External Ca^{2+} is required for the second phase of the LH release but is not critical for the induction of the LH response to GnRH. GnRH-induced PLC hydrolysis of phosphoinositides, PIP_2 in particular, is a rapid primary event that is sustained during continued GnRH action. Hormone activation of PLC has also been shown to require coupling to a G protein (for review see

Berridge, 1985). GnRH activation of PLC has been shown to be independent of Ca^{2+} (Kiesel et al., 1986; Andrews and Conn, 1986). The production of IP_3 may stimulate mobilization of intracellular (endoplasmic reticulum) stores of Ca^{2+} via specific IP_3 receptors. Ca^{2+} release from endoplasmic reticulum may also be mediated by a local G protein (Gill et al., 1986; Ueda et al., 1986). The other product of PLC hydrolysis, DG, serves as an activator of protein kinase C as well as a source of AA. GnRH-induced increases in PLA_2 activity also contribute to the stimulation of AA mobilization. AA or its metabolites may also play a role in mediating increases in intracellular Ca^{2+} levels. The observed increases in AA metabolism and PLC hydrolysis with GnRH action may in part explain the GnRH-induced increase in cGMP. IP_3 has been shown to increase cGMP formation (Europe-Finner and Newell, 1985), and elevation of cGMP levels has been shown to be secondary to AA metabolism (Harbon et al., 1983). Although Ca^{2+}/calmodulin-mediated pathways also participate in hormone-stimulation of LH release, the manner in which their role is coordinated with that of protein kinase C during GnRH action still remains to be clarified.

REFERENCES

Andrews, W. V., and Conn, P. M., 1986, Gonadotropin-releasing hormone stimulates mass changes in phosphoinositides and diacylglycerol accumulation in purified gonadotrope cell cultures, Endocrinology, 118:1148.

Bates, M. D., and Conn, P. M., 1984, Calcium mobilization in pituitary gonadotrope: relative roles of intra- and extra-cellular sources, Endocrinology, 115:1380.

Baukal, A. J., Guillemette, G., Rubin, R., Spat, A., and Catt, K. J., 1985, Binding sites for inositol trisphosphate in the bovine adrenal cortex, Biochem. Biophys. Res. Comm., 133:532.

Berridge, M. J., 1985, The molecular basis of communication within the cell, Amer. Sci., 253:142.

Berridge, M. J., and Irvine, R. F., 1984, Inositol triphosphate, a novel second messenger in cellular signal transduction, Nature, 312:315.

Catt, K. J., Loumaye, E., Katikineni, M., Hyde, C. L., Childs, G., Amsterdam, A., and Naor, Z., 1983, Receptors and actions of GnRH in pituitary gonadotrophs, in: "Role of Peptides in Control of Reproduction," S. M. McCann and D. Dhindsa, eds., p. 33, Elsevier, New York.

Catt, K. J., Loumaye, E., Wynn, P. C., Iwashita, M., Hirota, K., Morgan, R. O., and Change, J. P., 1985, GnRH receptors and actions in the control of reproductive function, J. Steroid Biochem., 23:677.

Catt, K. J., Loumaye, E., Wynn, P., Suarez-Quian, C., Kiesel, L., Iwashita, M., Hirota, K., Morgan, R., and Chang, J., 1984, Receptor-mediated activation mechanisms in the hypothalamic control of pituitary-gonadal function, in: "Endocrinology," F. Labrie and L. Proulx, eds., p. 57, Excerpta Medica, Amsterdam.

Chang, J. P., Graeter, J., and Catt, K. J., 1986, Coordinate actions of arachidonic acid and protein kinase C in gonadotropin-releasing hormone-stimulated secretion of luteinizing hormone, Biochem. Biophys. Res. Comm., 134:134.

Chang, J. P., McCoy, E. E., Graeter, J., Tasaka, K., and Catt, K. J., 1986, Participation of voltage-dependent calcium channels in the action of gonadotropin-releasing hormone, J. Biol. Chem., in press.

Conn, P. M., 1986, The molecular basis of gonadotropin-releasing hormone action, Endocrine Rev., 7:3.

Conn, P. M., Marian, J., McMillian, M., Stern, J., Rogers, D., Hamby, M., Penna, A., and Grant, E., 1981, Gonadotropin-releasing hormone action in the pituitary: a three step mechanism, Endocrine Rev., 2:174.

Cronin, M. J., Evans, W. S., Hewlett, E. L., and Thorner, M. O., 1984, LH release is facilitated by agents that alter cyclic AMP-generating system, Am. J. Physiol., 246:B44.

Europe-Finner, G. N., and Newell, P. C., 1985, Inositol 1,4,5-trisphosphate induces cyclic GMP formation in Dictyostelium discoideum, Biochem. Biophys. Res. Comm., 130:1115.

Gill, D. L., Ueda, T., Chueh, S. H., and Noel, M. W., 1986, Ca^{2} release from endoplasmic reticulum is mediated by a guanine nucleotide regulatory mechanism, Nature, 320:461.

Harbon, S., Leiber, D., and Vesin, M. F., 1983, Arachidonate metabolism in uterus. Selective implication of the cyclo-oxygenase and lipoxygenase pathways in the modulation of the cyclic AMP and cyclic GMP systems, in: "Hormones and Cell Regulation," J. E. Dumont, J. Nunez, and R. M. Denton, eds, Vol. 7, p. 39, Elsevier, New York.

Hirota, K., Hirota, T., Aguilera, G., and Catt, K. J., 1985, Hormone-induced redistribution of calcium-activated phospholipid-dependent protein kinase in pituitary gonadotrophs, J. Biol. Chem., 260:3243.

Holz, G. C. IV, Rane, S. G., and Dunlap, K., 1986, GTP-binding proteins mediate transmitter inhibition of voltage-dependent calcium channels, Nature, 319:670.

Hopkins, C. R., and Walker, A. M., 1978, Calcium as a second messenger in the stimulation of luteinizing hormone secretion, Mol. Cell. Endocrinol., 12:189.

Iwashita, M., and Catt, K. J., 1985, Photoaffinity labeling of pituitary and gonadal receptors for gonadotropin-releasing hormone, Endocrinology, 117:738.

Kiesel, L., Bertges, K., Rabe, T., and Runnebaum, B., 1986, Gonadotropin releasing hormone enhances polyphosphoinositide hydrolysis in rat pituitary cell, Biochem. Biophys. Res. Comm., 134:861.

Kiesel, L., and Catt, K. J., 1984, Phosphatidic acid and the calcium-dependent actions of gonadotropin-releasing hormone in pituitary gonadotrophs, Arch. Biochem. Biophys., 231:202.

Michell, B., 1983, Ca^{2+} and protein kinase C: two synergistic cellular signals, Trends Biochem. Sci., 8:263.

Morgan, R. O., and Catt, K. J., 1986, Novel aspects of gonadotropin-releasing hormone action on inositol polyphosphate metabolism, J. Biol. Chem., submitted.

Naor, Z., and Catt, K. J., 1980, Independent actions of gonadotropin releasing hormone upon cyclic GMP production and luteinizing hormone release, J. Biol. Chem., 225:342.

Naor, Z., Leifer, A. M., and Catt, K. J., 1980, Calcium dependent actions of gonadotropin releasing hormone on pituitary cGMP production and gonadotropin release, Endocrinology, 107:1438.

Naor, Z., Zer, J., Zakut, H., and Hermon, J., 1985, Characterization of pituitary calcium-activated, phospholipid-dependent protein kinase: redistribution by gonadotropin-releasing hormone, Proc. Natl. Acad. Sci. USA, 82:8203.

Parthasarathy, R., and Eisenberg, F. Jr., 1986, the Inositol phospholipids: a stereochemical view of biological activity, Biochem. J., 235:313.

Raymond, V., Leung, P. C. K., Veilleux, R., Lefevre, G., and Labrie, F., 1984, LHRH rapidly stimulates phosphatidylinositol metabolism in enriched gonadotrophs, Mol. Cell. Endocrinol., 36:157.

Schramm, M., Thomas, G., Towart, R., and Franckowiak, G., 1983, Novel dihydropyridines with positive inotropic action through activation of Ca^{2+} channels, Nature, 303:535.

Schrey, M. P., 1985, Gonadotropin releasing hormone stimulates the formation of inositol phosphates in rat anterior pituitary tissue, Biochem. J., 226:563.

Sen, K. K., and Menon, K. M. J., 1979, Dissociation of cyclic AMP accumulation from that of luteinizing hormone (LH) release in response to gonadotropin releasing hormone (GnRH) and cholera enterotoxin, Biochem. Biophys. Res. Comm., 87:221.

Snyder, G. D., and Bleasdale, J. E., 1982, Effect of LHRH on incorporation of [^{32}P]orthophosphate into phosphatidylinositol by dispersed anterior pituitary cells, Mol. Cell. Endocrinol., 28:55.

Towart, R., and Schramm, M., 1984, Recent advances in the pharmacology of the calcium channel, Trends Pharmacol. Sci., 5:111.

Triggle, D. J., 1984, Ca^{2+} mobilization: new sites and signals, Trends Pharmacol. Sci., 5:5.

Turgeon, J. L., Ashcroft, S. J. H., Waring, D. W., Milewski, M. A., Walsh, D. A., 1984, Characteristics of adenohypophyseal Ca^{2+}-phospholipid-dependent protein kinase, Mol. Cell. Endocrinol., 34:107.

Ueda, T., Chueh, S. H., Noel, M. W., and Gill, D. L., 1986, Influence of inositol 1,4,5-trisphosphate and guanine nucleotides on intracellular calcium release within the N1E-115 neuronal cell line, J. Biol. Chem., 261:3184.

Wolf, B. A., Turk, J., Sherman, W. R., and McDaniel, M. L., 1986, Intracellular Ca^{2+} mobilization by arachidonic acid comparison with myo-inositol 1,4,5-trisphosphate in isolated pancreatic islets, J. Biol. Chem., 261:3501.

PHOSPHOINOSITIDE TURNOVER, Ca^{2+} MOBILIZATION, AND PROTEIN KINASE C ACTIVATION IN GnRH ACTION ON PITUITARY GONADOTROPIN RELEASE

Zvi Naor

Department of Hormone Research
The Weizmann Institute of Science
Rehovot, 76100, Israel

Gonadotropin releasing hormone (GnRH) stimulates the biosynthesis and release of luteinizing hormone (LH) and follicle stimulating hormone (FSH) from pituitary gonadotrophs. A signal-transduction mechanism transfers the biological information from the activated receptor to the cell machinery culminating in exocytosis and gene expression. Since the major cyclic nucleotides (cAMP and cGMP) and the prostaglandins were ruled out as potential second messengers for the acute actions of GnRH on gonadotropin secretion (for review see Naor, 1982; Catt et al., 1983), alternative signal-transduction mechanism was investigated.

PHOSPHOINOSITIDE TURNOVER

The importance of phosphatidylinositol (PI) turnover in signal transduction was first recognized by Hokin and Hokin (1953) and later proposed as a more general mechanism for Ca^{2+}-mobilizing hormones (Michell, 1975; Michell et al., 1981) (Fig. 1). It is now thought that the activated receptor transfers the information via a GTP-binding protein (Okajima et al., 1985), to a specific phospholipase C. This particular enzyme can be activated even at very low Ca^{2+} concentrations (Wilson et al., 1984). The activated enzyme hydrolyses phosphatidylinositol-4,5-bisphosphate (PIP_2) to 1,2-diacylglycerol (DG) and myoinositol-1,4,5-trisphosphate (IP_3). The two products are regarded as 'second messengers' since DG activates the Ca^{2+}/phospholipid-dependent protein kinase (C-kinase; Nishizuka, 1984), and IP_3 mobilizes intracellular Ca^{2+} (Berridge and Irvine, 1984). DG is then phosphorylated to phosphatidic acid, and IP_3 is converted to inositol. The activated DG and inositol will produce PI, and phosphorylation of PI will result in the formation of phosphatidylinositol-4-phosphate (PIP) and PIP_2. It was recently suggested that while IP_3 is formed from PIP_2, most of the DG is produced from hydrolysis of PI (Wilson et al., 1985). Rapid changes in membrane phosphoinositides by the activated hormone-receptor complex will therefore furnish the second messengers needed for signal transduction for Ca^{2+} mobilizing hormones.

We have recently found that GnRH induced a rapid phosphodiester hydrolysis of polyphosphinositides in pituitary gonadotrophs (Naor et al., 1986a). Concomitant with the rapid hydrolysis of the polyphosphoinositides, GnRH stimulated the appearance of [^{3}H]myoinositol-1,4,5-trisphosphate (IP_3, 10 s), [^{3}H]myoinositol-1,4-bisphosphate (IP_2, 15 s), and [^{3}H]myoinositol-1-phosphate (IP_1, 1 min) (Fig. 2). The stimulatory effect of GnRH on

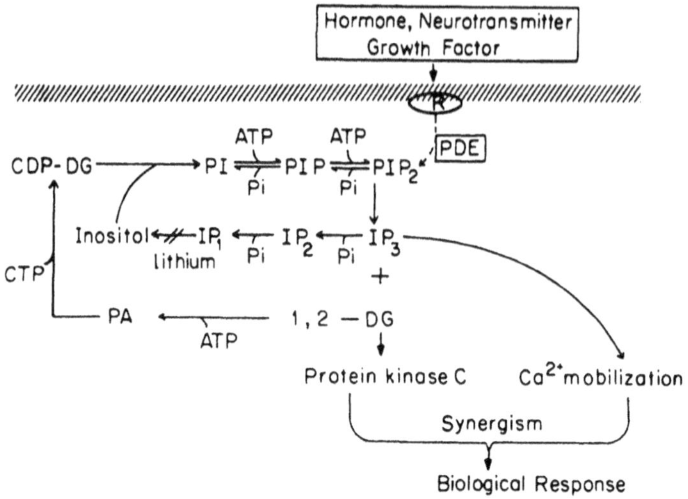

Fig. 1. Phosphoinositide turnover. Receptor (R) activation results in breakdown of phosphatidylinositol-4,5-bisphosphate (PIP_2) to 1,2-diacylglycerol (DG) and inositol trisphosphate (IP_3) by a specific phospholipase C (PDE). Phosphorylation by ATP and activation by diacylglycerol (DG) kinase forms phosphatidic acid (PA). Conjugation with CTP by phosphatidic acid: CTP cytidyltransferase forms CDP-diacylglycerol (CDP-DG). An exchange of the activated CDP with free inositol by CDP-diacylglycerol inositol phosphatidyltransferase forms PI. Multiple phosphorylation produces phosphatidylinositol-4-phosphate (PIP) and PIP_2. A series of inositol phosphatases converts IP_3 to free inositol which interacts with CDP-diacylglycerol to resynthesize PI. (Modified from Nishizuka, 1984; Berridge and Irvine, 1984; Wilson et al., 1985).

inositol phosphates (Ins-P) production was dose-related, and was blocked by a potent GnRH-antagonist. Since our working hypothesis claims that the enhanced phosphoinositide turnover induced by GnRH is responsible for Ca^{2+} mobilization, it was important to demonstrate that activation of phospholipase C is not secondary to Ca^{2+} mobilization. Indeed, elevation of cytosolic free Ca^{2+} levels ($[Ca^{2+}]_i$), by ionomycin and A23187 from intracellular or extracellular Ca^{2+} pools, respectively, had no significant effect on $[^3H]$-Ins-P production. Also, GnRH-induced $[^3H]$Ins-P production was not dependent on extracellular Ca^{2+}, and was noticed also after extracellular or intracellular Ca^{2+} mobilization by A23187 or ionomycin, respectively. We concluded that GnRH stimulates a rapid phosphodiester hydrolysis of polyphosphoinositides as also reported recently by others (Catt et al., 1984; Schrey, 1985; Kiesel et al., 1986). The stimulatory effect is not dependent on elevation of $[Ca^{2+}]_i$. Interestingly, the stimulatory effect of GnRH on $[^3H]$-Ins-P accumulation was not affected significantly by prior treatment of the cells with islet activating protein (IAP), pertussis toxin (Naor et al., 1986a). IAP is known to ADP-ribosylate the α-subunit of N_i and hence to inactivate the GTP-binding protein (Gilman, 1984). IAP-sensitive substrates were reported to communicate between Ca^{2+}-mobilizing receptors and phospholipase C (Okajima et al., 1985; Smith et al., 1986). We concluded that pituitary cells might utilize a different GTP-binding protein for signal-transduction mechanisms involved in GnRH (Naor et al., 1986a) and TRH (Martin et al., 1986) actions.

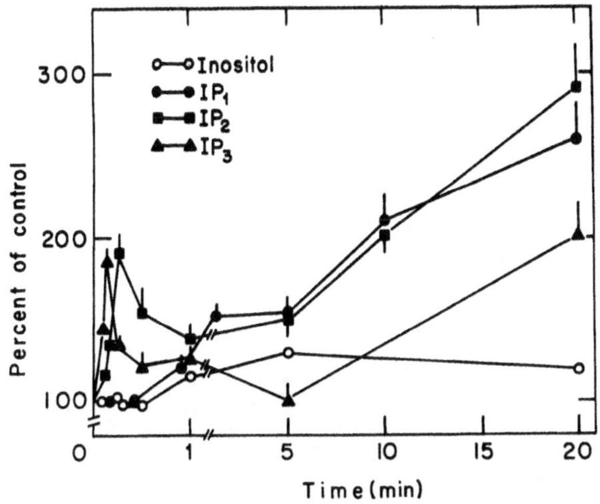

Fig. 2. Inositol phosphates production in GnRH-stimulated cells. Cultured rat pituitary cells were prelabeled for 3 days with myo-[2-^{3}H]-inositol and later stimulated with GnRH (100 nM) in the presence of Li^+ (10 mM) (from Naor et al., 1986a).

Ca^{2+} MOBILIZATION

The recent availability of enriched-fractions of pituitary gonaodtrophs, separated by centrifugal elutrition (Hyde et al., 1982), and the Quin-2 probe (Tsien et al., 1982), enabled us to measure changes in $[Ca^{2+}]_i$ in GnRH-responsive cells. GnRH induced a transient rise in $[Ca^{2+}]_i$, reaching maximal levels within 6-8 sec which was followed by a prolonged (∿15 min) plateau of elevated $[Ca^{2+}]_i$ (Naor et al., 1986b). GnRH elevates $[Ca^{2+}]_i$ only in gonadotroph-enriched cells, while thyrotropin-releasing hormone (TRH) and growth hormone-releasing factor (GHRF) elevate $[Ca^{2+}]_i$ in mammotrophs- and somatotrophs-enriched cells, respectively. The rapid first phase in $[Ca^{2+}]_i$ induced by GnRH was also observed in Ca^{2+}-free medium containing EGTA but this was terminated with 2 min. Readdition of Ca^{2+} to the medium induced a second slower rise in $[Ca^{2+}]_i$ ('plateau phase'). The rise in $[Ca^{2+}]_i$ in GnRH-stimulated gonadotrophs originates partly from intracellular pools and partly through influx across the cell membrane. The recent finding that GnRH stimulates the release of IP_3 (Catt et al., 1984; Schrey, 1985; Kiesel et al., 1986; Naor et al., 1986a) supports our finding that GnRH mobilizes Ca^{2+} also from intracellular stores. We have previously demonstrated the biphasic nature of GnRH-stimulated gonadotropin secretion in perifused cell-columns (Naor et al., 1982). The first phase of LH release is evident immediately upon GnRH stimulation and is followed by a second phase that is slower but more sustained. The kinetics of GnRH-induced elevation of $[Ca^{2+}]_i$ described here are in good agreement with the LH release profile. The two phases of GnRH-induced Ca^{2+} mobilization might therefore be involved in the biphasic release process of the gonadotropins.

PROTEIN KINASE C ACTIVATION

A Ca^{2+}-activated, phospholipid-dependent protein kinase (C-kinase) was discovered (for review see Nishizuka, 1984). C-kinase is activated by association with membrane phospholipids, in particular phosphatidylserine (PS). The enzyme is activated by Ca^{2+}, PS and diacylglycerol (DG). During enhanced

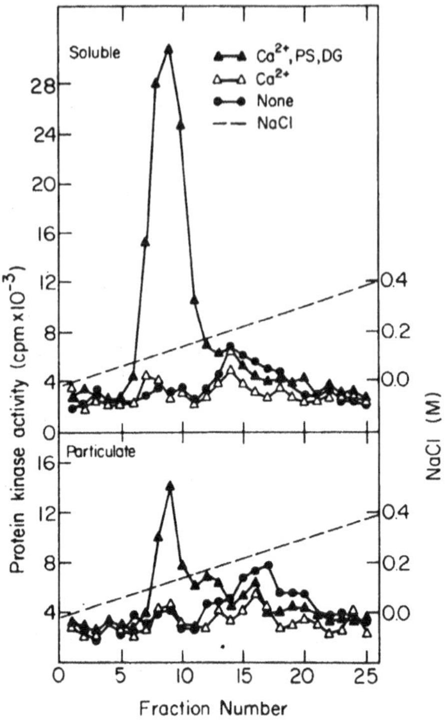

Fig. 3. DEAE-cellulose (DE-52) chromatography of pituitary protein kinase C.

PI turnover, DG is liberated and it activates C-kinase even in the presence of resting levels of Ca^{2+} providing PS is present in sufficient concentrations (Nishizuka, 1984). Brain C-kinase consists of a single polypeptide chain (80 kDa) composed of two domains. One is hydrophobic, most likely carrying the binding sites for Ca^{2+}, DG and PS. The second is hydrophilic (~50 kDa), known also as protein kinase M, and is the catalytic active fragment. C-kinase plays a role in signal transduction for hormones, neurotransmitters and growth-factors since it is linked to PI turnover and phosphorylates key proteins and enzymes. Since the tumor promoters phorbol esters directly activate C-kinase, it was suggested that the enzyme is also involved in proliferation and transformation (Nishizuka, 1984).

We, and others, have recently reported the presence of C-kinase in the pituitary (Turgeon et al., 1984; Hirota et al., 1985; Naor et al., 1985a). Pituitary C-kinase is mostly soluble (70%) and partly particulate (30%; Fig. 3). However, while the soluble form is recovered in an inactive state, the particulate form is found in a cofactor-insensitive state (Naor et al., 1985a). Therefore, the soluble form of the enzyme can also be detected in a crude cytosolic preparation, while the particulate enzyme is detectable only after solubilization and anion exchange chromatography (Hermon et al., 1986). Pituitary C-kinase is recruited from the cytosol to the membrane upon GnRH or TRH challenge (Hirota et al., 1985; Naor et al., 1985a; Drust and Martin, 1985; Fearon and Tashjian, 1985). Translocation of C-kinase to the membrane by GnRH or TRH following increased PI turnover will expose the inactive enzyme to DG and PS present in the membrane and permit activation even in the fact of resting levels of cellular concentrations of Ca^{2+}.

ARACHIDONIC ACID AND ITS METABOLITES

[^{3}H]Arachidonic acid (AA) release from prelabeled pituitary and granulosa cells was enhanced by treatment with GnRH or with a Ca^{2+} ionophore (Naor and Catt, 1981; Naor et al., 1985b; Minegishi and Leung, 1985). Diacylglycerol (DG), which is liberated during enhanced phosphoinositide turnover might be the source of the AA released by the neurohormone. Alternatively, phosphatidic acid or the major phospholipids might serve as the substrate for phospholipase A_2. The effect of GnRH on [^{3}H]AA release is Ca^{2+}-dependent, is mimicked by TPA and is not associated with prostaglandin production (Naor and Catt, 1981; Ojeda et al., 1979; Naor et al., 1985b). Interestingly, AA and 5-hydroxy-6,8,11,14-eicosa-tetraenoic acid (5-HETE) but neither 11-, 12-, nor 15-HETE stimulated LH release with ED_{50} values of 10-15 μM (Naor et al., 1985b). Snyder et al. (1983) have suggested that epoxygenase products of AA are involved in GnRH action since they could mimic the stimulatory effect of GnRH on LH release. Hulting et al. (1985) have recently demonstrated that leukotriene C_4 (LTC_4), but not LTB_4, stimulates LH release from cultured pituitary cells. Effective concentrations of LTC_4 were very low (10^{-14}-10^{-12} M) and the effect was observed mainly in the first 30 min of incubation. In the past we found only minor stimulation of LH release by LTC_4 (Naor et al., 1985b); however, with another preparation of LTC_4 (kindly provided by Dr. J. Rokach, Merck Frosst, Canada), we found stimulation of LH release (2-fold); albeit with ED_{50} of 10^{-10} M compared to ~10^{-14} M reported by Hulting et al. (1985). Various potencies of the different preparations of LTC_4 currently available might explain the discrepancies in the results.

We have previously shown that cyclooxygenase inhibitors do not interfere with GnRH-induced gonadotropin release, in agreement with the finding that PGE has no stimulatory effect on LH release (Ojeda et al., 1979). On the other hand, the lipoxygenase inhibitors nordihydroguaiaretic acid (NDGA), 5,8,11,14-eicosatetraynoic acid (ETYA) and 3-amino-1-[M-(trifluoromethyl)-phenyl]-2-pyrazoline (BW 755C) reduced GnRH-induced gonadotropin secretion (Naor et al., 1983; 1985b). At low concentrations, indomethacin potentiated GnRH action, perhaps by increasing the availability of AA for metabolism via the lipoxygenase pathway. The above results provide preliminary evidence for an intermediary role of lipoxygenase metabolites of arachidonic acid in the mechanism of action of GnRH on pituitary gonadotropin release. The implication of these findings is that lipoxygenase products, which appear to function as humoral mediators of the immune system, might also be involved in the regulation of endocrine systems. Indeed, the presence of lipoxygenase activity in pituitary gonadotrophs was demonstrated (Vanderhoek et al., 1984) providing further support to the proposal that AA metabolites are involved in GnRH action (Naor and Catt, 1981; Naor et al., 1983; 1985b; Snyder et al., 1983; Hulting et al., 1985; Minegishi and Leung, 1985).

INTEGRATION OF THE SIGNAL-TRANSDUCTION CASCADE

The earliest effect of the neurohormone GnRH is the stimulation of phosphoinositide turnover in a Ca^{2+}-independent mechanism (Catt et al., 1984; Schrey, 1985; Kiesel et al., 1986; Naor et al., 1986a). A GTP-binding protein most likely transfers the message from the activated receptor to phospholipase C, but unlike other systems (Okajima et al., 1985; Smith et al., 1986) it is not an IAP-sensitive substrate such as N_i (Naor et al., 1986a; Martin et al., 1986). The rapid stimulation of polyphosphoinositide hydrolysis will furnish the intracellular second messengers IP_3 and DG which are involved in signal transduction (Nishizuka, 1984; Berridge and Irvine, 1984).

Reconstitution of the full physiological response to GnRH in terms of LH release from cultured pituitary cells was achieved only when the two branches of the inositol-lipid cycle were activated (Naor and Eli, 1985; Chang et al.,

1986). The two second messengers (Ca^{2+} and C-kinase) might mediate the two phases of the LH release process. The elevated cytosolic free Ca^{2+} might activate phospholipase A_2 to generate AA and its active metabolites which might be needed for part of the exocytotic response via the lipoxygenase (Naor et al., 1983; 1985b) or the epoxygenase pathway (Snyder et al., 1983). In parallel, DG will activate C-kinase leading to phosphorylation of yet unidentified proteins which might be involved in a different phase of the exocytotic reaction mechanism. This is supported by recent findings that a combination of Ca^{2+} ionophore or AA and the C-kinase activator phorbol ester (TPA) could mimic the stimulatory effect of GnRH on LH release (Naor and Eli, 1985; Chang et al., 1986). Furthermore, lipoxygenase inhibitors blocked AA- but not TPA-stimulated LH release, and only a mixture of lipoxygenase and C-kinase inhibitors blocked completely the effect of GnRH on LH release (Chang et al., 1986). The results suggest that Ca^{2+} and C-kinase act in parallel in pituitary gonadotrophs rather than merge at a step proximal to the lipoxygenase pathway. This conclusion differs from that suggested by Halenda et al. (1985) and Volpi et al. (1985) for platelets and neutrophils, respectively, where they found that phorbol ester potentiates the stimulatory effect of Ca^{2+} ionophore on AA release and metabolite formation. Further studies are needed to clarify the synergistic interaction of the two branches of the phosphoinositide cycle. GnRH-induced LH release is therefore mediated by coordinated production of two second messengers Ca^{2+} and DG, which are derived from enhanced phosphoinositide turnover. The biochemical link to AA release and metabolism, and the coordinated mediation of the release phases of the gonadotropins should be further investigated.

REFERENCES

Berridge, M. J., and Irvine, R. F., 1984, Inositol trisphosphate, a novel second messenger in cellular signal transduction, Nature, 312:315.

Catt, K. J., Loumaye, E., Katikineni, M., Hyde, C. L., Childs, G., Amsterdam, A., and Naor, Z., 1983, Receptors and actions of gonadotropin-releasing hormone (GnRH) on pituitary gonadotrophs, in: "Role of Peptides and Proteins in Control of Reproduction," S. M. McCann, D. S. Dhindsa, eds., p. 33, Elsevier Press.

Catt, K. J., Loumaye, E., Wynn, P., Suarez-Quian, C., Kiesel, L., Iwashita, M., Hirota, K., Morgan, R., and Chang, J., 1984, Receptor mediated activation mechanisms in the hypothalamic control of pituitary-gonadal function, in: "Endocrinology", F. Labrie, L. Proulx, eds., Excerpta Medica, Amsterdam.

Chang, J. P., Graeter, J., and Catt, K. J., 1986, Coordinate actions of arachidonic acid and protein kinase C in gonadotropin releasing hormone stimulated secretion of luteinizing hormone, Biochem. Biophys. Res. Comm., 134:134.

Drust, D. S., and Martin, T. F. J., 1985, Protein kinase C translocates from cytosol to membrane upon hormone activation: effects of thyrotropin-releasing hormone in GH_3 cells, Biochem. Biophys. Res. Comm., 128:531.

Fearon, C. W., and Tashjian, A. H. Jr., 1985, Thyrotropin-releasing hormone induces redistribution of protein kinase C in GH_4C_1 rat pituitary cells, J. Biol. Chem., 260:8366.

Gilman, A. G., 1984, G protein and dual control of adenylate cyclase, Cell 36:577.

Halenda, S. P., Zavoico, G. B., and Feinstein, M. B., 1985, Phorbol esters and oleoyl acetoyl glycerol enhance release of arachidonic acid in platelets stimulated by Ca^{2+} ionophore A23187, J. Biol. Chem., 260:12484.

Hermon, J., Reiss, N., and Naor, Z., 1986, Phospholipid-dependent Ca^{2+} activated protein kinase (C-kinase) in the pituitary: further characterization and endogenous redistribution, Mol. Cell. Endocrinol., in press.

Hirota, K., Hirota, T., Aguilera, G., and Catt, K. J., 1985, Hormone-induced redistribution of calcium-activated phospholipid-dependent protein kinase in pituitary gonadotrophs, J. Biol. Chem., 260:3243.

Hokin, M. R., and Hokin, L. E., 1953, Enzyme secretion and the incorporation of [^{32}P] into phospholipids of pancreas slices, J. Biol. Chem., 203:976.

Hulting, A. L., Lindgren, J. A., Hökfelt, T., Eneroth, P., Werner, S., Patrono, C., and Samuelsson, B., 1985, Leukotriene C_4 as a mediator of luteinizing hormone release from rat anterior pituitary cells, Proc. Natl. Acad. Sci. USA, 82:3834.

Hyde, C. L., Childs (Moriarty), G., Wahl, L. M., Naor, Z., and Catt, K. J., 1982, Preparation of gonadotroph enriched cell population from adult rat anterior pituitary cells by centrifugal elutriation, Endocrinology, 111:1421.

Kiesel, L., Bertges, K., Rabe, T., and Runnebaum, B., 1986, Gonadotropin releasing hormone enhances polyphosphoinositide hydrolysis in rat pituitary cells, Biochem. Biophys. Res. Comm., 134:861.

Martin, T. F. J., Lucas, D. O., Bajjalieh, S. M., and Kowalchyk, J. A., 1986, Thyrotropin releasing hormone activates a Ca^{2+}-dependent polyphosphoinositide phosphodiesterase in permeable GH_3 cells. J. Biol. Chem., 261:2918.

Michell, R. H., 1975, Inositol phospholipids and cell surface receptor function, Biochem. Biophys. Acta, 415:81.

Michell, R. H., Kirk, C. J., Jones, L. M., Downes, C. P., and Creba, J. A., 1981, The stimulation of inositol lipid metabolism that accompanies calcium mobilization in stimulated cells: defined characteristics and unanswered questions, Phil. Trans. R. Soc. Lond. B., 296:123.

Minegishi, T., and Leung, P. C. K., 1985, Luteinizing hormone releasing hormone stimulates arachidonic acid release in rat granulosa cells, Endocrinology, 117:2001.

Naor, Z., 1982, Cyclic nucleotide production and hormonal control of anterior pituitary cells, in: "Multihormonal Regulation in Neuroendocrine Cells," A. Tixier-Vidal, P. Richards, eds., Inserm 110:395.

Naor, Z., and Catt, K. J., 1981, Mechanism of action of gonadotropin-releasing hormone: involvement of phospholipid turnover in luteinizing hormone release, J. Biol. Chem., 256:2226,

Naor, Z., and Eli, Y., 1985, Synergistic stimulation of luteinizing hormone release by protein kinase C activators and Ca^{2+}-ionophore, Biochem. Biophys. Res. Comm., 130:848.

Naor, Z., Katikineni, M., Loumaye, E., Garcia Vella, A., Dufau, M. L., and Catt, K. J., 1982, Compartmentalization of luteinizing hormone pools: dynamics of gonadotropin releasing hormone action in superfused pituitary cells, Mol. Cell. Endocrinol., 27:213.

Naor, Z., Vanderhoek, J. Y., Londner, H. R., and Catt, K. J., 1983, Arachidonic acid products as possible mediators of the action of gonadotropin releasing hormone, in: "Advances in Prostaglandins Thromboxane and Leukotriene Research," B. Samuelsson, R. Paoletti, P. Ramwell, eds., Vol. 12, p. 259, Raven Press, New York.

Naor, Z., Zer, J., Zakut, H., and Hermon, J., 1985a, Characterization of pituitary calcium-activated phospholipid-dependent protein-kinase: redistribution by gonadotropin releasing hormone, Proc. Natl. Acad. Sci. USA, 82:8203.

Naor, Z., Kiesel, L., Vanderhoek, J. Y., and Catt, K. J., 1985b, Mechanism of action of gonadotropin releasing hormone: role of lipoxygenase products of arachidonic acid in luteinizing hormone release, J. Steroid Biochem., 23:711.

Naor, Z., Azrad, A., Limor, R., Zakut, H., and Lotan, M., 1986a, Gonadotropin releasing hormone activates a rapid Ca^{2+}-independent phosphodiester hydrolysis of polyphosphoinositides in pituitary gonadotrophs, J. Biol. Chem., in press.

Naor, Z., Limor, R., and Hermon, J., 1986b, Phospholipid turnover, Ca^{2+} mobilization and protein kinase C activation in GnRH action on pituitary gonadotrophs, in: "Neuroendocrine Molecular Biology," G. Fink, A. J. Harmar, K. W. McKerns, eds., p. 113, Plenum Press, New York.
Nishizuka, Y., 1984, The role of protein kinase C in cell surface signal transduction and tumor promotion, Nature, 308:693.
Ojeda, S. R., Naor, Z., and Negro-Vilar, A., 1979, The role of prostaglandins in the control of gonadotropin and prolactin release, Prostag. Leukotriene Med., 5:249.
Okajima, F., Katada, T., and Ui, M., 1985, Coupling of the guanine nucleotide regulatory protein to chemotactic peptide receptors in neutrophil membranes and its uncoupling by islet-activating protein, pertussis toxin, J. Biol. Chem., 260:6761.
Schrey, M. P., 1985, Gonadotropin-releasing hormone stimulates the formation of inositol phosphates in rat anterior pituitary tissue, Biochem. J., 226:563.
Smith, C. D., Cox, C. C., and Snyderman, R., 1986, Receptor-coupled activation of phosphoinositide specific phospholipase C by an N protein, Science, 232:97.
Snyder, G. D., Capdevila, J., Chacos, N., Manna, S., and Falck, J. R., 1983, Action of luteinizing hormone releasing hormone: involvement of novel arachidonic acid metabolites, Proc. Natl. Acad. Sci. USA, 80:3504.
Tsien, R. Y,, Pozzan, T., and Rink, T. J., 1982, Calcium homeostasis in intact lymphocytes: cytoplasmic free calcium monitored with a new intracellularly trapped fluorescent indicator, J. Cell Biol., 94:325.
Turgeon, J. L., Ashcroft, S. J. H., Waring, D. W., Milewski, M. A., and Walsh, D. A., 1984, Characteristics of adenohypophyseal Ca^{2+}-phospholipid dependent protein kinase, Mol. Cell. Endocrinol., 34:107.
Vanderhoek, J. Y., Kiesel, L., Naor, Z., Bailey, J. M., and Catt, K. J., 1984, Arachidonic acid metabolism in gonadotroph enriched pituitary cells, Prostag. Leukotriene Med., 15:375.
Volpi, M., Molski, T. F. P., Naccache, P. H., Feinstein, M. B., and Shaafi, R. I., 1985, Phorbol 12-myristate, 13-acetate potentiates the action of the calcium ionophore in stimulating arachidonic acid release and production of phosphatidic acid in rabbit neutrophils, Biochem. Biophys. Res. Comm., 128:594.
Wilson, D. B., Bross, T. E., Hofman, S. L., and Majerus, P. W., 1984, Hydrolysis of polyphosphoinositides by purified sheep seminal vesicles phospholipase C enzymes, J. Biol. Chem., 259:11718.
Wilson, D. B., Neufeld, E. J., and Majerus, P. W., 1985, Phosphoinositide interconversion in thrombin stimulated human platelets, J. Biol. Chem., 260:1046.

THE OVARIAN GRANULOSA CELL AS A FOLLICLE-STIMULATING HORMONE TARGET TISSUE

Bruce Kessel, Xiao-Chi Jia, J. Benjamin Davoren, and Aaron J. W. Hsueh

Department of Reproductive Medicine
School of Medicine, M-025
University of California, San Diego
La Jolla, California, U.S.A. 92093

INTRODUCTION

The ovarian follicle is the basic functional unit of the ovary, and is comprised of an outer layer of theca cells, separated by a basement membrane from the inner layer of granulosa cells, which in turn surround the oocyte-cumulus cell complex. In response to the pituitary gonadotropins, follicle-stimulating hormone (FSH), luteinizing hormone (LH), and prolactin (PRL), the follicle is able to differentiate with regard to steroidogenesis and hormone receptor induction, as well as to produce a fertilizable ovum and adequate corpus luteum. The key anterior pituitary glycoprotein in the initiation of follicular maturation and granulosa cell differentiation is FSH. During reproductive life, FSH is released in an orchestrated cyclic fashion subject to endocrine control.

Although exogenous FSH treatment induces the maturation of multiple follicles, only a limited number of selected follicle(s) become dominant follicles and ovulate during each normal reproductive cycle, with the remainder undergoing atresia. Estrogen production by the selected follicle reduces FSH levels, thereby impeding the development of other follicles. Thus, the ability of a selected follicle to continue to develop, when neighboring follicles exposed to the same concentrations of FSH do not, is probably due to the actions of intra-ovarian hormones through paracrine mechanisms.

Whereas the ovarian effect of FSH is locally modulated by paracrine factors, the interaction of FSH with the ovary is also dependent upon the amount of FSH released from the anterior pituitary, and potentially also by the potency of the FSH molecule. Although the molecular heterogeneity of FSH is well known, the role of this FSH pleomorphism in physiologic situations is less clear due in part to the lack of an appropriate bioassay sensitive enough to detect circulating bioactive FSH.

We have studied the effect of steroids (estrogens, androgens and progestins) and selected growth factors (insulin and the insulin-like growth factors) in the regulation of FSH action upon cultured rat granulosa cells. This chapter examines these intragonadal modulators that augment FSH action, using key endpoints of FSH action - induction of aromatase activity and acquisition of LH receptors. Furthermore, we show how the studies of hormones involved in enhancing FSH action led to the development of a specific and sensitive in vitro bioassay for FSH. The newly developed Granulosa cell Aromatase Bioassay (GAB) allows the measurement of bioactive FSH levels in various physiologic and pathologic conditions.

FSH ACTION IN GRANULOSA CELL DIFFERENTIATION - IN VITRO STUDIES

The immature granulosa cells is endowed with FSH receptors that appear at the primary follicle stage (Richards and Midgley, 1976). This specific FSH binding is restricted to the granulosa cell component of the follicular complex, as demonstrated by autoradiography and in vivo uptake of radioiodinated hormones (Midgley, 1973; Richards et al., 1976). Through the actions of FSH a number of functional parameters are stimulated in cultured granulosa cells, including activation of the adenyl cyclase system, secretion of steroids and nonsteroidal cell products, induction of plasma membrane hormone receptors, and increased general cell function (Hsueh et al., 1984). Dorrington et al. (1975) first demonstrated the ability of FSH to induce aromatases in serum-free rat granulosa cell cultures. This effect is FSH dose dependent and occurs at physiologic levels with a lag period of more than 1 day. Furthermore, FSH induction of aromatases occurs in the relatively undifferentiated granulosa cells obtained from hypophysectomized, estrogen-primed rats, in which hCG and LH do not affect aromatase activity (Hsueh et al., 1983). FSH, at physiologic levels, also stimulates granulosa cell progestin biosynthesis by modulating steroidogenic enzyme activity.

The induction of plasma membrane receptors is also an important part of granulosa cell differentiation regulated by FSH. Granulosa cells are endowed with FSH receptors and FSH appears to have an autoregulatory role increasing its own receptors. In contrast, granulosa cells from preantral follicles show negligible binding of LH, with LH receptors found only in large preovulatory follicles. A number of studies have now demonstrated the ability of FSH to increase LH receptor numbers in granulosa cells in vivo and in vitro (Erickson et al., 1979; Zeleznik et al., 1976; Stouffer et al., 1976). Similarly, FSH treatment increases prolactin receptors in cultured granulosa cells (Wang et al., 1979).

It is generally believed that these FSH actions are mediated by cAMP as the second messenger. FSH stimulates adenyl cyclase activity, and increases cAMP production by cultured granulosa cells (Goff and Armstrong, 1977). Inhibition of cAMP metabolism by the addition of phosphodiesterase inhibitors also enhances FSH action. Furthermore, cAMP analogues such as $(Bu)_2cAMP$ and 8-Br-cAMP as well as cAMP-inducing agents such as cholera toxin and forskolin mimic FSH effects (Channing, 1970; Wang et al., 1982). These compounds stimulate aromatase activity, progestin biosynthesis, and will induce LH receptor formation. The FSH stimulation of cAMP production activates type II cAMP-dependent protein kinase in rat granulosa cells (Richards et al., 1983).

OVARIAN STEROIDS AS ENHANCERS OF FSH ACTION IN GRANULOSA CELLS

The ovary produces multiple steroids which may act locally as well as peripherally. Systemic and follicular fluid estrogen levels strongly correlate with the development of a healthy dominant follicle. Preovulatory increases in serum estrogen levels are involved in the triggering of the LH surge, and follicular fluid estrogen levels as well as estrogen/androgen ratios can be used to define a follicle as healthy or atretic. Following ovulation, luteal progestin production is necessary to prepare the uterus for implantation of the fertilized ovum.

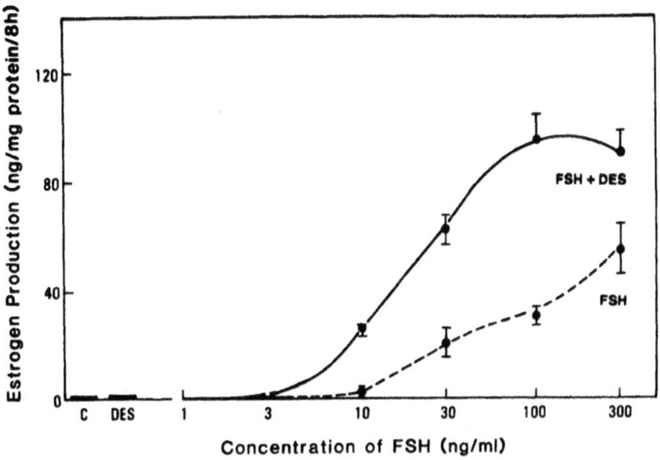

Fig. 1. Estrogen enhancement of FSH-stimulated aromatase activity in cultured granulosa cells. Rat granulosa cells were cultured for 3 days in the absence or presence of increasing concentration of FSH with or without DES. At the end of the 3 day culture period, cells were washed and further cultured in fresh media containing androstenedione for 8 h, and the media collected for measurement of estrogen content. (C, control; DES, diethylstilbestrol). Adapted with permission from Hsueh et al., 1984.

Role of Estrogens

Estrogen receptors are present in granulosa cells, and estrogen treatment stimulates granulosa cell proliferation. Furthermore, in vivo studies showed that estrogens enhance a number of FSH actions at the follicular level, including FSH-induced antrum formation and ovarian weight gain (Richards and Midgley, 1976). The actions of estrogens on two key functions of FSH in the granulosa cell, the induction of aromatase activity and LH receptor formation, were therefore examined in vitro.

For the examination of estrogen modulation of FSH-induced aromatase activity, granulosa cells were cultured for 3 days in serum-free media in the presence of increasing concentrations of FSH (1-300 ng/ml) in the presence or absence of the synthetic estrogen, diethylstilbestrol (DES) (Fig. 1). At the end of the 3 day culture period, cells were washed and further cultured in media containing 10^{-7}M androstenedione for 8 h, and estrogen accumulation measured. FSH stimulated aromatase activity in a dose dependent manner, and the concomitant treatment with DES (10^{-7} M) resulted in an increase in estrogen production, and a decrease in the half maximal effective dose of FSH from 55 to 20 ng/ml. In other studies, cells were cultured with low doses of FSH in the presence of increasing concentrations of estrogens. The minimal effective dose of estradiol-17β (3.7×10^{-10} M) capable of enhancing FSH-stimulated aromatase activity was well within reported follicular fluid concentrations (up to 10^{-7} M) (Adashi and Hsueh, 1982; Goff and Henderson, 1979). FSH is also known to stimulate progestin production in cultured rat granulosa cells in a dose-dependent fashion, and the concomitant addition of estrogens also enhanced FSH-stimulated progestin production (Welsh et al., 1983).

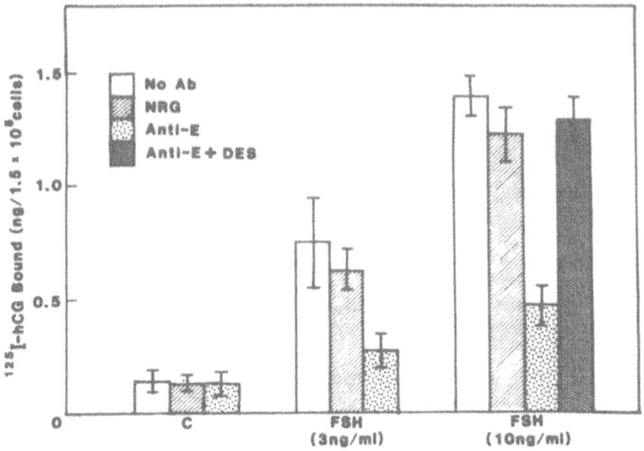

Fig. 2. Effect of treatment with estrogen antibodies on the FSH-induced LH/hCG receptor content. Granulosa cells were primed with FSH (10 ng/ml) for 3 days. At the end of this period, cells were washed and reincubated for 3 additional days in medium containing androstenedione (10^{-7} M) in the presence or absence of two different doses of FSH (3 and 10 ng/ml) with or without (No Ab) gamma globulin fractions obtained from estrogen antibodies (Anti-E; 1:100 dilution) or normal rabbit serum (NRG; 1:100 dilution). In some cells treated with the estrogen antiserum, 10^{-8} M diethylstilbestrol (DES) was also added. At the end of the incubation, cells were washed with buffer and the binding of ^{125}I-iodo-hCG to the granulosa cells were measured. Adapted from Kessel et al., 1985, with permission.

Induction of LH receptor formation is a key event in the differentiation of granulosa cells, allowing the dominant follicle to respond to the LH surge, ovulate, and produce an adequate corpus luteum. FSH has been shown to stimulate LH receptor formation in vivo and in vitro. We have recently shown that treatment with a number of synthetic and native estrogens enhance FSH-stimulated LH receptor formation, with an ED_{50} of approximately 3×10^{-9} M for estradiol-17β (Kessel et al., 1985). To further examine the local role of estrogens in LH receptor formation, specific estrogen antisera were used to block the action of endogenously produced estrogens. Granulosa cells were cultured with FSH for three days to induce aromatase activity. Cells were then washed and further cultured in media containing the aromatase substrate, androstenedione, in the presence of FSH, or FSH plus estrogen antisera (Fig. 2). FSH stimulated LH receptor formation; however, the presence of estrogen antisera partially blocked FSH induction of LH receptor formation. Furthermore, the addition of DES, which does not interact with the estrogen antisera, overcame the effect of the antisera.

Therefore, estrogens enhance both FSH-stimulated aromatase activity and LH receptor formation. In the context of the developing follicle during the follicular phase of the human menstrual cycle, the follicle achieving dominance is able to continue producing estrogens in the face of falling FSH levels due to the enhanced aromatase during the midfollicular phase of the cycle, and also to produce adequate LH receptors to respond to the upcoming midcycle gonadotropin surge.

Role of Androgens

Androgens, produced by theca and interstitial cells of the ovary, play a key role in follicular development. Androgens are available as aromatase substrates for granulosa cells, and also have specific stimulatory and inhibitory actions. These actions are presumed to be mediated through specific high affinity androgen receptors identified in granulosa cells (Schrieber et al., 1976). In the absence of gonadotropins, androgens stimulate follicular atresia and antagonize estrogen-induced ovarian weight gain in hypophysectomized immature rats. However, androgens also have obligate and stimulatory roles in follicular development. Granulosa cells, enriched in 17β-hydroxysteroid dehydrogenase, are lacking in 17α-hydroxylase and 17-20 desmolase. Therefore, granulosa cells cannot produce appreciable androgens *de novo*, and are dependent upon theca cell androgen production as a substrate for aromatase activity and estrogen production. Furthermore, although androgen treatment in the absence of gonadotropins induces atresia, androgen treatment in the presence of FSH augments steroidogenesis. *In vitro* studies demonstrate that androgens augment FSH-stimulated aromatase activity in cultured rat granulosa cells, and the observation that nonaromatizable androgens also augment FSH action supports the concept that androgens act not only as aromatase substrates, but also as an enhancer of the aromatase system (Hillier and DeZwart, 1981). Androgens also stimulate progestin biosynthesis in cultured granulosa cells (Lucky et al., 1977).

ROLE OF OVARIAN PEPTIDES IN ENHANCING FSH ACTION

Various regulatory peptides have been shown to have specific effects on the steroidogenic capacity of cultured rat granulosa cells. Both GnRH and epidermal growth factor have been found to be inhibitors of gonadotropin-stimulated steroidogenesis (Hsueh and Jones, 1981; Hsueh et al., 1982). Vasoactive intestinal peptide has been found to stimulate both estrogen and progestin production in the absence of gonadotropins. However, insulin and the insulin-like growth factors were found to enhance FSH action, and will therefore be discussed in more detail.

Role of Insulin and Insulin-like Growth Factors

A number of studies have implied that insulin or insulin-like growth factors may play a direct or indirect role in ovarian function. Experimentally induced insulin-deficient diabetes in female rats results in alterations of estrous cyclicity and ovarian function in female rats. Also, insulin has been used to supplement or replace serum in ovarian tissue culture experiments. To further dissect the possible direct ovarian role of insulin on granulosa cell differentiation, the *in vitro* effects of insulin on FSH-stimulated steroidogenesis were studied. Insulin at high doses was found to enhance FSH-stimulated progestin production and aromatase activity, with a two-fold increase in granulosa cell sensitivity to FSH in estrogen production (Davoren and Hsueh, 1984).

As the doses of insulin required to enhance FSH activity are above physiologic circulating levels, the possibility that the insulin action was mediated through other pathways than the insulin receptor was explored. The insulin-like growth factor I (IGF-I) shares 47% homology with insulin, and is believed to be the mediator of growth hormone (GH) action. Furthermore, growth hormone deficiency in rats is associated with a decrease of ovarian steroidal responsiveness to gonadotropins, and *in vitro* studies have shown that growth hormone treatment enhances FSH stimulated LH receptor formation (Jia et al., 1986). As these effects might be explained by changes in growth hormone-dependent changes in IGF-I levels, the possible role of IGF-I action on granulosa cell differentiation was examined (Fig. 3). Granulosa cells were cultured for 48 h with a physiologic dose of FSH in the presence or

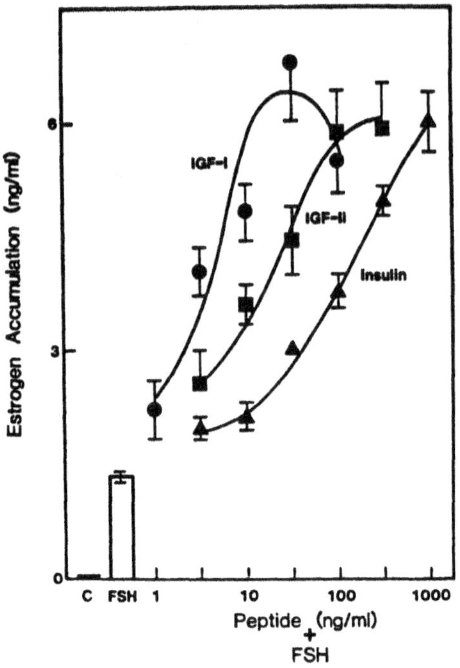

Fig. 3. Dose-dependent enhancement of FSH-stimulated estrogen production by insulin and insulin-like peptides. Granulosa cells were cultured for 2 days with FSH (3 ng/ml), androstenedione (10^{-7} M), and with or without increasing concentrations of IGF-I, IGF-II, or insulin. Medium estrogen content was determined by RIA.

absence of increasing concentrations of IGF-I, IGF-II, or insulin. FSH stimulated aromatase activity, and insulin or the insulin-like growth factors enhanced FSH-stimulated aromatase activity. The rank order of potency of these peptides was IGF-I > IGF-II > insulin.

To further assess the possible role in the ovary of IGF-I as a paracrine peptide and mediator of GH action, studies were done examining ovarian concentrations of immunoreactive IGF-I. Immature female rats were treated _in vivo_ with oGH for 0, 8, or 12 h, and ovarian tissue extracts were chromatographed, followed by IGF-I radioimmunoassay of the chromatography samples (Fig. 4) (Davoren and Hsueh, 1986). Immunoreactive IGF-I levels were increased 6.25 ± 1.9-fold over control levels after 8 h of GH treatment. These findings support the concept that there is GH-dependent local production of IGF-I. Because IGF-I can enhance FSH-stimulated granulosa cell differentiation, this peptide hormone may be a paracrine mediator of ovarian follicular development.

To more clearly understand the mechanism of IGF-I action at the ovarian level, studies were done to determine the existence of specific granulosa cell binding sites for insulin, IGF-I, and IGF-II. Granulosa cells were incubated with (^{125}I)-iodo-IGF-I with or without increasing concentrations of IGF-I, IGF-II, or insulin. Scatchard analysis of IGF-I binding data indicated that IGF-I bound to a single class of sites with a Kd of 1.36 ± 0.31 nM and the rank order competition potency of the peptides was IGF-I > IGF-II > insulin.

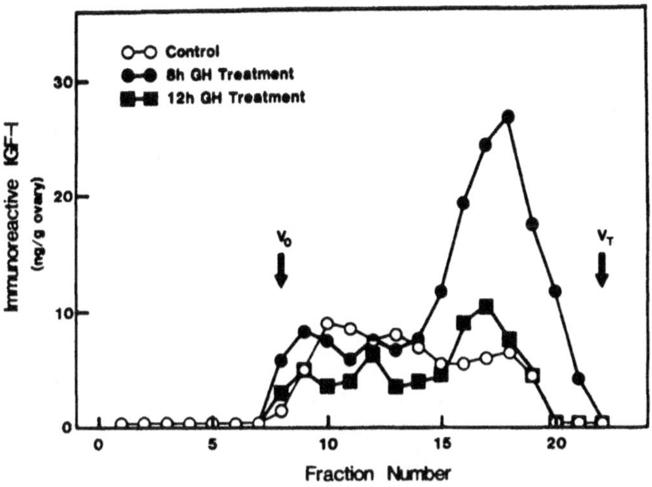

Fig. 4. Elution profile of ovarian Immunoreactive-IGF-I. Ovarian acid extracts were prepared and chromatographed on Sephadex G-50 columns. Adapted from Davoren and Hsueh, 1986, with permission.

Taken together, these studies indicate that IGF-I, as a locally produced mediator of GH action, may play a role as an intra-ovarian modulator of follicular development (Fig. 5). As an enhancer of FSH action, ovarian IGF-I may interact with specific granulosa cell receptors to enhance FSH-stimulated granulosa cell differentiation.

FSH MEASUREMENT BY RADIOIMMUNOASSAY AND BIOASSAY

The role of ovarian steroids and perhaps selected peptides in the paracrine control of follicular development is becoming clear. However, as intragonadal modulators, these factors are dependent upon the concentration of FSH and possibly the potency of the circulating FSH molecule. Whereas immunoreactive and bioactive LH levels have been well studied, the same is not true for FSH. Prior to the development of radioimmunoassays for FSH, large amounts of urine or plasma were necessary to detect the low concentrations of FSH. In 1966, sensitive radioimmunoassays for the quantitation of FSH were described, and since that time serum levels of immunoreactive FSH have been characterized in humans during puberty, the menstrual cycle, menopause, and many disease states (Styne and Grumbach, 1986; Ross et al., 1970; Crowley et al., 1985). At the same time, animal studies examining pituitary content of FSH were revealing that various molecular forms of FSH existed, and these forms could be altered by changing the hormonal milieu. Bogdanove reported that androgen treatment in orchidectomized rats resulted in a change in FSH molecular size (Bogdanove et al., 1974). Chappel has isolated six species of immunoreactive FSH based on differences in isoelectric point, and correlated the pI with sialic acid content of the glycoprotein (Chappel et al., 1983).

To further understand the physiologic importance of forms of FSH, it was also necessary to determine biological potency of FSH. The classical _in vivo_ FSH bioassay, the Steelman-Pohley method, was used for measuring pituitary and urinary FSH bioactivity; however, it was not sensitive enough for serum levels (Steelman and Pohley, 1953). Other _in vitro_ bioassay methods included employing the FSH stimulation of plasminogen activator activity by cultured granulosa cells, and the measurement of estrogen production in cultured Sertoli cells (Van Damme et al., 1979; Beers and Strickland, 1978). The plasminogen activator activity assay is hampered by a question of specificity as both FSH and LH control plasminogen activator activity, and may not be applicable to serum samples because serum contains plasminogen, plasminogen activators and various plasminogen activator inhibitors. The Sertoli cell bioassay was not sensitive enough for measurement of FSH in serum samples. Although these bioassays are selectively applicable to pituitary preparations there is clearly a need for a sensitive, specific FSH bioassay applicable to physiologic concentrations of FSH in serum samples.

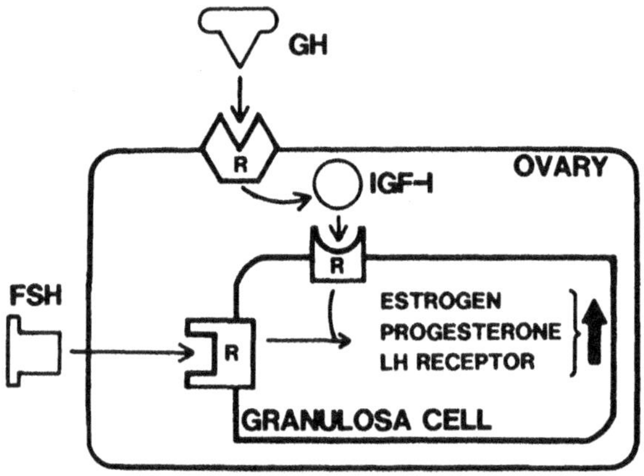

Fig. 5. Hypothetical scheme of the mechanism of growth hormone-enhanced granulosa cell differentiation.

DEVELOPMENT OF AN IN VITRO GRANULOSA CELL AROMATASE BIOASSAY (GAB) FOR FOLLICLE-STIMULATING HORMONE

As FSH is a potent stimulator of aromatase activity in granulosa cells, the possibility of using estrogen production as an endpoint for the measurement of biologically active FSH was examined. As discussed above, prior attempts at the development of a specific in vitro bioassay for FSH were often hampered by a lack of sensitivity, or could not be applied to serum samples due to inhibitory serum factors. Successful development of the GAB assay was therefore dependent upon increasing granulosa cell sensitivity to FSH, as well as pretreating serum samples to remove inhibitory substances. The many studies from this laboratory examining the possible paracrine modulators of FSH action in granulosa cells led to a number of hormones and factors that enhanced FSH induction of aromatase activity, and were therefore good candidates to examine with regard to increasing the sensitivity of granulosa cells to FSH. As granulosa cells are dependent upon androgen substrate for aromatization, and as it is well known that estrogens and androgens enhance FSH-induced aromatase activity, all cultures were performed in the presence of androstenedione and diethylstilbestrol. Since FSH actions are believed to be mediated by cAMP, the effect of minimizing cAMP breakdown by the addition of the phosphodiesterase inhibitor, 1-methyl-3-isobutyl-xanthine (MIX) on FSH induced aromatase activity was examined. MIX, at concentrations of 0.063 and 0.125 mM, was found to augment FSH-stimulated estrogen production as evidenced by an apparent decrease in the ED_{50} of 1.7- and 1.9-fold, respectively. Higher concentrations of MIX further increased FSH responsiveness; however, maximal estrogen production was inhibited. As our recent studies included the evaluation of insulin and insulin-like growth factors in FSH-stimulated steroidogenesis, the addition of insulin to cultures containing FSH, androstenedione, MIX, and DES was examined (Fig. 6). Combined treatments with MIX and insulin resulted in a synergistic augmentation of estrogen production by decreasing the minimal effective dose of FSH from 1 ng to 0.25 ng/culture. Although only low concentrations of LH receptors are present in granulosa cells obtained from preantral follicles of immature rats, preliminary results suggested that LH treatment is capable of enhancing FSH action in the presence of MIX. Addition of 30 ng/ml hCG to the hormones and factors described above resulted in a further decrease in the minimal effective dose of FSH from 0.25 ng to 0.06 ng/culture.

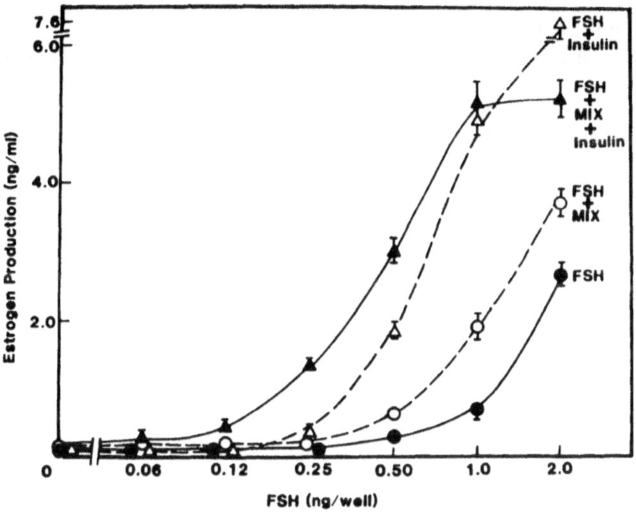

Fig. 6. Synergistic effect of treatment with MIX and insulin on FSH-stimulated estrogen production. Granulosa cells were cultured in medium containing 10^{-7} M androstenedione and 10^{-7} M DES in the presence of increasing doses of FSH with or without MIX (0.125 mM), insulin (1 μg/ml), or MIX plus insulin for 2 days.

This combination of hormones and factors now provided a highly sensitive in vitro bioassay for the measurement of FSH preparations; however, the addition of 4% gonadotropin-free serum to this culture system resulted in a substantial decrease in estrogen production at all doses of FSH tested (minimal effective dose of FSH of 0.5 ng/ml). Therefore, separation of FSH and inhibitory serum factors is necessary for optimal measurement of bioactive FSH in serum samples using the GAB method. Polyethylene glycol has dehydrating properties, and has been previously utilized to separate free peptide hormones from hormone-receptor complexes in radioreceptor assays. Polyethylene glycol pretreatment of serum samples was found to attenuate the inhibitory action of serum inhibitors. Treatment with increasing concentrations of polyethylene glycol (10-14%) dose-dependently increased the sensitivity of granulosa cells to FSH, with a minimal effective dose of FSH of 0.12 ng/culture in the presence of 12% polyethylene glycol-pretreated, gonadotropin free serum (Jia and Hsueh, 1985). At 12% polyethylene glycol,

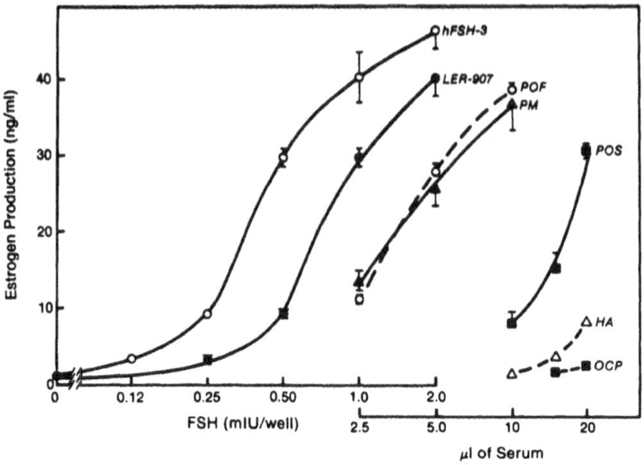

Fig. 7. Dose-dependent stimulation of estrogen production by pituitary FSH preparations and human serum from women during several clinical states. Granulosa cells were cultured as described for the GAB method with increasing concentrations of human pituitary FSH preparations (hFSH-3 or LER-907), or with increasing aliquots (2.5 to 20 μl) of human serum pretreated with 12% polyethylene glycol (POF = premature ovarian failure; PM = postmenopause; POS = preovulatory surge; HA = hypothalamic amenorrhea; OCP = oral contraceptive pill users). Medium estrogen content was measured by RIA. Adapted from Jia et al., 1986, with permission.

serum FSH was not precipitated, as evidenced by full recovery of exogenously added iodo-FSH or no loss in measured bioactivity of exogenously added standard FSH preparations.

The GAB assay was also examined as to hormone-specificity. Ovine GH (100 and 300 ng/culture), ovine PRL (100 and 300 ng/culture), rat TSH (30 mIU/culture), and human ACTH (30 mIU/culture) did not stimulate estrogen production by the cultured granulosa cells. hCG and LH stimulated estrogen production only at very high concentrations, probably due to FSH contamination or intrinsic FSH activity in the hormone preparations.

APPLICATION OF GAB METHOD

The GAB method has been successfully applied to the measurement of bioactive FSH levels in serum samples from rats and humans. Initially the FSH response to a single dose (5 μg) of GnRH given to adult male rats was examined. For these experiments, 10^{-6} M of a GnRH antagonist was included in all cultures to block the inhibitory effect of GnRH on granulosa cell estrogen production. (Inclusion of the antagonist alone does not affect the FSH stimulation of aromatase activity.) GnRH stimulated bioactive FSH levels within 30 min, with the maximal increase (2.8-fold) occurring between 60-120 min. Eight hours after hormone treatment, bioactive FSH levels declined to control values.

The GAB method was then applied to the measurement of bioactive FSH in human serum samples and pituitary FSH preparations (Jia et al., 1986; Kessel et al., 1986). All cultures were balanced with polyethylene glycol-

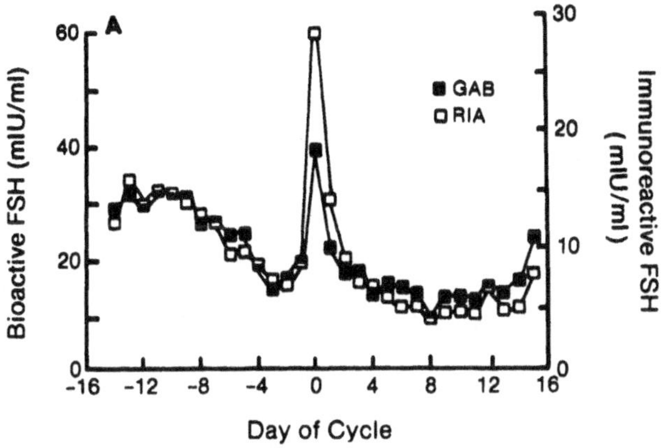

Fig. 8. Mean levels of bioactive and immunoreactive FSH throughout the menstrual cycle in seven regularly cycling women, with data centered around the midcycle LH surge.

pretreated, gonadotropin-free serum, to obtain a final serum concentration of 4%. Standard curves for the hFSH preparation had a working range of the assay between 0.12 - 2.0 mIU/culture for hFSH-3, and 0.25 - 2.0 mIU/culture for LER-907. Subsequent data is presented in units based on the biological potency of the LER-907 for comparison with the RIA results. FSH bioactivity as measured in serum from women in various physiologic or pathophysiologic states resulted in dose-response curves parallel to the standard curves; whereas serum from oral contraceptive pill users was unable to stimulate significant estrogen production at the highest aliquots used (Fig. 7).

Mean levels of serum FSH bioactivity in several clinical states representing a wide range of immunoreactive FSH levels were also consistent with the clinical presentation: undetectable in oral contraceptive pill users, intermediate in patients with hypothalamic amenorrhea (18.7 mIU/ml), and highest in postmenopausal women (191 mIU/ml) and patients with premature ovarian failure (163 mIU/ml). Serum FSH bioactivity was then determined in daily blood samples during ovulatory menstrual cycles in seven women. For purposes of data consolidation and analysis, the day of the LH peak was designated as day 0. The bioactive FSH levels exhibited a pattern closely resembling that of immunoreactive FSH. The bio- to immuno- (B:I) ratio throughout the menstrual cycle ranged from 1.4 to 3.4, with a mean of 2.5 (Fig. 8).

SUMMARY

FSH is a key gonadotropin inducer of ovarian follicular development, and stimulates granulosa cells to differentiate. The ability of a selected follicle(s) to become dominant while the majority of follicles become atretic cannot be explained on the basis of gonadotropin levels alone. Local modulatory factors therefore play an important role in follicular development. The paracrine role of steroids in the enhancement of FSH action is well established. Estrogen augments FSH stimulated aromatase activity, resulting in further increases in estrogen production, as well as enhancing FSH stimulated progestin production and LH receptor formation. Androgens, in the presence of FSH, not only serve as a substrate for aromatase activity, but augment FSH stimulation of estrogen and progesterone biosynthesis.

The physiologic role of ovarian peptides as paracrine modulators is less clear. Insulin, and the insulin-like growth factors, enhance FSH-stimulated steroidogenesis and LH receptor formation in vitro. Furthermore, GH increases IGF-I immunoreactivity in the ovary, and GH treatment results in enhanced FSH-stimulated LH receptor formation in granulosa cells. Taken together, the findings that GH stimulates gonadal IGF-I levels, and IGF-I enhances gonadotropin action, suggests that these peptides may be intra-ovarian modulators of FSH action.

The study of potential paracrine modulators that could enhance FSH action and therefore have potential physiologic roles in the development of the dominant follicle also led to the development of an in vitro bioassay for FSH. The combined action of enhancing hormones and factors resulted in granulosa cell assay highly sensitive to the aromatase inducing function of FSH. This assay is specific and sensitive, and has been applied to serum samples through the use of pretreatment of serum with polyethylene glycol to remove substances inhibitory in the granulosa cell culture. Preliminary applications of this assay to rat and human serum samples reveal that physiologic levels of FSH can be measured by this assay. Further applications of the GAB method in different animal species as well as to other physiologic and pathologic states in humans should be rewarding.

Acknowledgements

This work was supported by NIH Research Grants HD-14084 and HD-12303. AJWH is the recipient of Research Career Development Award HD-00375. BK is the recipient of NIH Physician Scientist Award HD-00653.

REFERENCES

Adashi, E. Y., and Hsueh, A. J. W., 1982, Estrogens augment the stimulation of ovarian aromatase activity by follicle-stimulating hormone in cultured rat granulosa cells, J. Biol. Chem., 257:6077.

Beers, W. H., and Strickland, S., 1978, A cell culture assay for follicle-stimulating hormone, J. Biol. Chem., 253:3877.

Bogdanove, E. M., Campbell, G. T., Blair, E. D., Mula, M. E., Miller, A. E., and Grossman, G. H., 1974, Gonado-pituitary feedback involves qualitative change: androgens alter the type of FSH secreted by the rat pituitary, Endocrinology, 95:219.

Channing, C. P., 1970, Influences of the in vivo and in vitro hormonal environment upon luteinization of granulosa cells in tissue culture, Recent Prog. Horm. Res., 26:589.

Chappel, S. C., Ulloa-Aguirre, A., and Coutifaris, C., 1983, Biosynthesis and secretion of follicle-stimulating hormone, Endocrine Rev., 4:179.

Crowley, W. F., Filicori, M., Spratt, D. I., and Santoro, N. F., 1985, The physiology of gonadotropin-releasing hormone (GnRH) secretion in men and women, Recent Prog. Horm. Res., 41:473.

Davoren, J. B., and Hsueh, A. J. W., 1984, Insulin enhances FSH-stimulated steroidogenesis by cultured rat granulosa cells, Mol. Cell. Endocrinol., 35:97.

Davoren, J. B., and Hsueh, A. J. W., 1986, Growth hormone increases ovarian levels of immunoreactive somatomedin C/insulin-like growth factor I in vivo, Endocrinology, 118:888.

Dorrington, J. H., Moon, Y. S., and Armstrong, D. T., 1975, Estradiol-17β biosynthesis in cultured granulosa cells from hypophysectomized immature rats: stimulation by follicle-stimulating hormone, Endocrinology, 97:1328.

Erickson, G. F., Wang, C., and Hsueh, A. J. W., 1979, FSH induction of functional LH receptors in granulosa cells cultured in chemically defined medium, Nature, 279:336.
Goff, A. K., and Armstrong, D. T., 1977, Stimulatory action of gonadotropins and prostaglandins on adenosine-3',5'-monophosphate production by isolated rat granulosa cells, Endocrinology, 101:1461.
Goff, A. K., and Henderson, K. M., 1979, Changes in follicular fluid and serum concentrations of steroids in PMS treated immature rats following LH administration, Biol. Reprod., 20:1153.
Hillier, S. G., and DeSwart, F. A., 1981, Evidence that granulosa cell aromatase induction/activation by follicle-stimulating hormone is an androgen receptor-regulated process in vitro, Endocrinology, 109:1303.
Hsueh, A. J. W., Adashi, E. Y., Jones, P. B. C., and Welsh, T. H., 1984, Hormonal regulation of the differentiation of cultured ovarian granulosa cells, Endocrine Rev., 5:76.
Hsueh, A. J. W., and Jones, P. B. C., 1981, Extrapituitary actions of gonadotropin-releasing hormone, Endocrine Rev., 2:437.
Hsueh, A. J. W., Erickson, G. F., Papkoff, H., 1983, Effect of diverse mammalian gonadotropins on estrogen and progesterone production by cultured rat granulosa cells, Arch. Biochem. Biophys., 225:505.
Hsueh, A. J. W., Welsh, T. H. Jr., and Jones, P. B. C., 1981, Inhibition of ovarian and testicular steroidogenesis by epidermal growth factor, Endocrinology, 108:2002.
Jia, X. -C., and Hsueh, A. J. W., 1985, Sensitive in vitro bioassay for the measurement of serum follicle-stimulating hormone, Neuroendocrinology, 41:445.
Jia, X. -C., Kalmijn, J., Hsueh, A. J. W., 1986, Growth hormone enhances FSH-induced differentiation of cultured rat granulosa cells, Endocrinology, 118:1401.
Jia, X. -C., Kessel, B., Yen, S. S. C., Tucker, E. M., Hsueh, A. J. W., 1986, Serum bioactive follicle-stimulating hormone during the human menstrual cycle and in hyper- and hypo-gonadotropic states: application of a sensitive granulosa cell aromatase bioassay (GAB), J. Clin. Endocrinol. Metab., 62:1243.
Kessel, B., Liu, Y. X., Jia, X. -C., and Hsueh, A. J. W., 1985, Autocrine role of estrogens in the augmentation of LH receptor formation in cultured rat granulosa cells, Biol. Reprod., 32:1038.
Kessel, B., Jia, X. -C., and Hsueh, A. J. W., 1986, Use of a rat granulosa cell aromatase bioassay (GAB) to measure serum levels of bioactive follicle-stimulating hormone (FSH): further validation of the method and analysis of the human menstrual cycle, 68th Endocrine Society meeting, Abstract No. 532.
Lucky, A. W., Schrieber, J. R., Hillier, S. G., Schulman, J. D., and Ross, G. T., 1977, Progesterone production by cultured preantral rat granulosa cells: stimulation by androgens, Endocrinology, 100:128.
Midgley, A. R. Jr., 1973, Autoradiographic analysis of gonadotropin binding to rat ovarian tissue section, Adv. Exp. Med. Biol., 36:365.
Richards, J. S., and Midgley, A. R. Jr., 1976, Protein hormone action: a key to understanding ovarian follicular and luteal cell development, Biol. Reprod., 14:82.
Richards, J. S., Ireland, J. J., Rao, M. C., Bernath, G. A., Midgley, A. R. Jr., and Reichert, L. E. Jr., 1976, Ovarian follicular development in the rat: hormone receptor regulation by estradiol, follicle-stimulating hormone and luteinizing hormone, Endocrinology, 99:1562.
Richards, J. S., Sehgal, A., and Tash, J. S., 1983, Changes in content and cAMP-dependent phosphorylation of specific proteins in granulosa cells of preantral and preovulatory ovarian follicles and in corpora lutea, J. Biol. Chem., 258:5227.

Ross, G. T., Cargille, C. M., Lipsett, M. B., Rayford, P. L., Marshall, J. R., Strott, C. A., and Rodbard, D., 1970, Pituitary and gonadal hormones in women during spontaneous and induced ovulatory cycles, Recent Prog. Horm. Res., 26:1.

Schrieber, J. R., and Hsueh, A. J. W., 1979, Progesterone "receptor" in rat ovary, Endocrinology, 105:915.

Steelman, S. L., and Pohley, F. M., 1953, Assay of the follicle stimulating hormone based on the augmentation with human chorionic gonadotropin, Endocrinology, 53:104.

Stouffer, R. L., Tyrey, L., and Schomberg, D. W., 1976, Changes in [^{125}I]-labeled human chorionic gonadotropin binding to porcine granulosa cells during follicle development and cell culture, Endocrinology, 99:516.

Styne, D. M., and Grumbach, M. M., 1986, Puberty in the male and female: Its physiology and disorders, in: "Reproductive Endocrinology, Physiology, Pathophysiology, and Clinical Management," S. S. C. Yen, R. B. Jaffe, eds., p. 313, W.B. Saunders Co., Philadelphia.

Van Damme, M. P., Robertson, D. M., Marana, R., Ritzen, E. M., and Diczfalusy, E., 1979, A sensitive and specific in vitro bioassay method for the measurement of follicle stimulating hormone actiivty, Acta Endocrinologica, 91:224.

Wang, C., Hsueh, A. J. W., and Erickson, G. F., 1979, Induction of functional prolactin receptors by follicle-stimulating hormone in rat granulosa cells in vivo and in vitro, J. Biol. Chem., 254:11330.

Wang, C., Hsueh, A. J. W., and Erickson, G. F., 1982, Role of cyclic AMP in the induction of estrogen and progestin synthesis in cultured granulosa cells, Mol. Cell. Endocrinol., 25:73.

Welsh, T. H. Jr., Zhuang, L. Z., Hsueh, A. J. W., 1983, Estrogen augmentation of gonadotropin-stimulated progestin biosynthesis in cultured rat granulosa cells, Endocrinology, 112:1916.

Zeleznik, A. J., Menon, K. M. J., Midgley, A. R. Jr., and Reichert, L. E. Jr., 1976, Induction of receptor for luteinizing hormone in rat granulosa cells in vivo and in vitro by follicle-stimulating hormone, Adv. Cyclic Nucleotide Res., 5:803.

INTRA-OVARIAN ACTIONS OF STEROIDS IN REGULATION OF FOLLICULAR STEROID BIOSYNTHESIS

D. T. Armstrong[1], S. A. J. Daniel, and R. E. Gore-Langton[2]

Medical Research Council of Canada
Group in Reproductive Biology
Departments of Physiology
Obstetrics and Gynaecology
University of Western Ontario
London, Ontario, Canada

The follicle is the principal functional unit of the mammalian ovary, the primary function of which is the production and release of a mature oocyte at the culmination of each ovarian cycle. A secondary role of the follicle is the biosynthesis and secretion of a group of steroid hormones which serve many important functions in regulation of peripheral (i.e extra-ovarian) cells and tissues.

In recent years it has become evident that the ovarian steroids also have important regulatory roles at intra-ovarian sites; in particular, exerting auto-regulatory actions on the cells within the follicle engaged in their biosynthesis.

There are two types of follicular somatic cells, of different embryological origins, which are responsible for the steroid output of the follicles at their various stages of differentiation and maturity. These are the theca interna cells (hereafter abbreviated as theca) and granulosa cells. Between them, these cells synthesize and secrete three major classes of steroids: progestins (C_{21}), androgens (C_{19}) and estrogens (C_{18}). The essential role of the pituitary hormones, follicle-stimulating hormone (FSH) and luteinizing hormone (LH) in regulating the activities of these cells has been recognized since the classical studies of P. E. Smith (1930) demonstrated the effects of hypophysectomy of the rat and replacement therapy with pituitary extracts, followed by the definitive demonstration by Greep et al. (1936, 1942) of two separate and distinct gonadotropic hormonal activities in these extracts.

The theca cells, under the influence of LH, convert sterol precursors to progesterone, and progesterone to androgens, but have only a limited capacity to aromatize androgens to estrogens. Granulosa cells, on the other hand, have abundant aromatase activity when stimulated by FSH, and under this stimulation, readily convert androgens to estrogens. Although they also synthesize progestins, to degrees which vary greatly at different stages of

[1]Career Investigator of the Medical Research Council of Canada.
[2]Career Scientist of the Ontario Ministry of Health.

their differentiation, granulosa cells are unable to convert C_{21}-steroids to androgens to a significant extent. Instead, androgens of theca cell origin diffuse into the granulosa cell compartment where they are used as substrates for conversion to estrogens. This cooperation between the theca and granulosa cells in the biosynthesis of steroids is the basis of the "two-cell, two gonadotropin" model for the control of steroidogenesis by the ovarian follicle (Armstrong and Dorrington, 1977).

More recently, evidence has been accumulating to indicate that the actions of the pituitary gonadotropins may be modulated or influenced by a number of regulatory substances produced and acting within the ovary itself, including steroids, prostaglandins, neurotransmitters, and small peptide molecules (reviewed by Gore-Langton and Armstrong, 1987). These compounds of intra-ovarian origin are agents of both paracrine and autocrine regulation. They regulate the activities of cells adjacent to their cells of origin, which they reach by diffusion through the interstitial fluid (including follicular fluid) bathing the cells, and of the cells in which they, themselves, are produced.

In this review we will consider the role of steroids of ovarian origin in local regulation of follicular steroid biosynthesis.

INTRA-OVARIAN REGULATION BY ESTROGENS

Morphologic Effects and Binding to Ovarian Cells

The stimulatory effects of estrogens on the follicle have been established (Pencharz, 1940; Williams, 1940; Simpson et al., 1941; Williams, 1944; Payne and Hellbaum, 1955; Payne and Runser, 1958; Bradbury, 1961; Smith, 1961). In experiments in which estrogen was administered to hypophysectomized immature rats, ovarian weight was maintained as a result of the ability of the estrogens to stimulate granulosa cell proliferation. In addition, the ovaries became more responsive to gonadotropic stimulation. This action has been demonstrated more recently to be due to an ability of estrogen to act synergistically with FSH in the induction of cell membrane receptors for both FSH and LH (Richards, 1980). Additional evidence for a synergism between estrogens and gonadotropins is provided by the observation that the ovarian weight response to gonadotropins is inhibited by an antiserum raised against estradiol (Reiter et al., 1972). Moreover, treatment with gonadotropins increased the number of atretic follicles in ovaries of hypophysectomized rats, and this could be partially reversed by administration of estrogens (Harman et al., 1975).

Specific uptake and retention of [^{3}H]estradiol *in vivo* by ovaries of immature rats (Saiduddin, 1971; Saiduddin and Milo, 1974), incorporation of estradiol into granulosa cells (Stumpf, 1969), and [^{3}H]estradiol binding to rat granulosa (Richards, 1975) and luteal cell nuclei (Richards, 1974) indicate the existence of estrogen binding sites in the ovary. Characterization of estrogen binding components in ovaries of immature rats indicates that ovarian estrogen binding sites are similar to specific, high-affinity estrogen receptors in extra-ovarian tissues such as the uterus (Saiduddin and Zassenhaus, 1977).

Effects on Androgen Biosynthesis and Metabolism

There is evidence to suggest that estrogen acts within the ovary to inhibit androgen production by the follicle. Treatment of immature rats with estradiol suppressed ovarian testosterone and 5α-dihydrotestosterone production. Administration of gonadotropins was unable to overcome this inhibition, indicating that the effect was not mediated by decreased circulating gonadotropins (Leung et al., 1978). Further evidence for a direct intra-ovarian action of estrogen was provided by experiments in which silastic implants containing estradiol were embedded under the ovarian bursa unilaterally. The LH-stimulated androgen content of the ovary in direct contact with the implant was considerably lower than that of the contralateral ovary (Leung et al., 1978). An inhibitory effect of estradiol on LH-induced androgen content of ovaries from immature hypophysectomized rats was reported as further evidence that this steroid was not acting by influencing secretion of a pituitary hormone (Leung et al., 1978).

Results of *in vitro* experiments have demonstrated similar effects of estrogen on ovarian androgen synthesis. Whole ovaries from immature intact or hypophysectomized rats administered estradiol *in vivo* responded to LH-stimulation *in vitro* with decreased androgen production when compared to that of ovaries obtained from rats that were not treated with estrogen. Dibutyryl cAMP was unable to increase testosterone production by cultured ovaries of estrogen pretreated rats (Leung and Armstrong, 1979a). Administration of estradiol to immature rats *in vivo* also decreased androgen secretion by isolated thecal tissue *in vitro* (Leung and Armstrong, 1980). In experiments with isolated porcine thecal tissue in organ culture (Leung and Armstrong, 1980) or as dispersed theca cell preparations (Hunter and Armstrong, 1986), addition of estrogens to the culture medium inhibited LH-stimulated androgen production in a dose-dependent manner, providing further evidence that the inhibitory action of estrogens is directly on theca cells.

Evidence available indicates that estrogen inhibits ovarian androgen synthesis at a site distal to cAMP production, probably at an enzymatic step or steps in the steroidogenic pathway between androgens and their C_{21}-steroid precursors. In support of this hypothesis, estrogen pretreatment of ovaries from intact immature rats *in vivo* was found to inhibit conversion of radioactively labeled progesterone to androgens (testosterone, androstenedione and androsterone) *in vitro*. On the other hand, incorporation into 3α-hydroxy-5α-pregnan-20-one was enhanced suggesting that estrogen may act by inhibiting the 17α-hydroxylase:C-17,20-lyase enzyme system or by diverting C_{21} substrates into an alternate pathway resulting in the formation of 5α-reduced pregnane compounds (Leung and Armstrong, 1979a). Treatment of immature rats with estradiol was also shown to suppress the stimulation by hCG of androstenedione, testosterone, 17α-hydroxyprogesterone and 17α-hydroxypregnenolone production by dispersed ovarian cells in culture (Magoffin and Erickson, 1982). Under these conditions, pregnenolone production was unchanged while progesterone production was markedly enhanced. Binding of hCG, hCG-stimulated cAMP synthesis and the viability of ovarian steroidogenic cells were not affected by estradiol. It was concluded that exogenous estradiol blocked ovarian androgen formation by reducing the activity of the 17α-hydroxylase enzyme. Further *in vitro* evidence with rat ovarian interstitial cell cultures indicates that estradiol causes a rapid inhibition of 17α-hydroxylase and C-17,20-lyase enzyme activities (Magoffin and Erickson, 1982).

Experiments of Eckstein and Nimrod (1977) have provided evidence that estrogens are capable of regulating metabolism of androgens by a direct action on 5α-reductase. In these experiments, an inhibitory effect of estradiol on 5α-reductase activity was demonstrated in microsomal preparations from immature rat ovaries. Since the minimal effective concentration of estradiol required to inhibit enzyme activity was in the range comparable to that present in follicular fluid, the authors suggested that estradiol may have a physiological role in the regulation of androgen metabolism in the follicle.

Administration of LH to intact immature rats has been shown to influence ovarian progesterone metabolism in a manner identical to that of estrogen. Ovarian androgen production *in vitro* was decreased and 3α-hydroxy-

5α-pregnan-20-one secretion increased (Leung and Armstrong, 1979a). Exposure of ovaries isolated from prepubertal rats to LH either in vivo before removal or in vitro by direct addition to culture media, alters progesterone metabolism favoring formation of 5α-reduced pregnane compounds and decreasing androgen and 5α-reduced androgen accumulation (Armstrong, 1979). In cultured preovulatory follicles from PMSG-treated immature rats, LH inhibited C-17,20-lyase activity. Addition of inhibitors of steroid biosynthesis prevented the inhibitory action of LH on the conversion of 17α-hydroxyprogesterone to androgens, but did not affect basal lyase activity. These experiments suggested that the inhibitory action of LH on androgen synthesis may be mediated by the action of another ovarian steroid. The possibility that this steroid is an estrogen is supported by experiments in which the aromatase inhibitor, 4-acetoxy-androstane-3,17-dione, was found to block the negative effect of LH on androgen synthesis by rat preovulatory follicles in vitro (Evans et al., 1981).

Androgen levels in ovarian tissue, follicular fluid, ovarian venous blood and serum initially rise after the preovulatory LH surge; this brief elevation is followed by a precipitous decline several hours later (Armstrong et al., 1976; Bahr, 1978; Goff and Henderson, 1979). This apparent inhibitory effect of LH on androgen production raises the possibility that the inhibition represents a physiological role of estrogens in the intrafollicular control of androgen biosynthesis. Further evidence for such a physiological role has been provided by experiments of Smith et al. (1975) and Kalra and Kalra (1974) showing that shortly after the LH surge on the day of proestrus, estradiol reaches a peak, and then declines rapidly. That the rise is dependent on the LH surge has been demonstrated in proestrous hamsters by blocking LH secretion with injections of phenobarbital (Saidapur and Greenwald, 1979). It may be that the surge of LH initially stimulates theca cells to produce androgens which are aromatized to estrogens by granulosa cells. The estrogens then inhibit the 17α-hydroxylase:C-17,20-lyase enzyme system in the theca cells, thereby limiting further synthesis of androgens and their subsequent use as substrates for aromatization. The ability of low doses of estradiol to enhance progesterone proudction by isolated bovine theca cells (Fortune and Hansel, 1979) may be a reflection of precursor accumulation following the inhibitory action of estrogens. This action may contribute to the transition of the follicle initiated by the LH surge from a structure primarily engaged in estrogen secretion to one whose main function is progesterone secretion.

Effects on Aromatase Activity

Another intrafollicular regulatory action of estrogens that has been clearly established is their ability to enhance ovarian estrogen production through direct actions on granulosa cells. Clomiphene citrate, a weak estrogen, increased estradiol and estrone synthesis from radiolabeled androstenedione by superfused canine ovaries in vivo (Engels et al., 1968). This compound, (Zhuang et al., 1982), as well as estradiol, estrone, hexestrol, moxestrol, ethinyl estradiol, chlorotrianisene, mestranol (Adashi and Hsueh, 1982) and triphenylethylene antiestrogens (Welsh et al., 1984) have also been reported to have similar effects on FSH-induced estrogen synthesis by cultured granulosa cells isolated from ovaries of immature, hypophysectomized, diethylstilbestrol (DES)-primed rats. Aromatase activity induced in rat granulosa cells by FSH was enhanced by in vitro addition of DES (Daniel and Armstrong, 1983). A good correlation was found between receptor binding affinity and biological potency of both natural and synthetic estrogens or antiestrogens. The stimulatory effect could not be accounted for by increased granulosa cell viability or protein mass (Adashi and Hsueh, 1982).

In support of the hypothesis that estrogens are physiological regulators of granulosa cell aromatase activity, the minimal effective dose of estradiol required to elicit a response (3.7×10^{-10} M) is well within the range of estradiol in antral fluid of preovulatory follicles (Adashi and Hsueh, 1982). Thus, estrogens may function within the ovary or in individual follicles as endproduct amplifiers to enhance FSH-induced aromatase.

Effects on Progesterone Biosynthesis

Estrogens have been shown to decrease progesterone secretion in a variety of ovarian preparations, including porcine (Thanki and Channing, 1976, 1978; Schomberg et al., 1976; Haney and Schomberg, 1978), bovine (Fortune and Hansel, 1979), rat (Hillier et al., 1977) and human (Bieszczad et al., 1982; Veldhuis et al., 1983) granulosa cells, and by large follicles from bovine ovaries (Shemesh and Ailenberg, 1977). The inhibitory action of estrogens on porcine granulosa cells was both time- and dose-dependent, and could be demonstrated in short-term but not long-term cultures, at estradiol concentrations similar to those found *in vivo* (Veldhuis, 1985). The inhibitory effect of estrogens on progesterone production in culture was independent of cell density and was not due to a cytotoxic effect. Instead, the action of estrogen appeared to limit the conversion of pregnenolone to progesterone, resulting in enhanced pregnenolone accumulation in culture. Increased pregnenolone production in the presence of estrogen was also the result of enhanced cholesterol side-chain cleavage activity (Veldhuis, 1985; Toaff et al., 1983) and mitochondrial content of P-450 (Toaff et al., 1983).

The inhibitory action of estrogen on progesterone production has been confirmed *in vivo*. Administration of estradiol for 3 days decreased the ability of LH to increase ovarian progesterone content in hypophysectomized, but not intact immature rats. Also, in ovaries of estradiol-treated hypophysectomized rats, dibutyryl cAMP, but not LH restored *in vitro* progesterone production to values comparable to those of ovaries from control animals. These findings suggested that estrogen inhibited progesterone synthesis by acting at a step or steps prior to cAMP generation (Leung and Armstrong, 1979b).

There are also reports of stimulatory effects of estrogen on progesterone secretion in rat (Bernard, 1975; Hillier et al., 1977; Welsh et al., 1983) and porcine granulosa cells (Goldenberg et al., 1972; Veldhuis et al., 1982). Unlike the inhibitory action of estrogen, the stimulatory effect on cultured porcine granulosa cells was demonstrable only in longer-term incubations and was found to be dependent on the density of granulosa cells in culture as well as on the maturational status of the follicle from which cells were isolated (Veldhuis, 1985). Granulosa cells from small but not large follicles responded to estrogen with increased progesterone secretion, an effect that was found to be due to enhanced activity of Δ^5-3β-hydroxysteroid dehydrogenase:Δ^{5-4}-isomerase. In addition, estrogen treatment increased pregnenolone accumulation as well as cholesterol side-chain cleavage activity and hydrolysis of endogenous cholesteryl esters.

Catechol estrogens, which may be synthesized within the follicle (Hammond et al., 1986), have also been shown to stimulate steroidogenesis in rat granulosa cells *in vitro* (Hudson and Hillier, 1985) and corpora lutea (Khan and Gibori, 1984).

INTRA-OVARIAN REGULATION BY ANDROGENS

Androgens and Follicular Growth

With the establishment of the essential role of androgens as precursors for estrogen biosynthesis, other possible roles of androgens in follicular function have not received much attention until recently. The antiandrogen, hydroxyflutamide had little or no effect on FSH-induction of enzyme activity or LH receptors in DES-primed, hypophysectomized, immature rats, suggesting that androgens are not essential for FSH to initiate development of antral follicles (Zeleznik et al., 1979). In support of this hypothesis, Neumann et al. (1970) demonstrated that another androgen antagonist, cyproterone acetate, did not disrupt estrous cycles or interfere with ovulation in adult rats. In addition, Lyon and Glenister (1974) reported that Tfm/O mice, a strain in which females carry a gene conferring androgen resistance, have normal reproductive cycles; follicular maturation, conception and pregnancy occur normally. On the other hand, more recent evidence of several types suggests a regulatory function of androgens in the follicle. In particular, the demonstration of specific androgen binding sites with characteristics similar to androgen receptors in ovaries from estrogen-primed, hypophysectomized, immature rats (Schreiber et al., 1976) indicates the likelihood of receptor-mediated actions of androgens in ovarian regulation. These receptors have been localized to the granulosa cell compartment of rat (Schreiber and Ross, 1976) and sheep (Campo et al., 1984) follicles. Similar androgen-binding proteins have been found in human ovarian cytosol (Milwidsky et al., 1980).

Androgens and Follicular Atresia

In healthy versus atretic follicles there is an inverse relationship between intrafollicular concentrations of androgens and estrogens. High androgen:estrogen ratios in follicular fluid have been associated with follicular atresia (McNatty et al., 1979; Richards, 1975; Carson et al., 1981). However, from these data it is uncertain whether the predominance of androgen over estrogen is the cause of atresia or merely a result of the process. Androgens (Hillier and Ross, 1979; Payne and Runser, 1958) as well as hCG (Louvet et al., 1975) which stimulate ovarian androgen synthesis, have been shown to promote atresia in rat ovaries. Concomitant administration of androgen antagonists, or of antiserum raised against androgen (Louvet et al., 1975) alleviated this effect of hCG. In addition, treatment of immature hypophysectomized, PMSG-treated rats with 5α-dihydrotestosterone also induced atresia (Bagnell et al., 1982). This latter effect could be at least partially overcome by estradiol. Recently, Opavsky and Armstrong (1985) showed an inhibitory effect of LH on the superovulatory response of immature rats to FSH, and this effect could be partially overcome by administration of the antiandrogen, hydroxyflutamide (Opavsky, M. A. and Armstrong, D. T., unpublished observation). Although there is considerable circumstantial evidence to implicate androgens in the process of follicular atresia, the mechanisms of this process and the specific role of androgens in the process are poorly understood.

In view of the considerable amount of evidence in favor of a negative role of androgens in follicular maturation, it is perhaps surprising to find that androgens also have positive effects on follicular growth. Ovarian degeneration occurs in androgen-resistant Tfm/O mice (Ohno et al., 1973), and an antiandrogen has been reported to accelerate atresia in preovulatory rat follicles (Peluso et al., 1979). Somewhat similarly, treatment of diestrous rats with the antiandrogen, flutamide, resulted in decreased growth and maturation of ovarian follicles (Kumari et al., 1978). An inhibitory action of androgen antisera on hCG-induced ovulation has also been reported in hypophysectomized rats (Mori et al., 1977).

In attempting to reconcile these apparently conflicting effects within the ovarian follicles, it is of interest that the antagonistic action of androgen on follicular events related to atresia appears only to affect those follicles at the preantral and early antral stages of development. The facilitatory effect of androgen may be reserved for those large follicles that have already entered the final stages of development (Tsafriri and Braw, 1984).

Androgens and Progesterone Biosynthesis

In culture, androgens have been shown to stimulate progesterone biosynthesis by intact follicles dissected from ovaries of cycling ewes (Moor et al., 1975) and cows (Shemesh and Ailenberg, 1977) and by granulosa cells isolated from pig (Schomberg et al., 1976; Haney and Schomberg, 1978), rat (Hillier et al., 1977; Lucky et al., 1977; Armstrong and Dorrington, 1976) and mouse (Corredor and Flickinger, 1983) ovaries. Moreover, administration of androgens to intact immature rats increased subsequent progesterone accumulation by their isolated ovarian cells *in vitro* (Leung et al., 1979). Since both aromatizable and nonaromatizable androgens were effective, the response appears to be a true androgen effect, rather than one that is dependent on aromatization of androgens to estrogens. This hypothesis is supported by findings in the pig in which implants of antiandrogens (flutamide or hydroxyflutamide) placed in the ovarian interstitium decreased progesterone secretion by subsequently isolated granulosa cells *in vitro* (Schomberg et al., 1978). Hydroxyflutamide and cyproterone acetate suppressed the stimulatory effect of testosterone on progesterone production by rat granulosa cell incubations (Hillier et al., 1977).

In addition to having their own stimulatory effects, androgens have been shown to enhance FSH-stimulated progestin synthesis (Nimrod and Lindner, 1976; Armstrong and Dorrington, 1976; Welsh et al., 1982; Nimrod, 1977). This action was blocked by hydroxyflutamide (Hillier and DeZwart, 1982), and the efficacy of various androgens appeared to be correlated with the extent to which they were converted to testosterone or 5α-dihydrotestosterone (Nimrod et al., 1980).

Depending on the animal model used, androgens have been found to act both before and after cAMP synthesis. Androstenedione has been found to enhance the stimulatory action of dibutyryl cAMP on progesterone production by isolated granulosa cells from immature hypophysectomized, estrogen-treated rats, but had no effect on $[^{125}I]$FSH binding to the cells, on FSH-stimulated cAMP production or on degradation of cAMP to 5'-AMP by the phosphodiesterase enzyme (Nimrod, 1977). On the other hand, with granulosa cells isolated from ovaries of intact immature rats, androgens enhanced FSH stimulation of cAMP production (Goff et al., 1979; Hillier and DeZwart, 1982; Daniel and Armstrong, 1984) and $[^{125}I]$FSH binding (Daniel and Armstrong, 1984; Knecht et al., 1984), and suppressed cAMP metabolism (Hillier and DeZwart, 1982).

Increased C_{21}-steroid production in the presence of androgen is not a reflection of decreased catabolism by 5α-reductase (Nimrod, 1977), although androgens have been reported to regulate catabolism of progesterone (Duleba et al., 1983; Moon et al., 1984) through inhibition of 20α-hydroxysteroid dehydrogenase (Moon et al., 1985). In cultured rat granulosa cells, androgens and FSH act synergistically to enhance lipoprotein utilization (Schrieber et al., 1983) although they have no effect on levels of free or esterified cholesterol in these cells, either alone or in combination with FSH (Nimrod, 1981). Both FSH and testosterone independently increase rate of conversion of cholesterol to pregnenolone indicating stimulatory action on cholesterol side-chain cleavage, and combined treatment resulted in synergism in this action (Nimrod, 1981; Jones and Hsueh, 1982; Welsh et al., 1982). Neither FSH nor androgen influence transport of cholesterol into mitochondria (Nimrod, 1981). Effects of androgens on 3α-hydroxysteroid dehydrogenase are equivocal. There are data to suggest that androgens act synergistically with FSH to increase conversion of pregnenolone to progesterone (Welsh et al., 1982), while others contend that androgens are without effect (Dorrington and Armstrong, 1979).

In contrast to the work using rat tissue, aromatizable androgens have a negative influence on progesterone production by human granulosa cells (Batta et al., 1980) and on FSH-stimulated progesterone accumulation by granulosa cells isolated form porcine ovaries (Evans et al., 1984; Lischinsky et al., 1983). Although estradiol had a similar inhibitory action and nonaromatizable androgens were ineffective, the effect of aromatizable androgens could not be accounted for by conversion to estrogens since the aromatase inhibitor, 4-acetoxy-4-androstene-3,17-dione, failed to prevent the testosterone-induced decrease in progesterone production (Evans et al., 1984).

Dibutyryl cAMP, like FSH, stimulated progesterone production by porcine granulosa cells. Addition of testosterone to cultures suppressed the stimulatory effect of dibutyryl cAMP indicating that testosterone acts at a site distal to cAMP generation (Lischinsky et al., 1983). Further studies revealed that androgens had no effect on progesterone metabolism (Evans et al., 1984). However, they did enhance pregnenolone synthesis in FSH-treated granulosa cell cultures (Lischinsky et al., 1983). Thus, decreased progesterone production in the presence of androgen appears to be due to restricted conversion of pregnenolone to progesterone through inhibition of Δ^5-3β-hydroxysteroid dehydrogenase:Δ^{5-4}-isomerase activity. A direct inhibitory action of testosterone on this enzyme has recently been demonstrated (Tan and Armstrong, 1984).

There is little known about possible intra-ovarian effects of androgens on steroidgenesis by theca cells. Androgen has been shown to enhance progesterone secretion by human thecal tissue in culture. The effect on granulosa cell progesterone synthesis was negative and in combined incubations of theca or stroma plus granulosa cells, no effect of androgen could be discerned (Batta et al., 1980).

Androgens and Estrogen Biosynthesis

The regulatory role of FSH in ovarian estrogen secretion through induction of aromatase activity in rat granulosa cells is now well established (Dorrington and Armstrong, 1979). Studies using cultured rat granulosa cells from intact immature rats have demonstrated that in addition to acting as substrates for FSH-stimulated aromatase, androgens also enhance FSH-induction of enzyme activity (Daniel and Armstrong, 1980). Both aromatizable and nonaromatizable androgens were effective in this regard, although nonaromatizable androgens (5α-dihydrotestosterone and androsterone) had only about half the activity of aromatizable androgens (testosterone and androstenedione). From these findings, it is clear that this action of androgens is not dependent upon their conversion to estrogens. This interpretation receives support from experiments showing that androgen enhancement of FSH-induced aromatase activity is suppressed by hydroxyflutamide (an androgen receptor blocker) (Armstrong et al., 1980; Hillier and DeZwart, 1981) and is not affected by 4-acetoxy-4-androstene-3,17-dione (an inhibitor of aromatase) or nafoxidine (an estrogen receptor blocker) (Daniel and Armstrong, 1983).

The action of androgens in influencing granulosa cell aromatase activity in the immature rat ovary appear to be exerted at a site before cAMP production. Although testosterone enhances FSH-induced aromatase activity, it has no effect on cAMP-induced estrogen synthesis. The responsiveness of cultured granulosa cells to FSH in the production of cAMP as well as cellular [^{125}I]-FSH binding is augmented by androgen (Daniel and Armstrong, 1984).

Aromatization of testosterone to estradiol by granulosa cells, isolated from the largest follicles in ovaries of rats showing diestrous II and proestrous vaginal smears, has been shown to be competitively inhibited by 5α-reduced androgens (Hillier et al., 1980a). A similar effect of 5α-reduced androgens was described using human follicles excised at all stages of maturity (Hillier et al., 1980b). Apart from alterations in aromatase enzyme activity itself, follicular estrogen biosynthesis may be influenced by the amount of C_{19} substrate available and variation in the ratio of aromatizable to nonaromatizable androgens as the result of changes in 5α-reductase activity. Considerable conversion of aromatizable androgens to 5α-reduced androgens has been found to occur in rat (Inabau et al., 1978) and human (Smith et al., 1974; McNatty et al., 1979) ovaries. Since high concentrations of estrogen in follicular fluid have been associated with healthy antral follicles and low estrogen with apparently degenerating follicles (McNatty et al., 1979), alterations in the capacity of granulosa cells to convert androgens to estrogens may be a physiologically important mechanism for regulating concentrations of intrafollicular steroid hormones and development of individual follicles.

Siiteri and Thompson (1975) have reported a 2½-fold increase in 5α-reductase activity and a 5-fold decrease in aromatase activity within a few hours of exposure of ovaries of PMSG-treated rats to hCG. Katz and Armstrong (1976) observed a similar decline in aromatase activity following LH treatment. The increased activity of 5α-reductase could contribute to the decreased estradiol secretion, which occurs dramatically following the LH surge, both by decreasing the intrafollicular levels of aromatizable androgens through increased 5α-reductase activity, and by increasing intrafollicular levels of 5α-reduced androgens which serve as competitive inhibitors of the aromatase enzyme system.

INTRA-OVARIAN REGULATION BY PROGESTERONE

Effect on Follicular Growth and Differentiation

It is uncertain whether progestins have any direct role in the intra-ovarian regulation of follicular function, and reports are often contradictory. There are reports that progesterone has an inhibitory effect on follicular development when given to intact animals (Jesel, 1970; Hori et al., 1973; Buffler and Roser, 1974; Beattie and Corbin, 1975) although other authors have failed to find an effect of administration of exogenous progesterone to estrogen-primed, hypophysectomized, immature rats on ovarian morphology (Richards, 1976; Saiduddin and Zassenhaus, 1978; Smith and Bradbury, 1966). That the effect of progesterone on follicular development might be mediated indirectly by depression of pituitary secretion of gonadotropins rather than by direct inhibition at the follicular level is suggested by the observation that retardation of follicular growth in intact animals by progesterone occurs only when plasma concentrations of both FSH and LH are significantly reduced (Beattie and Corbin, 1974). Goodman and Hodgen (1977) attempted to avoid this problem by placing progesterone directly in the monkey ovary and suggested that their results supported a direct inhibitory action of progesterone on follicular development.

Both serum LH concentrations and estradiol accumulation by the isolated follicles were decreased in prepubertal rats receiving progesterone implants, but surprisingly these implants facilitated the stimulatory effects of a low dose of hCG on growth of small antral follicles and on estrogen synthesis (Richards and Bogovich, 1982). Based on these observations, it was suggested that progesterone may facilitate LH action under physiological circumstances when basal LH is low. On the other hand, even in the presence of elevated serum progesterone as found during pregnancy, follicles undergo preovulatory maturation in response to small sustained increases in serum LH which occur at the end of pregnancy. These observations suggest that progesterone may have no direct inhibitory effect on follicle cell maturation (Bogovich et al., 1981).

Despite conflicting evidence concerning possible direct effects of progesterone on the follicle, specific progesterone receptors have been identified in rat ovary (Schreiber and Hsueh, 1979; Schreiber et al., 1983) and located in granulosa cells (Schreiber and Erickson, 1979; Naess, 1981). Similarly, progestin binding sites have been reported in ovaries of rabbits (Philibert et al., 1977), guinea pig (Pasqualini and Nguyen, 1980), cow (Jacobs and Smith, 1980) and women (Jacobs et al., 1980; Milwidsky et al., 1980).

When added directly to cultured rat granulosa cells, progesterone has been shown to enhance the ability of these cells to respond to FSH in the production of cAMP (Goff et al., 1979). In another study, a synthetic progestin (R5020) increased FSH-stimulation of progesterone and 20α-dihydroprogesterone synthesis in granulosa cells isolated from immature hypophysectomized, estrogen-treated rats. Similarly, R5020 enhanced LH-stimulated progestin production by cells in which prior exposure to FSH had induced LH-receptor formation. Furthermore, in the presence of cyanoketone, an inhibitor of Δ^5-3β-hydroxysteroid dehydrogenase, progesterone augmented the ability of FSH to stimulate pregnenolone synthesis (Fanjul et al., 1983). The physiological significance of this latter observation may be in doubt, as the concentration (1×10^{-6} M) of synthetic progestin required to elicit a response is at the extreme upper limit of the physiological progestin concentration. The possibility that the action of progestin was mediated by nonspecific binding to androgen receptors was considered; however, the apparent autoregulatory actions of progestins were not altered by treatment with antiandrogens suggesting that the effect was not due to binding of progestin to androgen receptors (Hsueh et al., 1984).

Effects of Progestin on Estrogen Biosynthesis

The effects of several progestins on FSH-stimulated estrogen production by cultured rat granulosa cells isolated from ovaries of DES-treated immature hypophysectomized rats have been examined (Schreiber et al., 1980). FSH-enhanced estrogen secretion was reduced as a result of treatment with progesterone, 20α-dihydroprogesterone, or R5020. The relative potencies of these three progestins in decreasing estrogen secretion was proportional to the relative abilities of the compounds to bind to ovarian progestin receptors, with R5020 being most effective and 20α-dihydroprogesterone least. From studies of the mechanism by which R5020 inhibits FSH-induction of aromatase activity, it was concluded that the synthetic progestin acts at a site distal to cAMP production and that it is not a competitive inhibitor of aromatase (Schreiber et al., 1981). In studies with rat granulosa cells isolated on the morning of proestrus from follicles of immature rats previously treated with PMSG (4 IU), progesterone had a slight suppressive effect on estradiol synthesis. However, it was evident that once aromatizing activity had been induced it was much less sensitive to inhibition by progesterone (Fortune and Vincent, 1983).

An inhibitory effect of progesterone on estrogen secretion has been demonstrated *in vivo* (Saidapur and Greenwald, 1979). Administration of progesterone to hamsters on the morning of the day of proestrus was followed by a fall in serum estradiol concentration without a detectable change in blood levels of gonadotropins. The fact that concomitant administration of testosterone did not reverse the effect of progesterone indicated that progesterone acted at the level of the aromatase enzyme system. These observations led the authors to speculate that the inhibitory effect of progesterone is one factor in the sharp decline in the serum concentration of estrogen that occurs after the LH surge in normally cycling hamsters (Greenwald, 1974).

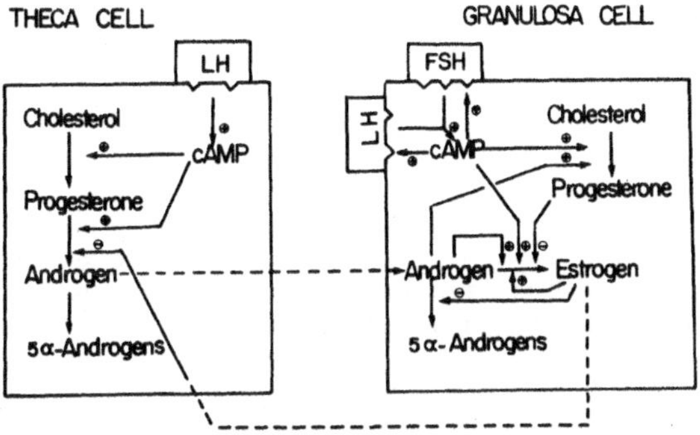

Fig. 1. Sites of action of steroids, and interactions with gonadotropins, in control of follicular steroid biosynthesis. This diagram localizes the major biosynthetic steps at which steroidogenesis is stimulated by FSH and LH (via cyclic AMP-mediated processes), and indicates those steps at which ovarian steroids have been shown to exert modulating actions of either a stimulatory (+) or inhibitory (-) nature.

Regulation of androgen production in the ovary by progestins has not been reported, but both progesterone (Mahajan and Samuels, 1975) and 5α-pregnane-3,20-dione (Brophy and Gower, 1974), are effective inhibitors of the C-17,20-lyase, thus indicating their potential in intra-ovarian regulation of androgen biosynthesis.

CONCLUDING COMMENTS

In this review, we have examined the evidence that steroids, of ovarian origin, act locally to influence the steroidogenic activity of the follicles, both in the absence and the presence of stimulation by the pituitary gonadotropins. As a result of such intra-ovarian actions, steroids represent one of the mechanisms by which follicular recruitment, growth, atresia, and ovulation may be influenced by local intra-ovarian factors. Although the pituitary gonadotropic hormones, FSH and LH are essential for progression of follicular development beyond the early antral stage, it has become increasingly evident that local intra-ovarian factors not only are involved in initiation of growth of follicles before they acquire gonadotropin responsiveness, but also influence the subsequent fate of these follicles, i.e. atresia or ovulation, under the primary stimuli of FSH and LH. Steroids are the best characterized of these intra-ovarian regulatory factors, and several sites at which their actions are exerted, have been identified, as summarized in Fig. 1.

Some of the intra-ovarian actions of steroids appear to be mediated through interactions with receptors similar to those well established in other steroid target organs, whereas others may be mediated via other mechanisms such as cell membrane effects or direct action on steroidogenic enzymes. Increased understanding of these basic mechanisms should enable development of important applications in the field of fertility and infertility, such as development of novel methods of contraception not dependent upon central suppression of gonadotropin secretion, and of new approaches to control ovulation rate for increasing fecundity of farm animals as well as of endangered wild species.

Acknowledgements

We are grateful to Mr. G. Barbe, Mrs. H. E. Ross and Ms. U. Williams for their diligent efforts in compiling and sorting references, typing and proof-reading during preparation of this manuscript.

REFERENCES

Adashi, E. Y., and Hsueh, A. J. W., 1982, Estrogens augment the stimulation of ovarian aromatase activity by follicle-stimulating hormone in cultured rat granulosa cells, J. Biol. Chem., 257:6077.

Armstrong, D. T., 1979, Alterations of progesterone metabolism in immature rat ovaries by luteinizing hormone, Biol. Reprod., 21:1025.

Armstrong, D. T., and Dorrington, J. H., 1976, Androgens augment FSH-induced progesterone secretion by cultured rat granulosa cells, Endocrinology, 99:1411.

Armstrong, D. T., and Dorrington, J. H., 1977, Estrogen biosynthesis in the ovaries and testes, in: "Regulatory Mechanisms Affecting Gonadal Hormone Action; Advances in Sex Hormone Research," J. A. Thomas, R. L. Singhal, eds., Vol. III, p. 217, University Park Press, Baltimore.

Armstrong, D. T., Daniel, S. A. J., Salhanick, A. R., and Sheela Rani, C. S., 1980, Hormonal interactions in regulation of steroid biosynthesis by the ovarian follicle, in: "Functional Correlates of Hormone Receptors in Reproduction," V. B. Makesh, T. G. Muldoon, B. B. Saxena, W. A. Sadler, eds., p. 245, Elsevier North Holland Inc., New York.

Armstrong, D. T., Dorrington, J. H., and Robinson, J., 1976, Effects of indomethacin and aminoglutethimide phosphate in vivo on luteinizing-hormone-induced alterations of cyclic adenosine monophosphate, prostaglandin F, and steroid levels in preovulatory rat ovaries, Can. J. Biochem., 54:796.

Bagnell, C. A., Mills, T. M., Costoff, A., and Mahesh, V. B., 1982, A model for the study of androgen effects of follicular atresia and ovulation, Biol. Reprod., 27:903.

Bahr, J. M., 1978, Simultaneous measurement of steroids in follicular fluid and ovarian venous blood in the rabbit, Biol. Reprod. 18:193.

Batta, S. K., Wentz, A. C., and Channing, C. P., 1980, Steroidogenesis by human ovarian cell types in culture: influence of mixing of cell types and effect of added testosterone, J. Clin. Endocrinol. Metab., 50:274.

Beattie, C. W., and Corbin, A., 1975, The differential effects of diestrous progesterone administration on proestrous gonadotrophin levels, Endocrinology, 97:885.

Bernard, J., 1975, Effect of follicular fluid and estradiol on the luteinization of rat granulosa cells in vitro, J. Reprod. Fertil., 45:453.

Bieszczad, R. R., McClintock, J. S., Pepe, G. J., and Dimino, M. J., 1982, Progesterone secretion by granulosa cells from different sized follicles of human ovaries after short term incubation, J. Clin. Endocrinol. Metab., 55:181.

Bogovich, K., Richards, J. S., and Reichert, L. E. Jr., 1981, Obligatory role of LH in the initiation of preovulatory follicular growth in the pregnant rat: specific effects of human chorionic gonadotropin and follicle-stimulating hormone on LH receptors and steroidogenesis in theca, granulosa and luteal cells, Endocrinology, 109:860.

Bradbury, J. T., 1961, Direct action of estrogen on the ovary of the immature rat, Endocrinology, 68:112.

Brophy, P. J., and Gower, D. B., 1974, Studies on the inhibition by 5-pregnane-3,20-dione of the biosynthesis of 16-androstenes and dehydroepiandrosterone in boar testis preparations, Biochim. Biophys. Acta, 360:252.

Buffler, G., and Roser, S., 1974, New data concerning the role played by progesterone in the control of follicular growth in the rat, Acta Endocrinol., 75:569.

Campo, S. M., Carson, R. S., and Findlay, K. K., 1984, Distribution and characterization of specific androgen-binding sites within the ovine follicle, 15th Annual Conference of the Australian Society for Reproductive Biology, Canberra, Abstract, p. 27.

Carson, R. S., Findlay, J. K., Clarke, I. J., and Burger, H. G., 1981, Estradiol, testosterone, and androstenedione in ovine follicular fluid during growth and atresia of ovarian follicles, Biol. Reprod., 24:105.

Corredor, A., and Flickinger, G. L., 1983, Hormonal regulation of progesterone secretion by cultured mouse granulosa cells, Biol. Reprod., 29:1142.

Daniel, S. A. J., and Armstrong, D. T., 1980, Enhancement of follicle-stimulating hormone-induced aromatase by androgens in cultured rat granulosa cells, Endocrinology, 107:1027.

Daniel, S. A. J., and Armstrong, D. T., 1983, Involvement of estrogens in the regulation of granulosa cell aromatase activity, Can. J. Physiol. Pharmacol., 61:507.

Daniel, S. A. J., and Armstrong, D. T., 1984, Site of action of androgens on follicle-stimulating hormone-induced aromatase activity in cultured rat granulosa cells, Endocrinology, 114:1975.

Dorrington, J. H., and Armstrong, D. T., 1979, Effect of FSH on gonadal functions, Rec. Prog. Horm. Res., 35:301.

Duleba, A. J., Takahashi, H., and Moon, Y. S., 1983, Androgenic modulation of progesterone metabolism by rat granulosa cells in culture, Steroids, 42:321.

Eckstein, B., and Nimrod, A., 1977, Properties of microsomal 3-ketosteroid 5α-reductase in immature rat ovary, Biochim. Biophys. Acta, 499:1.

Engels, J. A., Friedlander, R. L., and Eik-Nes, K. B., 1968, An effect in vivo of clomiphene on the rate of conversion of androstenedione-^{14}C to estrone-^{14}C and estradiol-^{14}C by the canine ovary, Metabolism, 17:189.

Evans, G., Leung, P. C. K., Brodie, A. M. H., and Armstrong, D. T., 1981, Effect of an aromatase inhibitor (4-acetoxy-4-androstene-3,17-dione) on the stimulatory action of luteinizing hormone on estradiol-17β synthesis by rat preovulatory follicles in vitro, Biol. Reprod., 25:290.

Evans, G., Lischinsky, A., Daniel, S. A. J., and Armstrong, D. T., 1984, Androgen inhibition of FSH-stimulated progesterone production by granulosa cells of prepubertal pig, Can. J. Physiol. Pharmacol., 62:840.

Fanjul, L. F., de Galarreta, R., and Hsueh, A. J. W., 1983, Progestin augmentation of gonadotrophin-stimulated progesterone production by cultured rat granulosa cells, Endocrinology, 112:405.

Fortune, J. E., and Hansel, W., 1979, The effects of 17β-estradiol on progesterone secretion by bovine theca and granulosa cells, Endocrinology, 104:1834.

Fortune, J. E., and Vincent, S. E., 1983, Progesterone inhibits the induction of aromatase activity in rat granulosa cells in vitro, Biol. Reprod., 28:1078.

Goff, A. K., and Henderson, K. M., 1979, Changes in follicular fluid and serum concentrations of steroids in PMS-treated immature rats following LH administration, Biol. Reprod., 20:1153.

Goff, A. K., Leung, P. C. K., and Armstrong, D. T., 1979, Stimulatory action of follicle-stimulating hormone and androgens on the responsiveness of rat granulosa cells to gonadotropins in vitro, Endocrinology, 104:1124.

Goldenberg, R. L., Bidson, W. E., and Kohler, P. O., 1972, Estrogen stimulation of progesterone synthesis by porcine granulosa cells in culture, Biochem. Biophys. Res. Comm., 48:101.

Goodman, A. L., and Hodgen, G. D., 1977, Systemic versus intra-ovarian progesterone replacement after lutiectomy in rhesus monkeys: differential patterns of gonadotropins and follicle growth, J. Clin. Endocrinol. Metab., 45:837.
Gore-Langton, R. E., and Armstrong, D. T., 1987, Follicular steroidogenesis and its control, in: "The Physiology of Reproduction," E. Knobil, J. D. Neill, eds., Raven Press, New York, in press.
Greenwald, G. S., 1974, Gonadotropin regulation of follicular development, in: "Gonadotropins and Gonadal Function," N. R. Moudgal, ed., p. 205, Academic Press, New York.
Greep, R. O., Fevold, H. L., and Hisaw, F. L., 1936, Effect of two hypophyseal gonadotrophic hormones on the reproductive system of the male rat, Anat. Rec., 65:261.
Greep, R. O., van Dyke, H. B., and Chow, B. F., 1942, Gonadotropins of the swine pituitary. I. Various biological effects of purified thylakentrin (FSH) and pure metakentrin (ICSH), Endocrinology, 30:635.
Hammond, J. M., Hershey, R. M., Walega, M. A., and Weisz, J., 1986, Catecholestrogen production by porcine ovarian cells, Endocrinology, 118:2292.
Haney, A. F., and Schomberg, D. W., 1978, Steroidal modulation of progesterone secretion by granulosa cells from large porcine follicles: a role for androgens and estrogens in controlling steroidogenesis, Biol. Reprod., 19:242.
Harman, S. M., Louvet, J. P., and Ross, G. T., 1975, Interaction of estrogen and gonadotrophins on follicular atresia, Endocrinology, 96:1145.
Hillier, S. G., and DeZwart, F. A., 1981, Evidence that granulosa cell induction/activation by follicle-stimulating hormone is an androgen receptor regulated process in vitro, Endocrinology, 109:1303.
Hillier, S. G., and DeZwart, F. A., 1982, Androgen/antiandrogen modulation of cyclic AMP-induced steroidogenesis during granulosa cell differentiation in tissue culture, Mol. Cell. Endocrinol., 28:347.
Hillier, S. G., and Ross, G. T., 1979, Effects of exogenous testosterone on ovarian weight, follicular morphology and intra-ovarian progesterone concentration in estrogen-primed hypophysectomized immature female rats, Biol. Reprod., 20:261.
Hillier, S. G., Knazek, R. A., and Ross, G. T., 1977, Androgenic stimulation of progesterone production by granulosa cells from preantral ovarian follicles: further in vitro studies using replicate cell cultures, Endocrinology, 100:1539.
Hillier, S. G., van den Boogaard, A. M. J., Reichert, L. E. Jr., and van Hall, E. V., 1980a, Alterations in granulosa cell aromatase activity accompanying preovulatory follicular development in the rat ovary with evidence that 5α-reduced C_{19} steroids inhibit the aromatase reaction in vitro, J. Endocrinol., 84:409.
Hillier, S. G., van den Boogaard, A. M. J., Reichert, L. E. Jr., and van Hall, E. V., 1980b, Intra-ovarian sex steroid hormone interactions and the regulation of follicular maturation: aromatization of androgens by human granulosa cells in vitro, J. Clin. Endocrinol. Metab., 50:640.
Hori, T., Kato, G., and Miyake, T., 1973, Acute effects of ovarian steroids upon follicular growth in the cycling rat, Endocrinol. Jpn., 20:475.
Hsueh, A. J. W., Adashi, E. Y., Jones, P. B. C., and Welsh, T. H., 1984, Hormonal regulation of the differentiation of cultured ovarian granulosa cells, Endocrine Rev., 5:76.
Hudson, K. E., and Hillier, S. G., 1985, Catechol estradiol control of FSH-stimulated granulosa cell steroidogenesis, J. Endocrinol., 106:R1.
Hunter, M. G., and Armstrong, D. T., 1986, Estrogens inhibit steroid production by dispersed porcine thecal cells, Biol. Reprod., 34 (Suppl. 1), Abstract No. 293, p. 196.
Inaba, T., Imori, T., and Matsumoto, K., 1978, Formation of 5α-reduced C_{19}-steroids from progesterone in vivo by 5α-reduced pathway in immature rat ovaries, J. Steroid Biochem., 9:1105.

Jacobs, B. R., and Smith, R. G., 1980, Evidence for a receptor-like protein for progesterone in bovine ovarian cytosol, Endocrinology, 106:1276.

Jacobs, B. R., Suchocki, S., and Smith, R. G., 1980, Evidence for human ovarian progesterone receptor, Am. J. Obstet. Gynecol., 138:332.

Jesel, L., 1970, Données nouvelles sur le contrôle eneece par le corps jaune sur la croissance folliculaire au debut du cycle oestral chez le Cobaye, C.R. Acad. Sci. [D] (Paris), 271:1693.

Jones, P. B. C., and Hsueh, A. J. W., 1982, Pregnenolone biosynthesis by cultured granulosa cells: modulation by follicle-stimulating hormone and gonadotropin-releasing hormone, Endocrinology, 111:713.

Kalra, S. P., and Kalra, P. S., 1974, Temporal interrelationships among circulating levels of estradiol, progesterone and LH during the rat estrous cycle: effects of exogenous progesterone, Endocrinology, 95:1711.

Katz, Y., and Armstrong, D. T., 1976, Inhibition of ovarian estradiol-17 secretion by luteinizing hormone in prepubertal, pregnant mare serum-treated rats, Endocrinology, 99:1442.

Khan, M. I., and Gibori, G., 1984, Catechol estrogens and their role in luteal steroidogenesis, Biol. Reprod., 30 (Suppl. 1), Abstract No. 194, p. 127.

Knecht, M., Darbon, J. M., Ranta, T., Baukal, A. J., and Catt, K. J., 1984, Estrogens enhance the adenosine 3',5'-monophosphate-mediated induction of follicle-stimulating hormone and luteinizing hormone receptors in rat granulosa cells, Endocrinology, 115:41.

Kumari, G. L., Datta, J. K., Das, R. P., and Roy, S., 1978, Evidence for a role of androgens in the growth and maturation ovarian follicles in rats, Horm. Res., 9:112.

Leung, P. C. K., and Armstrong, D. T., 1979a, Estrogen treatment of immature rats inhibits ovarian androgen production in vitro, Endocrinology, 104:1411.

Leung, P. C. K., and Armstrong, D. T., 1979b, A mechanism for the intra-ovarian inhibitory action of estrogen on androgen production, Biol. Reprod., 21:1035.

Leung, P. C. K., and Armstrong, D. T., 1980, Further evidence in support of a short-loop feedback action of estrogen on ovarian androgen production, Life Sci., 27:415.

Leung, P. C. K., Goff, A. K., and Armstrong, D. T., 1979, Stimulatory action of androgen administration in vivo on ovarian responsiveness to gonadotropins, Endocrinology, 104:1119.

Leung, P. C. K., Goff, A. K., Kennedy, T. G., and Armstrong, D. T., 1978, An intra-ovarian inhibitory action of estrogen and androgen production in vivo, Biol. Reprod., 19:641.

Lischinsky, A., Evans, G., and Armstrong, D. T., 1983, Site of androgen inhibition of FSH-stimulated progesterone production in porcine granulosa cells, Endocrinology, 113:1999.

Louvet, J. P., Harman, S. M., Schreiber, J. R., and Ross, G. T., 1975, Evidence for a role of androgens in follicular maturation, Endocrinology, 97:366.

Lucky, A. W., Schreiber, J. R., Hillier, S. G., Schulman, J. D., and Ross, G. T., 1977, Progesterone production by cultured preantral rat granulosa cells: stimulation by androgens, Endocrinology, 100:128.

Lyon, M. F., and Glenister, P. H., 1974, Evidence from Tfm/O that androgen is essential for reproduction in female mice, Nature (London), 247:366.

Magoffin, D.A., and Erickson, G. F., 1982, Direct inhibitory effects of estrogen on LH-stimulated androgen synthesis by ovarian cells cultured in defined medium, Mol. Cell. Endocrinol., 28:81.

Mahajan, D. K., and Samuels, L. T., 1975, Inhibition of 17,20(17-hydroxy-progesterone)-lyase by progesterone, Steroids, 25:217.

McNatty, K. P., Makris, A., Reinhold, V. N., DeGrazia, C., Osathanondoh, R., and Ryan, K. J., 1979, Metabolism of androstenedione by human ovarian tissues in vitro with particular references to reductase and aromatase activity, Steroids, 34:429.

McNatty, K. P., Moore Smith, D., Makris, A., Osathanondh, R., and Ryan, K. J., 1979, The microenvironment of the human antral follicle: interrelationships among the steroid levels in antral fluid, the population of granulosa cells, and the status of the oocyte in vivo and in vitro, J. Clin. Endocrinol. Metab., 49:851.

Milwidsky, A., Younes, M. A., Besch, N. F., Besch, P. K., and Kaufman, R. H., 1980, Receptor-like binding proteins for testosterone and progesterone in the human ovary, Am. J. Obstet. Gynecol., 138:93.

Moon, Y. S., Duleba, A. J., Kim, K. S., and Ho Yuen, B., 1985, Alterations of 20α-hydroxysteroid dehydrogenase activity in cultured rat granulosa cells by follicle-stimulating hormone and testosterone, Biol. Reprod., 32:998.

Moon, Y. S., Duleba, A. J., and Takahashi, H., 1984, Differential actions of LH and androgens on progesterone catabolism by rat granulosa cells, Biochem. Biophys. Res. Comm., 119:694.

Moor, R. M., Hay, M. F., and Seamark, R. F., 1975, The sheep ovary: regulation of steroidogenic, haemodynamic and structural changes in the largest follicle and adjacent tissue before ovulation, J. Reprod. Fertil., 45:595.

Mori, T., Suzuki, A., Nishimura, T., and Kambegawa, A., 1977, Evidence for androgen participation in induced ovulation in immature rats, Endocrinology, 101:623.

Naess, O., 1981, Characterization of cytoplasmic progesterone receptors in rat granulosa cells: evidence of nuclear translocation, Acta Endocrinol., 98:288.

Neumann, F., von Berswordt-Wallrabe, R., Elger, W., Steinbeck, K., Hann, J. D., and Kramer, M., 1970, Aspects of androgen-dependent events as studied by antiandrogens, Rec. Prog. Horm. Res., 26:337.

Nimrod, A., 1977, Studies on the synergistic effect of androgen on the stimulation of progestin secretion by FSH in cultured rat granulosa cells: a search for the mechanism of action, Mol. Cell. Endocrinol., 8:201.

Nimrod, A., 1981, On the synergistic action of androgen and FSH on progestin secretion by cultured rat granulosa cells: cellular and mitochondrial cholesterol metabolism, Mol. Cell. Endocrinol., 21:51.

Nimrod, A., and Lindner, H. R., 1976, A synergistic effect of androgen on the stimulation of progesterone secretion by FSH in cultured rat granulosa cells, Mol. Cell. Endocrinol., 5:315.

Nimrod, A., Rosenfield, R. L., and Otto, P., 1980, Relationship of androgen action to androgen metabolism in isolated rat granulosa cells, J. Steroid Biochem., 13:1015.

Ohno, S., Christian, L., and Attardi, B., 1973, Role of testosterone in normal female function, Nature (London), 243:119.

Opavsky, M. A., and Armstrong, D. T., 1985, The effectiveness of FSH in inducing superovulation is influenced by LH, Biol. Reprod., 32 (Suppl. 1), Abstract No. 67, p. 71.

Pasqualini, J. R., and Nguyen, B. J., 1980, Progesterone receptors in fetal uterus and ovary of the guinea pig: evolution during fetal development and induction and stimulation in estradiol-primed animals, Endocrinology, 106:1160.

Payne, R. W., and Hellbaum, A. A., 1955, The effect of estrogens on the ovary of the hypophysectomized rat, Endocrinology, 57:193.

Payne, R. W., and Runser, R. H., 1958, The influence of estrogen and androgen on the ovarian response of hypophysectomized immature rats to gonadotropins, Endocrinology, 62:313.

Peluso, J. J., Brown, I., and Steger, R. W., 1979, Effects of cyproterone acetate, a potent antiandrogen, on the preovulatory follicle, Biol. Reprod., 21:929.

Pencharz, R. I., 1940, Effect of estrogens and androgens alone and in combination with chorionic gonadotropin on the ovary of the hypophysectomized rat, Science, 91:554.

Philibert, D., Ojasoo, T., and Raynaud, J. P., 1977, Properties of the cytoplasmic progestin-binding protein in the rabbit uterus, Endocrinology, 101:1850.
Reiter, E. O., Goldenberg, R. L., Vaitukaitis, J. L., and Ross, G. T., 1972, Evidence for a role of estrogen in ovarian augmentation reaction, Endocrinology, 91:1518.
Richards, J. S., 1974, Estradiol binding to rat corpora lutea during pregnancy, Endocrinology, 95:1046.
Richards, J. S., 1975, Estradiol receptor content of rat granulosa cells during follicular development: modification by estradiol and gonadotropins, Endocrinology, 97:1174.
Richards, J. S., 1980, Maturation of ovarian follicles: actions and interactions of pituitary and ovarian hormones on follicular cell differentiation, Physiol. Rev., 60:51.
Richards, J. S., and Bogovich, K., 1982, Effect of human chorionic gonadotropin and progesterone on follicular development in the immature rat, Endocrinology, 111:1429.
Saidapur, S. K., and Greenwald, G. S., 1979, Regulation of 17β-estradiol synthesis in the proestrous hamster: role of progesterone and luteinizing hormone, Endocrinology, 105:1432.
Saiduddin, S., 1971, ^{3}H-estradiol uptake by the rat ovary, Proc. Soc. Exp. Biol. Med., 138:651.
Saiduddin, S., and Milo, G.E. Jr., 1974, Effect of hypophysectomy and pretreatment on uptake and retention of estradiol by the ovary, Proc. Soc. Exp. Biol. Med., 146:513.
Saiduddin, S., and Zassenhaus, H. P., 1977, Estradiol-17β receptors in the immature rat ovary, Steroids, 29:197.
Saiduddin, S., and Zassenhaus, H. P., 1978, Effect of testosterone and progesterone on the estradiol receptor in the immature rat ovary, Endocrinology, 102:1069.
Schomberg, D. W., Stouffer, R. L., and Tyrey, L., 1976, Modulation of progestin secretion in ovarian cells by 17β-hydroxy-5-adrostan-3-one (dihydrotestosterone): a direct demonstration in monolayer culture, Biochem. Biophys. Res. Comm., 68:77.
Schomberg, D. W., Williams, R. F., Tyrey, L., and Ulberg, L. C., 1978, Reduction of granulosa cell progesterone secretion in vitro by intra-ovarian implants of antiandrogen, Endocrinology, 102:984.
Schreiber, J. R., and Erickson, G. F., 1979, Progesterone receptor in the rat ovary: further characterization and localization in the granulosa cell, Steroids, 34:459.
Schreiber, J. R., and Ross, G. T., 1976, Further characterization of rat ovarian testosterone receptor with evidence for nuclear translocation, Endocrinology, 99:590.
Schreiber, J. R., Hsueh, A. J. W., and Baulieu, E. E., 1983, Binding of the antiprogestin RU-486 to rat ovary steroid receptors, Contraception, 28:77.
Schreiber, J. R., Nakamura, K., and Erickson, G. F., 1980, Progestins inhibit FSH-stimulated steroidogenesis in cultured rat granulosa cells, Mol. Cell. Endocrinol., 19:165.
Schreiber, J. R., Nakamura, K., and Erickson, G. F., 1981, Progestins inhibit FSH-stimulated granulosa estrogen production at a post-cAMP site, Mol. Cell. Endocrinol., 21:161.
Schreiber, J. R., Reid, R., and Ross, G. T., 1976, A receptor-like testosterone binding protein in ovaries from estrogen-stimulated hypophysectomized immature female rats, Endocrinology, 98:1206.
Schreiber, J. R., and Hsueh, A. J. W., 1979, Progesterone "receptor" in rat ovary, Endocrinology, 105:915.
Schrieber, J. R., Nakamura, K., and Weinstein, D. B., 1983, Androgen and FSH synergistically stimulate rat ovary granulosa cell utilization of rat and human lipoproteins, in: "Factors Regulating Ovarian Function," G. Greenwald, P. F. Terranova, eds., p. 311, Raven Press, New York.

Shemesh, M., and Ailenberg, M., 1977, The effect of androstenedione on progesterone accumulation in cultures of bovine ovarian follicles, Biol. Reprod., 17:499.

Siiteri, P. K., and Thompson, E. A., 1975, Studies of human placental aromatase, J. Steroid Biochem., 6:317.

Simpson, M. E., Evans, H. M., Fraenkel-Conrat, H. L., and Li, C. H., 1941, Synergism of estrogens with pituitary gonadotropins in hypophysectomized rats, Endocrinology, 28:37.

Smith, B. D., 1961, The effect of diethylstilbestrol on the immature rat ovary, Endocrinology, 69:238.

Smith, B. D., and Bradbury, J. T., 1966, Influence of progestins on ovarian responses to estrogen and gonadotrophins in immature rats, Endocrinology, 78:297.

Smith, M. S., Freeman, M. E., and Neill, J. D., 1975, The control of progesterone secretion during the estrous cycle and early pseudopregnancy in the rat: prolactin gonadotropin and steroid levels associated with rescue of the corpus luteum of pseudopregnancy, Endocrinology, 96:219.

Smith, O. W., Ofner, P., and Vena, R. L., 1974, In vitro conversion of testosterone-4-^{14}C to androgens of the 5α-androstane series by normal human ovary, Steroids, 24:311.

Smith, P. E., 1930, Hypophysectomy and a replacement therapy in the rat, Am. J. Anat., 45:205.

Stumpf, W. E., 1969, Nuclear concentration of ^{3}H-estradiol in target tissues: dry-mount autoradiography of vagina, oviduct, ovary, testis, mammary tumor, liver and adrenal, Endocrinology, 85:31.

Tan, C. H., and Armstrong, D. T., 1984, FSH-stimulated 3β-HSD activity of porcine granulosa cells: inhibition by androgens, Proceedings of the 3rd Joint Meeting of the British Endocrine Societies, Edinburgh, Abstract No. 12.

Thanki, K. H., and Channing, C. P., 1976, Influence of serum, estrogen, and gonadotropins upon growth and progesterone secretion by cultures of granulosa cells from small porcine follicles, Endocr. Res. Commun., 3:319.

Thanki, K. H., and Channing, C. P., 1978, Effects of follicle-stimulating hormone and estradiol upon progesterone secretion by porcine granulosa cells in tissue culture, Endocrinology, 103:74.

Toaff, M. E., Strauss III, J. F., and Hammond, J. M., 1983, Regulation of cytochrome P-450_{SCC} in immature porcine granulosa cells by FSH and estradiol, Endocrinology, 112:1156.

Tsafriri, A., and Braw, R. H., 1984, Experimental approaches to atresia in mammals, in: "Oxford Reviews of Reproductive Biology," R. E. Clarke, ed., Vol. 6, p. 226, Claredon Press, Oxford.

Veldhuis, J. D., 1985, Bipotential actions of estrogens on progesterone biosynthesis by ovarian cells. II. Relation of estradiol's stimulatory actions to cholesterol and progestin metabolism in cultured swine granulosa cells, Endocrinology, 117:1076.

Veldhuis, J. D., Klase, P. A., Sandon, B. A., and Kolp, L. A., 1983, Progesterone secretion by highly differentiated human granulosa cells isolated from preovulatory Graafian follicles, induced by endogenous gonadotropins and human chorionic gonadotropin, J. Clin. Endocrinol. Metab., 57:287.

Veldhuis, J. D., Klase, P. A., Strauss, J. F., and Hammond, J. M., 1982, Facilitative interactions between estradiol and luteinizing hormone in the regulation of progesterone production by cultured swine granulosa cells: relation to cholesterol metabolics, Endocrinology, 111:441.

Welsh, T. H. Jr., Jones, P. B. C., Ruiz de Galaretta, C. M., Fanjul, L. F., and Hsueh, A. J. W., 1982, Androgen regulation of progestin biosynthetic enzymes in FSH-treated rat granulosa cells in vitro, Steroids, 40:691.

Welsh, T. H. Jr., Jia, X. -C., Jones, P. B. C., Zhuang, L. -Z., and Hsueh, A. J. W., 1984, Disparate effects of triphenylethylene antiestrogens on estrogen and progestin biosyntheses by cultured rat granulosa cells, Endocrinology, 115:1275.
Welsh, T. H. Jr., Zhuang, L. -Z., and Hsueh, A. J. W., 1983, Estrogen augmentation of gonadotropin-stimulated progestin biosynthesis in cultured rat granulosa cells, Endocrinology, 112:1916.
Williams, P. C., 1940, Effect of stilbestrol on the ovaries of hypophysectomized rats, Nature (London), 145:388.
Williams, P. C., 1944, Ovarian stimulation by oestrogens: effects in immature hypophysectomized rats, Proc. R. Soc. Series B, 132:189.
Zeleznik, A. J., Hillier, S. G., and Ross, G. T., 1979, Follicle-stimulating hormone-induced follicular development: an examination of the role of androgens, Biol. Reprod., 21:673.
Zhuang, L. -Z., Adashi, E. Y., and Hsueh, A. J. W., 1982, Direct enhancement of gonadotropin-stimulated ovarian estrogen biosynthesis by estrogen and clomiphene citrate, Endocrinology, 110:2219.

INTERACTION BETWEEN THE OOCYTE AND THE GRANULOSA CELLS IN THE PREOVULATORY FOLLICLE

Nava Dekel

Department of Hormone Research
The Weizmann Institute of Science
Rehovot, Israel 76100

INTRODUCTION

The oocyte in the primordial follicle is surrounded by a single layer of follicular cells. These cells, originating from the germinal epithelium of the genital ridge, will proliferate to form the granulosa cell layer of the secondary follicle (Brambell, 1962; Franchi et al., 1962). Throughout folliculogenesis the granulosa layer is divided into two cellular subpopulations: the peripheral granulosa cells which become part of the follicular wall and the cumulus cells which remain attached to the oocyte to form the cumulus oophorus.

The cells of the innermost layer of the cumulus, the corona radiata, are intimately connected with the oocyte by long cytoplasmic projections which traverse the zona pellucida and intermingle with numerous microvilli from the oocyte surface (Paladino, 1890; Yamada et al., 1957; Sotelo and Porter, 1959; Anderson and Beams, 1960; Franchi, 1960; Odor, 1960; Tardini et al., 1960; Björkman, 1962; Zamboni and Mastroianni, 1966; Baca and Zamboni, 1967; Zamboni, 1974; Dekel et al., 1976; Dekel et al., 1978; Albertini, 1984). As uptake of uridine and certain amino acids by mouse, rat and sheep oocytes could only be obtained in the presence of the attached cumulus (Cross, 1973; Wassarman and Letourneau, 1976; Heller et al., 1981; Eppig, 1982; Brower and Schultz, 1982; Colonna and Mangia, 1983; Racowski, 1984; Moor et al., 1980), it appeared that these compounds are first taken up by the cumulus cells and subsequently enter the oocyte. Transfer of small molecules is apparently executed via gap junction present in the regions of contact between the cumulus cell projections and the colemma (Szöllösi, 1975; Amsterdam et al., 1976; Anderson and Albertini, 1976; Gilula et al., 1978). Interconnection by means of gap junctions were also described between adjacent cumulus as well as granulosa cells (Björkman, 1962; Merk et al., 1972; Albertini and Anderson, 1974; Zamboni, 1974; Szöllösi, 1975; Gilula et al., 1978). These specific structures were observed both between closely apposed cell membranes and at the regions of contact of microvilli and cell membranes.

The extensively developed communicating network established between the somatic cellular components of the follicle and the female gamete is probably of major functional importance. A likely physiological role for this type of heterologous cell communication is nutritional; follicle cells directly provide nutrients which are necessary for oocyte growth. A role for follicle cells in promoting growth is suggested by the observation that oocytes do not grow in the absence of follicle cells and the oocyte does not occur (Eppig, 1979). The findings that metabolic cooperation exists between granulosa cells and oocytes in culture systems which support oocyte growth (Heller et al., 1981) and that the rate of oocyte growth in vitro is positively correlated with the extent of metabolic coupling (Brower and Schultz, 1982) clearly support this idea. Junctional communication within the follicle could also mediate transmission of regulatory signals which are responsible for the control of the meiotic status of the follicular oocyte. The studies discussed in the present paper were carried out to explore the role of cell-to-cell communication in regulation of oocyte maturation.

CELL-TO-CELL COMMUNICATION AND REGULATION OF OOCYTE MATURATION

Meiosis of the mammalian oocyte is initiated during embryonic life. It proceeds up to the diplotene stage of the first prophase and is arrested at birth. Meiotically arrested oocytes are characterized by the presence of the nuclear structure known as germinal vesicle (GV). Meiotic arrest persists until the onset of puberty. In the sexually mature female mammal, a number of oocytes characeristic of the species reinitiate their first reduction division at each cycle. The GV in these oocytes disappears and the first polar body is extruded. The whole series of events initiated with GV breakdown (GVB) and completed by the formation of the first polar body leads to the production of a mature fertilizable oocyte. This process is therefore defined as oocyte maturation.

Oocyte maturation in vivo is clearly dependent upon the preovulatory surge of luteinizing hormone (LH, reviewed by Lindner et al., 1974; Tsafriri, 1978). However, when meiotically arrested oocytes are removed from the antral follicles they resume meiosis spontaneously. This observation, which was initially reported by Pincus and Enzmann (1935) for the rabbit, and later extended to other mammalian species (Edwards, 1965; Schuetz, 1974), led to the following conclusions: a) The antral follicle provides an inhibitory factor which maintains the oocyte in meiotic arrest. b) The oocyte is unable to generate the inhibitory factor on its own. c) The inhibitory action of the follicular factor is reversed as a result of LH action. The fact that unlike isolated oocytes, follicle-enclosed oocytes undergo maturation in vitro only upon exposure to gonadotropins (Tsafriri et al., 1973) stands in accordance with the above conclusions. The idea that the antral follicle generates a factor responsible for meiotic arrest gained further support when inhibition of spontaneous maturation by follicular components co-cultured with the isolated oocytes has been demonstrated (Chang, 1955; Foote and Thibault, 1969; Tsafriri and Channing, 1975; Leibfried and First, 1980; Meinecke and Meinecke-Tillman, 1981). Our studies were aimed at a) definition of the nature of the inhibitory factor for meiosis resumption; b) investigation of the mode of communication of this factor from the follicle to the oocyte; c) exploration of the mechanism by which LH terminates its action.

Almost forty years after the original observation of Pincus and Enzmann (1935), it has been demonstrated that the spontaneous maturation in vitro of mouse oocytes released from their follicles can be reversibly blocked by addition of either a membrane permeable derivative of cyclic AMP (cAMP) or a phosphodiesterase inhibitor (Cho et al., 1974). These findings, that were later extended to other species (Magnusson and Hillensjö, 1977; Dekel and Beers, 1978; Jagiello et al., 1981), suggested that cAMP could serve as a likely candidate for the follicular factor responsible for meiotic arrest. If cAMP is involved in meiotic arrest, incubation under conditions which will increase cAMP levels should result in inhibition of oocyte maturation. To elevate cAMP levels within the oocyte, we have recently used a highly active adenylate cyclase preparation which is elaborated by the bacteria of the genus Bordetella and can be internalized by mammalian cells (Confer and

Eaton, 1982). This invasive adenylate cyclase triggers the host cell to generate high amounts of intracellular cAMP using its own pool of ATP (Hanski and Farfel, 1985). The possible inhibition of spontaneous maturation of cumulus-free oocytes subjected to the invasive cyclase was examined and its correspondence to intra-oocyte cAMP levels tested. We found that elevation of cAMP levels within the oocytes corresponded with inhibition of their spontaneous maturation (Fig. 1). Persistent inhibition was obtained upon the continuous presence of the enzyme while its removal resulted in resumption of meiosis associated with a drop in intra-oocyte cAMP levels (Dekel et al., 1985). The results of this study suggest that elevated levels of intra-oocyte cAMP are responsible for the maintenance of meiotic arreast while a drop in cAMP levels is followed by meiosis resumption.

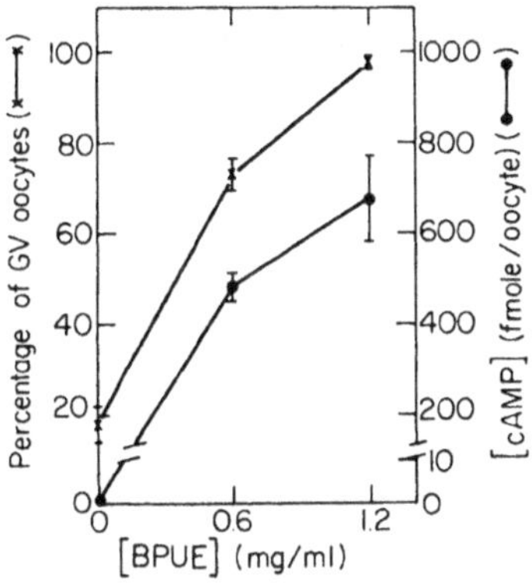

Fig. 1. Concentration dependence of the effect of _Bordetella_ pertussis urea extract (BPUE) on rat oocytes. Ooctyes were isolated from the ovaries of immature PMSG-primed rats and incubated with the indicated concentrations of BPUE for 11 h. Each group of oocytes was initially analyzed for the presence of GV by Nomarski interference contrast microscopy and further processed for cAMP determination. Determinations of cAMP were performed using NEN [^{125}I]-RIA kit.

Isolated oocytes administered _in vitro_ with an invasive adenylate cyclase cannot represent any physiological conditions. We therefore performed further determinations of cAMP in mature as compared to immature oocytes to establish the physiological role of cAMP in regulation of oocyte maturation. These experiments revealed that rat postovulatory mature oocytes contain lower levels of cAMP than follicular, GV oocytes (Table 1). Moreover, they also demonstrated that the spontaneous maturation _in vitro_ is preceded by a sharp drop in intra-oocyte cAMP (Fig. 2), while no decrease in cAMP levels is observed in oocytes maintained meiotically arrested by a

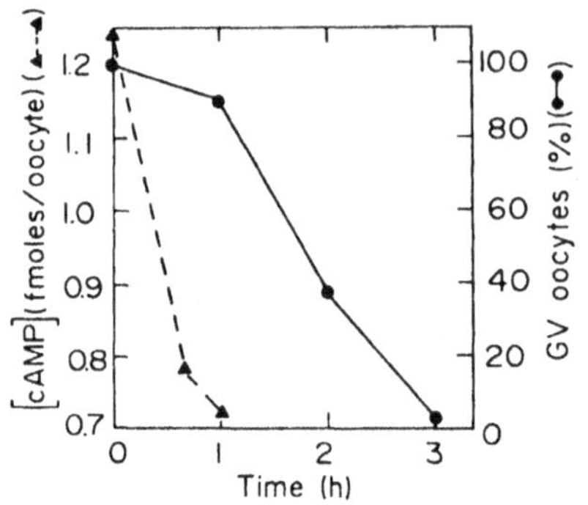

Fig. 2. Cyclic AMP levels in spontaneously maturing rat oocytes. Oocytes were released from the ovaries of immature PMSG-primed rats into L-15 Leibovitz's tissue culture medium. Determinations of cAMP and analysis for the presence of GV were performed in the oocytes at the indicated time points after their release from the ovarian follicles as described in the legend to Fig. 1.

Table 1. Cyclic AMP Levels in Immature as compared to Mature Rat Oocytes

Oocytes		[cAMP]
Treatment	Meiotic Status	fmoles/oocyte
Freshly isolated from ovary	immature	1.23 ± 0.3
Incubated for 2 h with a phosphodiesterase (PDE) inhibitor	immature	1.21 ± 0.1
Incubated for 2 h in absence of a PDE inhibitor	mature	0.80 ± 0.1
Freshly isolated from oviduct	mature	0.81 ± 0.2

Oocytes were isolated from the ovarian follicules of immature, PMSG-primed rats. Cyclic AMP determinations were performed by NEN [^{125}I]-RIA kit in either freshly isolated follicular oocytes or following 2 h of their incubation in the presence or absence of a phosphodiesterase inhibitor (0.2 mM methyl isobutyxanthine, MIX). Postovulatory oocytes were isolated from oviducts of the above rats following hCG administration. The levels of cAMP in the immature as opposed to the mature groups of oocytes are significantly different (student's t-test, $p < 0.01$).

Table 2. Cyclic AMP Levels in Oocytes exposed to Activators of Adenylate Cyclase

Treatment	[cAMP] (fmoles/oocyte)
None	1.23 ± 0.3
Forskolin (100 μM)	1.19 ± 0.4
Choleratoxin (5 μg/ml)	1.53 ± 0.1

Oocytes were isolated from the ovaries of immature PMSG-primed rats and incubated for 24 h in medium containing a phosphodiesterase inhibitor (0.2 mM MIX) with or without the indicated activators of adenylate cyclase. Determinations of cAMP were performed by NEN [^{125}I]-RIA kit.

phosphodiesterase inhibitor (Table 1). A clear correlation between the meiotic status and intra-oocyte levels of cAMP was also found in the mouse (Schultz et al., 1983; Vivarelli et al., 1983).

Administration of the invasive enzyme enabled the oocyte to generate cAMP bypassing its own adenylate cyclase. This experiment, however, could not give any indication as to whether or not rat oocytes can generate sufficient cAMP to maintain meiotic arrest. In an attempt to answer this question we have incubated cumulus-free oocytes in the presence of either cholera toxin or forskolin that interact, respectively, with the stimulatory GTP-binding regulatory component and the catalytic subunit to stimulate the adenylate cyclase. Both these activators of the adenylate cyclase system could not elevate intra-oocyte levels of cAMP (Table 2) and failed to inhibit the spontaneous maturation in vitro of rat oocytes (Dekel and Beers, 1980; Dekel et al., 1984a; Racowsky, 1984). It should be mentioned at this point that mouse oocytes apparently can respond to forskolin by generating cAMP which is accompanied by a transient inhibition of the spontaneous maturation (Schultz et al., 1983; Urner et al., 1983). Moreover, using a labeled precursor, Olsiewski and Beers (1983) could demonstrate that a certain amount of cAMP is apparently also generated by rat oocytes. However, as oocytes of all the mammalian species tested as yet undergo spontaneous maturation in vitro and as a drop in cAMP levels immediately follows their release from the follicle, it seems that our conclusion that the oocyte cannot generate by its own cAMP in levels which will reach the threshold required to maintain meiotic arrest is apparently valid.

The fact that the oocyte lacks the potential to generate high levels of cAMP while at the moment of isolation the oocyte does contain levels of cAMP which are sufficient to maintain meiotic arrest suggests that this nucleotide is probably provided to the oocyte by the follicular cells. The possibility that the oocyte is affected by cAMP secreted into the follicular fluid should be eliminated since, as opposed to cAMP derivatives, nonderivatized cAMP cannot penetrate cell membranes failing therefore to elicit an inhibitory action on maturation of isolated oocytes (Magnusson and Hillensjö, 1977; Dekel and Beers, 1978). Alternatively, cAMP can be transmitted from the follicle cells to the oocyte via cel-to-cell communication.

Table 3. Intracellular Levels of cAMP in Oocytes incubated with Forskolin in the Presence of the Cumulus Cells as compared to Oocytes similarly incubated in the Absence of the Cumulus Cells

	[cAMP] fmoles/oocyte
Cumulus-free oocytes	1.19 ± 0.47
Cumulus-enclosed oocytes	3.51 ± 0.53

Either cumulus-enclosed or cumulus-free oocytes recovered from immature PMSG-primed rats were incubated with 100 μM of forskolin in 0.1 mM MIX-containing medium for 1 h. Cumulus cells were removed from the cumulus-enclosed oocytes at the end of the incubation. Cyclic AMP in the oocytes was determined by RIA. Cumulus-enclosed oocytes contained significantly higher amounts of cAMP as compared to cumulus-free oocytes (student's t-test $p < 0.01$).

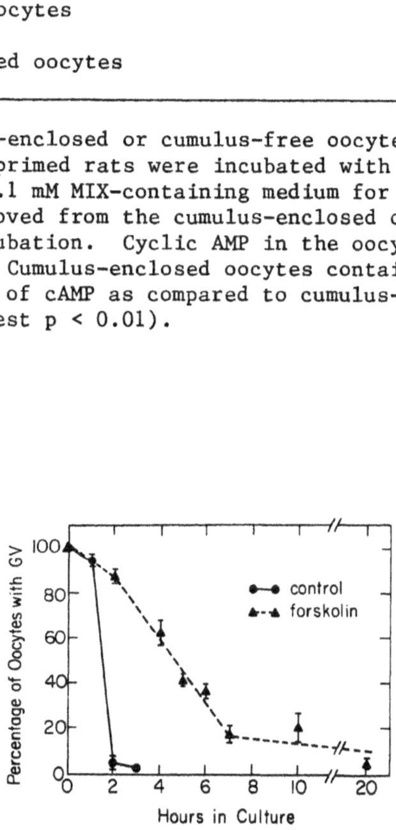

Fig. 3. Effect of forskolin on the time course of spontaneous maturation in rat cumulus-enclosed oocytes. Cumulus-oocyte complexes were isolated from the ovaries of immature PMSG-primed rats and incubated in the presence or absence of 100 μM of forskolin. The presence of GV in the oocytes was analyzed at the indicated time points and the fraction of GV oocytes out of the total examined is presented (from Dekel et al., 1984, Biol. Reprod. 31:244).

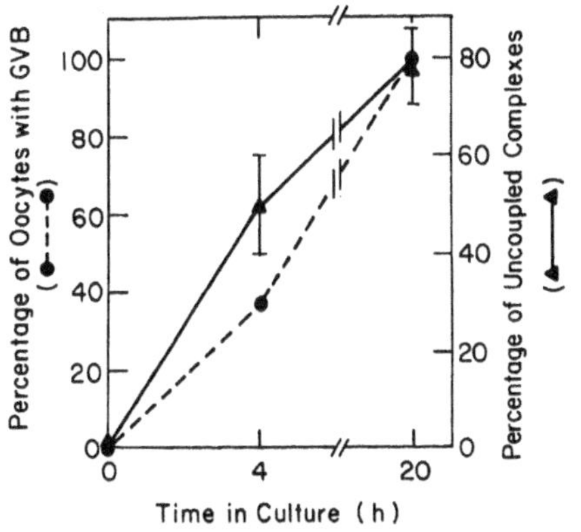

Fig. 4. Time course of EGF-induced uncoupling in the cumulus-oocyte complexes and maturation of the oocytes. Intact follicles isolated from immature PMSG-primed rats were incubated with EGF (100 ng/ml). After 20 h the follicles were incised and the cumulus-oocyte complexes recovered. Coupling was assessed by the transfer of labeled uridine from the cumulus cells to the oocyte. Maturation is expressed by the fraction of GVB oocytes at each time point (from Dekel and Sherizly, 1985, Endocrinology, 116:406).

As mentioned in the introduction to this paper, the oocyte communicates with the cumulus cell via gap junctions present at the regions of contact between these two cell types (Amsterdam, 1976; Anderson and Albertini, 1976; Gilula, 1978). Gap junctions were also observed between adjacent cumulus and granulosa cells (Bjorkman, 1962; Merk et al., 1972; Albertini and Anderson, 1974). Theoretically, according to its size cAMP can be easily communicated via gap junctions since these specialized regions in the membrane permit transcellular flow of ions and molecules which are not larger than 1000 daltons (Bennet, 1973). The actual transfer of cAMP from the cumulus cells to the oocyte has been recently demonstrated by us in rat cumulus-oocyte complexes. In these experiments we have shown that oocytes derived from forskolin-stimulated cumulus-oocyte complexes contain higher levels of cAMP than identically incubated cumulus-free oocytes (Table 3). We further demonstrated that as opposed to the cumulus-free, the cumulus-enclosed oocyte maturation *in vitro* is subsequently delayed by the presence of forskolin (Fig. 3) (Dekel et al., 1984a). Actual transfer of cAMP from the cumulus cells to the oocyte was also demonstrated by other investigators both in the rat and in the mouse (Racowsky, 1984; Bornslaeger et al., 1985; Salustri et al., 1985).

As mentioned earlier in this paper meiosis resumption *in vivo* is triggered by LH. Based on the fact that oocytes released from their follicles resume meiosis spontaneously, we have raised the hypothesis that LH probably does not provide a direct stimulatory signal for oocyte maturation but rather acts to relieve the inhibitory influence of the ovarian follicle. Since, as appears from the data presented so far, intra-oocyte elevated levels of cAMP are associated with arrest of meiosis, we assumed that any mechanism for LH-induced oocyte maturation should result in a drop of cAMP levels within the oocyte. We further predicted that if maintenance of meiotic arrest is dependent upon communication of the inhibitory cAMP, then uncoupling of the oocyte from the follicular cells could provide the appropriate conditions for resumption of meiosis. The effect of LH on communication in the cumulus-

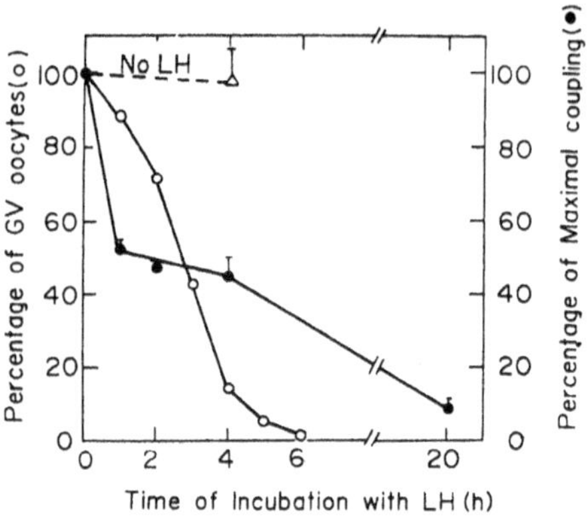

Fig. 5. Time-course of LH-induced oocyte maturation and uncoupling in rat cumulus-oocyte complexes. Cumulus-enclosed oocytes were isolated from follicles following culture in the presence or absence of 10 μg/ml of ovine LH, the oocytes were analyzed for the presence of GV. The extent of coupling is expressed as the fraction of incorporation of labeled uridine into the oocytes out of the total uptake by the corresponding cumulus cells.

oocyte complex was initially reported by Gilula et al. (1978). These investigators demonstrated that following hCG administration, coupling in the cumulus-oocyte complex is terminated. In a later study we have shown that follicle-enclosed oocytes induced to mature in vitro by LH are uncoupled from the cumulus cells (Dekel et al., 1981). These studies demonstrated that meiosis resumption and breakdown of communication are both subsequent to LH stimulation. The association between these two events is also suggested in our report that oocytes induced to mature by epidermal growth factor (EGF), are concomitantly uncoupled from the follicular cells (Fig. 4) (Dekel and Sherizly, 1985). Analysis of the temporal correlation between these two events was performed in an attempt to provide the answer to the question as to whether or not breakdown of communication could signal for oocyte maturation. In these studies, we have demonstrated that in rat follicle-enclosed cumulus-oocyte complexes metabolic coupling is decreased to 50% of its initial level following 1 h of incubation with LH (Fig. 5) (Dekel et al., 1984b). Almost all the oocytes are still meiotically arrested by this time point. Similar chronological relationships between uncoupling and reinitiation of meiosis are demonstrated in vivo in cumulus-oocyte complexes isolated from hCG-treated rats. These results clearly show that oocyte maturation follows uncoupling in the cumulus-oocyte complex. Junctional modulation in the rat cumulus-oocyte complex, which follows hCG administration in vivo, has also been demonstrated by Larsen et al. (1984). These investigators demonstrated a significant loss in the net area of cumulus gap junction membrane which occurs prior to GVB. Studies on intercellular coupling in the hamster cumulus-oocyte complex also revealed a decrease in communication during the early stages of meiosis (Racowski and Satterlie, 1985). All these studies provide support for the hypothesis that junctional disruption may signal meiotic resumption. On the other hand, the results reported by Moor et al. (1980) in sheep and Eppig (1982) in mice seem to suggest that reinitiation of maturation occurs in advance of any reduction in cumulus cell-oocyte coupling. Changes in intercellular coupling between pig oocytes and cumulus cells during maturation have been recently studied by Motlik et al. (1986). These investigators report that the contact of the cumulus-oocyte complex to the mural granulosa layers is distended prior to uncoupling of the oocyte from the cumulus cells. Since the granulosa cells serve probably as the major source for cAMP, while the cumulus cells provide mainly the channels for communication of this inhibitory signal, changes in the connections between the parietal granulosa and the cumulus cells may lead to a substantial decrease in transfer of cAMP to the oocyte. In those species in which decrease in coupling in the cumulus-oocyte complex could not be detected prior to onset of meiosis resumption, it is possible that LH interference with communication between the granulosa and the cumulus cells is the event leading to oocyte maturation.

The question whether or not communication breakdown in the cumulus-oocyte complex serves as the signal for oocyte maturation is still under controversy. There is no doubt, however, that LH uncouples the oocyte from the cumulus cells (Gilula et al., 1978; Moor et al., 1980; Dekel et al., 1981; Eppig, 1982). The mechanism by which LH interferes with communication in the cumulus-oocyte complex is not clear as yet. Cumulus cells communicate with the oocyte via cellular projections in which cytoskeletal elements, specifically microfilaments, are observed (Zamboni, 1974; Albertini, 1984). These projections are retracted following the preovulatory surge of LH (Sotelo and Porter, 1959; Odor, 1960; Dekel et al., 1976). We investigated the possible effect of gonadotropins on organization of microfilaments in the cumulus cells by staining the actin with fluorescein-conjugated NBD-phallacidin. We found that the nice bundles of filaments observed under control conditions disappeared following either gonadotropin treatment being replaced by aggregates of actin (Dekel, 1986). We concluded from this study that gonadotropins cause disassembly of the microfilaments in the cumulus cells. These findings may suggest that gonadotropin-induced retraction of the trans-zona projections of the cumulus cells is a result of their interference with the organization of the microfilaments. In the absence of the communicating channels, the oocyte is uncoupled from the cumulus cells and the transfer of an inhibitory signal is stopped. Under these conditions, the oocyte is free to reinitiate its meiotic division.

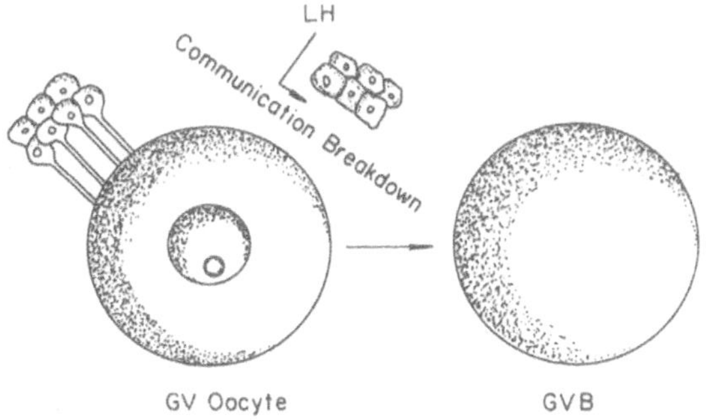

Fig. 6. A model for regulation of oocyte maturation by cell-to-cell communication between the oocyte and the follicle cells (for details see Conclusions).

CONCLUSION

Although it is clearly evident that resumption of meiosis in the follicular oocyte is under the control of gonadotropins, the mechanism responsible for regulation of oocyte maturation remained for many years one of the major puzzles of reproductive physiology. The experiments discussed in the present paper suggest the following model for regulation of oocyte maturation (Fig. 6). Cyclic AMP generated by the follicular cells is transferred to the oocyte via junctional communication with the cumulus cells to keep it in meiotic arrest. As a result of LH action, which terminates cell-to-cell communication, the transfer of cAMP is stopped, inhibition is relieved and the oocyte is allowed to resume meiosis. This model presents a unique system of heterologous cells regulated by alternation between established and interruption of cell-to-cell communication.

ACKNOWLEDGEMENTS

Studies from the author's laboratory were performed with the assistance of Mrs. D. Galiani, and supported by grants from the World Health Organization and the Israel Academy for Science and Humanities. I thank Mrs. R. Levin for secretarial assistance.

REFERENCES

Albertini, D. F., 1984, Novel morphological approaches for the study of oocyte maturation, Biol. Reprod., 30:13.

Albertini, D. F., and Anderson, E., 1974, The appearance and structure of intercellular connections during ontogeny of the rabbit ovarian follicle with particular references to gap junctions, J. Cell Biol., 63:234.

Amsterdam, A., Josephs, R., Lieberman, M. E., and Lindner, H. R., 1976, Organization of intermembrane particles in freeze-cleaved gap junctions of rat Graafian follicles: optical diffraction analysis, J. Cell Sci., 21:93.

Anderson, E., and Albertini, D. F., 1976, Gap junctions between the oocyte and the companion follicle cells in the mammalian ovary, J. Cell Biol., 71:680

Anderson, E., and Beams, H. W., 1960. Cytologic observations on the fine structure of the guinea pig ovary with special reference to the oogonium, primary oocyte and associated follicle cells, J. Ultrastruct. Res., 3:432.

Baca, M., and Zamboni, L., 1967, The fine structure of human follicular oocytes, J. Ultrastruct. Res., 19:354.

Bennet, M. W. L., 1973, Function of electronic junctions in embryonic and adult tissue, Fed. Proc., 32:65.

Björkmann, N., 1962, A study of the ultrastructure of the granulosa cells of the rat ovary, Acta Anat., 51:125.

Bornslaeger, E. A., and Schultz, R. M., 1985, Regulation of mouse oocyte maturation: effect of elevating cumulus cells cAMP on oocyte cAMP levels, Biol. Reprod., 33:698.

Brambell, F. W. R., 1962, Ovarian changes, in: "Marshall's Physiology of Reproduction," A. S. Parkes, ed., p. 397, Longmans, Green and Company, London.

Brower, P. T., and Schultz, R. M., 1982, Intercellular communication between granulosa cells and mouse oocytes: existence and possible nutritional role during oocyte growth, Develop. Biol., 90:144.

Chang, M. C., 1955, The maturation of rabbit oocytes in culture and their maturation, activation, fertilization and subsequent development in the fallopian tubes, J. Exp. Zool., 128:378.

Cho, W. K., Stern, S., and Biggers, J. D., 1974, Inhibitory effect of dibutyryl cAMP on mouse oocyte maturation in vitro, Exp. Zool., 187:383.
Colonna, R., and Mangia, F., 1983, Mechanism of amino acid uptake in cumulus-enclosed mouse oocytes, Biol. Reprod., 28:797.
Confer, D. L., and Eaton, J. W., 1982, Phagocyte impotence caused by an invasive bacterial adenylate cyclase, Science, 217:948.
Cross, P. C., 1973, Role of cumulus cells andserum in mouse oocyte maturation in vitro, J. Reprod. Fertil., 34:243.
Dekel, N., 1986, Hormonal control of ovulation, in: "Biochemical Actions of Hormones," G. Litwack, ed., Vol. 13, p. 57, Academic Press.
Dekel, N., Aberdam, E., and Hanski, E., 1985, Invasive bacterial adenylate cyclase maintains meiotic arrest in isolated rat oocytes, Proc. Fifth Ovarian Workshop, R. J. Ryan, D. O. Toft, eds., in press.
Dekel, N., Aberdam, E., and Sherizly, I., 1984a, Spontaneous maturation in vitro of cumulus-enclosed rat oocyte is inhibited by forskolin, Biol. Reprod., 31:244.
Dekel, N., and Beers, W. H., 1978, Rat oocyte maturation in vitro: relief of cyclic AMP inhibition by gonadotropins, Proc. Natl. Acad. Sci. USA, 75:4369.
Dekel, N., and Beers, W. H., 1980, Development of the rat oocyte in vitro: inhibition and induction of maturation in presence or absence of the cumulus oophorus, Develop. Biol., 75:247.
Dekel, N., Kraicer, P. F., Phillips, D. M., Ramon, S., and Segal, S. J., 1978, Cellular association in the rat oocyte-cumulus cell complex: morphology and ovulatory changes, Gamete Res., 1:47.
Dekel, N., Lawrence, T. S., Gilula, N. B., and Beers, W. H., 1981, Modulation of cell-to-cell communication in the cumulus-oocyte complex and the regulation of oocyte maturation by LH, Develop. Biol., 86:356.
Dekel, N., Messer, G., and Dabush, S., 1976, Effect of LH on the fine structure of rat preovulatory cumulus oophorus, Proc. 6th Europ. Cong. Electron Microscopy, p. 596.
Dekel, N., Sherizly, I., and Galiani, D., 1984b, Maturation of rat oocytes follows uncoupling in the cumulus-oocyte complex, Biol. Reprod., Suppl. 1, 30:144.
Dekel, N., and Sherizly, I., 1985, Epidermal growth factor induces maturation of rat follicle-enclosed oocytes, Endocrinology, 116:406.
Edwards, R. G., 1965, Maturation in vitro of mouse, sheep, cow, pig, rhesus monkey and human ovarian oocytes, Nature, 208:349.
Eppig, J. J., 1979, A comparison between oocyte growth in co-culture with granulosa cells and oocytes with granulosa cell-oocyte junctional contact maintained in vitro, J. Exp. Zool., 209:345.
Eppig, J. J., 1982, The relationship between cumulus cell-oocyte coupling, oocyte meiotic maturation, and cumulus expansion, Develop. Biol., 89:268.
Foote, W. E., and Thibault, C., 1969, Recherches experimentales sur la maturation in vitro des ovocytes de truie et de veau, Annls. Biol. Anim. Biochim. Biophys., 9:329.
Franchi, L. L., Mandl, A. M., and Zukerman, S., 1962, The development of the ovary and the process of oogenesis, in: "The Ovary," S. Zuckerman, ed., Vol. 1, p. 1, Academic Press, New York.
Gilula, N. B., Epstein, M. C., and Beers, W. H., 1978, Cell-to-cell communication and ovulation: a study of the cumulus-oocyte complex, J. Cell Biol., 78:58.
Hanski, E., and Farfel, Z., 1985, Bordetella pertussis invasive adenylate cyclase: partial resolution and properties of its cellular penetration, J. Biol. Chem., 290:5526.
Heller, D. T., Cahill, D. M., and Schultz, R. M., 1981, Biochemical studies of mammalian oogenesis: metabolic cooperativity between granulosa cells and growing mouse oocytes, Develop. Biol., 84:455.

Jagiello, G., Ducayen, M. B., and Goonan, W. D., 1981, A note on the inhibition of in vitro meiotic maturation of mammalian oocytes by dibutyryl cyclic AMP, J. Exp. Zool., 218:309.
Larsen, W. J., Wert, S. E., and Brunner, G. D., 1984, The disruption of cumulus cells gap junctions could provide a signal to the egg to resume meiotic maturation, J. Cell Biol., 99:345a.
Leibfried, L., and First, N. L., 1980, Follicular control of meiosis in the porcine oocyte, Biol. Reprod., 23:699.
Lindner, H. R., Tsafriri, A., Lieberman, M. E., Zor, U., Koch, Y., Bauminger, S., and Barnea, A., 1974, Gonadotrophin action on cultured Graafian follicles: induction of maturation division of the mammalian oocyte and differentiation of the luteal cell, Rec. Prog. Horm. Res., 30:79.
Magnusson, C., and Hillensjö, T., 1977, Inhibition of maturation and metabolism in rat oocytes by cyclic AMP, J. Exp. Zool., 201:139.
Meinecke, B., and Meinecke-Tillman, S., 1981, Induction and inhibition of meiotic maturation of follicle-enclosed porcine oocytes, Theriogenology, 216:205.
Merck, F. B., Botticelli, C. R., and Albright, J. J., 1972, An intercellular response to estrogen by granulosa cells in the rat ovary, an electron microscope study, Endocrinology, 90:992.
Motlik, J., Fulka, J., and Flechon, J. E., 1986, Changes in intercellular coupling between pig oocytes and cumulus cells during maturation in vivo and in vitro, J. Reprod. Fertil., 76:31.
Moor, R. M., Smith, M. W., and Dawson, R. M. C., 1980, Measurement of intercellular coupling between oocytes and cumulus cells using intracellular markers, Exp. Cell Res., 126:15.
Odor, L. D., 1960, Electron microscopic studies on ovarian oocytes and unfertilized tubal ova in the rat, J. Biophys. Biochem. Cytol., 7:567.
Olsiewski, P., and Beer, W. H., 1983, cAMP synthesis in rat oocyte, Develop. Biol., 100:287.
Paladino, G., 1890, Il ponte intercellulare tra l'uovo ovarico e la cellula follicolare e la formazion della zona pellucida, Anat. Anz., 15:254.
Pincus, G., and Enzmann, E. V., 1935, The comparative behaviour of mammalian eggs in vivo and in vitro. I. The activation of ovarian eggs, J. Exp. Med., 62:665.
Racowsky, C., 1984, Effect of forskolin on the spontaneous maturation and cyclic AMP content of rat oocyte-cumulus complexes, J. Reprod. Fertil., 72:107.
Racowsky, C., and Satterlie, R. A., 1985, Metabolic, fluorescent dye and electrical coupling between hamster oocytes and cumulus cells during meiotic maturation in vivo and in vitro, Develop. Biol., 108:191.
Salustri, A., Petrungaro, S., De Felici, M., Conti, M., and Siracusa, D., 1985, Effect of follicle-stimulating hormone on cyclic adenosine monophosphate level and on meiotic maturation in mouse cumulus cell-enclosed oocyte cultured in vitro, Biol. Reprod., 33:797.
Schuetz, A. W., 1974, Role of hormones in oocyte maturation, Biol. Reprod., 10:150.
Schultz, R. M., Montgomery, R. R., and Belanoff, J. R., 1983, Regulation of mouse oocyte meiotic maturation: implication of a decrease in oocyte cAMP and protein dephosphorylation in commitment to resume meiosis, Develop. Biol., 97:264.
Sotelo, J. R., and Porter, K. R., 1959, An electron microscope study of the rat ovum, J. Biophys. Biochem. Cytol., 5:327.
Szollosi, D., 1975, Ultrastructural aspects of oocyte maturation and fertilization in mammals, in: "La fecondation," C. Thibault, ed., p. 13, Masson et Cie, Paris.
Tardini, A. L., Vitali-Mozza, and Manzani, F. E., 1960, Ultrastruttura dell-ovocito umano maturo. 1. Rapporti fra cellule dela corona radiata, pellucida a ed ovoplasma, Arch. "de Vecchi" Anat. Pathol. Med. Clin., 33:281.

Tsafriri, A., 1978, Oocyte maturation in mammals, in: "The Vertebrate Ovary," R. E. Jones, ed., p. 409, Plenum Press, New York.

Tsafriri, A., and Channing, C. P., 1975, An inhibitory influence of granulosa cells and follicular fluid upon porcine oocyte meiosis in vitro, Endocrinology, 96:922.

Tsafriri, A., Lieberman, M. E., Barnea, A., Bauminger, S., and Lindner, H. R., 1973, Induction by luteinizing hormone of ovum maturation and steroidogenesis in isolated Graafian follicles of the rat: role of RNA and protein synthesis, Endocrinology, 93:1378.

Urner, F., Herrman, W. L., Baulieu, E., and Schorderet-Slatkine, S., 1983, Inhibition of denuded mouse oocyte meiotic maturation by forskolin, an activator of adenylate-cyclase, Endocrinology, 113:1170.

Vivarelli, E., Conti, M., De Felici, M., and Siracusa, G., 1983, Meiotic resumption and intracellular cAMP levels in mouse oocytes treated with compounds which act on cAMP metabolism, Cell Differentiation, 12:271.

Wassarman, P. M., and Letourneau, G. E., 1976, RNA synthesis in fully grown mouse oocytes, Nature (London), 261:73.

Yamada, E., Muta, T., Motamura, A., and Kaga, H., 1957, The fine structure of the oocyte in the mouse ovary studied with electron-microscope, Kurune Med. J., 4:148.

Zamboni, L., 1974, Fine morphology of the follicle wall and follicle cell-oocyte association, Biol. Reprod., 10:125.

Zamboni, L., and Mastroianni, L., 1966, Electron microscopic studies on rabbit ova. I. The follicular oocyte, J. Ultrastruct. Res., 14:95.

SECRETION OF OXYTOCIN BY THE CORPUS LUTEUM

AND ITS ROLE IN LUTEOLYSIS IN THE SHEEP

E. L. Sheldrick and A. P. F. Flint

AFRC Institute of Animal Physiology
and Genetics Research
Babraham, Cambridge, U.K. CB2 4AT

INTRODUCTION

The peptide hormone oxytocin has two well-defined functions: it stimulates uterine muscle contractions during parturition and facilitates the milk ejection response during lactation. The active posterior pituitary principle was named oxytocin (Gr. Oxys, sharp; Tokos, birth) as a result of its action on the uterus which was first described by Dale (1906, 1909) and Blair-Bell (1909). These two actions of oxytocin have been described comprehensively over several decades. In 1959, Armstrong and Hansel published data which tentatively pointed to another role for oxytocin in the control of luteal function in the cow: exogenous oxytocin given early in the estrous cycle caused premature regression of the corpus luteum.

Later it was shown that oxytocin stimulated secretion of the newly-identified uterine luteolysin prostaglandin $F_{2\alpha}$ (Sharma and Fitzpatrick, 1974). The more recent finding that oxytocin is synthesized and secreted by the corpus luteum of the sheep and cow (Flint and Sheldrick, 1982a; Ivell and Richter, 1984; Swann et al., 1984) has raised many important questions with regard to the involvement of oxytocin in ovarian function. Some of these questions will be discussed here with particular reference to the cyclic and early pregnant sheep.

LOCALIZATION OF OXYTOCIN IN CORPORA LUTEA

High concentrations of oxytocin are present in the corpora lutea of cyclic sheep (Table 1) (Wathes and Swann, 1982), cow (Wathes et al., 1984), and goat (Freeman and Currie, 1985). The identity of oxytocin contained within the corpus luteum has been confirmed by several methods including bioassay (Wathes and Swann, 1982; Sheldrick and Flint, 1984), radioreceptor assay (Sheldrick and Flint, 1985), radioimmunoassay, high-performance liquid chromatography, fast-atom bombardment mass spectrometry (Flint and Sheldrick, 1986), and sequence analysis (Watkins et al., 1985). Furthermore, the sequence of bovine oxytocin mRNA from the corpus luteum has been shown to be identical to that in the hypothalamus (Ivell and Richter, 1984). All of these techniques confirm that the oxytocic peptide contained in and secreted by the corpus luteum is identical to that of neurohypophyseal origin.

The ovine corpus luteum consists of three main cell types: the non-steroidogenic endothelial cells; large (> 22 μm) granulated, steroidogenic

Table 1. Concentrations of Oxytocin in Corpora Lutea from Sheep in which Luteal Function was prolonged beyond the Normal Time of Luteal Regression

Treatment	Days after estrus when corpora lutea were collected	No. of corpora lutea	Oxytocin ng/g wet weight
Untreated, cyclic ewes	4 - 12	14	1855 ± 500
	13 - 15	14	204 ± 44
Immunization against oxytocin	20	4	120 ± 39
Immunization against prostaglandin $F_{2\alpha}$	22	2	160 ± 50
Untreated, pregnant ewes	14 - 15	15	266 ± 63
	20 - 30	8	49 ± 15
	70 - 135	11	2.5 ± 0.5

All values: mean ± SEM

cells and the morphologically distinct small (< 22 μm) luteal cells which are also steroidogenic (Schwall and Niswender, 1985). The large luteal cells contain oxytocin and secrete it in vitro (Rodgers et al., 1983). Subcellular localization of oxytocin using a postembedding immunogold technique has shown the oxytocin to be contained within secretory granules in the large luteal cells (Theodosis et al., 1986). Oxytocin is co-localized, as in the posterior pituitary, with its associated neurophysin (Theodosis et al., 1986). This indicates that the hormone is synthesized and processed in a similar manner in the two glands. Immunocytochemical techniques have also been used to show that oxytocin and neurophysin are localized in the large luteal cells in the cow (Guldenaar et al., 1984).

PROPOSED MECHANISM OF ACTION OF LUTEAL OXYTOCIN IN THE CYCLIC EWE

Evidence that oxytocin plays some role in the process of luteal regression includes the observations that immunization against oxytocin delays luteolysis (Sheldrick et al., 1980; Schams et al., 1983), that oxytocin is released episodically towards the end of the cycle, together with its neurophysin (Moore et al., 1986), and that uterine oxytocin receptor concentrations rise at luteal regression (Roberts et al., 1976; Sheldrick and Flint, 1985). Despite this evidence, the mechanism of action of oxytocin at luteolysis is uncertain; the most frequent suggestion is that its secretion contributes to the pulsatile nature of the release of prostaglandin $F_{2\alpha}$, and evidence for this can be summarized as follows.

Luteal oxytocin is secreted episodically into the ovarian vein causing significant veno-arterial differences across the ovary of the sheep (Flint and Sheldrick, 1982b). No such differences occur across ovaries without corpora lutea. Oxytocin concentrations in the peripheral circulation of the sheep parallel those of progesterone during the estrous cycle (Sheldrick and Flint, 1981; Webb et al., 1981; Schams et al., 1982). During luteolysis, oxytocin is secreted in discrete episodes, each of which may represent total depletion of stored, active hormone (Sheldrick and Flint, 1986). These episodes frequently occur simultaneously with episodes of secretion of the ovine uterine luteolysin, prostaglandin $F_{2\alpha}$ (Flint and Sheldrick, 1983). To be most effective as a luteolysin, prostaglandin $F_{2\alpha}$ must be secreted in a pulsatile manner: pulses of approximately one hour duration usually occur at intervals of 5-6 h during a minimum period of 25 h (McCracken et al., 1984). It has been proposed that luteal oxytocin may control secretion of prostaglandin $F_{2\alpha}$ thus ensuring effective regression of the corpus luteum (Sheldrick and Flint, 1986).

For oxytocin to be effective in this action, the uterus, particularly the prostaglandin-secreting endometrium, must be sensitive to stimulation by oxytocin at the appropriate time. As in other species, uterine oxytocin receptor formation is thought to be under endocrine control in the sheep: its synthesis is stimulated by withdrawal of progesterone or by an increase in circulating estrogen (Sheldrick and Flint, 1985). Receptor concentrations can be raised in ovariectomized ewes by treatment with progesterone followed by estrogen, after which the uterus will respond to exogenous oxytocin by secreting prostaglandin $F_{2\alpha}$ (Sheldrick and Flint, 1986). During the mid-luteal phase of the estrous cycle when progesterone levels are high, oxytocin receptor concentrations are low in both myometrium and endometrium (Table 2). From the time of the initiation of luteolysis, receptor concentrations increase rapidly in both compartments reaching peak levels at estrus when peripheral plasma progesterone concentrations are low (Table 2). Thus, the uterus is sensitive to oxytocin during the time that prostaglandin $F_{2\alpha}$ is required to be most potent.

Administration of a luteolytic dose of a prostaglandin $F_{2\alpha}$ analogue, cloprostenol, to sheep in the midluteal phase of the estrous cycle results in a transient increase in oxytocin secretion by the corpus luteum (Flint and Sheldrick, 1982a). Because of the ability of oxytocin to stimulate prostaglandin secretion and vice versa, it has been suggested that a positive feedback loop may account for the pulsatile nature of the secretion of the two hormones (Flint and Sheldrick, 1983). Subsequently, it has been shown that following oxytocin administration there is a period of uterine refractoriness which is not associated with a reduction in oxytocin receptor concentration; this may also contribute to the episodic release of prostaglandin $F_{2\alpha}$ by limiting the period of uterine sensitivity to oxytocin (Sheldrick and Flint, 1986).

LUTEAL OXYTOCIN IN THE EARLY PREGNANT EWE

If oxytocin is directly involved in controlling luteolysis in the ewe, one would expect some changes in secretion and action to occur in early pregnancy, and indeed this does happen. The developing conceptus secretes an antiluteolytic substance, ovine trophoblastic protein-1 (Bazer et al., 1986) which is thought to act by inhibiting endometrial prostaglandin $F_{2\alpha}$ synthesis. The major episodes of prostaglandin $F_{2\alpha}$ secretion are blocked and there is increased secretion of prostaglandin E_2, which is thought to be luteotrophic (Lewis et al., 1978; Ellinwood et al., 1979). Ovine trophoblastic protein-1 is secreted by the trophectoderm between days 12 and 22 of gestation, but thereafter its production appears to cease, and this raises the question of the mechanisms maintaining pregnancy after day 22.

Another factor which may contribute to the maintenance of luteal function is a lack of sensitivity of the uterus to oxytocin. Receptor concentrations fall following fertile mating and remain low (cf. values in Table 2) during the time luteolysis would have occurred had the animal not been pregnant (Mean oxytocin receptor concentrations for days 14-17 of pregnancy: 6.4, 9.6 and 9.6 fmol/mg protein for caruncular endometrium,

Table 2. Uterine Oxytocin Receptor Concentrations during the Estrus Cycle in Sheep

Days of estrous cycle	No. of ewes	Mean oxytocin receptor conc. fmol/mg protein			Mean jugular venous progesterone concentrations ng/ml plasma
		Caruncular endometrium	Intercaruncular endometrium	Myometrium	
10	2	6.6	3.7	17.7	1.27
12	2	9.9	0.8	8.8	2.56
13	2	26.0	1.3	12.5	2.52
14	2	109.4	54.9	8.5	1.50
15	2	372.7	178.7	19.1	1.56
Estrus	2	749.6	1084.7	179.1	0.32
2	2	504.8	624.5	197.2	0.31
4	2	35.0	31.5	73.5	0.75

intercaruncular endometrium and myometrium, respectively, n = 5 ewes; see also Flint and Sheldrick, 1986). However, it is not certain whether this is a cause of maintained luteal function or occurs as a result of it.

A major change in luteal oxytocin metabolism which may contribute to pregnancy maintenance after day 22 is the cessation of oxytocin secretion which occurs at this time (Table 1; Sheldrick and Flint, 1983). Loss of oxytocin from the ovary is reflected in a drop in concentrations of circulating oxytocin (Schams and Lahlou-Kassi, 1984); it is not accompanied by detectable loss of large luteal cells (Sheldrick and Flint, 1984).

The three major changes involving oxytocin in pregnancy: suppression of prostaglandin $F_{2\alpha}$ secretion by ovine trophoblastic protein-1; loss of oxytocin from the corpus luteum; and lack of sensitivity of the uterus to oxytocin would be expected to act together to prevent the pulsatile secretion of prostaglandin $F_{2\alpha}$ necessary for luteolysis, thus causing a failure of the positive feedback loop.

OVARIAN OXYTOCIN AND EXTENDED LUTEAL FUNCTION

Active immunization against potentially important hormones has been widely used as method of investigating their effects. Immunization of ewes against oxytocin results in significant prolongation of the estrous cycle (Sheldrick et al., 1980; Schams et al., 1983). This action may be mediated through a reduced oxytocic stimulus for prostaglandin $F_{2\alpha}$ secretion as endogenous oxytocin is inactivated by circulating bodies.

Immunization against prostaglandin $F_{2\alpha}$ also leads to prolongation of luteal function, presumably by reducing the amount of prostaglandin available to the corpus luteum; a secondary effect may also be to reduce prostaglandin-stimulated oxytocin secretion by the corpus luteum (Sheldrick and Flint, 1984). Both mechanisms will have a disruptive effect on the proposed positive feedback loop. As in pregnancy, extended luteal function caused by immunization against oxytocin or prostaglandin $F_{2\alpha}$ leads to loss of oxytocin from the corpus luteum (Table 1), and thus it seems likely that a reduction in concentration of luteal oxytocin consistently accompanies prolonged luteal function. It is possible that a timing mechanism is involved in controlling the ability of the ovine corpus luteum to produce oxytocin. This may be reflected in temporal limitations on expression of the oxytocin gene within the corpus luteum such as those which appear to occur in the cow (Ivell and Richter, 1984) and the resulting reduction in secretion may reflect depletion of stores of unprocessed hormone. Once oxytocin is lost from the ovary it might be expected that the ability to stimulate episodes of prostaglandin $F_{2\alpha}$ secretion, which are necessary to initiate luteolysis, would also be lost. However, it is clear that this mechanism alone is insufficient to account for prolonged luteal maintenance, as immunized animals do eventually return to estrus (Sheldrick et al., 1980; Sheldrick and Flint, 1984).

CONTINUOUS INFUSION OF OXYTOCIN BLOCKS LUTEAL REGRESSION

A further insight into the consequences of the loss of luteal oxytocin following corpus luteum maintenance arises from experiments in which cyclic ewes were given continuous infusions of oxytocin.

Continuous intravenous infusion of oxytocin (3 μg/h) between days 13 and 21 after estrus delayed return of estrus by 7 days (mean cycle length 23.3 ± 0.6 days compared to 16.6 ± 0.2 in control ewes). Oxytocin infusion was ineffective at blocking estrus if administration commenced after luteolysis had begun (Flint and Sheldrick, 1985). Uterine oxytocin receptor concentrations were low in ewes with prolonged cycles (Table 3), and one

Table 3. Uterine Oxytocin Receptor Concentrations and Luteal Oxytocin Levels in Ewes receiving Continuous Oxytocin or Saline Infusion from Day 13 of the Estrus Cycle

Treatment	No. of ewes	Day of measurement	Oxytocin receptor conc. fmol/mg protein			Luteal oxytocin ng/g wet weight	Corpus luteum weight (mg)
			Caruncular endometrium	Intercaruncular endometrium	Myometrium		
Saline	5	estrus	675 ± 78	638 ± 190	130 ± 30	350 ± 69	367 ± 35
Oxytocin (3 μg/h) from day 13	5	17	76 ± 23	36 ± 5	9 ± 2	229 ± 37	652 ± 59
Oxytocin (3 μg/h) from day 13 plus Cloprostenol on day 15	5	estrus	649 ± 124	852 ± 76	109 ± 15	99 ± 13	310 ± 32

All values: mean ± SEM.

explanation for this effect of continuous oxytocin infusion is that this treatment down-regulates uterine oxytocin receptors and results in reduced secretion of prostaglandin $F_{2\alpha}$. This hypothesis gains some support from the observation that administration of the prostaglandin $F_{2\alpha}$ analogue cloprostenol will override this effect and induce estrus in ewes receiving oxytocin treatment. However, cloprostenol-induced estrus in ewes receiving continuous oxytocin infusion is accompanied by an increase in uterine receptor concentrations (Table 3). It seems that down-regulation of the receptor may require continued exposure of the uterus to progesterone; alternatively, oxytocin infusion may prevent withdrawal of progesterone by another mechanism and the low uterine oxytocin receptor concentrations on day 17 in treated ewes reflect, rather than cause, the failure in luteolysis.

As expected on the basis of observations in pregnant ewes and in those immunized against oxytocin or prostaglandin $F_{2\alpha}$, prolongation of luteal function was associated with depletion of luteal oxytocin (Table 3). However, it should be noted that spontaneous luteolysis closely followed withdrawal of oxytocin infusion, and therefore in these animals, as in those immunized against oxytocin, depletion of luteal oxytocin was not alone sufficient to lead to prolonged luteal function.

CONCLUSION

The discovery, in 1982, that the ovine corpus luteum contained and secreted oxytocin, a peptide hormone which was previously thought to be solely of neurohypophyseal origin, has reawakened interest in the involvement of oxytocin in reproductive function. Over the last five years, a good deal of information has been gained: we know that oxytocin is synthesized in a manner similar to that of the neurohypophysis; but unlike the neurohypophyseal peptide, luteal oxytocin production is limited to a particular time period of about 20 days regardless of the lifespan of the corpus luteum, and progesterone is secreted independently of oxytocin by corpora lutea in these circumstances.

Oxytocin has been shown to stimulate secretion of prostaglandin $F_{2\alpha}$ by the uterus, and the postulated role of oxytocin in controlling luteal function is based on this response. The proposed positive feedback loop in which oxytocin and prostaglandin $F_{2\alpha}$ each stimulate secretion of the other has been shown to undergo certain adaptations during the time of maternal recognition of pregnancy which ensure protection of the developing embryo. One of these, loss of luteal oxytocin following the prolongation of corpus luteum function beyond the normal time of luteolysis, may play an important role in the interruption of the positive feedback loop, and thereby contribute to the maintenance of the corpus luteum. However, in view of the occurrence of luteal regression in nonpregnant ewes in which luteal oxytocin has been depleted following treatments designed to delay luteolysis, it appears the peptide plays a facilitatory, rather than an obligatory role in stimulating uterine prostaglandin $F_{2\alpha}$ secretion at luteolysis.

In addition to the information gained on the actions of endogenous oxytocin, it has also been shown that exogenous oxytocin given in a physiological dose by continuous infusion into the jugular vein will delay luteal regression provided that the infusion is started before luteolysis has begun. This indicates that oxytocin has an antiluteolytic action as well as the luteolytic effect which first gave rise to interest in oxytocin as a hormone involved in ovarian function.

REFERENCES

Armstrong, D. T., and Hansel, W., 1959, Alteration of the bovine estrous cycle with oxytocin, J. Dairy Sci., 42:533.

Bazer, F. W., Vallet, J. L., Roberts, J. M., Sharp, D. C., and Thatcher, W. W., 1986, Role of conceptus secretory products in establishment of pregnancy, J. Reprod. Fert., 76:841.
Blair-Bell, W., 1909, The pituitary body and the therapeutic value of the infundibular extract in shock, uterine atony and intestinal paresis, Br. Med. J., 2:1609.
Dale, H. H., 1906, On some physiological actions of ergot, J. Physiol. (Lond.), 34:163.
Dale, H. H., 1909, The action of extracts of the pituitary body, Biochem. J., 4:47.
Ellinwood, W. E., Nett, T. M., and Niswender, G. D., 1979, Maintenance of the corpus luteum of early pregnancy in the ewe II. Prostaglandin secretion by the endometrium in vitro and in vivo, Biol. Reprod., 21:845.
Flint, A. P. F., and Sheldrick, E. L., 1982a, Ovarian secretion of oxytocin is stimulated by prostaglandin, Nature (Lond.), 297:587.
Flint, A. P. F., and Sheldrick, E. L., 1982b, Ovarian secretion of oxytocin in the sheep, J. Physiol., 330:61P.
Flint, A. P. F., and Sheldrick, E. L., 1983, Evidence for a systemic role for ovarian oxytocin in luteal regression in sheep, J. Reprod. Fert., 67:215.
Flint, A. P. F., and Sheldrick, E. L., 1985, Continuous infusion of oxytocin prevents induction of uterin oxytocin receptor and blocks luteal regression in cyclic ewes, J. Reprod. Fert., 75:623.
Flint, A. P. F., and Sheldrick, E. L., 1986, Ovarian oxytocin and the maternal recognition of pregnancy, J. Reprod. Fert., 76:831.
Freeman, L. C., and Currie, W. B., 1985, Variation in the oxytocin content of caprine corpora lutea across the breeding season, Theriogenology, 23:481.
Guldenaar, S. E. F., Wathes, D. C., and Pickering, B. T., 1984, Immunocytochemical evidence for the presence of oxytocin and neurophysin in the large cells of the bovine corpus luteum, Cell Tissue Res., 237:349.
Ivell, R., and Richter, D., 1984, The gene for the hypothalamic peptide hormone oxytocin is highly expressed in the bovine corpus luteum: biosynthesis, structure and sequence analysis, EMBO J., 3:2351.
Lewis, G. S., Jenkins, P. E., Fogwell, R. L., and Inskeep, E. K., 1978, Concentrations of prostaglandins E_2 and $F_{2\alpha}$ and their relationship to luteal function in early pregnant ewes, J. Anim. Sci., 47:1314.
McCracken, J. A., Schramm, W., and Okulicz, W. C., 1984, Hormone receptor control of pulsatile secretion of $PGF_{2\alpha}$ from the ovine uterus during luteolysis and its abrogation in early pregnancy, Anim. Reprod. Sci., 7:31.
Moore, L. G., Choy, V. J., Elliot, R. L., and Watkins, W. B., 1986, Evidence for the pulsatile release of $PGF_{2\alpha}$ inducing the release of ovarian oxytocin during luteolysis in the ewe, J. Reprod. Fert., 76:159.
Roberts, J. S., McCracken, J. A., Gavagan, J. E., and Soloff, M. S., 1976, Oxytocin-stimulated release of prostaglandin $F_{2\alpha}$ from ovine endometrium in vitro: correlation with estrous cycle and oxytocin-receptor binding, Endocrinology, 99:1107.
Rodgers, R. J., O'Shea, J. D., Findlay, J. K., Flint, A. P. F., and Sheldrick, E. L., 1983, Large luteal cells the source of oxytocin in sheep, Endocrinology, 113:2302.
Sharma, S. C., and Fitzpatrick, R. J., 1974, Effect of estradiol-17β and oxytocin treatment on prostaglandin F alpha release in the anestrous ewe, Prostaglandins, 6:97.
Schams, D., and Lahlou-Kassi, A., 1984, Circulating concentrations of oxytocin during pregnancy in ewes, Acta Endocr. Copenh., 106:277.
Schams, D., Lahlou-Kassi, A, and Glatzel, P., 1982, Oxytocin concentrations in peripheral blood during the estrous cycle and after ovariectomy in two breeds of sheep with low and high fecundity, J. Endocr., 92:9.

Schams, D., Prokopp, S., and Barth, D., 1983, The effect of active and passive immunization against oxytocin on ovarian cyclicity in ewes, Acta Endocr. Copenh., 103:337.
Schwall, R. H., and Niswender, G. D., 1985, Two types of steroidogenic luteal cells in the ewe: morphological and biochemical characteristics, in: "Implantation of the Human Embryo," R. G. Edwards, J. M. Purdy, P. C. Steptoe, eds., p. 31, Academic Press, London.
Sheldrick, E. L., and Flint, A. P. F., 1981, Circulating concentrations of oxytocin during the estrous cycle and early pregnancy in sheep, Prostaglandins, 22:631.
Sheldrick, E. L., and Flint, A. P. F., 1983, Luteal concentrations of oxytocin decline during early pregnancy in the ewe, J. Reprod. Fert., 68:477.
Sheldrick, E. L., and Flint, A. P. F., 1984, Ovarian oxytocin, in: "Gonadal Proteins and Peptides and their Biological Significance," M. R. Sairam, L. E. Atkinson, eds., World Scientific Publishing, Singapore.
Sheldrick, E. L., and Flint, A. P. F., 1985, Endocrine control of uterine oxytocin receptors in the ewe, J. Endocr., 106:249.
Sheldrick, E. L., and Flint, A. P. F., 1986, Transient uterine refractoriness after oxytocin administration in ewes, J. Reprod. Fert., 77:523.
Sheldrick, E. L., MItchell, M. D., and Flint, A. P. F., 1980, Delayed luteal regression in ewes immunized against oxytocin, J. Reprod. Fert., 59:37.
Swann, R. W., O'Shaughnessy, P. J., Birkett, S. D., Wathes, D. C., Porter, D. G., and Pickering, B. T., 1984, Biosynthesis of oxytocin in the corpus luteum, FEBS Lett., 174:262.
Theodosis, D. T., Wooding, F. B. P., Sheldrick, E. L., and Flint, A. P. F., 1986, Ultrastructural localization of oxytocin and neurophysin in the ovine corpus luteum, Cell Tissue Res., 243:129.
Wathes, D. C., and Swann, R. W., 1982, Is oxytocin an ovarian hormone?, Nature (Lond.), 297:225.
Wathes, D. C., Swann, R. W., and Pickering, B. T., 1984, Variations in oxytocin, vasopressin and neurophysin concentrations in the bovine ovary during the estrous cycle and pregnancy, J. Reprod. Fert., 71:551.
Watkins, W. B., Choy, V. J., Chaiken, I. M., and Spiess, J., 1985, Isolation and sequence analysis of oxytocin from the sheep corpus luteum, Neuropeptides, 7:87.
Webb, R., Mitchell, M. D., Falconer, J., and Robinson, J. S., 1981, Temporal relationships between peripheral plasma concentrations of oxytocin, progesterone and 13,14-dihydro-15-keto prostaglandin $F_{2\alpha}$ during the estrous cycle and early pregnancy in the ewe, Prostaglandins, 22:443.

CATECHOLAMINE EFFECTS ON LEYDIG CELL STEROIDOGENESIS

William H. Moger*, Onyeama O. Anakwe† and Paul R. Murphy+

Departments of Physiology & Biophysics
and Obstetrics & Gynaecology
Dalhousie University
Halifax, Nova Scotia B3H 4H7, Canada

Introduction

Luteinizing hormone (LH) is without question the most important regulator of Leydig cell steroidogenesis. In some species such as the ram there is concordance between episodes of LH secretion and episodes of testosterone secretion (Lincoln, 1976). In others, such as man and rat, the concordance is less precise. Ellis and Desjardins (1982) have, for example, suggested that two or more episodes of LH secretion must occur within 70 min of each other to induce an episode of testosterone secretion. LH not only acutely stimulates testosterone secretion but is important in maintaining the morphological and enzymatic attributes required for androgen production (Wing et al., 1984). There are, however, a number of situations where testosterone secretion appears to be altered without a corresponding change in LH secretion. Examples include "testitoxicosis", which is characterized by precocious puberty in boys with low serum gonadotropin concentrations (Weirman et al., 1985); the "testicular hemicastration response" in rats where the testosterone secretion rate of the remaining testis doubles within 24 h of hemicastration without a corresponding increase in serum LH concentrations (Frankel and Wright, 1982); the decline in serum testosterone concentrations in fetal male rats and mice late in gestation despite increasing serum LH concentrations (Pointis et al., 1980; Slob et al., 1980; Habert and Picon, 1982); and the reduction in serum testosterone concentrations induced by stress which has variable effects on LH secretion (Aona et al., 1976; Du Ruisseau et al., 1978; Gray et al., 1978; Tache et al., 1980). It must be noted that the dynamic nature of LH secretion makes it difficult to prove that small alterations in LH secretion have not occurred in situations where testosterone secretion appears to vary independent of changes in LH concentrations. Possible alterations in testicular blood flow or androgen metabolism also complicate interpretation. However, LH-independent testosterone secretion raises the possibility that factors other than LH influence Leydig

*Studies from the authors' laboratory were supported by the Medical Research Council of Canada.
†Department of Obstetrics & Gynecology, University of Michigan, Ann Arbor, Michigan, U.S.A.
+Department of Physiology, University of Manitoba, Winnipeg, Manitoba, Canada.

cell steroidogenesis. There is considerable evidence for direct effects of prolactin (Purvis et al., 1979), estrogens (Moger, 1980), vasopressin (Adashi et al., 1981, 1984) and, in some species, gonadotropin-releasing hormone (Sharpe, 1983) on Leydig cells and for indirect effects of follicle-stimulating hormone (FSH) acting on Leydig cells via an unidentified local testicular factor (Moger and Murphy, 1982). This review will focus on our recent studies that indicate that the catecholamines epinephrine and norepinephrine can directly stimulate Leydig cell steroidogenesis and thus may have physiologic or pathophysiologic roles in the LH-independent regulation of steroidogenesis.

Testicular Catecholamines

Norepinephrine and epinephrine are widely distributed via the sympathetic nervous system (principally norepinephrine) and via the circulation as a result of adrenal medullary secretion (principally epinephrine) and the escape of norepinephrine from sites of innervation. In addition, fetal tissues contain high concentrations of catecholamines as a result of extra-adrenal production (Phillippe, 1983). The mammalian testis receives sympathetic innervation via the superior and inferior spermatic nerves. The superior spermatic nerve originates in the spermatic ganglion with fibers reaching the testis in close association with the spermatic artery. The inferior spermatic nerve originates in the pelvic ganglion and innervates the vas deferens and epididymis but also sends fibers into the testis (Hodson, 1970). Within the testis histofluorescence studies indicate monoamine-containing neurons in association with blood vessels of the interstitial tissue. Although there is species variation in extent, small innervated blood vessels are seen adjacent to Leydig cells (Baumgarten et al., 1968; Norbert et al., 1967). However, nerve terminals directly associated with Leydig cells have only been reported in some non-mammalian species. Nerve fibers do not penetrate the wall of the seminiferous tubule (Hodson, 1970).

With the restriction of adrenergic nerve fibers to the interstitial area, which in most species comprises only a small percentage of testicular volume, it is not surprising that the reported concentrations of catecholamines in the adult testis are quite low. Concentrations of norepinephrine between 20 and 210 ng/g have been reported in testes from the rat, rabbit, guinea pig, cat, and human, whereas epinephrine concentrations are reportedly <50 ng/g (Eliasson and Risely, 1968; Baumgarten et al., 1968; Zieher et al., 1971; Srivastava and Singh, 1985). Higher concentrations of norepinephrine (5440 ng/g) have been observed in neonatal rat testes (Zieher et al., 1971) and we have found elevated norepinephrine (3200 ± 940 ng/g) and epinephrine (1020 ± 710 ng/g) concentrations in fetal mice at 17.5 days of gestation (Anakwe, Nance, and Moger, unpublished observation).

Stress and Androgen Secretion

Physical or psychological stress results in activation of the sympathetic nervous system including adrenal medullary secretion if the stress is severe enough. However, the role of catecholamines in the testicular response to stress is difficult to interpret since the secretion of other hormones, notably ACTH and glucocorticoids, are also affected. There is a large body of evidence documenting that a variety of stressors reduce testicular testosterone secretion in a variety of species (Bliss et al., 1972; Kreuz et al., 1972; Aono et al., 1976; Repcekova and Mikulaj, 1977; Gray et al., 1978; Goncharov et al., 1979; Tache et al., 1980; Charpenet et al., 1981). The effect of stress is not immediate but a decline in serum testosterone concentrations is usually seen within a few hours of the application of the stress. Chronic application of intermittent stress will also reduce serum testosterone concentrations.

The mechanism by which stress inhibits testosterone secretion remains elusive. Inhibition of LH secretion is frequently, but not consistently, observed (Aona et al., 1976; Du Ruisseau et al., 1978; Gray et al., 1978; Tache et al., 1980). In a series of studies, Collu and co-workers have investigated the mechanisms by which immobilization, a form of psychological stress, reduces serum testosterone concentrations in the male rat. In this model, where rats are restrained for 6 or more hours, serum testosterone concentrations decline approximately 70% by the end of the period of stress whereas mean LH levels are either unchanged, increased or decreased, in various seemingly identical experiments. This stress procedure also decreases the ability of low doses of hCG to increase serum testosterone concentrations (Charpenet et al., 1981). Hypophysectomy abolished the effects of stress in this model (Charpenet et al., 1982) in part by eliminating vasopressin secretion (Collu et al., 1984a). Vasopressin had previously been found to inhibit androgen secretion by cultured testicular cells (Adashi et al., 1981, 1984), and Collu et al. (1984a) demonstrated a small, transient, increase in serum vasopressin concentrations during stress. Adrenalectomy eliminated epinephrine and glucocorticoids from the circulation and attenuated the stress-induced rise in norepinephrine levels but did not prevent the decline in basal serum testosterone concentrations during stress nor prevent the reduced response to hCG (Gray et al., 1978; Tache et al., 1980; Collu et al., 1984b). However, adrenalectomy combined with guanethidine treatment (which depletes norepinephrine in peripheral neurons) prevented the stress-induced inhibition of both basal or hCG-stimulated testosterone concentrations. Further studies, however, showed that although β_2-receptor subtype antagonists prevented the effect of stress on the response to hCG, neither α- nor β-adrenergic receptor antagonists blocked the stress-induced inhibition of basal testosterone secretion (Collu et al., 1984b). Evidence for a role of catecholamines in the stress-induced decline in testosterone secretion, with the exception of a role in regulating the sensitivity of Leydig cells to hCG (and presumably LH), is thus ambiguous.

In Vivo Effects of Catecholamines on Testosterone Secretion

The putative role of catecholamines on Leydig cell function has also been investigated by the exogenous administration of catecholamines. Early studies of the effects of chronic epinephrine administration to rats and rabbits suggested that such treatment disrupted spermatogenesis and reduced sex accessory gland weights (Perry, 1941; VanDemark and Boyd, 1956). However these chronic effects of epinephrine were not always reproducible (Lamar, 1943; Armstrong and Hansel, 1958; Ewing et al., 1964). More recent studies have demonstrated that acute administration of catecholamines inhibits testosterone secretion. Levin et al. (1967) reported that infusion of epinephrine in 10 normal men over a 3 h period reduced plasma testosterone concentrations by 28%. An important aspect of this study was the measurement of the metabolic clearance rate (MCR) of testosterone in 4 subjects during the epinephrine infusion. The MCR was determined by the constant infusion of [^{3}H]-testosterone after a priming bolus injection. No effect of the epinephrine infusion on the MCR of testosterone was observed. Hence, the calculated production rate of testosterone (i.e. plasma testosterone concentration times the MCR) declined 28% in these subjects.

Infusions of norepinephrine or epinephrine for 20 min in male rats was also reported to reduce testosterone secretion (Damber and Janson, 1978). Damber and Janson also measured testicular blood flow at the end of these infusions using a radioactive microsphere method. Although both norepinephrine and epinephrine infusions increased arterial blood pressure, neither altered testicular blood flow. Taken together, the studies by Levin et al. (1967) and Damber and Janson (1978) indicate that the inhibitory effect of systemically administered catecholamines on testosterone secretion cannot be easily explained by vascular effects on either hepatic or testicular blood flow. These experiments do not necessarily indicate a testicular site of catecholamine action; serum LH concentrations, for example, were not determined by either group. Gotz et al. (1983) attempted to examine the roles of LH and the adrenal in the inhibitory effect of epinephrine on testosterone secretion. They administered epinephrine by daily injection to intact or adrenalectomized rats for 5 days. Plasma testosterone concentrations were reduced in samples from both intact and adrenalectomized animals 2 h after the last epinephrine injection. LH concentrations were significantly reduced in adrenalectomized, but not in intact rats. As a result of the interaction between adrenalectomy and the effect of epinephrine on LH concentrations, the contribution of altered LH secretion in the reduced testosterone concentrations cannot be assessed in this experiment.

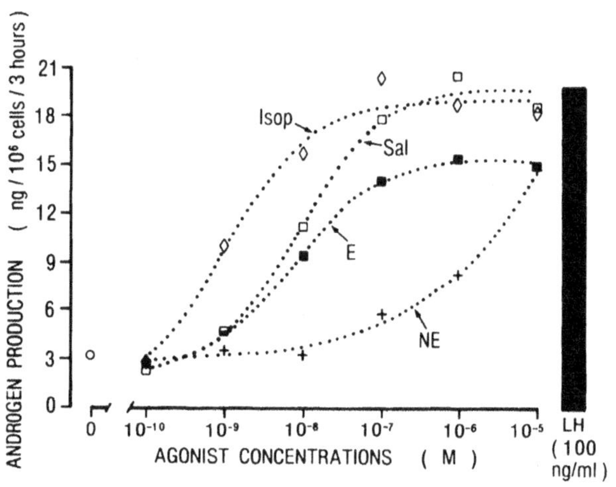

Fig. 1. Androgen production by mouse interstitial cells cultured for 48 h and then stimulated with the β-adrenergic agonists isoproterenol (Isop), salbutamol (Sal), epinephrine (Epi), or norepinephrine (Norepi) for 3 h. LH-stimulated androgen production is shown for comparison (black bar). (Reprinted with permission from Anakwe and Moger (1984a), Copyright 1984, Pergamon Journals Ltd.)

In Vitro Effects of Catecholamines on Androgen Secretion

In 1969, Eik-Nes made the surprising observation that addition of the β-adrenergic agonist isoproterenol to the arterial blood perfusing the dog testis *in situ* or in a metabolic chamber increased testosterone concentrations in the venous effluent and in the testicular tissue. Norepinephrine and epinephrine were also effective. Altered pituitary or adrenal secretion could not be involved in the stimulatory effect as the testicular venous blood was not recirculated. No change in overall blood flow was observed, although changes in regional flow could not be ruled out. The studies by Eik-Nes remained the only suggestion of a direct stimulatory effect of catecholamines on Leydig cell steroidogenesis until two groups (Moger et al., 1982; Cooke et al., 1982), working independently, reported that norepinephrine, epinephrine, and isoproterenol stimulated androgen production by mouse Leydig cells in primary culture (Fig. 1). We have recently reported a similar effect of these catecholamines on cultured rat Leydig cells (Anakwe et al., 1985; Anakwe and Moger, 1986).

Although some of our early studies used cultured testicular interstitial cells, it is reasonably certain that the response is a result of a direct effect of catecholamines on Leydig cells as mouse or rat Leydig cells purified on Percoll gradients respond to catecholamines (Cooke et al., 1982; Moger and Murphy, 1983; Anakwe and Moger, 1986).

It is also clear that the stimulatory effect of catecholamines on cultured Leydig cells is mediated via a β-adrenergic receptor mechanism. Isoproterenol, which is a β-receptor specific agonist that does not appreciably affect α-adrenergic receptors, is more potent than either epinephrine or norepinephrine which are β-selective and α- and β-nonselective agonists, respectively. The stimulatory effect of norepinephrine, which could be via either α- or β-receptors, is inhibited by the β-receptor antagonist propranolol, but is not inhibited by the α-receptor antagonists phentolamine or phenoxybenzamine (Anakwe and Moger, 1984a). The stimulatory effect of catecholamines on Leydig cells has been further resolved to be via the β_2 subtype of the adrenergic receptor. This conclusion is based on the ability of the β_2-selective agonist salbutamol to increase androgen production in mouse interstitial cell cultures and by the ability of the β_2-selective antagonist ICI 118,551 to inhibit isoproterenol-stimulated androgen production. The β_1-selective antagonists metoprolol and atenolol did not inhibit isoproterenol-stimulated steroidogenesis (Anakwe and Moger, 1984a).

Further evidence for β_2-receptors in Leydig cells comes from radioligand binding studies. In 1983, Poyet and Labrie reported in an abstract that rat interstitial cells contain specific β-adrenergic binding sites using [^{125}I]-cyanopindolol. As purified Leydig cells were not used in this study, it does not establish that the binding sites are on Leydig cells. We have recently characterized the binding of the β-receptor antagonist [^{3}H]CGP-12177 to purified rat and mouse Leydig cells (Anakwe et al., 1985). High affinity (K_D = 0.79 ± 0.22 nM) and low capacity binding (1716 ± 245 sites per cell) was observed with rat Leydig cells (Fig. 2). Mouse Leydig cells appeared to have about half the number of binding sites as rat Leydig cells. The characterization of the binding as a β_2 subtype was based on the order of potency of agonists (isoproterenol > epinephrine = salbutamol > norepinephrine) and antagonists (propranolol = ICI 118,551 >> atenolol) for competition with [^{3}H]CGP-12177 for the binding site.

Physiological Relevance

Despite the convincing evidence for a direct stimulatory effect of catecholamines on rodent Leydig cells in primary culture, there is considerable difficulty in interpreting this effect as being physiologically relevant. Catecholamines do not stimulate androgen production by freshly isolated Leydig cells from either mice (Cooke et al., 1982; Moger et al., 1982; Moger and Anakwe, 1983; Moger and Murphy, 1983) or rats (Anakwe et al., 1985; Anakwe and Moger, 1986). Mouse Leydig cells require 24 h of culture before they become responsive to catecholamines while responsiveness of rat Leydig cells begins to develop by 3 h of culture. The explanation for the failure of freshly isolated cells to respond to catecholamines has been elusive. The lack of response does not appear to be related to the type of culture medium or the presence or absence of the phosphodiesterase inhibitor

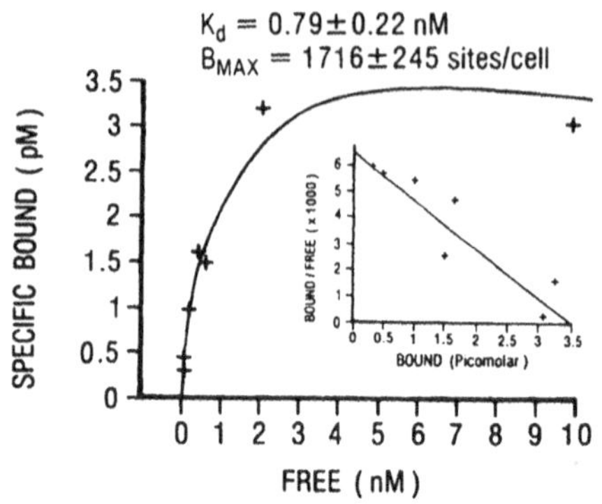

Fig 2. Saturation curve for [^{3}H]CGP-12177 binding to purified rat Leydig cells. The inset is a Scatchard analysis of the saturation curve. (Reprinted with permission from Anakwe et al., 1985.)

3-isobutyl-1-methylxanthine. Both Cooke et al. (1982) and Moger and Murphy (1983) concluded that down-regulation of the β-receptors was not involved (see below). Studies using whole decapsulated testes from postnatal rats or mice, which eliminates most of the steps in the preparation of interstitial cells, revealed only sporadic responses to isoproterenol with very small increases in androgen production (Moger et al., 1982; Anakwe and Moger, 1984b). Fetal rat and mouse testes consistently responded to isoproterenol *in vitro* but the increase in androgen production was again small (Anakwe and Moger, 1984b). This contrasts with the response of cultured cells where the response to catecholamines, although often less than that of LH, is of a similar magnitude.

Freshly isolated Leydig cells may not respond to catecholamines with increased androgen production but they do contain β-receptors. The characterization of [^{3}H]CGP-12177 binding described above was done with freshly isolated Leydig cells. In addition, both mouse (Cooke et al., 1982) and rat (Anakwe and Moger, 1986) Leydig cells when initially isolated produce cyclic AMP in response to isoproterenol (Fig. 3). Although the amount of cyclic AMP produced in response to isoproterenol is considerably less than that achieved with LH stimulation, the concentrations are sufficient to stimulate androgen production if invoked by low concentrations of LH (Cooke et al., 1982). An analogous situation has been reported where low concentrations of cholera toxin stimulated Leydig cell cyclic AMP production, but not testosterone production, whereas an equivalent amount of cyclic AMP produced by hCG stimulation did increase testosterone secretion (Dufau et al., 1978). The nature of this apparent compartmentalization of cyclic AMP or the cyclic AMP dependent protein kinase in Leydig cells is unknown.

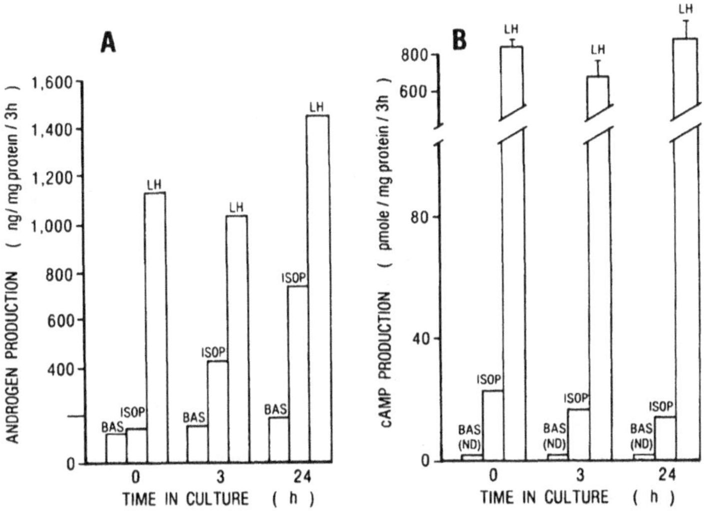

Fig. 3. Relationship between time in primary culture and androgen (panel A) and cAMP (panel B) production by rat Leydig cells. Bas = basal; ISOP = isoproterenol 10 μM; LH = luteinizing hormone 100 ng/ml; ND = not detectable. (Reprinted with permission from Anakwe and Moger, 1986.)

At this time we cannot rule out the possibility that the stimulatory effect of catecholamines on Leydig cell steroidogenesis is an artifact arising from placing the cells in primary culture. However, the presence of β-receptors functionally coupled to adenylate cyclase in freshly isolated cells suggests that the isolation procedure altered the link between cyclic AMP and steroidogenesis. Certain protocols for Leydig cell preparation have been reported to impair LH-stimulated cyclic AMP and testosterone formation (Aldred and Cooke, 1982, 1983; Molenaar et al., 1983). Further, Cooke et al. (1979) have reported that LH-stimulated testosterone secretion by freshly isolated rat Leydig cells was reduced by inhibitors of RNA synthesis. However, if the cells were preincubated for 3 h these inhibitors were no longer effective. This time course is similar to the development of the androgen response to catecholamines in rat cells and, with mouse cells, we have shown that the development of androgen responsiveness can be prevented by protein or RNA synthesis inhibitors (Moger and Murphy, 1983). The possibility should be considered that the current Leydig cell isolation procedures of collagenase dispersion and density gradient centrifugation induce a switch in gene expression leading to the synthesis of stress-proteins (Subjeck and Shyy, 1986) and thus altering Leydig cell function. During culture the cells may revert to normal gene expression and thus more nearly reflect the characteristics of Leydig cells in situ than do freshly isolated cells.

As mentioned in the Introduction, testosterone secretion increases independent of changes in LH secretion in the testicular response to hemicastration in the rat. Within 24 h of hemicastration and continuing

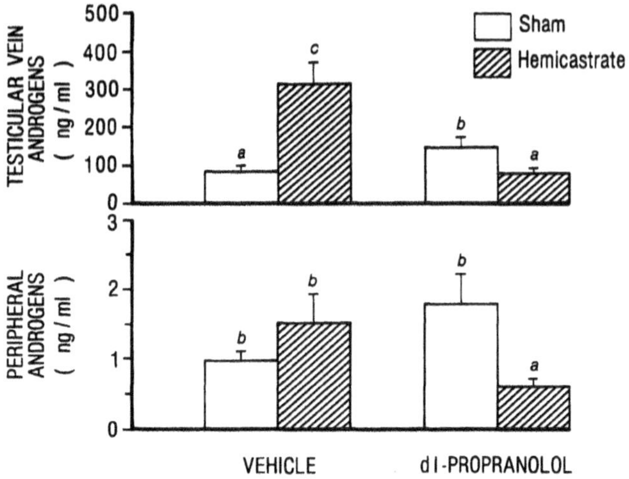

Fig. 4. Effect of intratesticular injections of the β-receptor antagonist propranolol on testicular vein (upper panel) and peripheral (lower panel) androgen concentrations in hemicastrated and sham operated animals. Within each panel bars with different letters are significantly different from each other. See Moger and Anakwe (1986) for additional details.

thereafter, peripheral testosterone concentrations are normalized as a result of a doubling of the testosterone secretion rate by the remaining testis (Lindgren et al., 1976; Mock and Frankel, 1982; Moger and Anakwe, 1986). Despite extensive investigation, there is no evidence for a compensatory increase in serum LH concentrations that would account for the increase in testosterone secretion (Howland and Skinner, 1975; Gomes and Jain, 1976; Frankel and Wright, 1982; Frankel and Mock, 1982). In 1984, Frankel et al. suggested a neural component to the testicular hemicastration response when they reported that severing the inferior spermatic nerve to the remaining testis partially prevented the increase in testosterone secretion by that testis.

To investigate whether β-adrenergic mechanisms might be involved in the testicular hemicastration response, we studied the effects of intratesticular administration of adrenergic antagonists (Moger and Anakwe, 1986). As shown in Fig. 4, treatment of the remaining testis with the β-receptor antagonist propranolol prevented the increase in testicular vein androgen concentrations and hence the normalization of peripheral androgen concentrations. Propranolol administration to sham operated testes did not decrease, and in fact slightly increased, testicular vein androgen concentrations.

Conclusions

The evidence for an inhibitory effect of stress or systemically administered catecholamines on testicular testosterone secretion is overwhelming but the mechanism(s) remains only partially resolved. In particular, there is no evidence for a direct inhibitory effect of catecholamines on Leydig cell steroidogenesis. If a direct inhibitory effect on Leydig cells steroidogenesis exists, it would most likely be via α-receptor mechanisms. Although α-receptor mediated effects on steroidogenesis have not been studied in detail, there is as yet no evidence for such effects (Collu et al., 1984b; Anakwe and Moger, 1984a). β-Receptor activation in Leydig cells is consistently stimulatory as would be anticipated from the knowledge that cyclic AMP is commonly the second messenger for this type of receptor. Studies on the interaction of catecholamines and LH on cultured mouse and rat Leydig cells indicate only additive effects at submaximum LH concentrations. With maximum stimulatory concentrations of LH and catecholamines, there are no additive or inhibitory effects, suggesting that both are acting via the same mechanism (Moger et al., 1982; Anakwe and Moger, 1986). Even when Leydig cells are desensitized by prolonged exposure to high concentrations of catecholamines, the desensitization is homologous. That is, the cells become refractory to catecholamine stimulation but do not become refractory to LH stimulation (Moger and Murphy, 1983; Anakwe and Moger, in preparation). The conclusion that catecholamines are unlikely to directly inhibit Leydig cells does not preclude the possibility of indirect inhibitory effects within the testis. Sertoli cells, for example, may contain β-receptors (Heindel et al., 1981; Kierszenbaum et al., 1985) and are likely to produce factors that influence Leydig cells (Sharpe, 1983).

Whether the direct stimulatory effect of catecholamines on steroidogenesis observed in cultured Leydig cells is physiologically relevant and if so in what situations are open questions. As hypophysectomy leads to the near total collapse of testicular function, the sympathetic nervous system obviously cannot replace the effects of gonadotropins. If catecholamine stimulation is physiologically relevant, it is likely to be in situations, such as the response to hemicastration, where the testis increases its secretion rate despite normal LH concentrations. In the rat, where not all episodes of LH secretion are followed by episodes of testosterone secretion (Ellis and Desjardins, 1982), increased activity of the adrenergic nerves of the testis could increase the correspondence between LH and testosterone episodes, thus increasing the testosterone secretion rate.

References

Adashi, E. Y., and Hsueh, A. J. W., 1981, Direct inhibition of testicular androgen biosynthesis revealing antigonadal activity of neurohypophyseal hormones, Nature, 293:650.

Adashi, E. Y., Tucker, E. M., and Hsueh, A. J. W., 1984, Direct regulation of rat testicular steroidogenesis by neurohypophyseal hormones, J. Biol. Chem., 259:5440.

Aldred, L. F., and Cooke, B. A., 1982, Tne deleterious effect of mechanical dissociation of rat testes on the functional activity and purification of Leydig cells using Percoll gradients, Int. J. Androl., 5:191.

Aldred, L. F., and Cooke, B. A., 1983, The effect of cell damage on the density and steroidogenic capacity of rat testis Leydig cells, using an NADH exclusion test for determination of viability, J. Steroid Biochem., 18:411.

Anakwe, O. O., and Moger, W. H., 1984a, β_2-Adrenergic stimulation of androgen production by cultured mouse testicular interstitial cells, Life Sci., 35:2041.

Anakwe, O. O., and Moger, W. H., 1984b, Ontogeny of rodent testicular androgen production in response to isoproterenol and luteinizing hormone in vitro, Biol. Reprod., 30:1142.

Anakwe, O. O., and Moger, W. H., 1986, Catecholamine stimulation of androgen production by rat Leydig cells. Interactions with luteinizing hormone and luteinizing hormone-releasing hormone, Biol. Reprod., in press.

Anakwe, O. O., Murphy, P. R., and Moger, W. H., 1985, Characterization of β-adrenergic binding sites on rodent Leydig cells, Biol. Reprod., 33:815.

Aono, T., Kurachi, K., Miyata, M., Nakasima, A., Koshiyama, K., Uozumi, T., and Matsumoto, K., 1976, Influence of surgical stress under general anesthesia on serum gonadotropin levels in male and female patients, J. Clin. Endocr. Metab., 42:144.

Armstrong, D. T., and Hansel, W., 1958, Effects of hormone treatment on testes development and pituitary function, Intern. J. Fert., 3:296.

Baumgarten, H. G., Falck, B., Holstein, A.-F., Owman, C., and Owman, T., 1968, Adrenergic innervation of the human testis, epididymus, ductus deferens and prostate: a fluorescence microscopic and fluorimetric study, Zeit. Zell. Mikros. Anat., 90:81.

Bliss, E. L., Frischat, A., and Samuels, L., 1972, Brain and testicular function, Life Sci., 11(Part 1):231.

Charpenet, G., Tache, Y., Bernier, M., Ducharme, J. R., and Collu, R., 1982, Stress-induced testicular hyposensitivity to gonadotropin in rats. Role of the pituitary, Biol. Reprod., 27:616.

Charpenet, G., Tache, Y., Forest, M. G., Hoar, F., Saez, J. M., Bernier, M., Ducharme, J. R., and Collu, R., 1981, Effect of chronic immobilization stress on rat testicular androgenic function, Endocrinology, 109:1254.

Collu, R., Gibb, W., Brichet, D. G., and Ducharme, J. R., 1984a, Role of arginine-vasopressin (AVP) in stress-induced inhibition of testicular steroidogenesis in normal and in AVP-deficient rats, Endocrinology, 115:1609.

Collu, R., Gibb, W., and Ducharme, J. R., 1984b, Role of catecholamines in the inhibitory effect of immobilization stress on testosterone secretion in rats, Biol. Reprod., 30:416.

Cooke, B. A., Janszen, F. H., van Driel, M. J., and van der Molen, H. J., 1979, Evidence for the involvement of lutropin-independent RNA synthesis in Leydig cell steroidogenesis, Mol. Cell. Endocr., 14:181.

Cooke, B. A., Golding, M., Dix, C. J., and Hunter, M. G., 1982, Catecholamine stimulation of testosterone production via cyclic AMP in mouse Leydig cells in monolayer culture, Mol. Cell. Endocr., 27:221.

Damber, J. E., and Janson, P. O., 1978, The effect of LH, adrenaline and noradrenaline on testosterone concentrations in anaesthetized rats, Acta Endocr., 88:390.

Du Ruisseau, P., Tache, Y., Brazeau, P., and Collu, R., 1978, Pattern of adenohypophyseal hormone changes induced by various stressors in female and male rats, Neuroendocrinology, 27:257.

Dufau, M. L., Horner, K. A., Hayashi, K., Tsuruhara, T., Conn, P. M., and Catt, K. J., 1978, Actions of choleragen and gonadotropin in isolated Leydig cells. Functional compartmentalization of the hormone-activated cyclic AMP response, J. Biol. Chem., 253:3721.

Eik-Nes, K. B., 1969, An effect of isoproterenol on rates of synthesis and secretion of testosterone. Am. J. Physiol., 217:1764.

Eliasson, R., and Risley, P. L., 1968, Adrenergic innervation of the male reproductive ducts of some mammals. III. Distributions of noradrenaline and adrenaline, Acta Physiol. Scand., 73:311.

Ellis, G. B., and Desjardins, C., 1982, Male rats secrete luteinizing hormone and testosterone episodically, Endocrinology, 110:1618.

Ewing, L. L., Noble, D. J., and Ebner, K. E., 1964, The effect of epinephrine and competition between males on the in vitro metabolism of rabbit testes, Can. J. Physiol. Pharmacol., 42:527.

Frankel, A. I., and Mock, E. J., 1982, A study of the first eight hours in the stabilization of plasma testosterone concentration in the hemicastrated rat, J. Endocr., 92:225.

Frankel, A. I., and Wright, W. W., 1982, The hemicastrated rat: definition of a model for the study of the regulation of testicular steroidogenesis, J. Endocr., 92:213.

Frankel, A. I., Mock, E. J., and Chapman, J. C., 1984, Hypophysectomy and hemivasectomy can inhibit the testicular hemicastration response of the mature rat, Biol. Reprod., 30:804.

Gomes, W. R., and Jain, S. K., 1976, Effect of unilateral and bilateral castration and crypt orchidism on serum gonadotropins in the rat, J. Endocr., 68:191.

Goncharov, M. P., Taranov, A. G., Antonichev, A. V., Gorlushkin, V. M., Aso, T., Cekan, S. Z., and Diczfalusy, E., 1979, Effect of stress on the profile of plasma steroids in baboons (Papio hamadryas), Acta Endocr., 90:372.

Gotz, F., Stahl, F., Rhohde, W., and Dorner, G., 1983, The influence of adrenaline on plasma testosterone in adult and newborn male rats, Exp. Clin. Endocr., 81:239.

Gray, G. D., Smith, E. R., Damassa, D. A., Ehrenkranz, J. R. L., and Davidson, J. M., 1978, Neuroendocrine mechanisms mediating the suppression of circulating testosterone levels associated with chronic stress in male rats, Neuroendocrinology, 25:247.

Habert, R., and Picon, R., 1982, Control of testicular steroidogenesis in fetal rat: effect of decapitation on testosterone and plasma luteinizing hormone-like activity, Acta Endocr., 99:466.

Heindel, J. J., Steinberger, A., and Strada, S. J., 1981, Identification and characterization of a β_1-adrenergic receptor in the rat Sertoli cell, Mol. Cell. Endocr., 22:349.

Hodson, N., 1970, The nerves of the testis, epididymis, and scrotum, in: "The Testis," A. D. Johnson, W. R. Gomes, and N. L. VanDemark, eds., Vol. 1, p. 47, Academic Press, New York.

Howland, B. E., and Skinner, K. R., 1975, Changes in gonadotropin secretion following complete or hemicastration in the adult rat, Horm. Res., 6:71.

Kierszenbaum, A. L., Spruill, W. A., White, M. G., Tres, L. L., and Perkins, J. P., 1985, Rat Sertoli cells acquire β-adrenergic response during primary culture, Proc. Natl. Acad. Sci. USA, 82:2049.

Kreuz, L. E., Rose, R. M., and Jennings, J. R., 1972, Suppression of plasma testosterone levels and psychological stress. A longitudinal study of young men in Officer Candidate School, Arch. Gen. Psychiat., 26:479.

Lamar, J. K., 1943, Epinephrine effects on young male rats, Anat. Rec. Abbr., 87:453.

Levin, J., Lloyd, C. W., Lobotsky, J., and Friedrich, E. H., 1967, The effect of epinephrine on testosterone production, Acta Endocr., 55:184.

Lincoln, G. A., 1976, Seasonal variation in the episodic secretion of luteinizing hormone and testosterone in the ram, J. Endocr., 69:213.

Lindgren, S., Damber, J.-E., and Carstensen, H., 1976, Compensatory testosterone secretion in unilaterally orchidectomized rats, Life Sci., 18:1203.

Mock, E. J., and Frankel, A. I., 1982, Response of testosterone to hemicastration in the testicular vein of the mature rat, J. Endocr., 92:231.

Moger, W. H., 1980, Direct effects of estrogens on the endocrine function of the mammalian testis, Can. J. Physiol. Pharmacol., 58:1011.

Moger, W. H., and Anakwe, O. O., 1983, Effects of forskolin on androgen production by mouse interstitial cells in vitro. Interactions with luteinizing hormone and isoproterenol, Biol. Reprod., 29:932.

Moger, W. H., and Anakwe, O. O., 1986, Propranolol inhibits the compensatory increase in androgen secretion after unilateral orchidectomy in rats, J. Reprod. Fert., 76:251.

Moger, W. H., and Murphy, P. R., 1982, Reevaluation of the effect of follicle stimulating hormone on the steroidogenic capacity of the testis: The effects of neuraminidase-treated FSH preparations, Biol. Reprod., 26:422.

Moger, W. H., and Murphy, P. R., 1983, β-Adrenergic agonist induced androgen production during primary culture of mouse Leydig cells, Arch. Androl., 10:135.

Moger, W. H., Murphy, P. R., and Casper, R. F., 1982, Catecholamine stimulation of androgen production by mouse interstitial cells in primary culture, J. Androl., 3:227.
Molenaar, R., Rommerts, R. F. G., and van der Molen, H. J., 1983, The steroidogenic activity of isolated Leydig cells from mature rats depends on the isolation procedure, Int. J. Androl., 6:261.
Norberg, K. A., Risley, P. L., and Ungerstedt, U., 1967, Adrenergic innervation of the reproductive tract of some mammals. I. The distribution of adrenergic nerves, Zeit. Zell. Mikros. Anat., 76:278.
Perry, J. C., 1941, Gonadal response of male rats to experimental hyperadrenalism, Endocrinology, 29:592.
Phillippe, M., 1983, Fetal catecholamines, Am. J. Obstet. Gynecol., 146:840.
Pointis, G., Latreille, M.-T., and Cedard, L., 1980, Gonadopituitary relationships in the fetal mouse at various times during sexual differentiation, J. Endocr., 86:483.
Poyet, P., and Labrie, F., 1983, Characterization of β-adrenergic receptors in dispersed rat Leydig cells, Biol. Reprod. Suppl. 1 Abbr., 28:59.
Purvis, K., Clausen, O. P. F., Olsen, A., Haug, E., and Hansson, V., 1979, Prolactin and Leydig cell responsiveness to LH/hCG in the rat, Arch. Androl., 3:219.
Repcekova, D., and Mikulaj, L., 1977, Plasma testosterone of rats subjected to immobilization stress and/or hCG administration, Horm. Res., 8:51.
Sharpe, R. M., 1983, Local control of testicular function, Quart. J. Exp. Physiol., 68:265.
Slob, A. K., Ooms, M. P., and Vreebury, J. T. M., 1980, Prenatal and early postnatal sex differences in plasma and gonadal testosterone and plasma luteinizing hormone in female and male rats, J. Endocr., 87:81.
Srivastava, A. K., and Singh, U. S., 1985, Effect of aflatoxin B_1 on the androgen receptor and catecholamines in the rat testis, IRCS Med. Sci., 13:46.
Subjeck, J. R., and Shyy, T.-T., 1986, Stress protein systems of mammalian cells, Am. J. Physiol., 250:C1.
Tache, Y., Ducharme, J. R., Charpenet, G., Saez, J., and Collu, R., 1980, Effect of chronic intermittent immobilization stress on hypophysiogonadal function of rats, Acta Endocr., 93:168.
VanDemark, N. L., and Boyd, L. J., 1956, The effect of epinephrine upon testicular function in rabbits, Intern. J. Fert., 1:245.
Weirman, M. E., Beardsworth, D. E., Mansfield, M. J., Badger, E. M., Crawford, J. D., Crigler, Jr., J. F., Bode, H. H., Loughlin, J. S., Kushner, D. C., Scully, R. E., Hoffman, W. H., and Crowley, Jr., W. F., 1985, Puberty without gonadotropins. A unique mechanism of sexual development, New Engl. J. Med., 312:65.
Wing, Y.-Y., Ewing, L. L., and Zirkin, B. R., 1984, Effects of luteinizing hormone withdrawal on Leydig cell smooth endoplasmic reticulum and steroidogenic reactions which convert pregnenolone to testosterone, Endocrinology, 115:2290.
Zieher, L. M., Debeljuk, L., Iturriza, F., and Mancini, R. E., 1971, Biogenic amine concentrations in testes of rats at different ages, Endocrinology, 88:351.

EVIDENCE FOR INTRATESTICULAR FACTORS WHICH MEDIATE THE RESPONSE OF LEYDIG CELLS TO DISRUPTION OF SPERMATOGENESIS

David K. Pomerantz and Gwenderlyn F. Jansz

Medical Research Council Group in Reproductive Biology
Departments of Physiology and Obstetrics & Gynecology
University of Western Ontario
London, Ontario, Canada N6A 5C1

INTRODUCTION

The traditional view of endocrine regulation of the testis placed the hypothalamic-pituitary unit in a commanding position such that the releasing and inhibiting hormones of the hypothalamus regulated the adenohypophysial secretion of luteinizing hormone (LH), follicle stimulating hormone (FSH), and prolactin (McCann, 1974; Neill, 1974). Studies of the action of these pituitary hormones led to the general acceptance that LH acts on the interstitial cells of Leydig to stimulate secretion of testosterone (Christensen, 1975). FSH was shown to bind to the seminiferous tubule, primarily the Sertoli cell, and to regulate the myriad of functions attributed to this cell (Means, 1977; Ritzen et al., 1981; Steinberger and Steinberger, 1977). The possibility remains that FSH may act upon additional testicular cell types; for example, testicular macrophages have been implicated (Yee and Hutson, 1985). Prolactin was shown to enhance the action of LH and was presumed to act at the level of the Leydig cell (Bartke et al., 1978). The role of prolactin in the maintenance of normal testicular function remains unclear. Negative feedback regulation of gonadal function is achieved by both steroidal and protein products of the testis acting on the hypothalamus and/or anterior pituitary (di Zerega and Sherins, 1981; Setchell, 1978).

While changes in gonadotropin secretion may help explain relatively long term changes in testicular function such as those occurring in association with puberty or seasonal breeding cycles, it has been suggested by others (Parvinen, 1982; Sharpe, 1983) that this is an incomplete view of testicular function. These workers have argued that descriptions of testicular regulation must also account for local intratesticular mechanisms. To appreciate this contention, a brief consideration of the spermatogenic cycle and its spatial characteristics is required.

Spermatogenesis is the process by which relatively undifferentiated diploid spermatogonia are transformed into highly differentiated, haploid, immature spermatozoa. Virtually the entire process occurs while the germ cells are in close contact with Sertoli cells. The process is not random, but tightly regulated with one "stage" following another in a highly ordered predictable fashion. Clermont (1972) identified 12 stages in the spermatogenic cycle of the rat. Each stage is characterized by a specific constellation of steps of the total spermatogenic process and occupies a small and

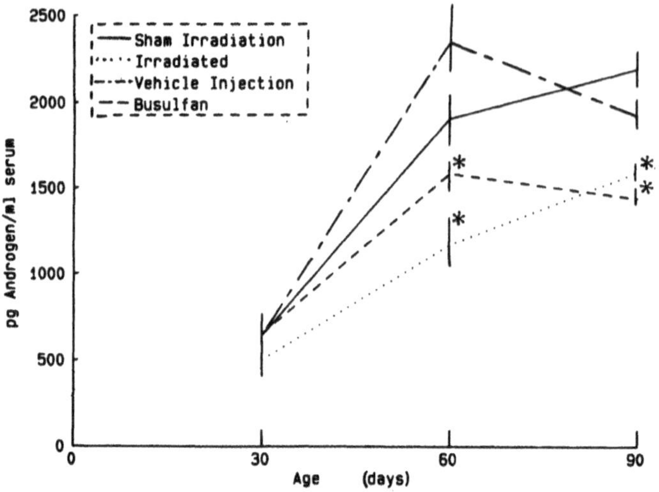

Fig. 1. Effects of age on androgen concentration in the sera of control and spermatogenically disrupted rats. All values are mean ± SEM of 10 to 20 animals. * indicates value is significantly lower than respective control. (Jansz and Pomerantz, 1984; 1985a)

continually changing portion of the seminiferous tubule. The hormonal requirements for the different stages are not identical. It was shown that hypophysectomy caused an increase in the rate of degeneration of midpachytene spermatocytes and spermatids at steps 7 and 19 (stage VII). LH could reverse this defect (Russell and Clermont, 1977). These three germ cell types are at very different stages of development and adjacent stages of spermatogenesis were not similarly sensitive. It was proposed that stage VII was particularly sensitive to testosterone, presumably mediated by the effect of the steroid on the Sertoli cell. These same authors noted that in normal animals the highest rate of spontaneous degeneration of germ cells also occurred at stage VII, suggesting that the availability of testosterone could be a limiting factor in the progression of germ cells past stage VII. Additional studies have shown that other testicular functionsl change with the spermatogenic cycle. Parvinen (1982) and coworkers studied the differences in physiologic function between small segments of seminiferous tubules, in which the stage of spermatogenesis had first been determined by transillumination. Among the observations reported are that the number of tubular receptors for FSH is highest at stage I and lowest at stages VI-VIII of the spermatogenic cycle (Parvinen, 1982). Ritzen et al. (1982) showed that the secretion of androgen-binding protein (ABP) is maximal in tubular segments in stage VII and VIII. Transferrin is secreted by the Sertoli cells (Skinner and Griswold 1980) and this occurs at a higher rate during stages IX-XII (Mather et al., 1983). The secretion of plasminogen activator, which may be involved in tissue remodeling and cell migration, is secreted at a 10-fold higher rate during stages VII and VIII compared to other stages of the spermatogenic cycle (Lacroix et al., 1981). The activity of adenyl cyclase in germ cells (Cusan et al., 1981) and Sertoli cells (Gordeladze et al., 1982) varies throughout the spermatogenic cycle.

The experiments of Clermont and Parvinen supported the ideas that the regional requirements for germ cell development within the testis and along the seminiferous tubule were not constant, nor was the metabolic activity of

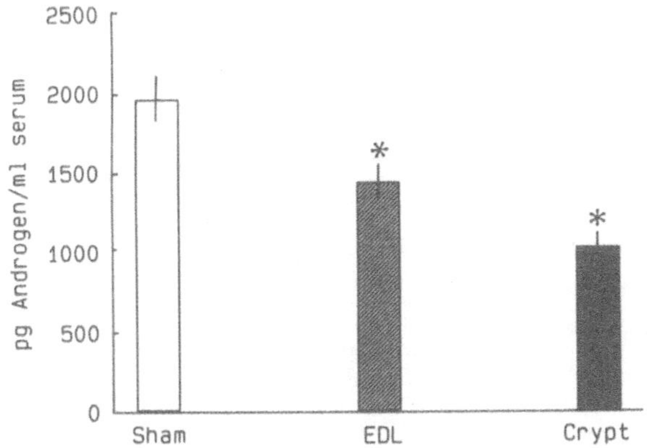

Fig. 2. Effect of efferent-duct ligation (EDL) and cryptorchidism (Crypt) on the androgen concentration in serum 4 weeks later. All values are mean ± SEM of 10 to 20 animals. * indicates value is significantly less than sham operated control. (Jansz and Pomerantz, 1986)

the Sertoli and germ cells. Because the secretory rate of gonadotropic hormones is relatively constant during the time required for the completion of one spermatogenic cycle, it was reasonable to propose that local intratesticular mechanisms of a paracrine nature were required to assure such that local adjustments to the milieu for gametogenesis could occur. More specifically, if the androgen requirement for ABP synthesis and germ cell development change during the spermatogenic cycle, then it is possible that a coordination of tubular and Leydig cell activity occurs.

In support of the idea that such local or paracrine mechanisms for tubular regulation of Leydig cell activity may exist, Aoki and Fawcett (1978) showed that implantation of substances which caused focal disruption of spermatogenesis also caused changes in Leydig cell morphology which were confined to the adjacent interstitium. Later, Bergh (1982; 1983) showed that the microscopic appearance of Leydig cells changed in concert with the progression of spermatogenesis in adjacent tubules. Such observations have led a number of laboratories, including our own, to a series of studies which are summarized in this review. Each addresses aspects of the hypothesis that the state of spermatogenesis is communicated to the Leydig cells and results in altered formation of androgens.

CURRENT STUDIES

Disruption of spermatogenesis is associated with changes in both gonadal and pituitary hormone secretion. The data in Fig. 1 represent the changes in the circulating concentration of androgen in rats of various ages that were exposed in utero to either ionizing radiation or busulfan to eliminate germ cells from the seminiferous epithelium (Jansz and Pomerantz, 1984; 1985a). It is evident that at 30 days of age the aspermatogenic and normal gonads produce equivalent amounts of androgen. Between 30 and 60 days of age, when the first wave of spermatogenesis is completed in normal rats (Clermont and Perey, 1957), the disruption of gametogenesis is associated with impaired production of androgens.

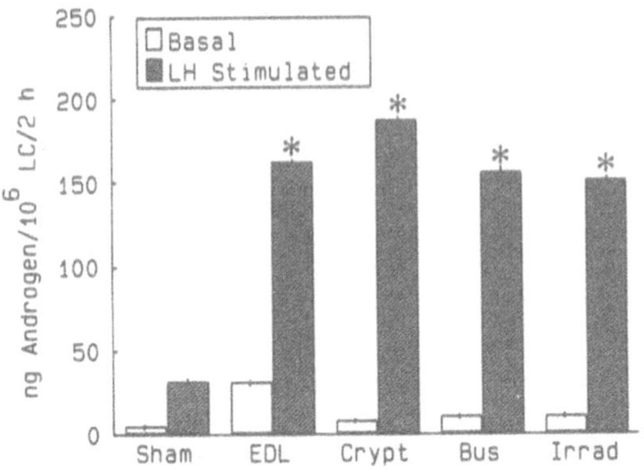

Fig. 3. Effect of efferent-duct ligation (EDL), cryptorchidism (Crypt), *in utero* busulfan treatment (Bus) or irradiation (Irrad) on the *in vitro* basal and LH-stimulated androgen production by Leydig cells (LC). All values are mean ± SEM of at least 5 determinations. * indicates LH-stimulated values significantly different from sham control. (Jansz and Pomerantz, 1984; 1985a; 1986)

Fig. 2 shows similar data from animals in which spermatogenesis was disrupted after the animals had reached sexual maturity. Cryptorchidism was surgically induced or the efferent ducts were bilaterally ligated and 4 weeks later blood samples were collected (Jansz and Pomerantz, 1986). Again, it is clear that the gametogenic disruption was associated with decreased androgen production by the lesioned gonads. In all cases where it was measured, the concentration of LH in serum had significantly increased whenever a treatment caused decreased production of androgen. These studies were in agreement with with data obtained by others (Amatayakul et al., 1971; Au et al., 1984; de Kretser et al., 1979; Hall and Gomes, 1975; Inano and Tamaoki, 1968; Rich and de Kretser, 1979; Rich et al., 1979).

In situations in which experimental interventions caused disruption to spermatogenesis, it was noted that the Leydig cells were hypertrophied and the appearance of their intracellular organelles suggested secretory hyperactivity (Kerr et al., 1979; Leeson and Leeson, 1970). In an attempt to better understand the functional characteristics of Leydig cells from gametogenically damaged gonads, we examined the ability of Leydig cells (from collagenase dispersed testes and identified by 3β-hydroxysteroid dehydrogenase histochemistry) to form androgens *in vitro*. The effect of various types of spermatogenic disruption were then studied. The data shown in Fig. 3 summarize the findings.

It is obvious that in all cases the basal and LH-stimulated androgen production was markedly elevated in Leydig cells obtained from damaged gonads. The reason for increased production of androgen per Leydig cell in the fact of decreased total testicular secretion is not clear at this point. Possible explanations are that the spermatogenically disrupted gonads have decreased blood flow (Setchell, 1978), increased intratesticular metabolism of androgens, or decreased numbers of steroidogenically active cells. In support of this latter possibility, we have found that the total binding of

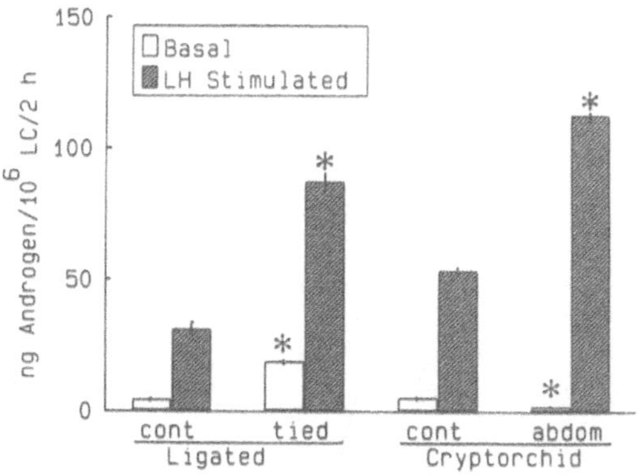

Fig. 4. Effect of unilateral efferent-duct ligation and unilateral cryptorchidism on the basal and LH-stimulated *in vitro* androgen production by Leydig cells (LC). All values are mean ± SEM of at least 5 determinations. * indicates value is significantly different from the respective contralateral control (cont). (Jansz and Pomerantz, 1986)

labelled hCG to testes from fetally irradiated rats was decreased compared to control, yet the binding per histochemically identified Leydig cell was increased (Jansz and Pomerantz, 1984).

The studies described thus far suggest that Leydig cells increased their androgen synthesizing capacity in response to gametogenic disruption, but provide little insight into how these changes are affected. It is possible that alterations in gonadotropin secretion or some other extratesticular response to damage of the germinal epithelium caused the increased responsiveness of Leydig cells to LH. While this remains a viable explanation for many of the published observations, this review is particularly concerned with local mechanisms and their potential contributions to testicular function. We have used an approach introduced by others and have rendered one gonad sterile in adult rats by surgically producing unilateral crytorchidism or ligating the efferent ducts of one testis. Because both gonads in such experimental models are exposed to similar blood concentrations of extragonadal "messengers", any difference in the functions of Leydig cells obtained from the spermatogenically disrupted gonad and its contralateral control organ must be due to intratesticular factors. A comparison of the androgen secretory capacity of Leydig cells from unilaterally cryptorchid or efferent-duct-ligated gonads and their respective controls are shown in Fig. 4. It is evident that the Leydig cells from both types of treated gonad are more responsive to LH than are the cells from the contralateral control. Thus, we contend that part of the increase in the responsiveness of Leydig cells in gametogenically damaged testes is due to a locally produced factor that can act within the gonad in which it is produced. While our studies (Jansz and Pomerantz, 1984; 1986) and that of Sharpe et al. (1984) utilized histochemically identified Leydig cells to evaluate the effects of tubular disruption, earlier workers had pioneered this same concept by showing that

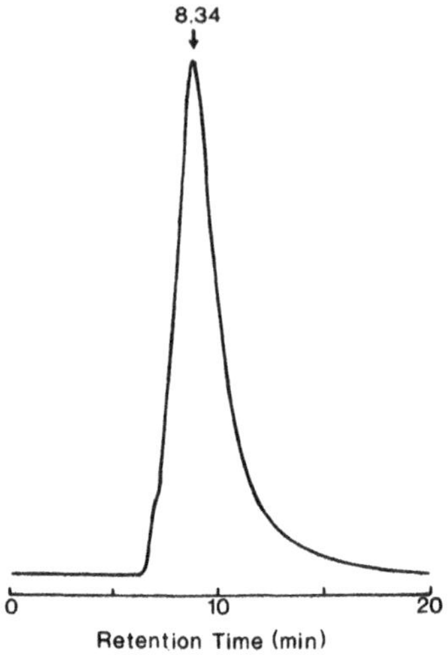

Fig. 5. HPLC elution profile of bioactive fraction obtained after Sephadex G-75 chromatography of testicular fluid obtained from bilaterally cryptorchid rats. Elution used a 0.1 M Tris and 0.1 M Na_2SO_4 buffer at a flow rate of 5 m./min through a double I^{125} protein column. The purity of this fraction was not tested; however, it possessed enhanced stimulatory activity when applied to normal Leydig cells.

androgen production by testicular tissue obtained from unilaterally treated gonads was significantly different from the contralateral control (Risbridger et al., 1981a, b).

If a factor produced locally in association with spermatogenesis participates in the control of androgen secretion by the Leydig cell, such a factor could arrive at the Leydig cell via the interstitial fluid of the testis. Sharpe and Cooper (1983) developed a technique that permits collection of testicular fluid rich in interstitial fluid. Utilizing this technique, four laboratories have provided information concerning the ability of such testicular fluid to modify the functions of the Leydig cell. Sharpe and Cooper (1984) found that testicular fluid contained a factor that could increase the steroidogenic capability of the rat Leydig cell. These workers found that the fluid obtained from cryptorchid testes was more effective in this regard than that from normal gonads. This laboratory (Jansz and Pomerantz, 1985b) confirmed this observation and attempted to further characterize the factor or factors. We found that this stimulatory material from unilaterally and bilaterally cryptorchid gonads increased the responsiveness of normal Leydig cells to a maximal dose of dibutyryl cAMP or LH, that the stimulatory material was thermolabile and its migration on columns of Sephadex G-75 suggested a molecular size greater than 45,000 daltons. Fig. 5 shows an elution profile from further HPLC of the active material obtained from the gonads of bilaterally cryptorchid animals. The material with a retention time of 8.34 min was enriched in potency, had stimulatory activity similar to that of unpurified fluid and had an apparent molecular size of 48,000-53,000 daltons.

We have been able to show that this stimulatory activity is increased in gonads subjected to other methods for disruption of spermatogenesis such as <u>in utero</u> exposure to busulfan or ionizing radiation and ligation of the efferent ducts (unpublished observations). We concluded that disruption of gametogenesis in the testis caused an increase in the amount of thermolabile factor or factors which acted to increase the androgen-secreting capacity of Leydig cells and acted, in part, distal to the intra-Leydig cell formation of cAMP. Finally, Rommerts et al. (1986) and Risbridger et al. (1986) have reported that testicular fluid from normal rats contained a factor which could increase the conversion of cholesterol to pregnenolone by Leydig cells and that this activity was independent of LH. A possible intratesticular source of this or similar material was suggested by Verhoeven and Cailleau (1985), who were able to show that spent medium from Sertoli cell cultures was able to stimulate androgen production by Leydig cells.

SUMMARY

Taken together, these recent studies provide evidence that the state of gametogenesis in the testis may control the production of "messages" by the seminiferous tubule which are able to act in a paracrine manner to control the local secretion of androgen. These messages appear to act on the Leydig cells in addition to the normal actions of LH. The physiologic role of such a factor or factors is unknown. One may speculate that during the process of spermatogenesis at a single point in the testis the required hormonal environment changes in that location. The factor(s) we have been studying may represent the means by which the seminiferous tubule communicates its metabolic requirements to the adjacent interstitium. The initial waves of spermatogenesis which occur in assciation with puberty and seasonally in some species are accompanied by changes in steroidogenesis by Leydig cells. The testicular factors we have begun to describe may provide the mechanism by which the seminiferous tubule heralds the accelerations of gametogenesis and induces the appropriate changes in testicular steroidogenesis.

REFERENCES

Amatayakul, K., Ryab, R., Uozomi, T., and Alberta, A., 1971, A reinvestigation of testicular-anterior pituitary relationships in the rat: I. Effects of castration and cryptorchidism, Endocrinology, 8:872.

Aoki, A., and Fawcett, D. W., 1978, Is there a local feedback from the seminiferous tubules affecting activity of the Leydig cell?, Biol. Reprod., 19:144.

Au, C. L., Robertson, D. M., and de Kretser, D. M., 1984, Relationship between testicular inhibin content and serum FSH concentrations in rats after bilateral efferent-duct ligation, J. Reprod. Fertil., 72:351.

Bartke, A., Hafiez, A. A., Bex, F. J., and Dalterio, S., 1978, Hormonal interactions in the regulation of androgen secretion, Biol. Reprod., 18:44.

Bergh, A., 1982, Local differences in Leydig cell morphology in the adult rat testis: evidence for a local control of Leydig cells by adjacent seminiferous tubules, Int. J. Androl., 5:325.

Bergh, A., 1983, Paracrine regulation of Leydig cells by the seminiferous tubules, Int. J. Androl., 6:57.

Christensen, A. K., 1975, Leydig cells, in: "Handbook of Physiology, Section 7: Endocrinology, vol. V, Male Reproductive System," R. O. Greep and E. B. Astwood, eds., p. 57, American Physiology Society, Washington, D. C.

Clermont, Y., 1972, Kinetics of spermatogenesis in mammals; seminiferous epithelium cycle and spermatogonial renewal, Physiol. Rev., 52:198.

Clermont, Y., and Perey, B., 1957, Quantitative study of the cell population of the seminiferous tubules in immature rats, Am. J. Anat., 100:241.

Cusan, L., Gordeladze, J. O., Parvinen, M., Clausen, O. P. F., and Hansson, V., 1981, Protein carboxylmethylase and germ cell adenylyl cyclase at specific stages of the spermatogenic cycle of the rat, Biol. Reprod., 25:915.

de Kretser, D. M., Sharpe, R. M., and Swanston, I. A., 1979, Alterations in steroidogenesis and human chorionic gonadotropin binding in the cryptorchid testis, Endocrinology, 105:135.

di Zerega, G. S., and Sherins, R. J., 1981, Endocrine control of adult testicular function, in: "The Testis," H. Burger and D. M. de Kretser, eds., p. 127, Raven Press, New York.

Gordeladze, J. O., Parvinen, M., Clausen, O. P. F., and Hansson, V., 1982, Stage-dependent variation in Mn^{2+}-sensitive adenylyl cyclase (AC) activity in spermatids and FSH-sensitive AC in Sertoli cells, Arch. Androl., 8:43.

Hall, R. W., and Gomes, W. R., 1975, The effect of artificial cryptorchidism on serum oestrogen and testosterone levels in the adult rat, Acta Endocr., 80:583.

Inano, H., and Tamaoki, B., 1968, Effect of experimental bilateral cryptorchidism on testicular enzymes related to androgen formation, Endocrinology, 83:1074.

Jansz, G. F., and Pomerantz, D. K., 1984, Fetal irradiation increases androgen production by dispersed Leydig cells of the rat, J. Androl., 5:344.

Jansz, G. F., and Pomerantz, D. K., 1985a, The effect of prenatal treatment with busulfan on in vitro androgen production by testes from rats of various ages, Can. J. Physiol. Pharmacol., 63:1155.

Jansz, G. F., and Pomerantz, D. K., 1985b, Cryptorchidism increases the content of a "factor" in testicular interstitial fluid which stimulates Leydig cell androgen production, Biol. Reprod., 32:85, Suppl. 1.

Jansz, G. F., and Pomerantz, D. K., 1986, A comparison of Leydig cell function after unilateral and bilateral cryptorchidism and efferent-duct ligation, Biol. Reprod., 34:316.

Kerr, J. B., Rich, K. A., and de Kretzer, D. M., 1979, Alterations of the fine structure and androgen secretion of the interstitial cells in the experimentally cryptorchid rat testis, Biol. Reprod., 20:409.

Lacroix, M., Parvinen, M., and Fritz, I. B., 1981, Localization of testicular plasminogen activator in discrete portions (stages VII and VIII) of the seminiferous tubule, Biol. Reprod., 25:143.

Leeson, T. S., and Leeson, C. R., 1970, Experimental cryptorchidism in the rat. A light and electron microscope study, Invest. Urol., 8:127

Mather, J. P., Gunsalus, G. L., Musto, N. H., Cheng, C. Y., Parvinen, M., Wright, W., Perez-Infante, V., Margioris, A., Liotta, A., Becker, R., Krieger, D. T., and Bardin, C. W., 1983, The hormonal control of Sertoli cell secretion, J. Steroid Biochem., 19:41.

McCann, S. M., 1974, Regulation of secretion of follicle-stimulating hormone and luteinizing hormone, in: "Handbook of Physiology, Section 7: Endocrinology, vol. IV, Part 2," R. O. Greep and E. B. Astwood, eds., p. 489, American Physiology Society, Washington, D. C.

Means, A. R., 1977, Mechansims of action of follicle-stimulating hormone (FSH), in: "The Testis," A. D. Johnson and W. R. Gomes, eds., p. 163, Academic Press, New York.

Neil, J. D., 1974, Prolactin: its secretion and control, in: "Handbook of Physiology, Section 7: Endocrinology, vol. IV, Part 2," R. O. Greep and E. B. Astwood, eds., p. 469, American Physiology Society, Washington, D. C.

Parvinen, M., 1982, Regulation of the seminiferous epithelium, Endocr. Rev., 3:404.

Rich, K. A., and de Kretser, D. M., 1979, Effect of fetal irradiation on testicular receptors and testosterone response to gonadotrophin stimulation in adult rats, Int. J. Androl., 2:343.

Rich, K. A., Kerr, J. B., and de Kretser, D. M., 1979, Evidence of Leydig cell dysfunction in rats with seminiferous tubule damage, Mol. Cell. Endocrinol., 13:123.

Ritzen, E. M., Hansson, V., and French, F. S., 1981, The Sertoli cell, in: "The Testis," H. Burger and D. de Kretser, eds., p. 171, Raven Press, New York.
Ritzen, E. M., Boitani, C., Parvinen, M., French, F. S., and Feldman, M., 1982, Stage-dependent secretion of ABP by rat seminiferous tubules, Mol. Cell. Endocrinol., 25:25.
Risbridger, G. P., Kerr, J. B., Peake, R. A., and de Kretser, D. M., 1981a, An assessment of Leydig cell function after bilateral or unilateral efferent duct ligation: further evidence for local control of Leydig cell function, Endocrinology, 109:1234.
Risbridger, G. P., Kerr, J. B., and de Kretser, D. M., 1981b, An evaluation of Leydig cell function and gonadotropin binding in unilateral and bilateral cryptorchidism. Evidence for local control of Leydig cell function by the seminiferous tubule, Biol. Reprod., 24:534.
Risbridger, G. P., Jenkin, G., and de Kretser, D. M., 1986, The interaction of hCG, hydroxysteroids and interstitial fluid on rat Leydig cell steroidogenesis in vitro, J. Reprod. Fertil., 77:239.
Rommerts, F. F. G., Hoogerbrugge, J. W., and van der Molen, H. J., 1986, Stimulation of steroid production in isolated rat Leydig cells by unknown factors in testicular fluid differs from the effects of LH or LH-releasing hormone, J. Endocrinol., 109:111.
Russell, L. D., and Clermont, Y., 1977, Degeneration of germ cells in normal, hypophysectomized and hormone-treated hypophysectomized rats, Anat. Rec., 187:347.
Setchell, B. P., 1978, The Mammalian Testis, Paul Elek, London.
Sharpen, R. M., 1983, Local control of testicular function. Quart. Rev. Exp. Physiol., 68:265.
Sharpe, R. M., and Cooper, I., 1983, Testicular interstitial fluid as a monitor for changes in the intratesticular environment in the rat, J. Reprod. Fertil., 69:125.
Sharpe, R. M., and Cooper, I., 1984, Intratesticular secretion of a factor(s) with major stimulatory effects on Leydig cell testosterone secretion in vitro, Mol. Cell. Endocrinol., 37:159.
Sharpe, R. M., Cooper, I., and Doogan, D. G., 1984, Increase in Leydig cell responsiveness in the unilaterally cryptorchid rat testis and its relationship to the intratesticular levels of testosterone, J. Endocrinol., 102:319.
Skinner, M. K., and Griswold, M. D., 1980, Sertoli cells synthesize and secrete transferrin-like protein, J. Biol. Chem., 255:9523.
Steinberger, A., and Steinberger, E., 1977, The Sertoli cells, in: "The Testis," A. D. Johnson and W. R. Gomes, eds., p. 371, Academic Press, New York.
Verhoeven, G., and Cailleau, J., 1985, A factor in spent media from Sertoli cells enriched cultures that stimulates steroidogenesis in Leydig cells, Mol. Cell. Endocrinol., 40:57.
Yee, J. B., and Hutson, J. C., 1985, Effects of testicular macrophage-conditioned medium on Leydig cells in culture, Endocrinology, 116:2682.

hCG/LH-INDUCED CHANGES IN TESTICULAR BLOOD FLOW, MICROCIRCULATION AND VASCULAR PERMEABILITY IN ADULT RATS

Anders Bergh and Jan-Erik Damber

Departments of Pathology, Urology,
and Andrology
University of Umea
S-90187 Umea, Sweden

INTRODUCTION

Control of testicular blood flow and vascular permeability are important aspects of testicular physiology. Vascular factors are crucial both for the supply of substances necessary for spermatogenesis and for the secretion of androgens. Testicular blood vessels are situated in the interstitial tissue, often close to groups of Leydig cells. The seminiferous tubules, which constitute up to 90% of testicular volume, are avascular. Substances from the blood must transverse the interstitium before they reach the tubules. The passage of substances into the tubules is restricted also by the blood-testis barrier situated partly in the tubular wall and partly at junctional complexes between adjacent Sertoli cells. The interstitial space, particularly in the rat, is composed mainly of large lymphatic sinusoids that completely surround the tubules and Leydig cells secrete their products into this lymph (Clark, 1976; Fawcett et al., 1973). Studies during recent years have shown that paracrine factors are secreted into this fluid, factors that locally modulate the function both in Sertoli and in Leydig cells (Sharpe, 1984; Saez et al., 1985). Testicular interstitial fluid (IF) thus serves as a medium for communication between tubules, Leydig cells and blood (for review see Sharpe, 1984). The control of IF composition and volume are therefore of importance for testicular function (Sharpe, 1984).

The factors involved in the physiological control of testicular blood flow and vascular permeability are unfortunately almost unknown. Treatment of adult rats with human chorionic gonadotropin (hCG) results in a delayed increase in blood flow occurring 16-24 h after treatment (Sertoli and Sharpe, 1981; Damber et al., 1981). Since testosterone secretion is maximally stimulated within 1 h, it is not likely that the blood flow stimulating factor is testosterone itself. In the normal rat there are up to 6 spontaneous LH peaks per 8 h (Ellis and Desjardins, 1982); the role of changes occurring after 16 h is therefore somewhat obscure. The mechanism behind the delayed increase in blood flow and its physiological relevance are consequently unknown. Apart from LH/hCG, prostaglandins, catecholamines, ACTH and serotonin have been reported to influence testicular blood flow (Setchell, 1978; Damber, 1978), but whether these substances are involved in the physiological control of testicular blood flow is still unknown.

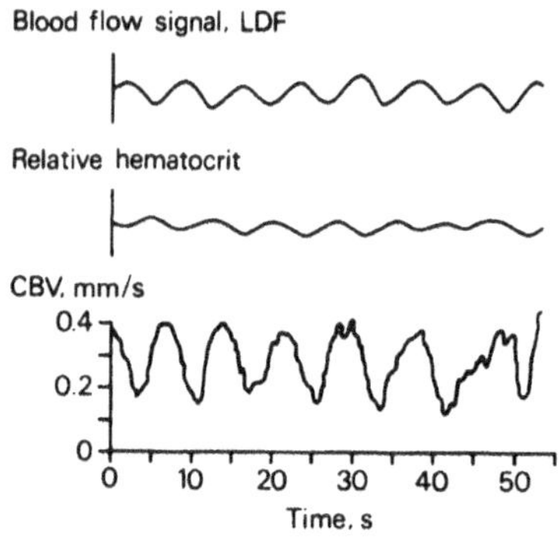

Fig. 1. Testicular microcirculation as recorded by videophotometric capillaroscopy (capillary blood cell velocity, CBV) and laser doppler flowmetry (LDF) in a control rat. Changes in the relative hematocrit were registered simultaneously and are shown in the middle section of the figure. Simultaneous analysis of CBV, LDF and relative hematocrit showed rythmical oscillations of the three parameters with the same frequency. With permission reproduced from Damber et al., 1986a, Acta Physiologica Scandinavia.

hCG treatment results in an increase in testicular interstitial fluid (IF) volume occurring 6-8 h after treatment (Setchell and Sharpe, 1981; Sharpe, 1984; Widmark et al., 1986a). The mechanism behind this increase has not been established, but it is apparently not related to an increase in flow since that was not increased until at least 8 h later. Setchell and Sharpe (1981, and see Sharpe 1984 for review) suggested that the increase in IF volume was caused by an hCG-induced increase in capillary permeability and that this was of physiological importance for the uptake of substances to the testis during hormonal stimulation.

Apart from hCG, one additional factor influences IF volume. Treatment of hypophysectomized (hypox) or intact rats with LHRH-agonists increase or decrease IF volume (Sharpe et al., 1983). An LHRH-like peptide is probably produced in the tubules, Leydig cells contain LHRH-receptors, and this peptide may modulate Leydig cell function (Sharpe, 1984 for review). The seminiferous tubules may thus be involved in the control of vascular permeability. If so, this would not be surprising since most of the metabolic activity in the testis occurs inside the tubules.

From this summary, it is evident that our knowledge of the physiological regulation of testicular blood flow and vascular permeability is far from

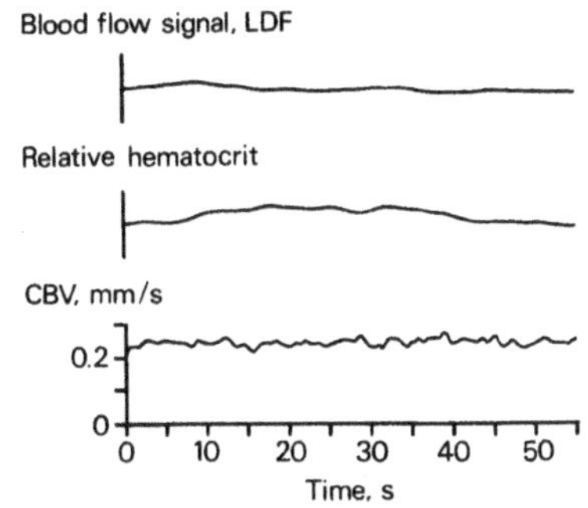

Fig. 2. Testicular blood flow as recorded by videophotometric capillaroscopy (capillary blood cell velocity, CBV) and laser doppler flowmetry (LDF) in a rat treated with 200 IU hCG 16-20 h prior to experiment. The relative hematocrit was registered simultaneously and is shown in the middle section of the figure. In comparison with Fig. 1, there were no rhythmical oscillations in blood flow (CBV, LDF) or in relative hematocrit. With permission reproduced from Damber et al., 1986a, Acta Physiologica Scandinavia.

satisfactory. In a series of papers, we have studied blood flow and permeability in the rat testis and our observations are summarized in this mini-review.

LOCAL VARIATIONS IN TESTICULAR BLOOD FLOW

Testicular blood flow has usually been studied using clearance techniques or with radioactive microspheres (Damber, 1978). Such methods can, however, not be used to study flow over a time period. Blood flow can be measured continuously without interference using laser doppler flowmetry. The laser probe, which is fixed just above the testicular surface, measures erythrocyte flux in fine caliber vessels in a tissue volume of approx. 1 mm^3 (Damber et al., 1982; 1983 for references). When studied in this way, testicular blood flow shows large rhythmical variations with approximately 6-8 peaks per min (Damber et al., 1982; 1983). By studying the microcirculation in vivo with capillaroscopy simultaneously with laser doppler flowmetry, we found that the variations in laser signal were due to local variations in erythrocyte velocity (Fig. 1, Damber et al., 1986a). In a group of capillaries, supplied by the same arterioli, blood flow was synchronized and periods of high erythrocyte velocity alternate with periods of very slow or no flow. These variations are apparently due to rhythmical changes in myogenic activity in precapillary sphincters (called vasomotion). This type

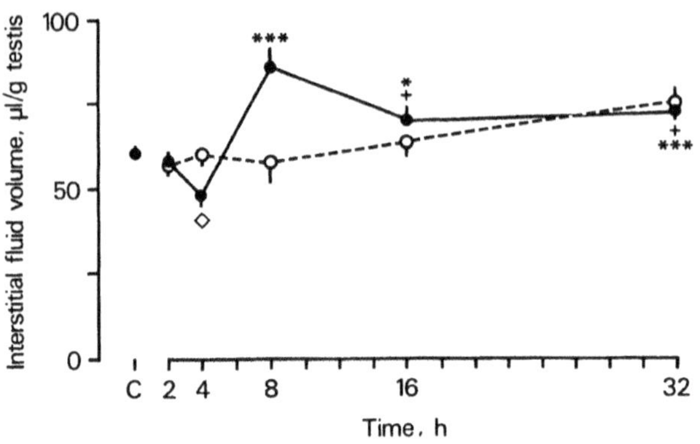

Fig. 3. Time dependent changes of testicular interstitial fluid (IF) after s.c. injection of 12.5 IU (o) and 50 IU (o) hCG. Each point represents the mean ± SEM of 10-15 rats. , $p < 0.01$ Significantly smaller than control values; *, $p < 0.05$, ***, $p < 0.02$ significantly larger than control values; +, $p < 0.05$ significantly smaller than values obtained 8 h after hCG (Mann-Whitney U test). With permission reproduced from Widmark et al., 1986a, Journal of Endocrinology.

of microcirculation is not unique for the testis, a similar blood flow pattern has been described in other organs (Eggert and Weiss, 1980; Fagrell et al., 1980; Intaglietta, 1981). Such variations in erythrocyte velocity are probably involved in transvascular exchange and in the formation of interstitial fluid in other organs. During periods with high flow there is a net filtration of plasma from the vasculature but during stops the net flow is reversed (Intaglietta, 1981; Intaglietta and Gross, 1982). The pronounced vasomotion in the testis could thus facilitate transvascular exchange, and be involved in the formation of testicular interstitial fluid. The factors regulating tonus in testicular precapillary sphincters are, however, unknown; spontaneous variations in myogenic activity, nerves and hormones could all be involved.

hCG treatment, within 4-8 h results in a change in blood flow pattern. The flow is now continuous (Figs. 1 and 2, Damber et al., 1986a). This is apparently due to relaxation of precapillary sphincters. A normal type of flow pattern then returns 24 h after treatment. As could be suspected, continuous flow favors filtration of plasma from the vasculature and IF volume was increased simultaneously with the change in flow pattern (Fig. 3, Widmark et al., 1986a). From these observations, we conclude that the hCG-induced increase in IF volume could be caused by increased filtration and is not necessarily due to increased capillary permeability as suggested earlier. hCG treatment apparently results in the secretion of factors causing dilatation of precapillary sphincters, but the nature of these factors remains unknown. Testicular microcirculation changes during development; the flow is continuous in immature animals (our own unpublished observations). This

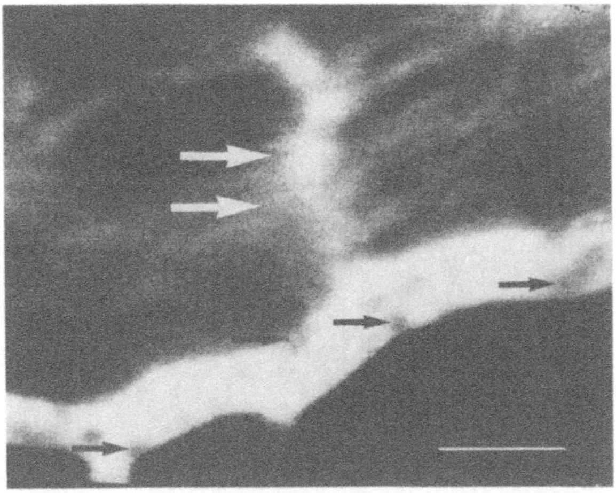

Fig. 4. Photomicrograph from a videorecording showing a leaking postcapillary venule 4 h after treatment with 50 IU hCG s.c. Adhering leukocytes (black arrows) and macromolecular dextran (white arrows) leaking from the blood vessel are observed. Bar 50 μm. With permission from Bergh et al., 1986a, Journal of Reproduction and Fertility.

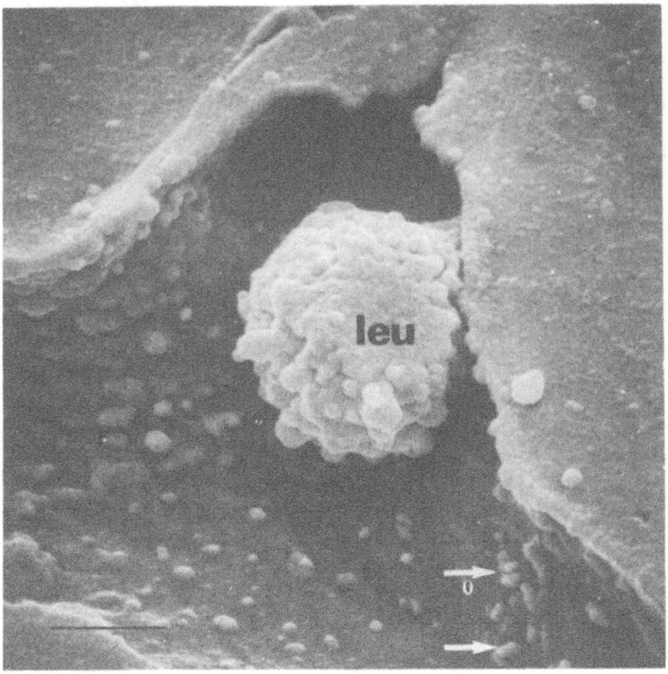

Fig. 5. Scanning EM showing the inside of a postcapillary venule 4 h after treatment with 50 IU hCG s.c. Despite vascular perfusion with saline followed by the fixative, an adhering leukocyte (leu) is observed. Microvilli-like structures are seen on the endothelial surface (arrows). Bar 5 μm.

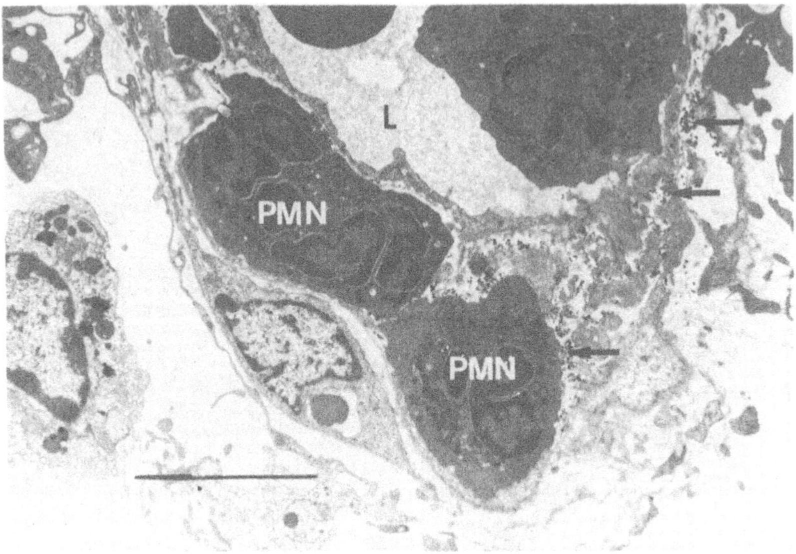

Fig. 6. Electron micrograph showing migrating leukocytes (PMN) and subendothelial deposits of carbon particles (arrows) in a postcapillary venule (lumen = L) 4 h after treatment with 50 IU hCG s.c. Bar 5 μm. With permission reproduced from Bergh et al., 1986a, Journal of Reproduction and Fertility.

observation and the effect of hCG (and LH/LHRH, see below) suggest that local variations in testicular microcirculation are of physiological significance, but their precise role has yet to be determined.

REGULATION OF VASCULAR PERMEABILITY - ROLE OF LEUKOCYTES

The hCG-induced increase in IF volume could thus be due to hemodynamic changes causing increased filtration, but changes also in permeability cannot be excluded. Vascular permeability was therefore studied in vivo after intravascular injection of FITC-labelled dextran (150,000 daltons) using a fluorescence microscope connected to a videorecorder. Under basal conditions macromolecules of this size remain in the blood stream. However, 4 h after hCG treatment the injected dextran was rapidly leaking into the interstitium (Bergh et al., 1986a). This leakage occurred only in small postcapillary venules and not in capillaries. The leakage sites were similar to those described in other tissues after administration of inflammation mediators such as histamine or leukotrienes (Björk et al., 1982; Persson and Svensjö, 1985). When the leaking venular segments were examined in higher magnification, we observed numerous adhering leukocytes (Figs. 4 and 5). Adhering leukocytes were not observed in control testes.

Vascular permeability was also studied using the colloidal carbon labelling technique. A suspension of carbon particles is injected intravascularly. After approximately 1 h, the injected carbon is cleared from the circulation by the reticulo-endothelial system. Carbon particles, which are too large to penetrate through pores or channels in endothelial cells, may however penetrate open junctions between adjacent endothelial cells. They

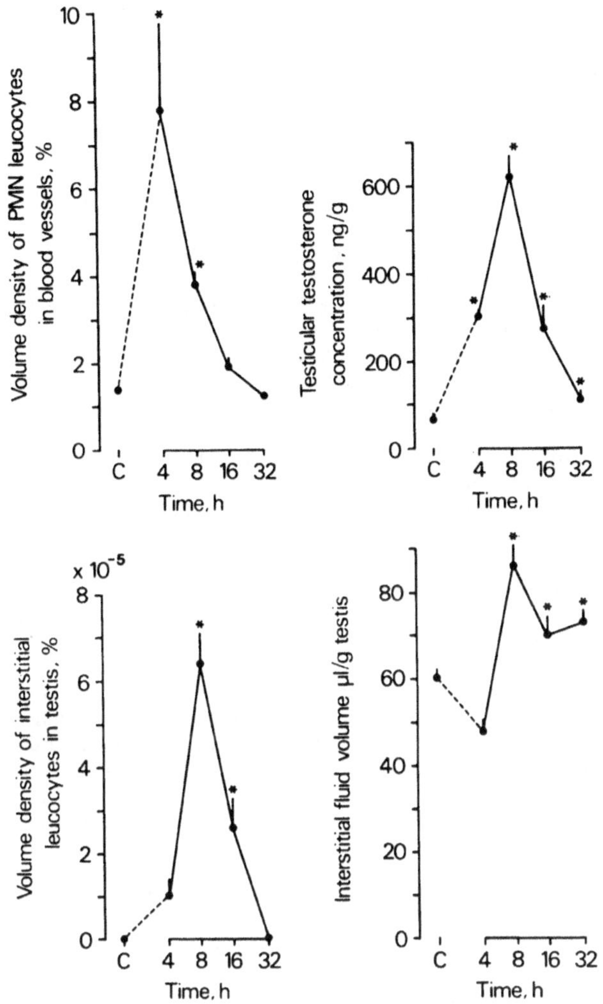

Fig. 7. Volume density of leukocytes in testicular blood vessels (upper left) and in the interstitial space (lower left), testicular testosterone concentration (upper right) and IF volume (lower right) at different times after treatment with 50 IU hCG. *, Significantly different from control value ($p < 0.05$), n = 5 animals, mean ± SEM. With permission reproduced from Bergh et al., 1986b, Endocrinology.

are then stopped by the underlying basement membrane. Leaking blood vessels are thus black-labelled and can be localized later during dissection or in sections (Cotran and Majno, 1964). Blood vessels were not labelled by carbon in control testes. In contrast, 4 h after hCG treatment segments of postcapillary venules were heavily black stained. The leakage sites were examined by electron microscopy and we confirmed that they only occurred in postcapillary venules and that they were caused by interendothelial cell gaps. We could also observe neutrophil polymorphonuclear leukocytes (PMNs) migrating through such openings (Fig. 6, Bergh et al., 1986a). hCG treatment apparently resulted in the formation of gaps between adjacent endothelial cells in postcapillary venules and an increased permeability for macromolecules. The hCG-induced increase in IF volume could thus be caused both by increased filtration (see above) and an increased venular permeability.

We were initially surprised to observe adhering leukocytes in the leaking venules. We therefore quantified the volume density of PMNs inside blood vessels and inside the interstitial space in tissue sections at different times after hCG treatment (Bergh et al., 1986b). Under basal conditions a small number of PMNs was observed inside blood vessels but not in the interstitial space. Four hours after hCG treatment, there was a 6-fold increase in the number of PMNs inside blood vessels (Fig. 7, Bergh et al., 1986b). This increase occurred mainly in postcapillary venules and the PMNs were mainly of the neutrophil type. At 8 and 16 h, leukocytes appeared in the interstitial space. Thirty-two hours after treatment, intravascular and and interstitial PMN numbers had returned to basal values. The volume of IF was also measured and we found that the increase in IF was preceded by intravascular accumulation of PMNs and paralleled with migration of leukocytes into the interstitial space (Fig. 7). Our observations suggest that hCG treatment results in secretion of a chemotactic factor for leukocytes in the testis. The nature and cellular source of this factor is unknown. Leydig cells may, however, secrete leukotriene B 4 (Cocke et al., 1984), one of the most potent leukotactic factors known (Björk et al., 1982; Williams, 1985; Persson and Svensjö, 1985). The testicular interstitium contains numerous macrophages (Bergh, 1985) and macrophages in general secrete leukotactic factors when activated (Takemura and Werb, 1984).

Dilatation of precapillary sphincters, leukocyte adhesion and migration, formation of interendothelial cell gaps in postcapillary venules, increased permeability for macromolecules and tissue edema are well-known characteristics of acute inflammation (Williams, 1985). hCG treatment of the testis thus results in a situation with striking similarities to inflammation and the increase in IF could perhaps be considered as an inflammatory exudate! The tissue edema in inflammation is caused by two parallel but independently regulated events (Williams, 1985): (1) Dilatation of precapillary sphincters, caused by secretion of vasodilators (principally prostaglandins) resulting in increased hydrostatic pressure in capillaries and postcapillary venules. Interestingly, the concentration of prostaglandins is increased in the testis 6-8 h after hCG treatment (Haour et al., 1979). (2) An increase in venular permeability, caused by secretion of mediators that directly or indirectly result in the formation of interendothelial cell gaps (Williams, 1985; Persson and Svensjö, 1985). These mediators are, for instance, histamine with a direct effect on venular endothelium, or chemotactic factors like complement and leukotrienes that cause gaps indirectly by attracting leukocytes. The substances causing gaps in testicular venules are unknown but histamine is probably not involved since infusion of histamine, in contrast to most other organs, does not induce venular leakage in the testis (Gabbiani et al., 1970). The testis apparently secretes a leukotactic factor in response to hCG and adhering leukocytes are able to open interendothelial cell junctions (Wedmore and Williams, 1981; Williams, 1985; Persson and Svensjö, 1985). Leukocytes play a crucial role for the supply of macromolecules to the inflamed tissue; perhaps they could serve a similar role in the hormonally stimulated testis.

In conclusion, our results suggest that leukocytes could be involved in mediating the increase in venular permeability. It can, however, not be excluded that leukocyte migration is only a passive reflection of an increased permeability caused by other factors. To test this, we examined the effects of hCG in leukopenic rats. Such rats were obtained by pretreatment with an antiserum against neutrophil PMNs (ANS), causing an approximately 90% reduction in the number of circulating neutrophils. In such rats hCG did not cause a significant increase in IF volume and testicular venules could not be labelled with carbon (Fig. 8, Widmark et al., 1986c; Bergh et al., 1986c). These observations suggest that the increase in IF is caused mainly by a leukocyte-mediated increase in venular permeability. In contrast ANS treatment had no effect on IF testosterone content or on the hCG-induced change in blood flow pattern (Widmark et al., 1986c; Bergh et al., 1986c).

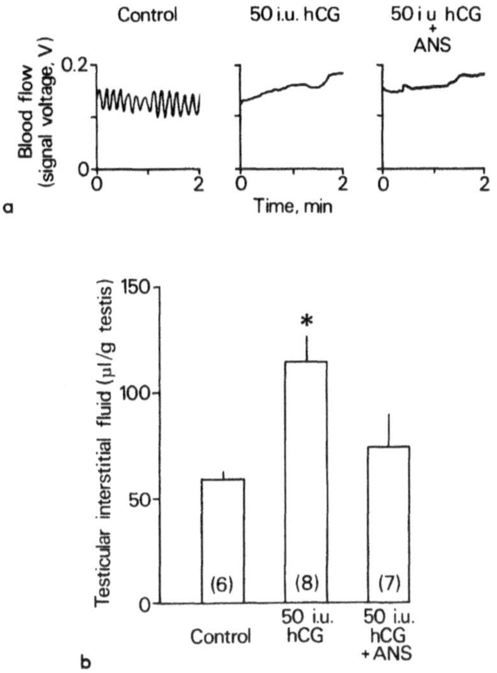

Fig. 8. Testicular blood flow measured with laser doppler flowmetry (a), and testicular IF volume (b) in rats treated with 50 IU hCG 8 h prior to experiment, with and without pretreatment with antineutrophil serum (ANS). *, Significantly different from controls, $p < 0.01$, Mann-Whitney U test, n = 6-8 animals, mean ± SEM.

This observation supports our hypothesis that the hCG-induced effect on arteries is differently regulated from that on venules. The finding that IF volume was not significantly increased by the change in blood flow pattern suggests that increased hydrostatic pressure and filtration is only of minor importance compared to the increase in venular permeability for the hCG-induced change in IF volume. However, it does not exclude the possibility that rhythmical variations in blood flow and the change induced by hCG could be of importance for transvascular exchange in the testis.

WHICH CELLS IN THE TESTIS ARE INVOLVED IN THE CONTROL OF BLOOD FLOW AND PERMEABILITY?

Changes in blood flow and permeability have only been observed using substances with effects on Leydig cells (LH/hCG and LHRH). Leydig cells must therefore play a central role and depletion of Leydig cells by EDS (a Leydig cell toxin, see Setchell and Rommerts, 1985) abolish the hCG-induced effects on IF volume, leukocyte accumulation and total blood flow (Setchell and Rommerts, 1985; our own unpublished observations) and inhibition of Leydig cells by estradiol impair the hCG-induced increase in blood flow (Damber et al., 1981). The mechanisms by which a stimulated Leydig cell function results in changes in blood flow and permeability are, however, unknown. Blocking Leydig cell prostaglandin synthesis by indomethacin or androgen-estrogen synthesis by aminoglutethimide does not prevent the hCG-induced increase in IF volume (Veiola and Raijaniemi, 1985). These observations and the delay in response after hCG treatment suggest that additional cells in the testis could be involved; cells that are stimulated by the increased Leydig cell activity, i.e. testicular macrophages, peritubular or Sertoli cells.

Destruction of seminiferous tubules by different means results in a decrease in testicular blood flow that is parallel to the degree of tubular damage (Setchell and Galil, 1983; Wang et al., 1983). Setchell and coworkers therefore suggested that blood flow is in some way related to tubular function. A role of the tubules is also indicated by the observation that LHRH agonists influence IF volume. Intratesticular injection of LHRH in a low dose, not influencing LH, decreased IF volume. A higher dose, resulting also in an increase in LH, initially decreased but later increased IF volume (Sharpe et al., 1983). Using the same LHRH-agonist, we observed that the low dose decreased (2 h after treatment) but the large dose increased total testicular blood flow (Widmark et al., 1986b). The increase in flow cannot be explained only by the increase in LH, since LH treatment on its own does not increase blood flow until 12 h after treatment (Damber et al., 1986b) and the LHRH effect was observed already 2 h after treatment (Widmark et al., 1986b). A large dose of LHRH given s.c. also increased blood flow in the testis 2-4 h after treatment in hypophysectomized rats (Damber et al., 1984). It also results in a change in blood flow pattern and in migration of PMNs into the interstitial space (our own unpublished observations). An LHRH-like peptide is probably produced in the tubules; it may modulate Leydig cell activity and testicular LHRH content is increased after hCG treatment (Sharpe, 1984). It appears that this peptide (considerably more rapid than LH/hCG, see below) augments the secretion of Leydig cell products that directly or indirectly influence testicular blood vessels. Apart from the LHRH-like peptide, Sertoli cells also produce other substances with modulatory effects on Leydig cell function (Saez et al., 1985), but whether they influence testicular microcirculation remains unknown. Particularly tubules in stages VII-VIII stimulate adjacent Leydig cells (Bergh, 1983a) but if this results in local effects on adjacent blood vessels has not been studied.

If the tubules in some way are involved in blood flow regulation, directly or indirectly via Leydig cells, it could be of interest to study the effects of hCG in unilaterally cryptorchid rats (in testes with a primary damage to the seminiferous tubules, Bergh, 1983b). We found that hCG did not increase blood flow in abdominal as in scrotal testes (Fig. 9, Damber et al., 1985). Vascular permeability was subnormal in abdominal testes during basal conditions but after hCG treatment permeability and IF volume increased considerably (Fig. 10, Damber et al., 1985). These observations suggest that

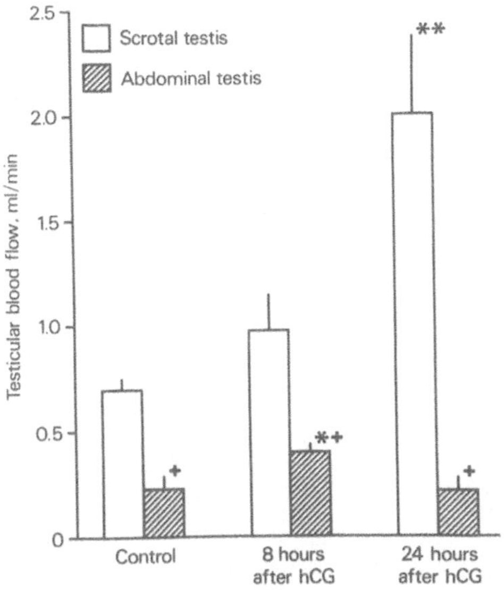

Fig. 9. Testicular blood flow in unilaterally cryptorchid rats. Each bar represents the mean ± SE of eight rats. Some rats were treated with 200 IU hCG 8 or 24 h before the experiment. *, Significantly higher ($p < 0.05$) than the abdominal control testis and abdominal testis 24 h after hCG treatment; **, significantly higher ($p < 0.001$) than scrotal testis in control group; +, significantly smaller ($p < 0.001$) than the scrotal testis in the same group. With permission reproduced from Damber et al., 1985, Endocrinology.

blood flow and permeability are differently regulated, and that abdominal testes may lack the substance stimulating flow but not the one increasing permeability. Whether this blood flow stimulating factor is produced only when tubules are intact remains unknown. The altered function in testicular macrophages in abdominal testes (Bergh, 1985) may, however, also be of importance since macrophages in general secrete vasoactive substances (Takemura and Werb, 1984). In this experiment, we also measured testosterone secretion into the testicular vein and testosterone concentration in IF. IF testosterone increased comparably in scrotal and abdominal testes after hCG treatment, but the increase in testicular venous blood was considerably lower at the abdominal than at the scrotal side (Damber et al., 1985). The hCG-stimulated secretion of testosterone from the testis (product of blood flow and concentration in the vein) was positively correlated to IF testosterone content in scrotal ($r = 0.65$, $p < 0.01$) but not in abdominal testes ($r = 0.34$, $p < 0.05$, Damber et al., 1985). This lack of correlation between local production and secretion can partly be explained by the inability to increase blood flow at the abdominal side. These observations in cryptorchid

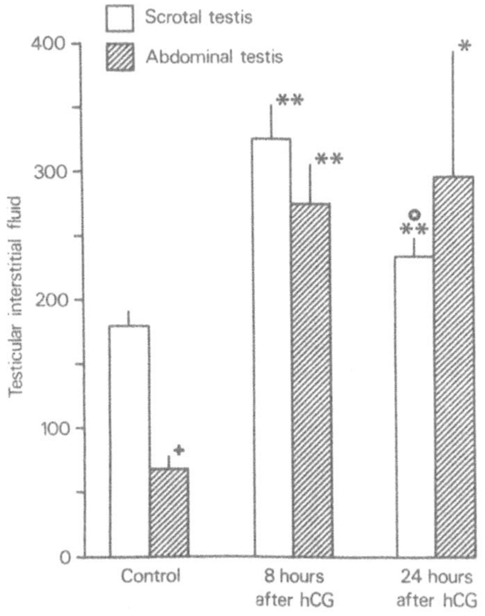

Fig. 10. The amount of testicular interstitial fluid obtained from scrotal and abdominal testes from unilaterally cryptorchid rats. Each bar represents the mean ± SE of 6-8 rats. Some of the rats were treated with 200 IU hCG s.c. 8 or 24 h before the experiment. *, Significantly higher ($p < 0.05$) than the abdominal testis in the control group; **, significantly higher ($p < 0.001$) than the scrotal and abdominal testes in the control group; +, significantly smaller ($p < 0.001$) than the scrotal testis in the control group; o, significantly smaller ($p < 0.05$) than the scrotal testis in the group treated with hCG 8 h before the experiment. With permission reproduced from Damber et al., 1985, Endocrinology.

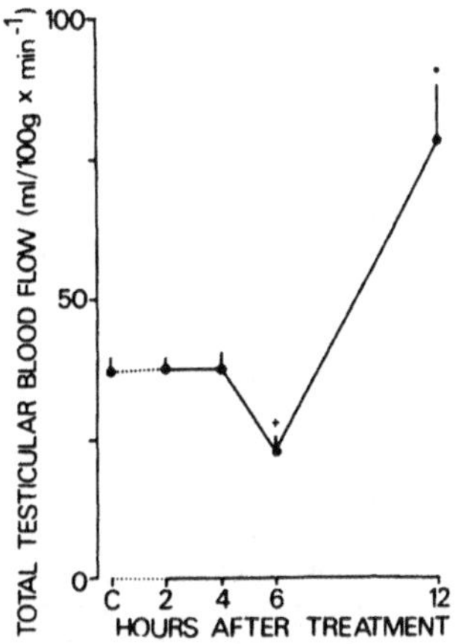

Fig. 11. Total testicular blood flow (TBF) as measured with radioactive microspheres at different times after 25 µg ovine LH given s.c. Six hours after treatment there was a significant ($p < 0.02$, Mann-Whitney U test) decrease in TBF followed by a significant ($p < 0.01$) increase 12 h after LH. Each point represents the mean of 6-8 rats and the SEM is indicated.

rats clearly suggest that testicular endocrine function cannot be evaluated properly without considering blood flow and permeability. Our results, together with those of Setchell and coworkers, also suggest that tubules are involved in the control of testicular blood flow, but the mechanisms are still unknown.

PHYSIOLOGICAL ROLE OF hCG/LH-INDUCED CHANGES IN TESTICULAR MICROCIRCULATION - OBSERVATIONS AND SPECULATIONS

In the previous experiments, we used hCG doses (50-200 IU) giving a maximal testosterone response and such doses were also used by Setchell and Sharpe (Sharpe, 1984). It can however be argued that the testis is never maximally stimulated during physiological conditions. We therefore tested lower doses. However, in our hands, lower doses did not change IF volume or blood flow pattern, and did not result in leukocyte accumulation and migration (i.e. 12.5 IU giving approximately half the maximal testosterone response was without effect, Fig. 3, Widmark et al., 1986a; Bergh et al., 1986b). Inflammation-like changes in testicular microcirculation may occur only during supraphysiological stimulation. If so, this is a serious warning to all investigators using hCG doses of this magnitude (doses of this magnitude have generally been used). On the other hand, microcirculatory changes induced by lower doses may have escaped our attention since; IF volume is a crude way to estimate changes in filtration and permeability, we do not know whether the laser doppler is able to monitor more discrete changes in blood flow pattern, and a small increase in the number of open endothelial cell gaps, although difficult to detect, may nevertheless be of physiological importance. Recent studies suggest that the physiological control of macromolecular permeability in general takes place at interendothelial cell junctions in small postcapillary venules. By controlling the size of these "variable large pores", permeability is regulated (Grega et al., 1986; Crone, 1986). What we observe in the testis after hCG may be an indication that mcaromolecular permeability, also in the testis, is controlled by variable pores between endothelial cells in postcapillary venules. Testicular blood vessels are freely permeable to albumin under basal conditions (for reference see Setchell, 1978). This observation is difficult to understand since morphologically testicular capillaries appear impermeable to macromolecules (Setchell, 1978) but could perhaps be explained by semi-open venular gaps also under basal conditions. Observations in the ovary also suggest a physiological role of inflammation-like microvascular changes in the gonads. Mating results in an LH peak that induces ovulation; this peak is followed by intravascular and perifollicular leukocyte accumulation and edema, and antiinflammatory drugs prevent ovulation (see Espey, 1980, for review).

The effect of LH on testicular blood and vascular permeability has not been studied systematically. We therefore treated adult rats with 25 μg ovine LH s.c. (a dose giving a maximal testosterone response) and studied blood flow and microcirculation at different times after treatment (Damber et al., 1986b). LH caused a decrease in blood flow at 6 h and an increase 12 h after treatment (Fig. 11). IF volume was increased 6 h after treatment but was unchanged at all other times studied. LH treatment induced a change in blood flow pattern (similar to that after hCG) observed 4 h and 6 h after treatment. Leukocyte concentration was increased in testicular blood vessels between 3 and 6 h, and at 4 h and 6 h leukocytes were also observed in the interstitial space. The overall effect of LH is thus similar to that of hCG, but some important differences are noted. The blood flow stimulating effect of LH is more rapid. LH induces an early decrease in flow; whether this occurs also after hCG stimulation is not fully established but results from Wang et al. (1984) suggest that hCG may have a similar early effect. The effect of submaximal doses of LH on testicular blood flow and microcirculation have not been studied yet and before that, it is not possible to know whether the observed gonadotropin-induced changes may be of physiological significance or not. It should be noted that there are up to 6 spontaneous LH pulses every 8 h in normal rats, but 8 h periods with no LH pulse can also be observed (Ellis and Desjardins, 1982). It may thus be argued that rapidly occurring changes in flow and permeability are probably of larger physiological significance than those occurring later. Before the physiological role of LH/hCG-induced vascular changes can be evaluated, it is also of importance to study the effects of multiple injections. To what extent is the secretion of factors regulating blood vessels influenced by desensitization? It thus remains to be shown whether endogenous LH pulses influence testicular microcirculation. In any case, depletion of gonadotropins by hypophysectomy results in a decrease in testicular blood flow that can be normalized after hCG treatment, suggesting that one role of LH is at least the maintenance of blood flow (Daehlin et al., 1985).

SUMMARY

LH, hCG and LHRH treatment, probably via effects on Leydig cells, influence testicular blood flow and vascular permeability. Both inhibitory and stimulatory effects can be observed depending on dose and time after treatment. The physiological role and significance of these changes are, however, still unknown.

In testes with damaged tubules (cryptorchidism), hCG is almost unable to increase blood flow but not permeability, suggesting that flow and permeability are differently controlled and that the seminiferous tubules may be involved in the control of testicular blood flow.

The increase in testicular IF volume after hCG/LH treatment is mainly caused by an inflammation-like, leukocyte-mediated increase in venular permeability. This observation may indicate a physiological role for leukocytes in regulating macromolecular permeability in the testis. Alternatively, it may be a warning not to use supraphysiological doses of LH/hCG when studying testicular physiology.

ACKNOWLEDGEMENTS

This work was supported by grants from the Swedish Medical Research Council (proj 5935 and 5653) and the Maud and Birger Gustavsson Foundation.

REFERENCES

Bergh, A., 1983a, Paracrine regulation of Leydig cells by the seminiferous tubules, Int. J. Androl., 6:57.

Bergh, A., 1983b, Early morphological changes in the abdominal testes in immature unilaterally cryptorchid rats, Int. J. Androl., 6:73.

Bergh, A., 1985, Effect of cryptorchidism on the morphology of testicular macrophages - evidence for a Leydig cell-macrophage interaction in the rat testis, Int. J. Androl., 8:86.

Bergh, A., Rooth, P., Widmark, A., and Damber, J-E., 1986a, Human chorionic gonadotropin treatment induces inflammation-like changes in testicular microcirculation, J. Reprod. Fertil., accepted.

Bergh, A., Widmark, A., Damber, J-E., and Cajander, S., 1986b, Are leukocytes involved in the human chorionic gonadotropin-induced increase in testicular vascular permeability?, Endocrinology, in press (Aug. issue).

Bergh, A., Widmark, A., Rooth, P., Damber, J-E., and Smedjegard, G., 1986c, Leukocytes mediate the hCG-induced increase in venular permeability in the testis, Proc. 4th European Workshop on the Molecular and Cellular Endocrinology of the Testis, Elsevier, Amsterdam, in press.

Björk, J., Hedqvist, P., and Arfors, K. E., 1982, Increase in vascular permeability induced by leukotriene B4 and the role of polymorphonuclear leukocytes, Inflammation, 6:189.

Clark, R. V., 1976, Three-dimensional organization of the testicular interstitial tissue and lymphatic space in the rat, Anat. Rec., 184:203.

Crone, C., 1986, Modulation of solute permeability in microvascular endothelium, Fed. Proc., 45:77.

Cooke, B. A., Dix, C. J., Habberfield, A. D., and Sullivan, M. H. F., 1984, Control of steroidogenesis in Leydig cells: role of Ca and lipoxygenase products in LH and LHRH agonist action, Ann. NY Acad. Sci., 438:269.

Cotran, R. S., and Majno, G., 1964, The delayed and prolonged vascular leakage in inflammation. I. Topography of the leaking vessels after thermal injury, Am. J. Path., 45:261.

Daehlin, L., Damber, J-E., Selstam, G., and Bergman, B., 1985, Effects of human chorionic gonadotropin, estradiol and estromustine on testicular blood flow in hypophysectomized rats, Int. J. Androl., 8:58.

Damber, J-E., 1978, Testicular blood flow. Methodological and functional studies, Umea University Medical Dissertation, New Series No. 39.

Damber, J-E., Selstam, G., and Wang, J. M., 1981, Inhibitory effect of estradiol-17β on human chorionic gonadotropin-induced increment of testicular blood flow and plasma testosterone concentration in rats, Biol. Reprod., 25:555.

Damber, J-E., Lindahl, O., Selstam, G., and Tenland, T., 1982, Testicular blood flow measured with a laser doppler flowmeter: acute effects of catecholamines, Acta Physiol. Scand., 115:209.

Damber, J-E., Lindahl, O., Selstam, G., and Tenland, T., 1983, Rhythmical oscillations in rat testicular microcirculation as recorded by laser doppler flowmetry, Acta Physiol. Scand., 118:117.

Damber, J-E., Bergh, A., and Daehlin, L., 1984, Stimulatory effects of an LHRH agonist on testicular blood flow in hypophysectomized rats, Int. J. Androl., 7:236.

Damber, J-E., Bergh, A., and Daehlin, L., 1985, Testicular blood flow, vascular permeability, and testosterone production after stimulation of unilaterally cryptorchid adult rats with human chorionic gonadotropin, Endocrinology, 117:1906.

Damber, J-E., Bergh, A., Fagrell, B., Lindahl, O., and Rooth, P., 1986a, Testicular capillary circulation in the rat studied by videophotometric capillaroscopy, fluorescence microscopy and laser doppler flowmetry, Acta Physiol. Scand., 126:371.

Damber, J-E., Widmark, A., and Bergh, A., 1986b, The effect of LH on testicular microcirculation and vascular permeability, 4th European Workshop on Molecular and Cellular Endocrinology of the Testis, Capri, Italy, Abstract No. F2.

Eggart, P., and Weiss, C., 1980, Periodic microflow pattern measured with a new microflow probe within the rat kidney cortex, Pflugers Archiv., 383:223.

Ellis, G. B., and Desjardins, C., 1982, Male rats secrete luteinizing hormone and testosterone episodically, Endocrinology, 110:1618.

Espey, L. L., 1980, Ovulation as an inflammatory process - an hypothesis, Biol. Reprod., 22:73.

Fagrell, B., Intaglietta, M., and Ostergren, J., 1980, Relative hematocrit in human skin capillaries and its relation to capillary blood flow velocity, Microvasc. Res., 20:327.

Fawcett, D. W., Neaves, W. B., and Flores, M. N., 1973, Comparative observations on the intertubular lymphatics and the organization of the interstitial tissue of the mammalian testis, Biol. Reprod., 9:500.

Gabbiani, G., Badonnel, M. C., and Majno, G., 1970, Intra-arterial injections of histamine, serotonin, or bradykinin: a topographic study of vascular leakage, Proc. Soc. Exp. Biol. Med., 135:447.

Grega, G. J., Adamski, S. W., and Dobbins, D. E., 1986, Physiological and pharmacological evidence for the regulation of permeability, Fed. Proc., 45:96.

Haour, F., Kouznetzova, B., Dray, F., and Saez, J. M., 1979, hCG-induced prostaglandin E_2 and F_2 release in adult rat testis: role in Leydig cell desensitization to hCG, Life Sci., 24:2151.

Intaglietta, M., 1981, Vasomotion activity, time-dependent fluid exchange and tissue pressure, Microvasc. Res., 21:153.

Intaglietta, M., and Gross, J. F., 1982, Vasomotion, tissue fluid flow and the formation of lymph, Int. J. Microcirc: Clin. and Exp., 1:55.

Persson, C. G. A., and Svensjö, E., 1985, Vascular responses and their suppression: drugs interfering with venular permeability, in: "The Pharmacology of Inflammation," I. L. Bonta, M. A. Bray, M. J. Parnham, eds., pp. 61-82, Elsevier, Amsterdam.

Saez, J. M., Tabone, E., Perrard-Sapuri, M. H., and Rivarola, M. A., 1985, Paracrine role of Sertoli Cells, Medical Biol., 64:225.

Setchell, B. P., 1978, The Mammalian Testis, Paul Elek, London.

Setchell, B. P., and Sharpe, R. M., 1981, The effect of human chorionic gonadotropin on capillary permeability, extracellular fluid volume and flow of lymph and blood in the testes of rats, J. Endocrinol., 91:245.

Setchell, B. P., and Galil, K. A. A., 1983, Limitations imposed by testicular blood flow on the function of Leydig cells in rats in vivo, Aust. J. Biol. Sci., 36:285.
Setchell, B. P., and Rommerts, F. F. G., 1985, The importance of Leydig cells in the vascular response to hCG in rats, Int. J. Androl., 8:436.
Sharpe, R. M., Doogan, D. G., and Cooper, I., 1983, Direct effects of a luteinizing hormone-releasing hormone agonist on intratesticular levels of testosterone and interstitial fluid formation in intact male rats, Endocrinology, 113:1306.
Sharpe, R. M., 1984, Intratesticular factors controlling testicular function, Biol. Reprod., 30:29.
Takemura, R., and Werb, A., 1984, Secretory products of macrophages and their physiological function, Am. J. Physiol., 246:C1.
Veiola, M., and Rajaniemi, H., 1985, The hCG-induced increase in hormone uptake and interstitial fluid volume in the rat testis is not mediated by steroids, prostaglandins or protein synthesis, Int. J. Androl., 8:69.
Wang, J. M., Galil, K. A. A., and Setchell, B. P., 1983, Changes in testicular blood flow and testosterone production during aspermatogenesis after irradiation, J. Endocrinol., 98:35.
Wang, J. M., Gu, C. H., Oian, Z. M., and Jing, G. W., 1984, Effect of gossypol on testicular blood flow and testosterone production in rats, J. Reprod. Fertil., 71:127.
Wedmore, C. V., and Williams, T. J., 1981, Control of vascular permeability by polymorphonuclear leukocytes in inflammation, Nature (Lond.), 289:646.
Widmark, A., Damber, J-E., and Bergh, A., 1986a, The relationship between human chorionic gonadotropin-induced changes in testicular microcirculation and the formation of testicular interstitial fluid, J. Endocrinol., 109:, in press.
Widmark, A., Damber, J-E., and Bergh, A., 1986b, Testicular vascular resistance in the rat after treatment with an LHRH-agonist, Int. J. Androl., in press.
Widmark, A., Bergh, A., and Damber, J-E., 1986c, Treatment of rats with antineutrophil serum prevents the hCG-induced increment in testicular interstitial fluid volume, 4th European Workshop on the Molecular Endocrinology of the Testis, Capri, Italy, Abstract No. F3.
Williams, T. J., 1985, Vascular responses and their suppression: vasodilatation and edema, in: "The Pharmacology of Inflammation," I. L. Bonta, M. A. Bray, M. J. Parnham, eds., pp. 49-59, Elsevier, Amsterdam.

MORPHOLOGY OF NORMAL AND ABNORMAL TESTICULAR DESCENT AND THE REGULATION OF THIS PROCESS

C. J. G. Wensing

Department of Anatomy
School of Veterinary Medicine
State University Utrecht
The Netherlands

INTRODUCTION

The mechanisms involved in testicular descent are not completely understood despite all the attention the subject has had from many investigators of various disciplines, since the classical description of the morphology of the process by John Hunter, in 1972. Several theories have been put forward to explain the changes in morphology involved in the movement of the testis from its site of origin in the genital ridge to its final resting place in the scrotum. Especially opinions on the role of the gubernaculum testis in the process differ (Backhouse, 1964, Wensing, 1968; Wensing et al., 1980; Gier and Marion, 1969; Hadziselimovic, 1983).

In this account, attention will first be focussed on the morphology of the process followed by a description of some aberrations of gubernacular development resulting in abnormal testicular descent. In the second part of this contribution, attention will be focussed on aspects of the hormonal regulation of the process. Only species with a strip-like cremaster muscle (for example, ungulates, carnivores and man) will be dealt with.

MORPHOLOGY OF TESTICULAR DESCENT

The gubernaculum is a mesenchymal structure which, before the migration of the testis actually starts, runs mainly within the abdomen in a peritoneal fold that extends from the testis across the mesonephros to the inguinal area; there is a small extra-abdominal extension (Fig. 1). This mesenchymal structure ends in a knob-like expansion between the differentiating internal and external oblique abdominal muscles. The peritoneum invades the extra-abdominal part of the gubernaculum in a more or less semicircular fashion at the level of the future inguinal ring. This evagination, which can be considered as the beginning of the vaginal process, divides the gubernaculum into three parts (Fig. 2):

a) proper gubernaculum: the intra-abdominal part and the extra-abdominal part suspended by the visceral peritoneal layer of the vaginal process;
b) vaginal part of the gubernaculum: the extra-abdominal part surrounding the parietal layer of the vaginal process externally;

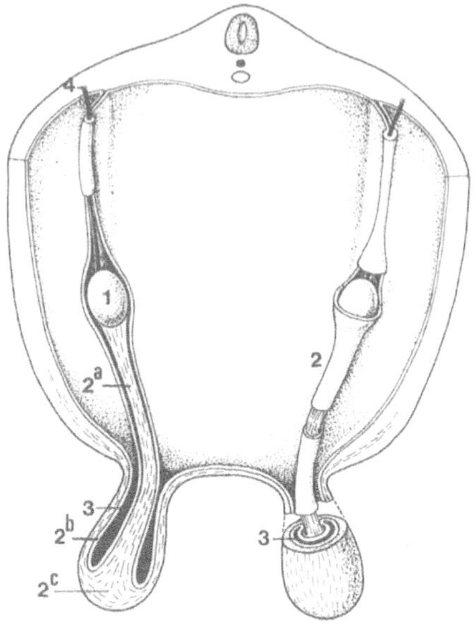

Fig. 1. Schematic drawing of testis and gubernaculum. On the left hand side the gubernaculum is sectioned longitudinally: 1. testis; 2. gubernaculum, 2a. gubernaculum proper, 2b. infravaginal part of gubernaculum, 2c. vaginal part of gubernaculum; 3. cavity of the vaginal process; 4. testicular artery.

c) infravaginal part of the gubernaculum: the caudal end of the gubernaculum which has not been invaded by the vaginal process.

In the pig, there is no organized connection between the caudal tip of the gubernaculum and the area of the future scrotum during the early stages of the testicular descent (Wensig, 1968; Heyns and De Klerk, 1985). The ratio of the intra-abdominal and extra-abdominal parts of the gubernaculum alters substantially during testicular migration, for while the total length of the gubernaculum (relative to the crown-rump length) hardly increases, the extra-abdominal part forms a progressively greater proportion of the whole as the testis approaches the internal ring. At the same time mainly the extra-abdominal part increases in bulk (Fig. 3) and its caudal tip extends into the region of the scrotum. In the 15-25 cm range pig fetus, the gubernaculum presents itself as a globular-ended mass of jelly lying free in the inguinal area or scrotum. In the vaginal part of the gubernaculum, myoblasts develop that in the inguinal area are closely related to the myoblasts that give origin to the internal abdominal oblique muscle. At a later stage, it can be said that the muscle fibers of the cremaster mainly arise from the caudal border of the internal abdominal oblique.

After passage of the testis through the inguinal canal, the gubernaculum extends as far caudally as the scrotal floor and apart from the skin it is surrounded only by the external spermatic fascia that extends from the external inguinal ring to the scrotal wall.

The outgrowth of the gubernaculum is mainly confined to that part covered by the visceral layer of the processus vaginalis, i.e. the extra-abdominal part of the gubernaculum proprium (Fig. 2).

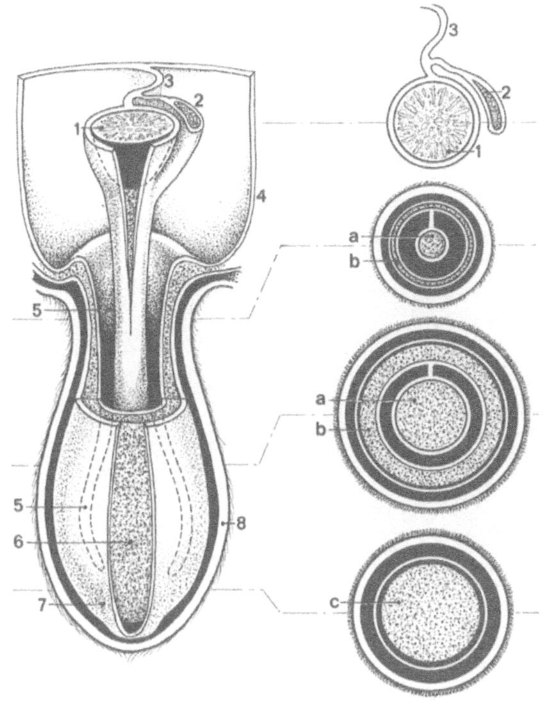

Fig. 2. Schematic drawing illustrating the three parts of the gubernaculum: a) gubernaculum proper; b) vaginal part; c) infravaginal part; 1. testis; 2. epididymis; 3. mesorchium; 4. parietal peritoneum; 5. vaginal process; 6. gubernaculum; 7. external spermatic fascia; 8. scrotum (Baumans et al., 1981, by permission).

In the dog, the intra-abdominal testicular migration during the outgrowth of the gubernaculum is far less pronounced (Baumans et al., 1981). In this species the outgrowth of the extra-abdominal part is clear but not as spectacular as in ungulates. In the male human fetus, the changes in the dimensions of the gubernaculum and the transabdominal migration of the testis (Backhouse, 1964; Heyns and De Klerk, 1985) more or less resemble the situation in ungulates and carnivores. The size of the gubernaculum in the human fetus relative to the testis is smaller than in the pig fetus.

Gubernacular outgrowth is caused partly by active cell division but also to a substantial degree by the increase of extracellular substance caused mainly by an increase in the total amount of glycosaminoglycans. The increase in weight and volume continues for some time after the passage of the testis through the inguinal canal. The last part of the enlargement is due to widening of the structure since its length actually decreases. The structure stays at its maximum weight for some time before reduction commences.

The vaginal process closely follows the outgrowth of the gubernaculum, and the fundus of the cavity is never more than 1-2 mm distant from the gubernacular tip. In consequence, the vaginal process is an elongated cleft whose visceral and parietal walls are closely opposed at the time of passage

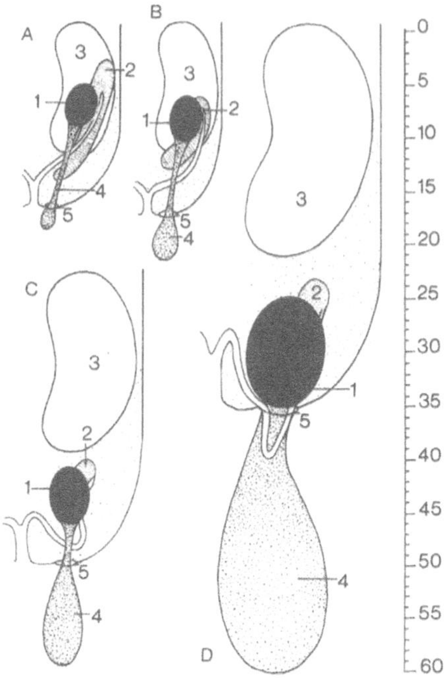

Fig. 3. Schematic drawings of the changes in gubernacular morphology during the outgrowth reaction in pig fetuses of A. 45 days post coitum; B. 52 days p.c.; C. 63 days p.c.; and D. 80 days p.c. (scale in centimeters). 1. testis; 2. mesonephros; 3. metanephros; 4. gubernaculum testis; 5. internal inguinal ring. The intra-abdominal migration of the testis towards the inguinal ring is evident.

of the testis through the inguinal canal. Shortly before this event takes place, the gubernaculum adjacent to testis-epididymis reaches a diameter which is as wide as or wider than the testis itself and in this way dilates the inguinal canal (Fig. 3). The elongation and outgrowth distal to the deep inguinal ring apparently exerts traction upon the abdominal part of the gubernaculum proper, shortening it and thus gradually pulling the testis towards the deep inguinal ring (Fig. 4).

With the subsequent regression of the gubernaculum proper and shift in the relative positions of the superficial and deep inguinal rings, the testis migrates within the peritoneal fold that carries the gubernaculum proper (Figs. 1, 2).

During regression, a large part of the extracellular matrix gradually disappears leading to a condensation of fibrous material and a sharp increase in cell density (Wensing, 1973a; Heyns and De Klerk, 1985). In the pig, the

Fig. 4. In these drawings, the extra-abdominal part of the gubernaculum is compared with a balloon. Inflation of the balloon with its neck enclosed in a restricted passage results in traction being exerted on the neck of the balloon. A mechanism like this might be responsible for the intra-abdominal migration of the testis.

transformation of the mucoid structure into a small fibrous structure takes place between 100-120 days p.c. (birth is at 114 p.c.). In the calf, this takes place in the 18-20th fetal week, in the human fetus shortly before birth, and in the dog in the neonatal period (Wensing, 1968; Baumans et al., 1981; Hadziselimovic, 1983).

The testis passes into the area formerly occupied by the gubernaculum proper and finally resides in the scrotal compartment (Fig. 5), still unattached to the external spermatic fascia.

The invasion of the vaginal process into the gubernaculum establishes an extension of the peritoneal cavity, keeping the testis and epididymis visceral organs after they have occupied the position formerly occupied by the mesenchyme of the gubernaculum proper. In man, the neck of the cavity of the vaginal process is normally closed after descent. In the domestic mammals this does not happen.

According to Backhouse and Butler (1960) the initial movement of the testis from a position lateroventral to the metanephros to the deep inguinal ring can be explained by the degeneration of the mesonephros and the growth of the tesits which expands to fill the space vacated by the former organ. This supposition is not very probable since the degeneration of the mesonephros shows pronounced interspecific differences in its timing and especially in its relation to testicular migration (Wensing, 1968). It seems more likely that this intra-abdominal migration is brought about by the outgrowth of the extra-abdominal part of the gubernaculum (Wensing, 1968).

The passage through the inguinal canal is believed to be produced by an increase in intra-abdominal pressure, facilitated by the generous dimensions of the inguinal canal which is dilated by the gubernaculum. In our opinion, this assumption, first proposed by Backhouse and Butler (1960), is correct.

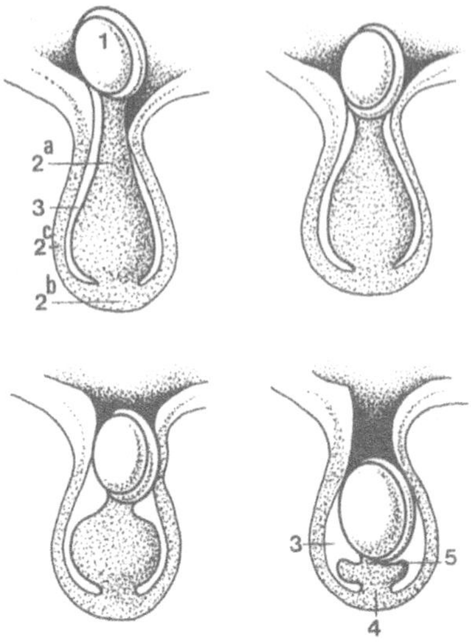

Fig. 5. In these schematic drawings the changes in gubernacular morphology during the regression reaction are depicted. Due to the regression of the gubernaculum proper, the retroperitoneal migration of the testis can take place. 1. testis and epididymis; 2. gubernaculum testis, 2a. gubernaculum proper, 2b. infravaginal part, 2c. vaginal part; 3. vaginal cavity; 4. caudal epididymal ligament; 5. proper testicular ligament.

The role of the gubernaculum in this phase is passive, ensuring that the canal is sufficiently dilated to offer no obstruction to the progress of the testis.

The most important factor in the last phase of testicular descent is, in our opinion, the regression and the transformation of the gubernaculum proprium and the infravaginal part. This regression process allows the further migration of the testis within the peritoneal fold that formerly carried the gubernaculum proper.

The hypothesis put forward by Bedford (1978) and Hadziselimovic (1983) claiming a role for the epididymis in normal descent has not been substantiated.

MORPHOLOGY OF ABNORMAL TESTICULAR DESCENT

According to the theory in which changes in gubernacular morphology are of great importance for a normal testicular descent, it can be postulated that an abnormality of the gubernaculum can affect testicular descent, namely through:

1. absolute or relative failure of outgrowth,
2. aberrant growth causing the gubernaculum to extend into an unusual position, and
3. excessive growth and absent or delayed regression.

Absolute or Relative Failure of Outgrowth

Complete absence of the outgrowth of the gubernaculum has never been observed, but substantial underdevelopment, usually unilateral, has been occasionally noticed. In these animals, insufficient outgrowth of the gubernaculum either on one or on both sides results in insufficient migration of the testis when compared to fetuses of the same age (bilateral insufficiency) or to the contralateral, normal side (unilateral insufficiency) (Wensing, 1973b).

Aberrant Growth

In a survey of several thousand pig fetuses close to birth, 42 fetuses with one or both gubernacula in abnormal locations were found (Wensing, 1973b). By selective breeding using a boar with a unilateral maldescended testis that was based on an abnormally-located gubernaculum, it appeared to be easy to produce newborn piglets with such abnormalities. Abnormal location of the gubernaculum takes three forms (Fig. 6) which may be considered seriatim:

a) the extra-abdominal part of the gubernaculum does not expand beyond the inguinal canal. After outgrowth within the confined space, it later thrusts back into the abdominal cavity. As a consequence, the site of invagination of the vaginal process is lifted away from the deep inguinal ring and carried cranially (Fig. 6A). The abdominal migration normally brought about by the outgrowth of the extra-abdominal part is now absent, thus the testis fails to leave its original position caudal to the kidney. In these animals, the sac formed by the external spermatic fascia is empty and remains narrow. This type of abnormality has also been observed in dogs (Wensing et al., 1980).

b) The outgrowth of the extra-abdominal part of the gubernaculum takes place mainly within the inguinal canal; the structure fails to pass through the external inguinal ring (Fig. 6B). The result is a dilation of the inguinal canal and sometimes an extension of the gubernaculum between the aponeurosis of the external and the fleshy part of the internal oblique abdominal muscle. The abdominal migration of the testis during the outgrowth phase is usually only slightly disturbed. The migration of the testis in conjunction with gubernacular regression is variable. Low abdominal or inguinal cryptorchidism can be the result. In a few cases complete, but delayed, descent is possible. During this regression phase of the gubernaculum the vaginal process in the inguinal canal becomes unusually wide predisposing to inguinal hernia.

c) Outgrowth of the gubernaculum takes place mainly, but not entirely, in the inguinal canal (Fig. 6C). In these cases, dilation of the inguinal canal is evident. With the regression of the gubernaculum, further descent of the testis and concurrently the development of an inguinal hernia can take place. Incomplete descent resulting in low abdominal or inguinal cryptorchidism is also possible (Wensing and Colenbrander, 1973).

Excessive Growth and Absent or Delayed Regression

Abnormalities caused by excessive outgrowth of the extra-abdominal part of the gubernaculum were regularly found, mostly unilaterally. By excessive outgrowth, the gubernaculum exceeds the normal size for that age by a substantial margin. The gubernaculum develops in the normal direction within the inguinal canal and external spermatic fascia. Since the outgrowth

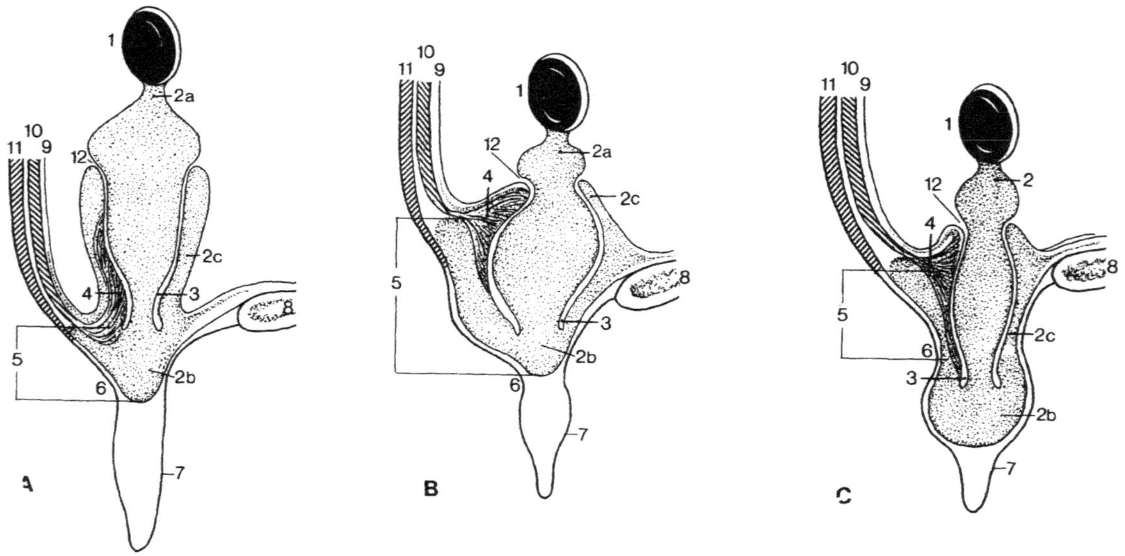

Fig. 6. Schematic drawings of gubernacular outgrowth at abnormal locations: A. outgrowth mainly within the abdominal cavity (reversed reaction); B. outgrowth in the inguinal region; C. outgrowth partly within the inguinal canal and partly beyond the superficial inguinal ring. 1. testis; 2. gubernaculum; 3. cavity of the vaginal process; 4. parietal peritoneum; 5. internal oblique abd. m.; 6. external oblique abd. m.; 7. cremaster muscle; 8. space confined by the external spermatic fascia (Wensing, 1973b, by permission).

continues over a longer period and the regression does not appear to commence until several days after birth, passage of the testis through the inguinal canal is delayed if it takes place at all. The initial large size of the gubernaculum results in a vaginal process with an unusually wide neck, which predisposes to inguinal hernia. In a follow-up study, it became evident that in several of these animals an inguinal hernia did in fact develop after or during the descent of the testis through the inguinal canal (Wensing and Colenbrander, 1973).

HORMONAL REGULATION OF TESTICULAR DESCENT

The factors responsible for gubernacular outgrowth and regression are still obscure. Dependence of the process of testicular descent on androgens and/or gonadotropins is suggested (Engle, 1932; Hamilton, 1938; Raynaud, 1958; Backhouse, 1964; Hadziselimovic and Herzog, 1976; Rajfer and Walsh, 1977; Radhakrishnan et al., 1979; Elder et al., 1982). An intact hypothalamo-pituitary-gonadal axis is also claimed to be required for a normal descent (Hadziselimovic, 1983). Hormonal therapy of testicular maldescent either by hCG, LHRH or testosterone is based on these assumptions. Clear scientific proof on these matters is, however, still lacking.

Initial experiments either to initiate gubernacular outgrowth in females by androgen administration or to depress the reaction in males by administration of antiandrogen were unsuccessful (Frostberg et al., 1968; Wensing, 1973c; Elger et al., 1971; Elger et al., 1977; Richter, 1973). These results suggest that testicular descent, at least the outgrowth of the gubernaculum, is not androgen-dependent. A clear indication in that direction can also be derived from data obtained on testicular feminized males (Wensing et al., 1975; Fentener van Flissingen et al., 1984; Hutson, 1986). The more or less normal gubernacular outgrowth that occurs in these mammals, while known androgen-dependent structures do not differentiate, speaks for itself.

Gubernacular regression important in the second phase of testicular descent does not take place or is insufficient in testicular feminized males (Wensing et al., 1975; Hutson, 1986). The regression can also be retarded by orchidectomy which results in a retarded and incomplete descent of the remaining epididymis (Baumans et al., 1982). The retarded regression after orchidectomy can be successfully counteracted by the supplementation of testosterone in such orchidectomized dogs (Baumans et al., 1983); in these animals testosterone was found to induce gubernacular regression.

In the last decade attention has also been focussed on a possible role of the Müllerian Inhibiting Factor (M.I.F.) in the regulation of testicular descent (Wensing, 1973c; Donahoe et al., 1977; Josso and Tran, 1979; Hutson, 1986). Persisting Müllerian ducts in otherwise relatively normal male development in combination with maldescent is a condition described in man (Brook et al., 1973) and in the dog (Brown et al., 1976; Marshall et al., 1982). This combination of abnormalities has been used to support a possible M.I.F. dependence of testicular descent. The failure to block descent by using antibodies against M.I.F. presents an argument against this hypothesis (Picard et al., 1983).

An intact hypothalamo-pituitary-gonadal axis is claimed to be a prerequisite for normal testicular descent. In decapitated pig fetuses in which no LH and FSH could be assayed, normal gubernacular outgrowth took place indicating that gubernacular outgrowth and thus the first phase of testicular descent are independent of a functional pituitary (Colenbrander et al., 1979). It cannot be excluded that in these animals the regression of the gubernaculum is retarded to some extent.

From castration experiments carried out during testicular descent, it has become clear that the presence of testes is required for outgrowth as well as for regression of the gubernaculum (Colenbrander et al., 1982; Baumans et al., 1982). Bilateral orchidectomy before the start of gubernacular outgrowth completely stopped the outgrowth reaction while bilateral orchidectomy just before the onset of gubernacular regression resulted in retarded regression of the gubernacula.

A clear indication that the testes are essential in provoking the gubernacular reaction can be derived from porcine freemartins. In these genotypically females of which the circulatory systems have been interconnected with that of male co-twins during fetal life, gubernacular reaction takes place resulting in a complete male-like vaginal process and a well-developed cremaster muscle despite complete absence of a gonad. In this experiment of nature, gubernacular outgrowth is brought about by (a) male hormone(s) that are either transferred through the vascular connections from the male to the female circulation or produced by the transforming gonads (Colenbrander and Wensing, 1975; Wensing and Colenbrander, 1986).

The conclusion of this survey in our opinion can only be that more detailed studies are needed to elucidate the relationship between fetal hormones and testicular descent.

REFERENCES

Backhouse, K. M., 1964, The gubernaculum testis Hunteri, testicular descent and maldescent, Ann. R. Coll. Surg. Engl., 35:227.

Backhouse, K. M., and Butler, H., 1960, The gubernaculum testis of the pig, J. Anat., 94:107.

Baumans, V., Dijkstra, G., and Wensing, C. J. G., 1981, Testicular descent in the dog, Zentbl. Vet. Med., C10:97.

Baumans, V., Dijkstra, G., and Wensing, C. J. G., 1982, The effect of orchidectomy on gubernacular outgrowth and regression in the dog, Int. J. Androl., 5:387.

Baumans, V., Dijkstra, G., and Wensing, C. J. G., 1983, The role of non-androgenic testicular factor of the process of testicular descent in the dog, Int. J. Androl., 6:541.

Bedford, M., 1978, Anatomical evidence for epididymis as a prime mover in the evolution of the scrotum, Am. J. Ant., 152:483.

Brook, C. G. D., Wagner, H., Zachmann, M., Prader, A., Armendares, S., Frenk, S., Aleman, P., Najjar, S. S., Slim, M. S., Genton, N., and Boric, C., 1973, Familial occurrence of persistent Müllerian structures in otherwise normal males, Br. Med. J., i:771.

Brown, T. T., Burek, J. D., and McEntee, K., 1976, Male pseudohermaphroditism, cryptorchidism and Sertoli cell neoplasia in three Miniature Schnauzers, J. Am. Vet. Med. Ass., 169:821.

Colenbrander, B., Macdonald, A. A., and Elsaesser, F., 1982, The effect of fetal castration on gubernacular development in the pig, Proceedings of the 2nd European Workshop on Molecular and Cellular Endocrinology, pp. E4.

Colenbrander, B., van Rossum-Kok, C. M. J. E., van Straaten, H. W. M., and Wensing, C. J. G., 1979, The effect of fetal decapitation on the testis and other endocrine organs in the pig, Biol. Reprod., 20:198.

Colenbrander, B., and Wensing, C. J. G., 1975, Studies on phenotypically female pigs with hernia inguinalis and ovarian aplasia. I. Morphological aspects, Proc. K. Ned. Akad. Wet., C78:33.

Donahoe, P. K., Ho, Y., Morikawa, Y., and Hendren, H. W., 1977, Müllerian Inhibiting Substance in human testis after birth, J. Pediatr. Surg., 12:322.

Elder, J. S., Issacs, J. T., and Walsh, P. C., 1982, Androgenic sensitivity of gubernaculum testis: evidence for hormonal/mechanical interactions in testicular descent, J. Urol., 127:170.
Elger, W., Neumann, F., and von Berswordt-Wallrabe, R., 1971, The influence of andorgen antagonists and progestagens on the sex differentiation of different mammalian species, in: "Hormones in Development," M. Hamburgh and E. J. W. Barrington, eds., pp. 641, Appleton-Century-Crofts, New York.
Elger, W., Richter, J., and Korte, R., 1977, Failure to detect androgen dependence of the descensus testiculorum in fetal rabbits, mice and monkeys, in: "Maldescensus Testis," J. R. Bierich, K. Rager, and M. B. Ranke, eds., pp. 187, Urban and Schwarzenberg, Baltimore.
Engle, E. T., 1932, Experimentally-induced descent of the testis in the Macadus monkey by hormones from the anterior pituitary and pregnancy urine, Endocrinology, 16:513.
Fentener van Flissingen, J. M., Colenbrander, B., Verbruggen, A. J. E. P., and Wensing, C. J. G., 1984, Testicular feminized males (TFM) in Nyctereutes procyonoides (Raccoon dog), in: "Développements récents de l'endocrinologie du testicule," J. M. Saez, M. G. Forest, A. Dazord, and J. Bertrand, eds., pp. 325, Colloque INSERM 132.
Frosberg, J. G., Jacobsohn, D., and Nordgren, A., 1968, Modifications of reproductive organs in male rats influenced prenatally or pre- and postnatally by an "antiandrogenic" steroid (cyproterone), Anat. Embryol., 126:175.
Gier, H. T., and Marion, G. B., 1969, Development of mammalian testis and genital ducts, Biol. Reprod., 1:1.
Hadziselimovic, F., 1983, Embryology of testicular descent and maldescent, in: "Cryptorchidism, Management and Implications, F. Hadziselimovic, ed., pp. 11, Springer Verlag, New York.
Hadziselimovic, F., and Herzog, B., 1976, The meaning of the Leydig cell in relation to the etiology of cryptorchidism, J. Pediatr. Surg., 11:1.
Hamilton, J. B., 1938, The effect of male hormone upon the descent of the testis, Anat. Rec., 70:533.
Heyns, C. F., and De Klerk, D. P., 1985, The gubernaculum during descent in the pig fetus, J. Urol., 133:694.
Hutson, J. M., 1986, A biphasic model of the hormonal control of testicular descent, Lancet, in press.
Josso, N., and Tran, D., 1979, Biochemical aspects of prenatal testicular development, in: "Cryptorchidism. Diagnosis and Treatment," J. C. Job, ed., pp. 37, Karger, New York.
Marshall, L. S., Oehlert, M. L., Haskins, M. E., Selden, J. R., and Patterson, D. F., 1982, Persistent Müllerian duct syndrome in Miniature Schnauzers, J. Am. Vet. Med. Ass., 181:798.
Picard, J. Y., Tran, D., Vigier, B., and Josso, N., 1983, Maintien des canaux de Müller chez la lapin mâle par immunisation passive contre l'hormone antimüllerienne pendant la vie foetale, C.r. Acad. Sci. Paris, 297:567.
Radhakrishnan, J., Morikawa, Y., Donahoe, P. K., and Hendren, W. H., 1979, Observations on the gubernaculum during descent of the testis, Invest. Urol., 16:365.
Rajfer, J., and Walsh, P. C., 1977, Hormonal regulation of testicular descent: experimental and clinical observations, J. Urol., 118:985.
Raynaud, A., 1958, Inhibition, sous l'effet d'une hormone oestrogene du développement du gubernaculum du foetus mâle de souris, C.r. Acac. Sci. Paris, 246:176.
Richter, J., 1973, Die hormonale Steuerung des Descensus testiculorum, Untersuchungen mit Androgenen, Antiandrogenen und Oestrogenen an männlichen und weiblichen Mäuseföten, Dissertation, Freie Universität, Berlin.
Wensing, C. J. G., 1968, Testicular descent in some domestic mammals, I. Anatomical aspects of testicular descent, Proc. Kon. Ned. Akad. Wetensch. C., 71:423.

Wensing, C. J. G., 1973a, Testicular descent in some domestic mammals, II. The nature of the gubernacular changes during the process of testicular descent in the pig, Proc. Kon. Ned. Akad. Wetensch. C., 76:190.

Wensing, C. J. G., 1973b, Abnormalities of testicular descent, Proc. Kon. Ned. Akad. Wetensch. C., 76:373.

Wensing, C. J. G., 1973c, Testicular descent in some domestic mammals, III. Search for the factors that regulate the gubernacular reaction, Proc. Kon. Ned. Akad. Wetensch. C., 76: 196.

Wensing, C. J. G., and Colenbrander, B., 1973, Cryptorchidism and inguinal hernia, Proc. Kon. Ned. Akad. Wetensch. C., 76:489.

Wensing, C. J. G., and Colenbrander, B., 1986, Normal and abnormal testicular descent, in: "Oxford Reviews of Reproductive Biology," Vol. 8, pp. 130, Oxford.

Wensing, C. J. G., Colenbrander, B., and Bosma, A. A., 1975, Testicular feminisation syndrome and gubernacular development in the pig, Proc. Kon. Ned. Akad. Wetensch. C., 78:402.

Wensing, C. J. G., Colenbrander, B., and van Straaten, H. W. M., 1980, Normal and abnormal testicular descent in some mammals, in: "Descended and Cryptorchid Testis," E. S. E. Hafez, ed., pp. 125, Nijhoff, Boston.

SEX DIFFERENTIATION

Nathalie Josso

Unité de Recherches sur l'Endocrinologie
du Développment, INSERM
Hôpital des Enfants-Maldes
75743, Paris cedex 15, France

Introduction

Sex differentiation can be defined as the process leading to the development of phenotypically dimorphic individuals within a species. This allows sexual reproduction, with its attached evolutionary advantages of meiosis and syngamy which increase individual variability, upon which natural selection acts. Although to achieve this purpose, sex differentiation could be limited to the gonads, in higher organisms it is a far more extensive process, affecting the morphology and function of the internal genital tract, bodily appearance, neuroendocrine relationships and social behavior. However, gonadal differentiation is the most important step, being the only one subject to direct genetic regulation, all subsequent steps are induced by hormones produced by the gonads themselves. This statement must be qualified, however: ovarian hormones do play an important role in postnatal sex differentiation; however, prior to birth, sex orientation is determined by the presence or absence of male-determining genes or hormones, which act to impose masculinity upon an organism which otherwise would become female (Jost, 1972). Thus is verified the theory put forward by Aristotle (cited by Chan and O, 1981): "A male is a male in virtue of a particular ability, and a female is a female in virtue of a particular inability."

This chapter will deal exclusively with the events of sex differentiation affecting the reproductive organs of mammals, with emphasis on the human species, during prenatal life: sex differentiation of the brain, psychosocial sex differentiation, and pubertal hormonal sex differentiation will not be discussed. In order not to inflate the reference list, recent reviews have been favored over original publications. A more detailed bibliography can be found in Josso (1981) and Josso and Picard (1986).

GONADAL SEX DIFFERENTIATION

Morphological and Functional Aspects

Initially, the first recognizable gonadal primordium appears, in human embryos 4-5 mm, as a thickening of the coelomic epithelium on the medial aspect of the mesonephros, immediately after the coelomic lining in this region has transformed itself into an epithelium. Opinions differ as to the origin of gonadal somatic cells: Zamboni and his associates (Zamboni and

Upadhyay, 1982) believe they are donated by the mesonephros, while Merchant-Larios (1984) emphasizes the contribution of the coelomic epithelium. The gonadal primordium is invaded by primitive germ cells, originating near the allantois stalk, and travelling by amoeboid movements along the gut wall and mesentery, to reach the gonadal ridge. The presence of germ cells is not required for initial gonadal development: few germ cells reach the gonad in mice with Steel mutation (McCoshen, 1982) or in fetuses from mothers treated with the alkylating drug Busulfan (Merchant, 1975), but lack of germ cells does not impair initial gonadal organogenesis in these animals.

Testicular differentiation is detectable in the human fetus by the 15 mm stage (approximately 5-6 weeks) (Wartenberg, 1978). The somatic elements of the gonad organize into seminiferous cords, which appear first at the anterior pole of the gonad, close to the mesonephros. Prior to that event, somatic cells acquire a large, clear cytoplasm, interdigitations (Magre and Jost, 1980), and the capacity to produce anti-Müllerian hormone (AMH) (Tran et al., 1977), characteristic of fetal Sertoli cells. These orient themselves along a basal membrane containing laminin and fibronectin (Pelliniemi et al., 1984) and surround germ cells, quenching their spontaneous tendency towards meiotic maturation (Luciani et al., 1977). According to Wartenberg (1978), the human fetal testis contains two types of Sertoli cells, those with a clear cytoplasm, which originate from the coelomic epithelium repress meiosis, while a minority of cells with a dark cytoplasm, which are of mesonephric origin have the opposite effect. Leydig cells develop from mesenchymal cells in the interstitium at the 30 mm stage (8 weeks). Their differentiation is marked by the acquisition of abundant smooth endoplasmic reticulum, a characteristic of steroid-producing cells. Testosterone production by Leydig cells does not initially require stimulation by gonadotropins (George et al., 1978a); however, hCG is necessary to ensure continued secretion, and the ontogeny of testosterone production by the human testis closely follows that of hCG (Winter et al., 1977). Fetal Leydig cells have few estrogen receptors (Huhtaniemi et al., 1982), thus they do not undergo desensitization by hCG (Leinonen and Jaffe, 1984) and they are able to produce high amounts of testosterone during the prolonged period necessary to the organogenesis of the male genital tract.

Ovarian differentiation is marked by the proliferation of germ cells at the periphery of the gonad, and subsequently by their organization in ovigerous cords, which differ from seminiferous cords by the undifferentiated state of their somatic elements, and by the absence of a basal membrane. The most striking feature of ovarian differentiation is found in its germ cell component. In contrast to male germ cells, which enter meiosis only at puberty, female germ cells undergo meiotic prophase up to the diplotene stage, resuming meiosis at ovulation. Germ cells in the medulla, close to the mesonephros, are the first to undergo this evolution, which spreads eventually towards the cortex. At the seventh month, the human fetal ovary no longer contains oogonia: all cells have either degenerated, or entered the méiotic prophase. Follicular cells condense around oocytes at the diplotene stage, and the first ovarian follicles are found at 11 to 12 weeks. In the absence of germ cells, no follicles form, therefore the ovary of Steel mutants or offspring of Busulfan-treated mothers is progressively transformed into a "streak gonad," incapable of sex steroid production (McCoshen, 1982; Merchant-Larios and Centeno, 1981). Germ cells with only one X apparently fall victims to a mechanism selectively destroying those cells unable to achieve pairing of sex chromosomes (Burgoyne and Baker, 1985); this explains the progressive disappearance of germ cells from the ovaries of 45,X individuals, and the pathogenesis of Turner's syndrome. Although ovarian steroids play no role in sex differentiation of the genital tract, they are produced by the fetal ovary at the same time that testosterone is secreted by the fetal testis (George et al., 1978b).

Genetic Factors of Gonadal Differentiation

Tdy, Bkm and H-Y. In mammals, the presence of a Y chromosome induces testicular differentiation of the gonad, irrespective of the number of associated X chromosomes, indicating that the Y plays an important part in persuading the fetal gonad to develop in a male direction. Jones and Singh (1981) have isolated from a snake, the banded krait, repetitive DNA sequences apparently involved in the differentiation of the gonad of the heterogametic sex. These sequences, named Bkm (banded krait minor) hybridize *in situ* with a portion of the Y chromosome of the mouse, located near the centromere. In man, however, no Bkm sequences are present in the Y chromosome (Jones and Singh, 1981; Kiel-Metzger et al., 1986); the Y chromosome genes responsible for testicular differentiation, designated Tdy, correspond to single-copy DNA sequences normally present in the short arm of the Y, in the vicinity of the X-Y pairing region, and identified in the genome of some, but not all, XX males (Guellaen et al., 1984; Vergnaud et al., 1986; Seboun et al., 1986).

The relationship between Tdy and the Hy gene coding for the minor transplantation antigen H-Y is still a matter for heated controversy. Eichwald and Silmser (1955) have demonstrated the existence of a male-specifc transplantation antigen, responsible for the rejection of male skin grafts by syngeneic females sharing the same histocompatibility determinants. This transplantation antigen, which has been well conserved during evolution, triggers a cytotoxic response in T cells, which can be conveniently studied *in vitro* (Simpson et al., 1983). This transplantation antigen was at first considered identical to the male-specific antigen evoking a serological response in females immunized with syngeneic male cells (SDM). However, recently discrepancies between the expression of the transplantation antigen defined by the cytotoxic T cell response and serologically defined antigen have become apparent (Simpson et al., 1982; Melvold et al., 1977), and it has been suggested that H-Y transplantation antigen and SDM are separate entities (Silvers et al., 1982). On the opposite hand, Ohno (1985) points out that many of the discrepancies could well be due to the contamination of H-Y antiserum by unrelated auto-antibodies, due to the fact that H-Y is a weak antigen.

Ohno (1985) still staunchly defends the "H-Y theory," first put forward by Wachtel et al. (1975), and reviewed in Ohno et al. (1979). According to this theory, gonadal cells are endowed with receptors for H-Y antigen, which lead them to associate and form seminiferous tubules, when these receptors are occupied. If no receptors are present, testicular organogenesis cannot take place, and this could explain why ovaries can form in individuals expressing H-Y. Conversely, receptors are present in ovarian cells. Zenzes et al. (1978a) and Ohno et al. (1979) have claimed that a fetal ovary can be masculinized by H-Y antigen; however, histological proof of this is flimsy and the effect of H-Y antigen on the fetal ovary has not been confirmed by other investigators (Benhaim et al., 1981).

The Sxr Mutation. The Sxr mutation, originally described by Cattanach et al. (1971), produces sex reversal in XX mice and is genetically transmitted in a manner suggesting autosomal dominance. However, recently Singh and Jones (1982) have shown that the Sxr mutation consists in the duplication of the segment of the Y chromosome containing male-determining sequences. This duplication is followed by translocation to a site distal to the pairing region, allowing its transfer to the X chromosome at meiosis (Evans et al. 1982) (Fig. 1). Such genes, termed pseudo-autosomal (Burgoyne, 1982) are probably involved in other instances of genetically transmitted sex-reversal, observed for instance in goats (Hamerton et al., 1969) or in XX human males (de la Chapelle, 1981).

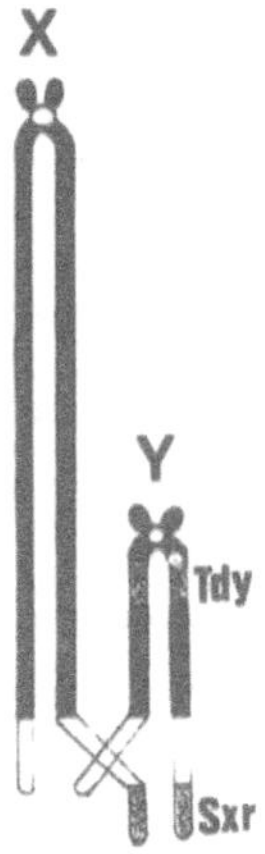

Fig. 1. Meiosis in a X/Y Sxr carrier male: the X is shown at the left, and the Y at the right. Most nonhomologous regions are shown in black, and homologous ones in white. The shaded regions on the Y are Sxr itself, attached distal to the pairing segment and the subcentric region which by duplication furnished Sxr. (From Ohno, 1985, with permission.)

X/X Sxr mice usually develop as males, and express H-Y antigen (Bennett et al., 1977), implying either that both Hy and Tdy have been translocated, or that the two genes are in fact one and the same. In view of the fact that random inactivation of one X chromosome takes place in somatic cells, one could expect male differentiation to occur only in those individuals in whom the active X chromosome in most cells happens to be the one carrying Sxr. However, Sxr is usually protected against inactivation by the barrier of the euchromatic pairing segment of the X chromosome (Burgoyne, 1982), therefore normally all X/X Sxr individuals develop as males. Notable exceptions are mice in whom the partner of the Sxr X is affected by the Searle translocation, T16H (Lyon et al., 1964). In T16H/X Sxr individuals, the X bearing the translocation remains active in all cells, and conversely, the X bearing Sxr is always inactivated. Inactivation may spread to the Sxr gene, causing mice of identical chromosomal constitution to develop either as males, females, or true hermaphrodites (Cattanach et al., 1982). Study of T16H/X Sxr females (McLaren and Monk, 1982) has thrown light upon the relationship between Tdy- and H-Y: although fertile, these females usually express H-Y transplantation antigen (Simpson et al., 1984). This has been interpreted as a proof of dissociation between testicular differentiation and expression of H-Y antigen; however, Ohno (1985) counters with the proposal that the gene for the plasma membrane antigen which normally associates with the putative H-Y receptor on the membrane of gonadal cells resides on chromosome 16 close to the break position.

The Sxr' Mutation. Expression of H-Y by fertile females proves only that H-Y does not prevent ovarian differentiation, but does not indicate whether this antigen is necessary for testicular differentiation. Interesting information in this respect can be gained from the study of an exceptional T16H/X Sxr female not expressing H-Y antigen (McLaren et al., 1984). Because this female gave birth to a son which did not express H-Y either, it was postulated that in this female, the Sxr segment of the Y translocated to the X contained Tdy, but not Hy, and that therefore, these are different entities. The hypothesis put forward by Ohno (1985) to explain this discrepancy, namely that the Hy gene was inactivated, is contradicted by the fact that XO Sxr' males, in whom obviously the single X is not inactivated, are also H-Y negative (Burgoyne et al., 1986). Spermatogenesis does not proceed normally in such males, suggesting that H-Y may play an important role in spermatogenesis, if not in testicular differentiation (Burgoyne et al., 1986).

Other Genes involved in Sex Differentiation. In parallel with the controversy concerning sex-determining genes on the Y chromosome, it has become apparent that the Y chromosome is not the only one involved in testicular differentiation. Fredga et al. (1977) have discovered that the wood-lemming produces two types of females, XX and XY: in the latter, a mutant gene on the X suppresses the male-determining ones on the Y. More recently, Eicher at al. (1982) and Johnson et al. (1982) have described yet another type of fertile XY female expressing H-Y antigen. In this case, the Y gene of Mus poschiavinus has been introduced into the genome of C57BL/6J mice, and suppression of its male determining properties has ensued. Another gene, Tas, located on mouse chromosome 17 and part of the T/t complex, inhibits testicular differentiation when present in XY Tas/+ mice (Washburn and Eicher, 1983). Finally, Chandra (1985a, b) has proposed that noncoding DNA sequences on the X chromosome may play a determining part in gonadal sex differentiation, by regulating the structural Tdy gene. As suggested by Seboun et al. (1986), mutations of non-Y linked genes may be responsible for those cases of human XX males which do not carry Y-specific DNA sequences on their paternally derived X chromosome. This group of patients is characterized by a high incidence of external sexual ambiguity, and familial transmission of the disease.

Tissular Interactions and Gonadal Sex Differentiation

Interaction between Germ Cells and Somatic Cells of the Gonad. As mentioned above, germ cells are not initially necessary to gonadal organogenesis. Neither do they influence the sex orientation of the gonad, which is determined essentially by the genetic endowment of its somatic component (Zenzes, 1981; Ford et al., 1975; Evans et al., 1977).

In contrast, germ cell sex is controlled by the somatic elements of the gonad, or rather, germ cells develop along female lines unless directed otherwise by Sertoli cells. Germ cells developing outside the gonad invariably develop as oocytes and undergo meiotic prophase (Uphadhyay and Zamboni, 1982; McLaren, 1983), as do XY cells developing within the ovary; whereas XX germ cells within a testis usually become spermatogonia (Cattanach et al., 1971; McLaren, 1981). However, the genetic makeup of the germ cells does control their ultimate fate. With only one X chromosome, a germ cell can adopt either a male or a female phenotype, according to its environment. With two X chromosomes, both of which are active in germ cells (Monk and McLaren, 1981; McLaren and Monk 1981), its choice is restricted to female development within an ovary, or death in the testis: XX spermatogonia can initially differentiate in a testicular environment, but rapidly degenerate and cannot support spermatogenesis.

Interaction between the Mesonephros and Fetal Gonads. Byskov and Grinstead (1981) have observed both a morphological and a functional dedifferentiation of the fetal mouse testis cultured in the presence of the mesonephros. Germ cells entered meiosis, thus supporting the earlier claim of Byskov (1974) and of O and Baker (1976) for the secretion of a meiosis-stimulating factor by the mesonephros. According to Byskov (1978), the effect of this factor (MIS) is opposed by a meiosis-preventing factor (MPF) secreted by young Sertoli cells.

Interaction between Testicular and Ovarian Tissue. Several natural or experimental models show that testicular cells are able to induce XX cells to differentiate along male lines. When heterosexual chimeras are constructed, the majority develop as males (McLaren, 1984). As prettily put by McLaren (1985), "sex differentiation of the gonad appears to follow a 'majority vote system' rather than proportional representation ... and the voting system is biased in favor of the heterogametic sex, at least in mammals and birds." Fetal ovaries exposed to fetal testes after sex differentiation become sterile and undergo variable degrees of masculization, depending upon experimental protocol and timing (review in Taketo et al., 1985). The testis can also act upon the fetal ovary at a distance: seminiferous tubules capable of producing anti-Müllerian hormone (Vigier et al., 1984a) develop in the ovary of freemartin bovine fetuses, females united to a male twin by vascular placental anastomoses (Jost et al., 1972). The means by which testicular cells act to virilize the fetal ovary is not known. H-Y antigen, a product of fetal Sertoli cells according to Zenzes et al. (1978b), has been suggested. Presence of H-Y has been detected serologically in the serum of freemartin fetuses (Wachtel et al., 1980).

Effect of Purified Anti-Müllerian Hormone upon Rat Fetal Ovaries. H-Y antigen is not the only protein produced by fetal Sertoli cells; in fact, its very secretion by Sertoli cells has been challenged recently (Gore-Langton et al., 1983). Therefore, it seemed reasonable to investigate the effect of other testicular proteins upon the development of the fetal ovary. Among these, anti-Müllerian hormone appeared to be a likely candidate, since ovarian stunting and oocyte loss in the freemartin gonad are chronologically correlated with Müllerian regression (Jost et al., 1972). Vigier and Josso (1986) cultured fetal rat ovaries, 14 days p.c., 5 to 10 days in the presence of purified anti-Müllerian hormone at a concentration of 0.75 to 3 μg/ml. In all cases, a time- and dose-dependent decrease of germ cell number was observed. Furthermore, somatic cells tended to acquire some morphological characteristics of Sertoli cells, and to arrange themselves in structures resembling seminiferous tubules (Fig. 2).

SEX DIFFERENTIATION OF THE GENITAL TRACT

The Undifferentiated Genital Tract

Shortly after gonadal differentiation, the genital tract of male or female embryos consists of unipotenital Wolffian and Müllerian ducts, and bipotential sinusal and external genital primordia. Wolffian ducts, the primordia for male accessory organs, are originally the excretory canals of the primitive kidney, or mesonephros, and are incorporated into the genital system when renal function is taken over by the definitive kidney. Müllerian ducts originate from a cleft in the coelomic epithelium, which burrows downwards in the mesenchyme, lateral and parallel to the Wolffian ducts, which act as guides (Didier, 1973). When the growing Müllerian duct reaches the pelvis, it crosses the Wolffian duct ventrally, and fuses with its opposite. The elevation of the dorsal wall of the urogenital sinus caused by the protrusion of the caudal tips of the paired Müllerian ducts is called the Müllerian tubercle, although it consists of sinusal tissue. Situated at the limit between the vesico-urethral canal and the urogenital sinus proper, it

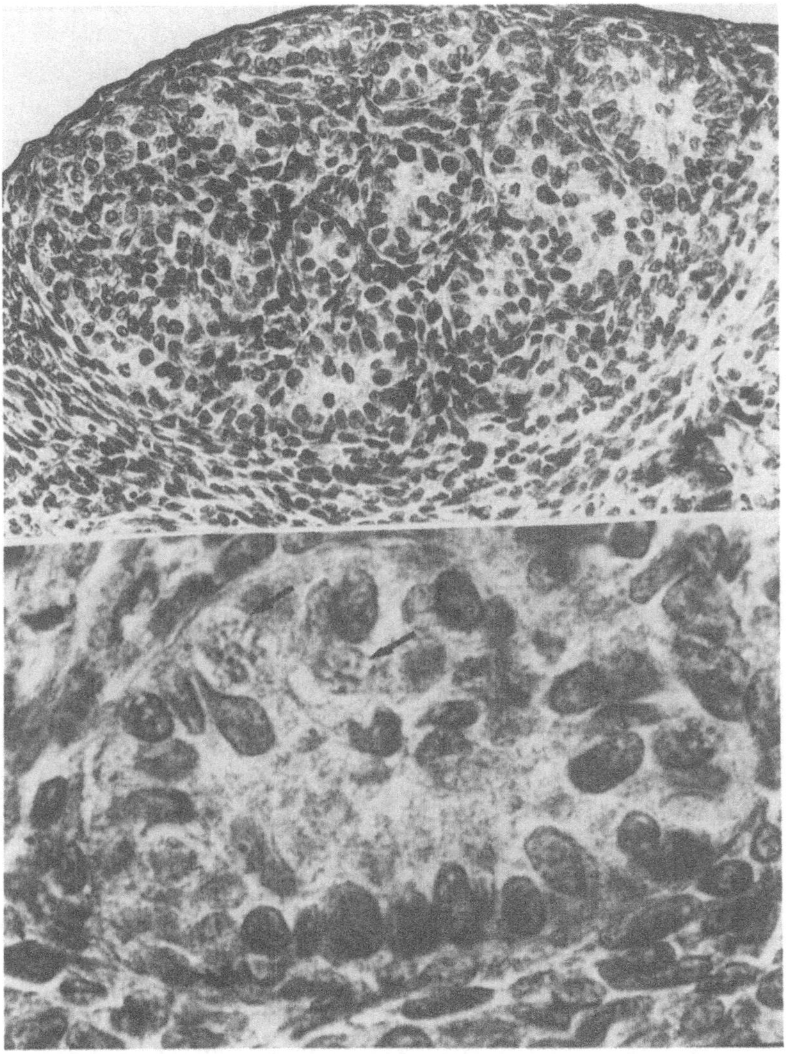

Fig. 2. Histological aspect of fetal rat ovary, cultured from 14 to 24 days post-coitum in the presence of purified anti-Müllerian hormone, 3 µg/ml. Top: Low power view (× 500), showing formation of tubular structures and a connective tissue layer at the surface of the gonad resembling a tunica albuginea. Bottom: High power view (× 1250) of a tubular structure, showing orientation of somatic cells along a basal membran. Two germ cells in meiotic prophase are visible (arrows).

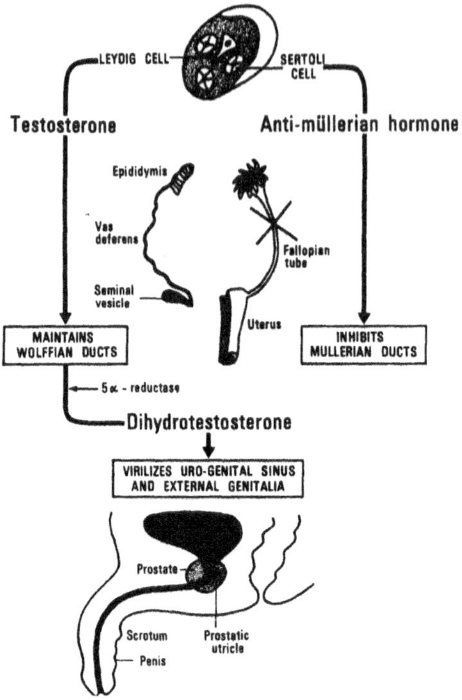

Fig. 3. Male differentiation of the genital tract. (From Josso, 1981, "The Intersex Child", Karger, Basel, with permission.)

is flanked on both sides by the openings of the Wolffian ducts. The undifferentiated external genitalia consist of the genital tubercle, surrounded laterally by the labioscrotal swellings.

Constitutive Events: Female Sex Differentiation

As demonstrated by Jost (1947), virilization of the genital tract is due to the secretion by the fetal testis of two distinct hormones, testosterone and the Müllerian inhibitor (Fig. 3). In the absence of testicular hormones, the genital tract differentiates along female lines. Wolffian ducts degenerate, Müllerian ducts differentiate into tubes, uterus and the upper part of the vagina. The lower part of the vagina is of sinusal origin. The Müllerian tubercle fuses with the sinovaginal bulbs which surround it laterally, to form the vaginal plate. This is followed by extensive caudal growth of this structure, and by its canalization, so that the vagina acquires a separate opening on the perineum. The polygonal epithelium, which lines the vagina initially, is progressively replaced by a stratified squamous epithelium of sinusal origin. This process is inhibited in females prenatally exposed to diethylstilbestrol, which are at increased risk for vaginal adenosis and cancer (Ulfelder and Robboy, 1974). Externally the labioscrotal swellings do not fuse, and give rise to the labia majora, while the genital folds remain separate and form the labia minora. The genital tubercle develops into the clitoris.

Hormonally-mediated Events: Male Sex Differentiation

Testosterone-mediated Events. Testosterone acts upon the Wolffian ducts, which differentiate into epididymis, vas deferens and seminal vesicle, and upon the urogenital sinus and external genitalia. Prostatic rudiments appear at the site of the Müllerian tubercle, first as epithelial buds, which progressively acquire a lumen. The prostatic utricle develops in its center, from the rudiments which in the female participate in vaginal organogenesis. The perineal and penile urethra are formed respectively from the fusion of the labioscrotal folds and of the urethral groove, which develops at the ventral surface of the growing genital tubercle. Penile organogenesis is completed in the human male at 10 weeks, but until the 16th week, its size is similar to that of the clitoris (Feldman and Smith, 1975).

All these events are mediated by testosterone secreted by the fetal Leydig cells. Testosterone acts directly upon Wolffian ducts derivatives, but must undergo reduction into dihydrotestosterone (DHT) in order to virilize the urogenital sinus and external genitalia. This reduction is achieved by a tissular enzyme, 5α-reductase, which is normally present in the urogenital sinus, but not in the Wolffian ducts at the time of sex differentiation. Deficiency or instability of the enzyme results in lack of fetal virilization, and patients with this disorder are often considered females at birth. At puberty, however, they become virilized, in spite of the fact that their enzyme deficiency is unchanged. Hodgins (1982) has suggested that this is because at that time, they are no longer exposed to progesterone, which acts as a competitor for binding to the androgen receptor. 5α-Reductase deficiency was first described in an islet of the Dominican Republic, where the patients commonly change their sex at puberty (Imperato-McGinley et al., 1979). An experimental form of the disorder has recently been produced in rats, using an inhibitor of the enzyme (Imperato-McGinley et al., 1985).

The reason for the increased virilizing efficiency of testosterone over DHT must be sought in the characteristics of the androgen receptor, a protein present in the cell cytoplasm, which translocates into the nucleus after binding to androgens. The binding affinity of the androgen receptor is greater for DHT (Wilson and French, 1976), but the molecule is able to accommodate testosterone also, in the event no 5α-reductase is present in the tissue to generate DHT. Deficiency or abnormalities of the androgen receptor are responsible for another type of male pseudohermaphroditism, known by the name of testicular feminization (Tfm), which is found also in mice. It is transmitted as a sex-linked disorder, since the gene coding for the androgen receptor is located on the X chromosome (Meyer et al., 1975).

Events mediated by Anti-Müllerian Hormone. Regression of the Müllerian primordia is the first sign of male differentiation of the genital tract. It is already detectable histologically in the human fetus at 8 weeks, at the area nearest to the caudal pole of the testis. Once initiated, Müllerian regression extends caudally, sparing the caudal tip, which participates in the formation of the prostatic utricle. The histological pattern of Müllerian regression has been subjected to close scrutiny in the rat fetus, since the rat fetal Müllerian duct has been used as end-organ to study anti-Müllerian activity of various tissues or media (Fig. 4). In contrast with the usual models of programmed cell death, which occur in the course of normal fetal development, Müllerian regression is not accompanied by cell necrosis. Dissolution of the basal membrane of the duct is followed by the migration of epithelial cells in the mesenchymal compartment, and by the formation of a tight ring of fibroblasts around the regressing duct (Trelstad

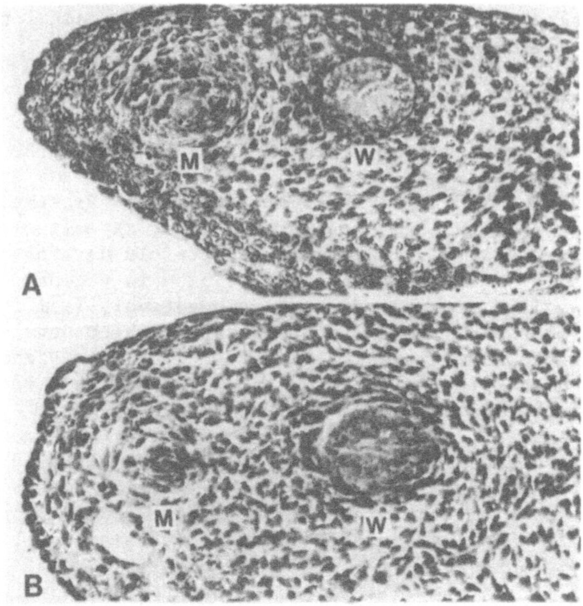

Fig. 4. Aspect of fetal rat Müllerian duct cultured from 14 to 17 days post-coitum in the presence of anti-Müllerian hormone, at a concentration of approximately 1 μg/ml, purified from A: incubation medium of fetal calf testicular tissue, and B: ovarian follicular fluid from adult cows. Note the development of a tight fibroblastic ring around the epithelial component of the duct, characteristic of the effect of anti-Müllerian hormone upon its target organ (× 350). (From Vigier et al., 1984b, with permission.)

et al., 1982). This histological pattern (Fig. 4) is rather similar to the one exhibited by fetal mammary rudiments exposed to testosterone in vitro (Kratochwil, 1977; Drews and Drews, 1977), and suggests that, as has been demonstrated in the latter organ (Heuberger et al., 1982), receptors to the regressor are in the mesenchyme, and act through epithelium-mesenchyme interaction.

The fetal testis mediates the regression of Müllerian primordia through anti-Müllerian hormone (AMH), a 145,000 dimeric glycoprotein (Picard et al., 1978, 1984, 1986; Budzik et al., 1983) produced by Sertoli cells (Tran and Josso, 1982; Hayashi et al., 1984), but also by postnatal ones (review in Vigier et al., 1983) and by granulosa cells of the mature ovary (Vigier et al., 1984b). The cDNA (Picard et al., 1986) and genes (Cate et al., 1986) coding for bovine and human anti-Müllerian hormone have been cloned recently, the C-terminal domain of the protein is highly conserved and shows marked homology with human transforming growth factor β and the β chain of porcine inhibin (Cate et al., 1986). In human intersexes, lack of Müllerian regression may occur as an isolated defect, i.e. in otherwise normally virilized males (Sloan and Walsh, 1976). It is genetically transmitted, suggesting that a biochemical abnormality affecting either synthesis or peripheral action of the hormone is responsible for the syndrome. In contrast, persistence of Müllerian duct derivatives may occur in incompletely virilized patients. Male pseudohermaphroditism with signs of both AMH and testosterone deficiency is usually ascribed to testicular dysgenesis (discussion in Josso et al., 1983). Whatever the cause of the persistence of Müllerian ducts, tesicular descent is usually severely impaired, most probably due to the close attachment of the testes to the abdominal Müllerian derivatives. Persistence of Müllerian derivatives in male rabbits has been obtained experimentally, by passive immunication of the fetuses against AMH (Tran et al., 1986).

References

Benhaim, A., Gangnerau, M., Bettane-Casanova, M., Fellows, M., and Picon, R., 1982, Effects of H-Y antigen on morphologic and endocrine differentiation of gonads in mammals, Differentiation, 22:53.

Bennett, D., Mathieson, B. J., Scheid, M., Yanagisawa, K., Boyse, E. A., Wachtel, S., and Cattanach, B. M., 1977, Serological evidence of H-Y antigenin Sxr, XX sex-reversed phenotypic males, Nature, 265:255.

Budzik, G. P., Powell, S. M., Kamagata, S., and Donahoe, P. K., 1983, Müllerian inhibiting substance fractionation by dye affinity chromatography, Cell, 34:307.

Burgoyne, P. S., 1982, Genetic homology and corssing over in the X and Y chromosomes of mammals, Human Genet., 61:85.

Burgoyne, P. S., and Baker, T. G., 1985, Perinatal oocyte loss in XO mice and its implications for the aetiology of gonadal dysgenesis in XO women, J. Reprod. Fertil., 75:633.

Burgoyne, P. S., Levy, E. R., and McLaren, A., 1986, Spermatogenic failure in male mice lacking H-Y antigen, Nature, 320:170.

Byskov, A. G., 1974, Does the rete ovarii act as a trigger for the onset of meiosis, Nature, 252:396.

Byskov, A. G., 1978, Regulation of initiation of meiosis in female gonads, Int. J. Androl., 1 Suppl. 2:29.

Byskov, A. G., and Grinsted, J., 1981, Feminizing effect of mesonephros on cultured differentiating mouse gonads and ducts, Science, 212:817.

Cate, R. L., Mattaliano, R. J., Hession, C., Tizard, R., Farber, N. M., Cheung, A., Ninfa, E. G., Frey, A. Z., Gash, D. J., Chow, E. P., Fisher, R. A., Bertonis, J. M., Torres, G., Wallner, B. P., Ramachandran, K. L., Ragin, R. C., Manganaro, T. F., MacLaughlin, D. T., and Donahoe, P. K., 1986, Isolation of the bovine and human genes for Mullerian inhibiting substance and expression of the human gene in animal cells, Cell, 45, in press.

Cattanach, B. M., Evans, E. P., Burtenshaw, M. D., and Barlow, J., 1982, Male, female and intersex development in mice of identical chromosome constitution, Nature, 300:445.

Cattanach, B. M., Pollard, C. E., and Hawkes, S. G., 1971, Sex-reversed mice: XX and XO males, Cytogenetics, 10:318.

Chan, S. T. H., O, W. S., 1981, Environmental and nongenetic mechanisms in sex determination, in: "Mechanisms of Sex Differentiation in Animals and Man", C. A. Austin and R. G. Edwards, eds., p. 55, Academic Press, London.

Chandra, H. S., 1985a, Sex determination: a hypothesis based on noncoding DNA, Proc. Natl. Acad. Sci., 82:1165.

Chandra, H. S., 1985b, Is human X chromosome inactivation a sex-determining device?, Proc. Natl. Acad. Sci., 82:6947.

De la Chapelle, A., 1981, The etiology of maleness in XX men, Human Genet., 58:105.

Didier, E., 1973, Recherches sur la morphogenese du canal de Müller chez les oiseaux, II Etude experimentale, Wilhelm Roux Arch, 172:287.

Drews, U., and Drews, U., 1977, Regression of mouse mammary gland anlagen in recombinants of tfm and wild-type tissues: testosterone acts via the mesenchyme, Cell, 10:401.

Eicher, E. M., Washburn, L. L., Whitney, J. B., and Morrow, K. E., 1982, Mus poschiavinus Y chromosome in the C57BL:6J murine genome causes sex reversal, Science, 217:535.

Eichwald, E. J., and Silmser, C.R., 1955, Communication, Transplant Bull., 2:148.
Evans, E. P., Burtenshaw, M. D., and Cattanach, B. M., 1982, Meiotic crossing-over between the X and Y chromosomes of male mice carrying the sex-reversing (Sxr) factor, Nature, 300:443.
Evans, E. P., Ford, C. E., and Lyon, M. F., 1977, Direct evidence of the capacity of the XY germ cell in the mouse to become an oocyte, Nature, 267, 430:431.
Feldman, K. W., and Smith, D. W., 1975, Fetal phallic growth and penile standards for newborn male infants, J. Pediatr., 86:395.
Ford, C. E., Evans, E. P., Burtenshaw, M. D., Clegg, H. M., Tuffrey, M., and Barnes, R. D., 1975, A functional "sex-reversed" oocyte in the mouse, Proc. R. Lond. B., 190:187.
Fredga, K., Gropp, A., Winking, H., and Fritz, F., 1977, A hypothesis explaining the exceptional sex ratio in the wood lemming, Hereditas, 85:101.
George, F. W., Catt, K. J., Neaves, W. B., and Wilson, J. D., 1978a, Studies on the regulation of testosterone synthesis in the fetal rabbit testis, Endocrinology, 102:665.
George, F. W., Milewich, L., and Wilson, J. D., 1978b, Oestrogen content of the embryonic rabbit ovary, Nature, 274:172.
Gore-Langton, R. E., Tung, P. S., and Fritz, I. B., 1983, The absence of specific interactions of Sertoli-cell-separated proteins with antibodies directed against H-Y antigen, Cell, 32:289.
Guellaen, G., Casanova, M., Bishop, C., Geldwerth, D., André, G., Fellous, M., and Weissenbach, J., 1984, Human XX males with Y single-copy DNA fragments, Nature, 307:172.
Hamerton, J. L., Dickson, J. M., Pollard, C. E., Grieves, S. A., and Short, R. V., 1969, Genetic intersexuality in goats, J. Reprod. Fertil. (Suppl.), 7:25.
Hayashi, H., Shima, H., Hayashi, K., Trelstad, R. L., and Donahoe, P. K., 1984, Immunocytochemical localization of Müllerian-inhibiting substance in the rough endoplasmic reticulum and Golgi apparatus in Sertoli cells of the neonatal calf testis, using a monoclonal antibody, J. Histochem. Cytochem., 32:649.
Heuberger, B., Fitzka, I., Wasner, G., and Kratochwil, K., 1982, Induction of androgen receptor formation by epithelium-mesenchyme interaction in embryonic mouse mammary gland, Proc. Natl. Acad. Sci., 79:2957.
Hodgins, M. B., 1982, Binding of androgens in 5α-reductase deficient human genital skin fibroblasts: inhibition by progesterone and its metabolites, J. Endocrinol., 94:415.
Huhtaniemi, I. T., Nozu, K., Warren, D. W., Dufau, M. L., and Catt, K. J., 1982, Acquisition of regulatory mechanisms for gonadotropin receptors and steroidogenesis in the maturing rat testis, Endocrinology, 111:1721.
Imperato-McGinley, J., Binienda, Z., Arthur, A., Mininberg, D. T., Vaughan, E., D. J. R., and Quimby, F. W., 1985, The development of a male pseudohermaphroditic rat using an inhibitor of the enzyme 5α-reductase, Endocrinology, 116:807.
Imperato-McGinley, J., Peterson, R. E., Gautier, T., and Sturla, E., 1979, Androgens and the evolution of male gender identity among male pseudohermaphrodites with 5α-reductase deficiency, New Engl. J. Med., 300:1233.
Johnson, L. L., Sargent, E. L., Washburn, L. L., and Eicher, E. M., 1982, XY female mice express H-Y antigen, Develop. Genet., 3:247.
Jones, K. W., and Singh, L., 1981, Conserved repeated DNA sequences in vertebrate sex chromosomes, Human Genet., 58:46.
Josso, N., 1981, Differentiation of the genital tract: stimulators and inhibitors, in: "Mechanisms of Sex Differentiation in Animals and Man", C. R. Austin and R. G. Edwards, eds., p. 165, Academic Press, London.
Josso, N., Feketé, C., Cachin, O., Nezelof, C., and Rappaport, R., 1983, Persistence of Müllerian ducts in male pseudohermaphroditism, and its relationship to cryptorchidism, Clin. Endocrinol., 19:247.

Josso, N., and Picard, J. Y., 1986, Anti-Müllerian hormone, Physiol. Rev., in press.
Jost, A., 1947, Recherches sur la différenciation sexuelle de l'embryon de lapin, III. Rôle des gonades foetales dans la différenciation sexuelle somatique, Arch. Anat. Micr. Morph. Exp., 36:271.
Jost, A., 1972, A new look at the mechanisms controlling sex differentiation in mammals, Johns Hopkins Med. J., 130:38.
Jost, A., Vigier, B., and Prepin, J., 1972, Freemartins in cattle: the first steps of sexual organogenesis, J. Reprod. Fertil., 29:349.
Kiel-Metzger, K., Warren, G., Wilson, G. N., and Erickson, R. P., 1986, Evidence that the human Y chromosome does not contain clustered DNA sequences (Bkm) associated with heterogametic sex determination in other vertebrates, N. Engl. J. Med., 313:242.
Kratochwil, K., 1977, Development and loss of androgen responsiveness in the embryonic rudiment of the mouse mammary gland, Dev. Biol., 61:358.
Leinonen, P. J., and Jaffe, R. B., 1985, Leydig cell desensitization by human chorionic gonadotropin does not occur in the human fetal testis, J. Clin. Endocrinol. Metab., 61:234.
Luciani, J. M., Devictor, M., and Stahl, A., 1977, Preleptotene chromosome condensation stage in human foetal and neonatal testes, J. Embryol. Exp. Morph., 38:175.
Lyon, M. F., Searle, A. G., Ford, D. E., and Ohno, S., 1964, A mouse translocation suppressing sex linked variegation, Cytogenetics, 3:306.
Magre, S., and Jost, A., 1980, The initial phases of testicular organogenesis in the rat; an electron microscopy study, Arch. Anat. Micr. Morph. Exp., 69:297.
McCoshen, J. A., 1982, In vivo sex differentiation of congeneic germinal cell aplastic gonads, Am. J. Obstet. Gynecol., 142:83.
McLaren, A., 1981, The fate of germ cells in the testis of fetal sex-reversed mice, J. Reprod. Fertil., 61:461.
McLaren, A., 1983, Studies on mouse germ cells inside and outside the gonad, J. Experiment. Zool., 228:167.
McLaren, A., 1984, Chimeras and sexual differentiation, in: "Chimeras in Developmental Biology", N. Ledouarin and A. McLaren, eds., p. 381, Academic Press, London.
McLaren, A., 1985, Relation of germ cell sex to gonadal differentiation, in: "The Origin and Evolution of Sex", p. 289, Alan R. Liss, New York.
McLaren, A., and Monk, M., 1981, X-chromosome activity in the germ cells of sex reversed mouse embryos, J. Reprod. Fertil., 63:533.
McLaren, A., and Monk, M., 1982, Fertile females produced by inactivation of an X chromosome of "sex-reversed" mice, Nature, 300:446.
McLaren, A., Simpson, E., Romonari, K., Chandler, P., and Hogg, H., 1984, Male sexual differentiation in mice lacking H-Y antigen, Nature, 312:552.
Melvold, R. W., Kohn, H. I., Yerganian, G., and Fawcett, D. W., 1977, Evidence suggesting the existence of two H-Y antigens in the mouse, Immunogenetics, 5:33.
Merchant, H., 1975, Rat gonadal and ovarian organogenesis with and without germ cells, an ultrastructural study, Develop. Biol., 44:1.
Merchant-Larios, H., 1984, Germ and somatic cell interactions during gonadal morphogenesis, in: "Ultrastructure of Reproduction", J. Van Blerkom and P. M. Motta, eds., p. 19, Martinus Nijhoff Publishers, Boston.
Merchant-Larios, H., and Centeno, B., 1981, Morphogenesis of the ovary from the sterile W/Wv mouse, in: "Advances in the Morphology of Cells and Tissues", p. 383, Alan R. Liss, New York.
Meyer, W. J., Migeon, B. R., and Migeon, C. J., 1975, Locus on the X chromosome for dihydrotestosterone receptor and androgen insensitivity, Proc. Natl. Acad. Sci. USA, 72:1469.
Monk, M., and McLaren, A., 1981, X-chromosome activity in foetal germ cells of the mouse, J. Embryol. Exp. Morph., 63:75.

O, W. S., and Baker, T. G., 1976, Initiation and control of meiosis in hamster gonads in vitro, J. Reprod. Fertil., 48:399.
Ohno, S., 1985, The Y-linked testis determining gene and H-Y plasma membrane antigen gene: are they one and the same?, Endocrine Rev., 6:421.
Ohno, S., Nagai, Y., Ciccarese, S., and Smith, R., 1979, In vitro studies of gonadal organogenesis in the presence and absence of H-Y antigen, In Vitro, 15:11.
Pelliniemi, L. J., Paranko, J., Grund, S. K., Fröjdman, K., Foidart, J. M., and Lakkala-Paranko, T., 1984, Extracellular matrix in testicular differentiation, in: "Hormone Action and Testicular Function", NY Acad. Sci., 438, 405:416.
Picard, J. Y., Benarous, R., Guerrier, D., Josso, N., Kahn, A., 1986, Cloning and expression of cDNA for anti-Müllerian hormone, Proc. Natl. Acad. Sci., in press.
Picard, J. Y., Goulut, C., Bourrillon, R., and Josson, N., 1986, Biochemical analysis of bovine testicular anti-Müllerian hormone, FEBS Lett., 195:73.
Picard, J. Y., and Josso, N., 1984, Purification of testicular anti-Müllerian hormone allowing direct visualization of the pure glycoprotein and determination of yield and purification factor, Mol. Cell. Endocrinol., 34:23.
Picard, J. Y., Tran, D., and Josso, N., 1978, Biosynthesis of labelled anti-Müllerian hormone by fetal testes: evidence for the glycoprotein nature of the hormone and for its disulfide-bonded structure, Mol. Cell. Endocrinol., 12:17.
Seboun, E., Leroy, P., Casanova, M., Magenis, E., Boucekkine, C., Disteche, C., Bishop, C., Fellows, M., 1986, A molecular approach to the study of the human Y chromosome and anomalies of sex determination in man, in: "Homo sapiens", Cold Spring Harbor Symposia, in press.
Silvers, W. K., Gasser, D. L., and Eicher, E. M., 1982, H-Y antigen, serologically detectable male antigen and sex determination, Cell, 28:439.
Simpson, E., Chandler, P., Washburn, L. L., Bunker, H. P., and Eicher, E. M., 1983, H-Y typing of karyotypically abnormal mice, Differentiation, 23:S116.
Simpson, E., McLaren, A., and Chandler, P., 1982, Evidence for two male antigens in mice, Immunogenetics, 15:609.
Simpson, E., McLaren, A., Chandler, P., and Tomonari, K., 1984, Expression of H-Y antigen by female mice carrying Sxrl, Transplantation, 37:17.
Singh, L., and Jones, K. W., 1982, Sex reversal in the mouse (Mus musculus) is caused by a recurrent nonreciprocal crossover involving the X and an aberrant Y chromosome, Cell, 28:205.
Sloan, W. R., and Walsh, P. C., 1976, Familial persistent müllerian duct syndrome, J. Urol., 115:459.
Taketo, T., Koide, S. S., and Merchant-Larios, H., 1985, Gonadal sex differentiation in mammals, in: "Origin and Evolution of Sex", H. O. Halvorson, A. Monroy, eds., p. 271, Alan R. Liss, New York.
Tran, D., and Josso, N., 1982, Localization of anti-Müllerian hormone in the rough endoplasmic reticulum of the developing bovine Sertoli cell using immunocytochemistry with a monoclonal antibody, Endocrinology, 111:1562.
Tran, D., Meusy-Dessole, N., and Josso, N., 1977, Anti-Müllerian hormone is a functional marker of foetal Sertoli cells, Nature, 269:411.
Tran, D., Picard, J. Y., Vigier, B., Berger, R., Josso, N., 1986, Persistence of Müllerian ducts in male rabbits passively immunized against bovine anti-Müllerian hormone during fetal life., Dev. Biol., in press.
Trelstad, R. L., Hayashi, A., Hayashi, K., and Donahoe, P. K., 1982, The epithelial mesenchymal interface of the male rat Müllerian duct: loss of basement membrane integrity and ductal regression, Dev. Biol., 92:27.
Ulfelder, H., and Robboy, S. J., 1976, Embryological development of human vagina, Am. J. Obstet. Gynecol., 126:769.

Upadhyay, S., and Zamboni, L., 1982, Ectopic germ cells - Natural model for the study of germ cell sexual differentiation, Proc. Natl. Acad. Sci., 79:6584.
Vergnaud, G., Page, D. C., Simmler, M. C., Brown, L., Rouyer, F., Noel, B., Botstein, D., de la Chapelle, A., Weissenbach, J., 1986, A deletion map of the human Y chromosome based on DNA hybridization, Am. J. Human Genet., 38:109.
Vigier, B., and Josso, N., 1986, Use of monoclonal antibodies to anti-Müllerian hormone in the study of sexual differentiation, in: "Monoclonal Antibodies: Basic Principles, Experimental and Clinical Applications in Endocrinology", G. Forti, M.O. Serio, M. B., Lipsett, eds., p. 309, Raven Press, New York.
Vigier, B., Picard, J. Y., Tran, D., Legeai, L., and Josso, N., 1984b, Production of anti-Müllerian hormone: another homology between Sertoli and granulosa cells, Endocrinology, 114:1315.
Vigier, B., Tran, D., Du Mesnil du Buisson, F., Heyman, Y., and Josso, N., 1983, Use of monoclonal antibody techniques to study the ontogeny of bovine anti-Müllerian hormone, J. Reprod. Fertil., 69:207.
Vigier, B., Tran, D., Legeai, L., Bezard, J., and Josso, N., 1984a, Origin of anti-Müllerian hormone in bovine freemartin fetuses, J. Reprod. Fertil., 70:473.
Wachtel, S. S., Hall, J. L., Muller, U., and Chaganti, R. S. K., 1980, Serum-borne H-Y antigen in the fetal bovine freemartin, Cell, 21:917.
Wachtel, S. S., Ohno, S., Koo, G. C., and Boyse, E. A., 1975, Possible role for H-Y antigen in the primary determination of sex, Nature, 257:235.
Wartenberg, H., 1978, Human testicular development and the role of the mesonephros in the origin of a dual Sertoli cell system, Andrologia, 10:1.
Washburn, L. L., and Eicher, E. M., 1983, Sex reversal in XY mice caused by dominant mutation on chromosome 17, Nature, 303:338.
Wilson, E. M., and French, F. S., 1976, Binding properties of androgen receptors, Evidence for identical receptors in rat testis, epididymis, and prostate, J. Biol. Chem., 251:5620.
Winter, J. S. D., Faiman, C., and Reyes, F. I., 1977, Sex steroid production by the human fetus: its role in morphogenesis and control by gonadotropins, in: "Morphogenesis and Malformation of the Genital System", R. Blandau, D. Bergsma, eds., p. 41, Alan R. Liss, New York.
Zamboni, L., and Upadhyay, S., 1982, The contribution of the mesonephros to the development of the sheep fetal testis, Am. J. Anat. 165:339.
Zenzes, M. T., 1981, H-Y antigen negative germ cells in gonadal sex organization in vitro, A Morphological study, Differentiation, 18:169.
Zenzes, M. T., Muller, U., Aschmoneit, I., and Wolf, U., 1978b, Studies on H-Y antigen in different cell fractions of the testis during pubescence, immature germ cells are H-Y antigen negative, Human Genet., 45:297.
Zenzes, M. T., Wolf, U., Gunther, E., and Engel, W., 1978a, Studies on the function of H-Y antigen: dissociation and reorganization experiments on rat gonadal tissue, Cytogenet. Cell. Genet., 20:365.

THE PLACENTAL LACTOGEN GENE FAMILY:

STRUCTURE AND REGULATION

M. L. Duckworth, M. C. Robertson, H. G. Friesen

Department of Physiology
University of Manitoba
Winnipeg, Manitoba

The placenta has been recognized as a major endocrine gland for well over half a century. Its repertoire of secretory products includes both steroids, proteins, and peptides, several of which are secreted in large amounts. Another notable feature of placental endocrinology is the great species variation of specific hormones and pattern of hormones secreted. This feature makes studies of placental endocrinology more fascinating as well as more complex and challenging. Excellent reviews on placental endocrinology should be consulted for a more detailed account of the subject (Forsyth, 1974; Blank et al., 1977; Talamantes et al., 1980).

In this chapter we wish to focus particularly on research on placental lactogen. The first report on luteotropic activity of the rat placental extracts appeared almost 50 years ago. In this report, Astwood and Greep (1938) stated that serum and placental extracts contained luteotropic activity from day 10 to 14 of gestation with maximal activity occurring at day 12, and none after day 15. Subsequent reports using other assays confirmed and extended these findings. Renewed interest in placental luteotropins was exhibited following the report by Josimovich and MacLaren (1962) that human placentas contained a protein which crossreacted with antibodies to human growth hormone (hGH). Subsequent studies led to the isolation and amino acid sequence analysis (Niall et al., 1973) and ultimately the cloning of genes for human placental lactogen (hPL) or human chorionic somatomammotropin (hCS) (Miller and Eberhart, 1983). Structural analysis demonstrated extensive homology between hGH and hPL. Although hGH and hPL exhibit greater than 80% amino acid sequence homology, the somatotropic activity of hPL is less than 1% that of hGH. This result proved somewhat disappointing at a time when an alternative to hGH for therapeutic use was widely sought.

A resurgence of interest in placental lactogens was triggered by the development of radioreceptor assays for prolactin and growth hormone (Shiu et al., 1973). The RRA's were able to detect PL's from a variety of sources and appeared to be class specific. As a result, placental lactogens were identified and isolated from several additional species (Kelly et al., 1976) including sheep, goats, cows, mice, guinea pigs, etc.

Using 3T3 cells, Linzer and Nathans reported that a number of genes were induced when 3T3 cells were stimulated to proliferate upon the addition of serum. One of these proliferation-inducible genes, when cloned and sequenced, exhibited extensive homology to PRL/GH (Linzer and Nathans, 1984).

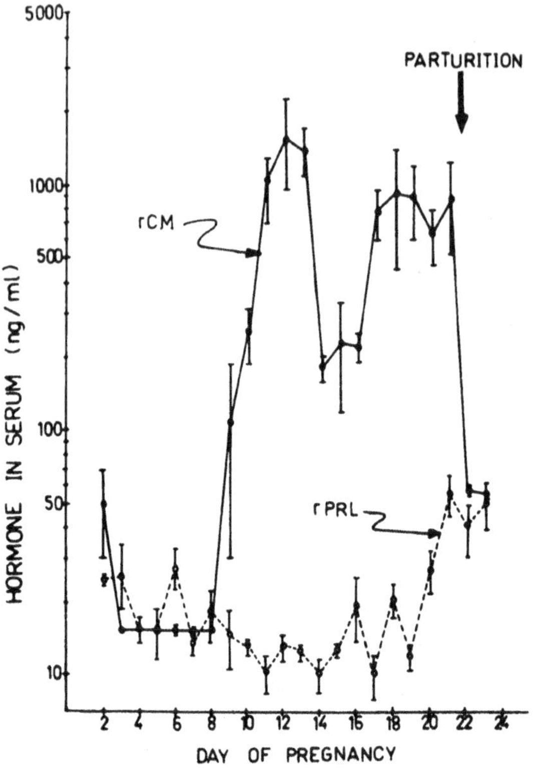

Fig. 1. Serum concentrations of rPL (rat chorionic mammotropin, rCM) were determined by radioreceptor assay on blood samples taken daily throughout pregnancy. The rabbit mammary gland membranes were used for the receptor assay and values are expressed as ovine prolactin (25 IU/mg) equivalents. The mean±SE values are shown for 5 animals at each point. Serum prolactin concentrations were measured by radioimmunoassay on aliquots of the same sample. The first peak of rCM occurs at day 12 and is attributable to rPL-I in the circulation. The second peak beginning at day 14 is due to rPL-II. Using an RIA for rPL-II, no crossreaction with rPL-I is noted. Moreover, antibodies to rPL-II which neutralize all activity of rPL-II when measured by bioassay fail to crossreact and to neutralize rPL-I.

They aptly named the gene, proliferin, and subsequently demonstrated it was expressed by the placenta along with a proliferin-related protein (Linzer and Nathans, 1985a, b). The functional role of proliferin remains to be defined.

In this chapter, we wish to focus especially on our studies on rat placental lactogen. In the study of any hormone, there are three principal features which must be addressed; namely, the structure of the hormone, the factors controlling hormone synthesis and secretion, and finally the function and mechanism of action of a hormone. These last two subjects remain

Table 1. Percentage Comparison of Amino Acid Sequences of rPL-II versus Other Members of the PRL, GH Gene Family[a]

Proteins	Identical	Related	Total
rPLII vs rPRL	78 (35%)	38 (17%)	116 (52%)
rPLII vs rGH	42 (19%)	33 (15%)	75 (34%)
rPLII vs hPRL	87 (38%)	31 (13%)	118 (51%)
rPLII vs hGH	44 (20%)	32 (14%)	76 (34%)
rPLII vs hPL	41 (19%)	32 (15%)	73 (34%)

[a]The related amino acids were considered to be: L = I = V = M, S = T, D = E, K = R, Q = N, Y = F.

poorly defined for virtually all placental hormones. More detailed information on the structure of placental protein and polypeptide hormones has emerged in the past decade.

When rat serum samples were measured by RRA for lactogens, it was apparent that contrary to the initial report by Astwood and Greep two peaks of lactogenic activity were identified; the first peak consistent with the initial report occurred at day 12 of gestation and the second towards term at days 18 to 20 (Fig. 1). Subsequently, we established that rPL-I (the first peak) and rPL-II (the second peak) differed in M_r, half-time disappearance rate, antigenic determinants, etc. (Robertson et al., 1982). An RIA for rPL-II failed to detect rPL-I, suggesting substantial structural differences between rPL-I and rPL-II.

To elucidate the structure of the members of the rPL gene family, a cDNA library was developed from mRNA isolated from day 18 placenta. One of the clones identified, hybrid selects a mRNA which translates _in vitro_ to a protein of 25,000 daltons. In the presence of pancreatic microsomes, the protein is processed to a 22,000 dalton protein which is immunoprecipitated by antiserum to rPL-II or antiserum to ovine PRL. On the basis of these findings and other evidence, we concluded that a cDNA clone for rPL-II had been isolated (Duckworth et al., 1986a). The cDNA clone for rPL-II hybridizes to rPRL, hPRL but not to rGH or hPL probes. Nucleotide sequence analysis of rPL-II cDNA reveals that rPL-II is more homologous to members of the PRL family than to members of the GH family. At an amino acid level, rPL-II is 52% homologous to rPRL, but only 34% related to the amino acid sequence of rGH (Table 1). A comparison of the structural features of members of the PRL, PL, Proliferin gene family is summarized in Fig. 2.

The cysteines at amino acid positions 56, 172, 189, 197 and the two tryptophans at positions 89 and 148 of rPRL are in similar locations within the coding regions of rPL-II. The highly conserved region from amino acids 58 to 71 found in all members of the PRL-GH gene family is present in rPL-II. Restriction enzyme analysis of rat genomic DNA shows that the rPL-II gene is distinctly different from the rPRL-gene. The rPL-II and rPRL gene are approximately 10 Kb in length compared to the GH gene family which is 2 Kb in length. Both the sequence and genomic DNA data support the conclusion that unlike hPL rPL-II is evolutionarily more closely related to PRL than to GH.

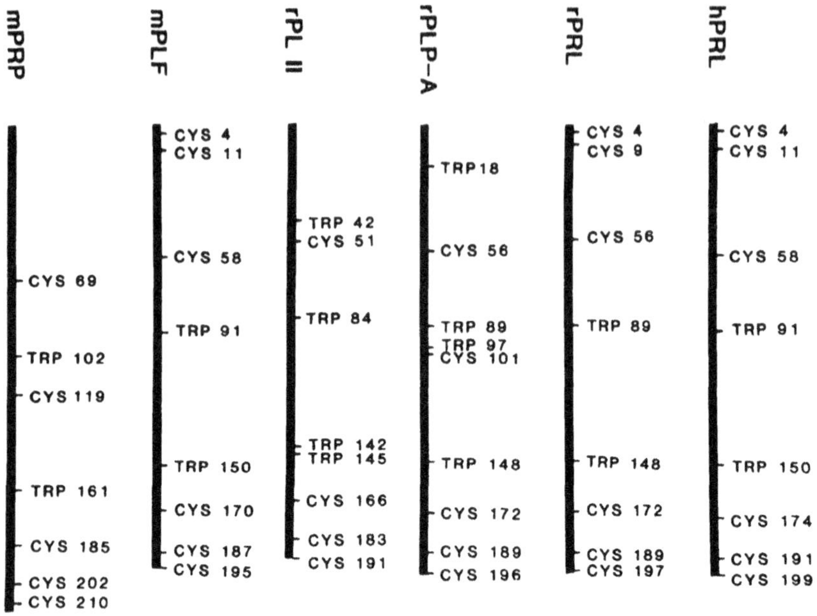

Fig. 2. Comparison of tryptophans and cysteines in members of the rat prolactin, placental lactogen, proliferin gene family. The tryptophan and cysteine residues of the mature rPLP-A are compared with those found in hPRL, rPRL, rPL-II, mPLF and mPRP. The signal peptide for rPLP-A is taken to be 31 amino acids. The rPLP-A contains several tryptophans and cysteines in conserved locations compared with other PRL family members. It also contains an extra cysteine at position 101 and extra tryptophans at positions 18 and 97. rPLP-A, rat placental lactogen-like protein-A; mPLF, mouse proliferin; mPRP, mouse proliferin related protein.

Four additional cDNA clones have been isolated from a day 18 rat placental library which show varying degrees of hybridization homology to rPRL and hPRL clones and to one another. Each of these clones is uniquely different based on restriction enzyme mapping, protein specified, pattern of mRNA induction during pregnancy (Fig. 3) and genomic DNA restriction enzyme pattern. One of these clones, rPLP-A which shows good hybridization to rPRL, hPRL and rPL-II cDNA clones at 55°C but not to rGH or hPL, has been studied in greater detail (Duckworth et al., 1986b). This clone codes for an mRNA which translates _in vitro_ to a protein of approximately 25K molecular weight. In the presence of dog pancreatic microsomes, it changes to approximately 27K possibly due to glycosylation. The mRNA corresponding to this clone first appears at day 14 of pregnancy and peaks at day 18. The nucleotide sequence of this clone confirms the relationship to prolactin. There is 50% homology between it and rPRL and hPRL at the nucleotide level and 43% homology at the amino acid level. The prolactins and this protein have essentially identical locations for tryptophan at positions 89 and 148, and cysteine residues at positions 56, 172, 189 and 197 (Fig. 2). A region corresponding to amino acids 58-71 of rPRL seems to be very highly conserved amongst a number of other prolactin-like proteins including rPLP-A, rPL-II and the more recently described growth-associated mouse proliferin (mPLF)

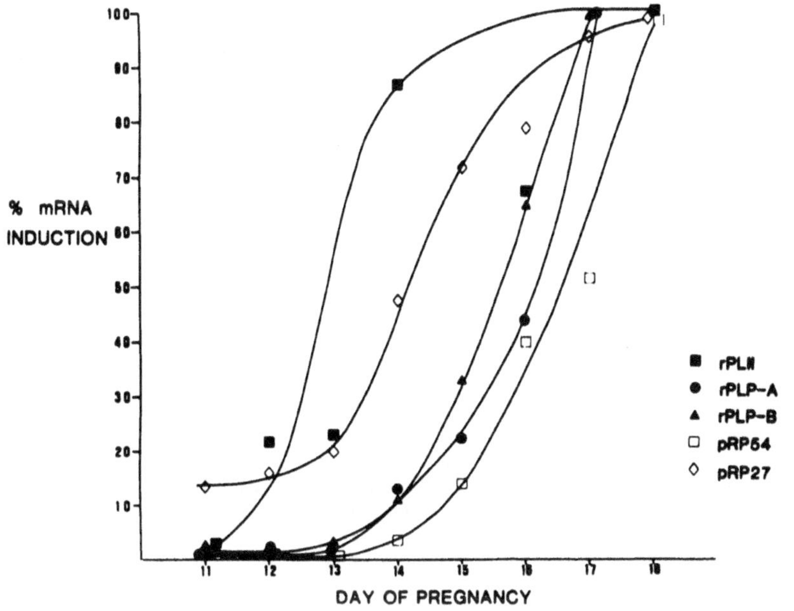

Fig. 3. The induction pattern of 5 members of the rPL-II gene family. The steady state mRNA levels in placental RNA preparations were determined by dot blot hybridization analysis using specific cDNA clones isolated from a day 18 placental library. The mRNA induction pattern throughout gestation was expressed as a percentage of the maximum level determined for each clone. It is apparent that each mRNA has a unique temporal pattern of induction. The half maximal level of induction of rPL-II precedes by 2 to 3 days the half maximal level observed for rPLP-A.

from stimulated 3T3 cells (Linzer and Nathans, 1984). Our studies of rPLP-A and other placental clones appear to suggest the existence of a family of PRL-related genes expressed by the rat placenta.

A third placental clone (rPLP-B) shows moderate hybridization homology with rPRL at 55°C (Duckworth et al., 1986c). The clone hybridizes with a 1.0 Kb mRNA transcript present in low levels as early as day 11 of pregnancy but which is highly expressed from day 14 to day 18. Hybrid select translation shows only one protein of 25,000 daltons. The amino acid sequence deduced from the nucleotide sequence also demonstrates significant homology with rat PRL and PRL-like placental proteins. In particular, there is conservation of cysteine residues equivalent to positions of 56, 172, 189, and 197 on rat PRL and tryptophans equivalent to positions 89 and 148.

As mentioned earlier, Linzer and Nathans (1984) have identified a PRL-related gene which is induced by a mitogenic signal in 3T3 cells. In subsequent studies they have also reported on the expression of a proliferin-related gene product in midterm placentae in the mouse (Linzer and Nathans, 1985).

Table 2. Members of the Human Placental Lactogen and GH Gene Family

Gene	hGH-N	hCS-L	hCS-A	hGF-V	hCSB
Product	22k	-	22k	22k Variant (glycosylated)	22k
Tissue	Pituitary	-	Placenta	Placenta	Placenta

<-------------------55Kb------------------------->
DNA
on Chromosome 17

In the mouse at least two distinct forms of placental lactogen have been identified (Soares et al., 1982); an initial midterm peak of PL(mPL-I) which is immunologically different from the second peak of PL(mPL-II) occurring later in pregnancy. These investigators have been successful in purifying mPL-I and developing a radioimmunoassay for it (Colosi et al., 1986). Again, as in the rat, they observed no crossreaction of mPL-II in an RIA for mPL-I.

Finally, it should be noted that in addition to the growing list of members of the placental lactogen gene family, Jayatilak et al. (1985) reported the appearance of a distinct and unique luteotropin in decidual tissue between days 7 to 10 of pregnancy. The characterization of this factor is still at an early stage, but unlike placental lactogen the luteotropic activity is more readily detected by an RRA employing rat ovarian receptors rather than mammary gland or liver receptors. Human decidua has been shown to synthesize authentic hPRL (Golander et al., 1978). However, in the rat, decidual cells appear to be capable of secreting luteotropic activity which differs from rPRL (Herz et al., 1986).

It is evident that substantial progress has been made in the identification, isolation and characterization of members of the placental lactogen gene family. In humans there are two hPL genes which are translated into a single protein (Barrera et al., 1983). These two genes, hCS-A and hCS-B, are located on the growth hormone gene cluster on either side of the HGH-V gene on chromosome 17 (Table 2) (Parks, 1986). The hPL genes are closely related structurally to growth hormone but functionally they exhibit predominantly prolactin-like activity. In the rodent there are two major classes of placental lactogen PL-I and PL-II with at least five related members of the PL-II gene family. Table 3 summarizes the different types of placental lactogens and the time of appearance of each during pregnancy. Bovine placental lactogen also appears more closely related to the PRL gene family (Schuler and Hurley, 1985). Thus in the primate, PL seems closely linked to the GH gene family whereas in several other species PL's seem structurally more closely related to PRL's.

Control of Placental Lactogen

Before detailed studies of the factors regulating hormone secretion in pregnancy can be launched, it is imperative that the normal secretory pattern of PL throughout gestation be defined and for a number of species this has been done (Kelly et al., 1976) (Fig. 4). Too often it has been assumed that the definition of the secretory pattern is synonymous with an understanding of the regulatory factors determining the temporal sequence of hormone synthesized and secreted at specific stages of pregnancy. But what are the factors that determine specific gene expression of individual members of the PL gene family? Are there extraplacental control mechanisms which influence

Table 3. The Appearance of Placental Lactogens, Decidual Prolactin and Proliferin during Pregnancy in the Rat

Day 7 - 9	Day 10 - 14	Day 12 - 20
PRL-like (decidua)	rPL-I	rPL-II
	Proliferin Proliferin-related Proteins	5 related genes

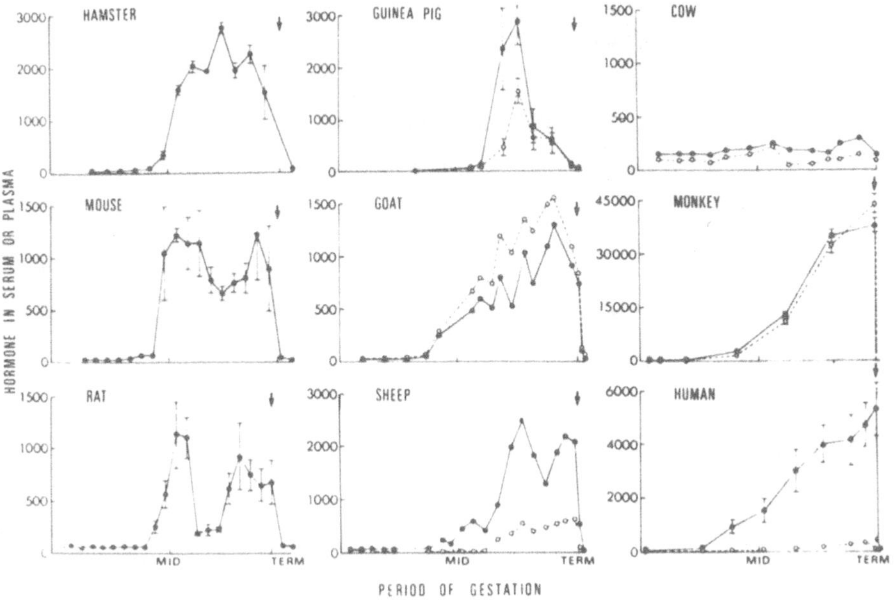

Fig. 4. Pattern of placental lactogen (PL; ● --- ●) and growth hormone-like activity (GHLA; O --- O) in serum or plasma samples from 9 different species. The amount of hormone activity in ng/ml equivalents determined by 2 radioreceptor assays is indicated on the ordinate A. A separate assay for PL and GHLA was carried out for each species, and, for each assay, equivalent quantities of male serum from the species being assayed was added to the standards to compensate for any nonspecific interference due to serum proteins. The abscissa scale has been normalized and is indicated as midpregnancy and term. The duration of pregnancy in days for the 9 species was 15 (hamster); 20 (mouse); 21 (rat); 65 (guinea pig); 148 (goat); 148 (sheep); 280 (Cow); 170 (monkey); and 280 (human). The values illustrated are the mean±SEM for 4 animals at each point in the case of the hamster, guinea pig, and monkey, and 5 animals each for the mouse, rat, and human. The curves for the goat, sheep, and cow are representative patterns of a single animal sample throughout pregnancy. The arrow (↓) indicates the day of parturition.

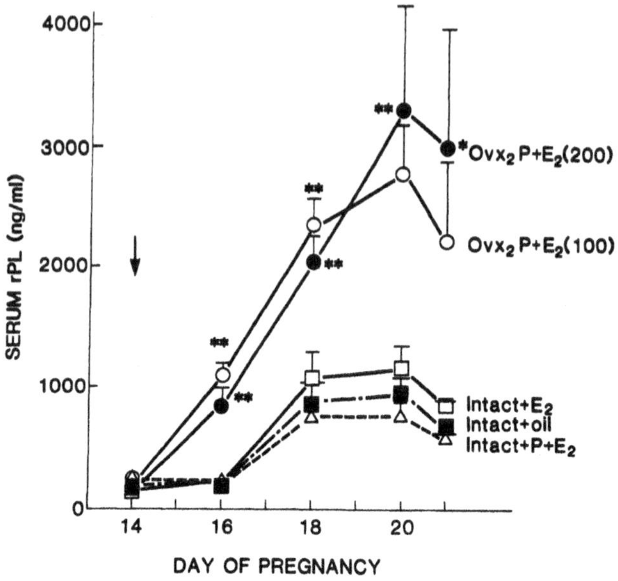

Fig. 5. Serum rPL levels in Ovx_2 or intact pregnant rats treated with P and/or E_2. Operations and/or steroid treatment began on day 14. Serum levels of rPL were compared to levels found in intact pregnant controls receiving daily injections of oil. Mean levels were significant at **, $P < 0.01$; or *, $P < 0.05$. E_2 was administered at either 100 or 200 ng/rat/day. The number of animals in each treatment group ranged from 3 to 7.

PL secretion? Studies of the regulation of rPL secretion are becoming even more complicated given the growing number of members that constitute the gene family. The necessity of establishing specific assays that measure individual members of the gene family becomes self evident. In the case of rPL, it is clear that there are two major peaks of PL activity as determined by RRA. Subsequent studies revealed that rPL-I is secreted from day 10 to 14 and rPL-II from day 12 to term (Fig. 1). A specific RIA for rPL-II enabled specific measurement of rPL-II, while a bioassay using Nb_2 lymphoma cells in the presence of antibodies to rPL-II allowed specific measurement of rPL-I.

To examine whether any extraplacental factors influence PL secretion, a series of endocrine ablation experiments were performed to examine the potential influence of the pituitary, ovaries and fetuses on PL secretion. Hypophysectomy of the dam at midpregnancy resulted in significant elevation of rPL-II during late pregnancy (Robertson et al., 1984a). Adrenalectomy or unilateral ovariectomy had no significant effect on serum rPL levels. Bilateral ovariectomy of day 14 pregnant rats led to a rapid increase in serum rPL levels (Fig. 5). Adrenalectomy combined with ovariectomy led to sustained elevated levels of serum rPL-II which were greater than those seen with bilateral ovariectomy alone. When progesterone or 17β-estradiol were administered to ovariectomized pregnant rats, serum rPL-II levels remained elevated.

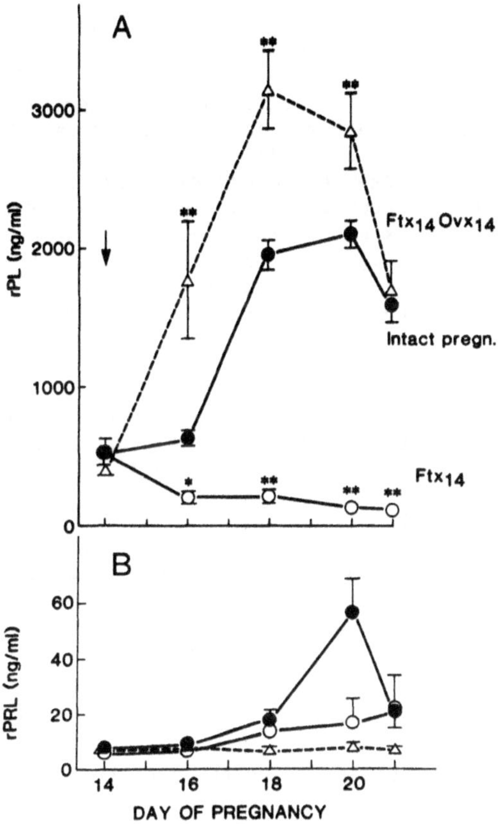

Fig. 6. The effect of fetectomy with or without ovariectomy on levels of rPL-II (A) and rPRL (B) in serum from pregnant rats. The operations were performed on day 14 of gestation. Serum rPL-II and rPRL concentrations were compared to levels found in intact pregnant controls. Mean levels were different at *, $P < 0.05$ or **, $P < 0.01$.

These data indicate that the ovary and adrenal influence secretion either directly or indirectly. Our data also show that the increase in rPL after ovariectomy was independent of the day of ovariectomy, the level of steroids administered, or the placental mass. Indeed, both types of assay used to measure rPL, that is the Nb_2 lymphoma cell bioassay or RIA revealed a large increase of rPL-II after ovariectomy. Since neither progesterone nor estrogen were able to reduce serum levels in ovariectomized animals, the data suggests that undetermined ovarian factors influence rPL-II secretion. At this time we have no indication of the nature of these factors. Additional studies are required to determine the nature of these factors and to delineate their role in the regulation of rPL-II.

Removal of fetuses at day 14 of gestation (Ftx_{14}) in the pregnant rat leads to marked suppression of serum levels of rPL-II (Robertson et al., 1984b). One might attribute this to compromised placental growth in the absence of a fetus. However, if ovariectomy and fetectomy (Ftx_{14} Ovx_{14}) are carried out at the same time, a great increase in serum rPL-II levels is seen (Fig. 6). This occurs despite a significant decrease in placental weight. When Ftx_{14} was performed on day 14 and Ovx was delayed 1, 2 or 3 days, the expected large increase in serum rPL-II was progressively attenuated compared to that seen when Ftx and Ovx were carried out simultaneously. Daily administration of 17β-estradiol (4 μg/rat/day) to Ftx_{14} Ovx_{14} pregnant rats resulted in a significant suppression of rPL-II and elevation of rPLR levels, a reversal of what is seen for those hormones in untreated Ftx_{14}, Ovx_{14} animals.

These and other previous studies suggest that a number of factors are involved in the regulation of rPL-II; namely, the ovary, fetus, and maternal pituitary. Initially, after implantation, the placenta grows rapidly with the early form, rPL-I suppressing the secretion of rPRL by the pituitary. When rPL-I disappears at day 14, this function appears to be taken over by rPL-II. The ovary appears to be a dominant factor in control of growth of the placenta and also influences the secretion of rPL. At day 16 to 17, however, fetal factors which overcome the inhibitory action of the ovary appear to dominate. Just before parturition, levels of rPL-II decline as rPRL levels rise. Our studies would indicate that the fetus may be involved in this change. Further studies are necessary to clarify the interrelationship of these factors in the regulation of rPL-II. In summary, the pituitary and ovary appear to release inhibitors and the fetus stimulators of rPL-II secretion. The nature of these factors remains to be defined.

A second experimental approach has been used to examine extraplacental factors influencing rPL secretion. Placental tissue fragments with or without pituitary fragments were transplanted under the kidney capsule into recipients. The recipients were intact, ovariectomized or hypophysectomized female rats. Following transplantation, serum rPL levels were measured daily. The peak concentrations of rPL were detected 2 days after transplantation when serum rPL-I levels reached values of 200 ng/ml in hypox rats compared to 20 ng/ml in intact and 40 ng/ml in ovariectomized rats. When pituitary and placental fragments were co-transplanted under the kidney capsule of hypox rats, the peak serum rPL-I concentrations were reduced to 18 ng/ml compared to 205 ng/ml in the hypox rats bearing only placental transplants. These results also demonstrate that the pituitary secretes an inhibitor of rPL-I secretion.

It is evident that a number of extraplacental endocrine factors importantly influence rPL secretion. These are derived from the pituitary, the ovary, and the fetus. The nature and mechanism of action of each of these factors remains to be defined.

Function and Mechanisms of Action of Placental Lactogen

Determining the function of placental lactogens is difficult because the number of placental lactogen and prolactin related molecules has increased. Secondly, it is evident that the specific secretory pattern of PL varies among species. Thirdly, the control of reproductive function varies greatly among species. An excellent example is the rodent where prolactin and/or placental lactogen is luteotropic, whereas in other species it appears not to be. Fourth, it is difficult to remove PL selectively to determine what effects the absence of PL might have on pregnancy and on the fetus. An exception to this statement is found in experiments of nature where selective gene deletion of hPL has been reported to have had no deleterious effect on pregnancy or on the fetus at the time of delivery (Parks et al., 1984). It is evident from these reports that the absence of hPL is not obviously deleterious but of course this does not preclude the possibility that hPL has a function but that other compensatory mechanisms can mask any deficiency of hPL. A fifth reason why little is known of the function of PL's is that the amount of PL available for research has been very limited. With the differences in somatogenic and lactogenic activity of individual PL's, it is imperative that examination of the biological role of any PL must be examined using homologous systems including PL and tissues or animals from the same species. For example, hPL has very little somatogenic activity when assayed by RRA for GH. On the other hand, oPL is equipotent with hGH in RRA's for GH and PRL using rabbit liver and mammary gland receptors. However, when RRA's are used employing ovine, liver oPL has predominantly somatogenic activity and very little lactogenic activity (Chan et al., 1978). Most recently, Freemark and Handwerger (1986) have provided evidence for a unique oPL receptor and therefore possibly unique function for oPL. They reported that there were unique binding sites for oPL in liver and that only oPL was active in stimulating glucose uptake in ovine fetal hepatocytes.

In conclusion, since the first report of luteotropic activity of rat placental extracts by Astwood and Greep (1938) significant advances have occurred. A growing number of PRL/PL related genes and proteins have been identified and characterized from a number of species. It is evident that not only the placenta but also the decidua is able to synthesize PRL or PRL-like proteins. In the case of proliferin and perhaps of PRL and other PRL-related proteins, other cell types not traditionally regarded as endocrine cells (3T3-fibroblasts, and most recently, lymphocytes) have been reported to be sites of synthesis of these molecules. It is evident that the list of PRL-like molecules synthesized by the placenta and/or fetal tissues is incomplete. It is also clear that the factors controlling the secretion of these PL's is ill defined. Lastly, the major biological role and function of PL throughout pregnancy for the most part is as enigmatic as it was fifty years ago. Clearly, the challenge to students of reproductive physiology is to rectify this situation.

References

Astwood, E. B., and Greep, R. O., 1938, A corpus luteum stimulating substance in the rat placenta, Proc. Soc. Exp. Biol. Med., 38:713.

Barrera-Saldena, H. A., Seeburg, P. H., and Saunders, G. F., 1983, Two structurally different genes produce the same human placental lactogen hormone, J. Biol. Chem., 258:3787.

Blank, M. S., Chan, J. S. D., and Friesen, H. G., 1977, Placental lactogens; new developments, J. Steroid Biochem., 8:403.

Chan, J. S. D., Robertson, H. A., and Friesen, H. G., 1978, Distribution of binding sites for ovine placental lactogen in the sheep, Endocrinology, 102:632.

Colosi, P., Ogren, L., and Talamantes, F., 1986, RIA and gestational serum profile of mouse midpregnancy lactogen, in: Endocrine Society Abstracts 68th Annual Meeting, Anaheim, Abstract No. 194.
Duckworth, M. L., Kirk, K. L., and Friesen, H. G., 1986a, Isolation and identification of a cDNA clone of rat placental lactogen II, J. Biol. Chem., in press.
Duckworth, M. L., Peden, L. M., and Friesen, H. G., 1986b, Isolation of a novel prolactin-like cDNA clone from developing rat placenta, J. Biol. Chem., in press.
Duckworth, M. L., Peden, L. M., and Friesen, H. G., 1986c, The characterization of two closely related prolactin-like cDNA clones from rat placenta: One gene or two, in: Endocrine Society Abstracts 68th Annual Meeting, Anaheim, Abstract No. 739.
Forsyth, I. A., 1974, The comparative study of placental lactogenic hormones: a review in lactogenic hormones, in: "Fetal Nutrition & Lactation", J. B. Josimovich, M. Reynolds, E. Cobo, eds., Vol. 2, p. 49, John Wiley and Sons, New York.
Freemark, M., and Handwerger, S., 1986, The glycogenic effects of placental lactogen and growth hormone in ovine fetal liver are mediated through binding to specific fetal ovine placental lactogen receptors, Endocrinology, 118:613.
Golander, A., Hurley, T., Barrett, J., Hizi, A., Handwerger, S., 1978, Prolactin synthesis by human chorionic-decidual tissue: a possible source of prolactin in amniotic fluid, Science, 202:311.
Herz, A., Khan, J., Jayatilak, P. G., and Gibori, G., 1986, Evidence for the secretion of decidual luteotropin: a prolactin-like hormone produced by rat decidual cells, Endocrinology, 118:2203.
Jayatilak, P. G., Glaser, L. A., Basuray, R., Kelly, P. A., Gibari, G., 1985, Identification and partial characterization of a prolactin-like hormone produced by the rat decidual tissue, Proc. Natl. Acad. Sci., 82:217.
Josimovich, J. B., and MacLaren, J. A., 1962, Presence in the human placenta and term serum of a highly lactogenic substance immunologically related to pituitary growth hormone, Endocrinology, 71:202.
Kelly, P. A., Tsushima, T., Shiu, R. P. C., and Friesen, H. G., 1976, Lactogenic and growth hormone-like activities in pregnancy determined by radioreceptor assays, Endocrinology, 99:765.
Linzer, D. I. H., Lee, S. J., Ogren, L., Talamantes, F., and Nathans, D., 1985, Identification of proliferin mRNA and protein in mouse placenta, Proc. Natl. Acad. Sci., 82:4356.
Linzer, D. I. H., and Nathans, D., 1984, Nucleotide sequence of a growth-related mRNA encoding a member of the prolactin-growth hormone family, Proc. Natl. Acad. Sci., 81:4255.
Linzer, D. I. H., and Nathans, D., 1985, A new member of the prolactin-growth hormone gene family expressed in mouse placenta, EMBO J., 4:1419.
Miller, W. L., and Eberhardt, N. O., 1983, Structure and evolution of the growth hormone gene family, Endocrine Rev., 4:97.
Niall, H. D., Hogan, M. L., Tregear, G. W., Segre, G. V., Heuang, P., and Friesen, H. G., 1973, The chemistry of growth hormone and the lactogenic hormones, in: "Recent Progress in Hormone Research", E. B. Astwood, ed., 29:397, Academic Press, New York.
Parks, J. S., 1986, Organization and function of the growth hormone gene cluster in human growth hormone, in: "Human Growth Hormone", S. Raiti, R. A. Tolman, eds., p. 199, Plenum Book Company, New York.
Parks, J. S., Nielsen, P. V., Sexton, L. A., and Jorgensen, E. H., 1984, An effect of gene dosage on production of human chorionic somatomammotropin. J. Clin. Endocrinol. Metab., 60:994.
Robertson, M. C., Gillespie, B., and Friesen, H. G., 1982, Characterization of the two forms of rat placental lactogen (rPL): rPL-I and rPL-II, Endocrinology, 111:1862.

Robertson, M. C., Owens, R. E., Klindt, J., and Friesen, H. G., 1984a, Ovariectomy leads to a rapid increase in rat placental lactogen secretion, Endocrinology, 114:1805.
Robertson, M. C., Owens, R. E., McCoshen, J. A., and Friesen, H. G., 1984b, Ovarian factors inhibit and fetal factors stimulate the secretion of rat placental lactogen, Endocrinology, 114:22.
Shuler, L. A., and Hurley, W. L., 1985, Evolution of bovine placental lactogen includes insertion 5' to the second exon, in: Endocrine Society Abstracts of 67th Annual Meeting, Baltimore, Abstract No. 528.
Shiu, R. P. C., Kelly, P. A., and Friesen, H. G., 1973, Radioreceptor assay for prolactin and other lactogenic hormones, Science, 180:968.
Soares, M. J., Colosi, P., and Talamantes, F., 1983, Identification and partial characterization of a lactogen from the midpregnant mouse conceptus, Endocrinology, 112:1313.
Talamantes, F., Ogren, L., Markoff, E., Woodward, S., and Madrid, J., 1980, The phylogenetic distribution, regulation of secretion and the prolactin like effects of placental lactogens, Fed. Proc., 39:2582.

HORMONAL INFLUENCES ON FETAL AND PERINATAL WATER METABOLISM

Anthony M. Perks and Sidney Cassin

Department of Zoology
University of British Columbia
Vancouver, B.C., Canada, and
Departments of Physiology and Pediatrics
University of Florida
Gainesville, Florida, U.S.A.

INTRODUCTION

Interest in the fluids and membranes which surround the fetus has persisted for many centuries. Hippocrates was probably the first to recognize that amniotic fluid was mainly fetal urine (Reynolds, 1972). However, Aristotle believed that the outer chorion contained residual seminal fluid (Harvey, 1651). By the time of Fabricius, opinions had changed, and to quote Harvey: "-the fluid within the amnion, wherein the foetus swims, consists of sweat; and that within the chorion of urine."

With his usual clarity, William Harvey (1651) quarreled with these ideas. Firstly, he noticed a "limpid and pure watery fluid" within the early amnion, long before there was a recognizable fetus, much less skin or kidneys. Secondly, he saw fluid present in "-unfruitful conceptions where nothing like a fetus was discoverable." In addition, Harvey pointed out that many animals, such as dogs and cats, never sweat, and he believed that "birds, serpents and fishes" did not sweat, or make urine.

Harvey was able to obtain fetal urine by pressure on the bladder. However, he was unwilling to accept the idea that the fetus should "-float about in its own excrement." He believed that wastes arose secondarily from materials left behind after the baby had extracted nutritional substances from the fluid, "-just as wine becomes poor and tasteless when the spirit has evaporated." Harvey believed that nutrition was the main use of the external fluids. He writes: "-the fluid of the amnion is usually consumed by the time of birth; so that it is probable the fetus seeks its exist on account of deficiency of nutriment." However he also recognized other functions: "(the fluids exist)-for the purpose, no doubt, of protecting the body of the foetus from coming in contact with the adjacent parts when the mother runs, jumps, or uses violent exertion-" and he thought the fluids facilitated birth.

Today, we recognize Harvey's concept of protection. The amniotic fluid supports and protects the fetus from abrasion, allows it to move in relative weightlessness, and provides thermal insulation (Reynolds, 1972). Centuries later, there has been a rediscovery of Harvey's observation that fluid precedes a recognizable fetus, as tacit support for secretory activity by the amniotic membrane (Adolph, 1967). However, Harvey's concept of nutrition is mostly lost, except in one sense; it is possible that the fluid forms a buffer for water and ions, which pass into the sac when in excess, but are taken back, perhaps by fetal drinking, when necessary (Seeds, 1965; Cassin and Perks, 1982).

Today, we regard the amniotic fluid as part of a complex and dynamic system. Over the long term, it shows changes in volume and tonicity. In most species, the volume increases throughout most of pregnancy, although later deviations suggest possible uses and the presence of controls; close to birth, the volume usually declines (chick, Adolph, 1967; cow, horse, pig, cat, Arthur, 1969; sheep, Barnes, 1976; guinea pig, Pisani and Perks, unpublished observations; human, Elliot and Inman, 1961; Kerpel-Fronius, 1970). Until midterm, the fluid remains isotonic and resembles a dialysate of plasma (Seeds, 1965; Lind, 1969; Lind and Hytten, 1972). However, around the time the skin keratinizes, the composition of the fluid becomes more independent of the fetus, and progressively more hypotonic; this recalls the transition from "salt" to "fresh" water during evolution (Seeds, 1965; Lind, 1969; Lind and Hytten, 1972; Perks et al., 1978).

In the short term, the amniotic fluid undergoes a rapid circulation and exchange. In the guinea pig it is renewed every hour (Flexner and Gellhorn, 1942), and in the human every three hours (Vosburgh et al., 1948). In late gestation, it receives hypotonic fetal urine at a high rate (sheep, 200-500 ml/day, Barnes, 1976), and comparisons have been made with adults with diabetes insipidus (Seeds, 1965). However, even in renal agenesis, considerable fluid may be found; usually the volume is reduced, but cases of polyhydramnios have also been reported (Jeffcoate and Scott, 1959; Bain and Scott, 1960; Wagner and Tygstrup, 1963). Part of this fluid is probably from the lungs, which, near term, secrete at a rate almost equal to production by the kidneys (goat, 320 ml/day; sheep, 295 ml/day; Perks and Cassin, 1985a, b). In addition, there may be minor contributions from the salivary glands, and other sources (Arthur, 1969; Reynolds, 1972). In terms of bulk flow, the possible contributions of the fetal skin and amnion are still uncertain (Bourne, 1962; Kerpel-Fronius, 1970; Seeds, 1972; Barnes, 1976). However, the amnion is clearly permeable to water (Seeds, 1965) and an active secretion has been suggested many times. Evidence comes from the presence of fluid in the early amnion, as noted by Harvey (1651) and Adolph (1967), and from the histology and ultrastructure of the membrane (Danforth and Hull, 1958; Bourne, 1962; cf. Diamond, 1971; see Seeds, 1965).

In terms of bulk flow, the fluid which enters the amniotic cavity forms part of a circulation; the fetus drinks back fluid at about the same rate as urine is produced (500 ml/day; sheep, Bradley and Mistretta, 1973; human, Jeffcoate and Scott, 1959).

Isotopes, which demonstrate diffusional movements of water rather than bulk flow, also indicate a circulation of water around and through the fetus. Water passes rapidly between the amniotic cavity, mother and fetus, in all possible directions (Hutchinson et al., 1959; Reynolds, 1972; Seeds, 1972). However, the net fluxes suggest a general circulation along the route mother-fetus-amniotic fluid-mother. Although the exact pathways are not clear, the evidence for a large direct route between the amniotic fluid and the mother could implicate the amnion, which surrounds the fluid, and covers the placental disc and umbilical cord (Bourne, 1962; Reynolds, 1972).

This complex system can malfunction. Both polyhydramnios and oligohydramnios can indicate fetal anomalies (Benirschke and McKay, 1953; Jeffcoate and Scott, 1959). Lack of fluid can also cause fetal damage (Arthur, 1969), and complete loss around midterm can lead to fetal death (Watson, 1906).

Despite the importance of understanding these problems, the mechanisms which accumulate fluid, and those which both control the system and change it fundamentally at birth, are little understood (Adolph, 1967; Kerpel-Fronius, 1970; Barnes and Seeds, 1972). However, recent studies in our laboratories have raised a number of interesting possibilities. These were based on work on the fetal neurohypophysis, and the purpose of this review is to bring together these studies with our later work on the possible functioning of the system.

THE FETAL NEUROHYPOPHYSIS

There is increasing evidence that the fetal neurohypophysis is well developed, and probably functional through much of fetal life. The system develops early. In the human, the sac-like infundibular process contacts Rathke's pouch at 4 weeks of gestation, and forms a solid gland by 12 weeks (Benirschke and McKay, 1953). Characteristic Gomori-positive neurosecretion is seen first in the hypothalamic nuclei at 19-20 weeks, and later in the pituitary at 23 weeks. After this, the secretion becomes denser around the blood vessels, and Herring bodies are formed, although they are smaller than in the adult (Benirschke and McKay, 1953). Parallel changes are found in many other mammals, and, except in the rat, the system appears structurally mature at birth (Yakovleva, 1965).

Despite the structural integrity of the fetal system, early studies showed distinct differences between the hormones of the fetus and the adult. Firstly, there was an unusual predominance of vasopressor over oxytocic activity in the neural lobe of fetal cats, dogs and humans, and this was also found in some newborns (cats, dogs, rats and guinea pigs) (Dicker and Tyler, 1953a, b; Acher and Fromageot, 1957; Levina, 1968). The V/O ratio (vasopressor/oxytocic activity) in midterm human fetuses was so high (28/1) that it was difficult to explain in terms of AVP alone (Dicker and Tyler, 1953a). Secondly, it was suggested that these ratios were only artifacts, caused by an unusual solubility of the oxytocic factor in the acetone used to dry the glands (Heller and Lederis, 1959). Either of these characteristics could have indicated the presence of neurohypophysial principles different from those of the adult.

Accordingly, from 1965 onwards a series of pharmacological and chemical studies were carried out in our laboratory. These used lyophilised neural lobes, where possible extraction by acetone was avoided. Studies were carried out on glands from sheep, seals, pigs and guinea pigs, since these could give rapidly dissected material through many stages of gestation. Previous work had not followed changes sequentially through development, except in acetone-dried tissues from the human, where irregular changes in the V/O ratios suggested that postmortem degeneration could have affected the results (Dicker and Tyler, 1953a; Levina, 1968). The glands were extracted in 0.25% acetic acid, and assayed for oxytocic, vasopressor and antidiuretic activities (isolated rat uterus, without magnesium, Holton, 1948; Munsick, 1960; rat vasopressor, Sawyer, 1961 (synthetic lysine vasopressin was used as standard for pigs only); ethanol-anesthetized rat antidiuretic assay, Sawyer, 1961). The main conclusions from this study are reviewed below:

Changes in Biological Activities in the Neural Lobe through Gestation

These studies showed that the high V/O ratios in the fetus were not artifacts, but reflected a predominance of AVP in the pituitary during fetal life.

Vasopressor and oxytocic activities were detected in all fetal pituitaries examined, even in the youngest yet studied, the fetal seal (*Callorhinus ursinus*) at 0.19 of term (V = 115.7, O = 30.3 mU/mg dry powder;

2.2 and 0.7% of adult levels, respectively) (Perks and Vizsolyi, 1973; Vizsolyi and Perks, 1976a). As expected, activities increased steadily with age in all four species, but vasopressor activity always predominated. After midterm, vasopressor activities, expressed as percentages of adult values, rose from 8 to 26% in the guinea pig, 8 to 82% in the pig, 23 to 52% in the seal, and 52 to 105% in the sheep (periods represented: guinea pig, 0.77-0.92 of term; term = 65 days; pig, 0.57-0.89 of term, term = 114 days; seal, 0.56-0.93 of term, term = 240 days, implantation to delivery; sheep, 0.59-0.96 of term, term = 147 days). In the sheep, these results were confirmed by antidiuretic assays. Corresponding oxytocic activities were always lower. There was a rise of 2 to 7% in the guinea pig, 2 to 30% in the pig, 4 to 35% in the seal, and 3 to 10% in the sheep (for detailed values, see: sheep, Vizsolyi and Perks, 1976a; other species, Perks and Vizsolyi, 1973).

These results showed that the vasopressin-producing system was well developed in fetal life, with most of the storage of oxytocin taking place after birth. Surprisingly, there was a relatively slow maturation in the guinea pig, a species well known for its maturity at birth. This was also found by Burton and Forsling (1972), and it may be related to the preponderance of AVP retained in the adult gland (adult V/O ratio: 2.9, Burton and Forsling, 1972; 3.4, Perks and Vizsolyi, 1973). In contrast, the fetal sheep was notably mature at birth, at least in terms of AVP.

In the sheep, results over the period of birth suggested a powerful depletion of the fetal pituitary over parturition (Vizsolyi and Perks, 1976a). The average vasopressor activity in the neural lobe of the late-term fetuses (0.96 of term) was 1721 ± 217 mU/mg dry tissue, slightly above the adult value (1645 ± 151 mU/mg). However, over the perinatal period this fell by 54%, to reach 801 ± 79.5 mU/mg in 12 day newborns. This fall was confirmed in acetone-dried tissues. When these results were combined with those for a fetus in early natural labor, and also for one presenting, the fall appeared to have taken place between early labor and delivery. These results agree with later immunological assays of plasma AVP by other groups, and suggest a large depletion of the fetal gland during birth (sheep, Starke et al., 1977, 1979; human, Pohjavuori and Fyhrquist, 1980). This change was important in our later studies of the lungs.

In all species investigated, the V/O ratios were higher in the fetus than the adult, despite the absence of acetone (Perks and Vizsolyi, 1973; Vizsolyi and Perks, 1976a). This was particularly striking in the guinea pig and sheep. The maximum individual value of 20.8 was found in a fetal sheep at 0.61 of term. In the seal, the only species in which early tissues could be obtained, the ratio remained stable, close to 4, until midterm. After midterm, all species showed a steady decline in values, although they remained high in newborn guinea pigs and sheep. The change is shown clearly in the sheep, where average values at midterm, late term, term, in the neonate and adult were 16.3, 10.7, 5.4, 5.2 and 0.93, respectively. Direct comparisons with acetone dried glands showed closely similar results, except the overall values in acetone-dried glands were slightly lower. Clearly, the high V/O ratios in the fetus were not artifacts due to acetone, but reflections of a predominance of AVP in fetal life.

Histological studies showed that the probable explanation of the predominance of AVP was early maturation of the supraoptic (SO) nucleus in the hypothalamus (Perks and Vizsolyi, 1973). In the adult, this nucleus is thought to produce mainly AVP, while oxytocin-producing neurones are mainly in the paraventricular (PV) nucleus (Heller, 1961, 1966). Histological studies in fetal seals showed no stainable neurosecretion in either nucleus at 0.3 of term (aldehyde fuchsin and Alcian blue-Schiff's orange techniques). However, by 0.5 of term, granules were seen in 52% of the cells of the SO nucleus, but in only 14% of those in the PV nucleus. In the neonate

(1-3 days pn), granules were present in 92% of the SO cells, but in only 85% of the PV neurones. Estimates of the total amount of neurosecretory material in the nuclei at different ages showed close parallels between that in the SO nucleus and the vasopressor activity in the neural lobe, and between that in the PV nucleus and the oxytocic activity of the pituitary.

The lack of stainable granules in the nuclei of the early fetus (0.3 of term), at a time when biological activities were present in the pituitary, is not altogether surprising, since this has been seen by other authors (sheep, Alexander et al., 1973). It may reflect different sensitivities of the methods. However, Levina (1968) has commented that "oxytocin activity in the human fetal hypophysis may be associated with its transitory disappearance from the hypothalmus", so that it is possible that the early nuclei send intermittent pulses of material to the neural lobe.

These results suggest that the supraoptic nucleus, with its production of AVP, becomes active earlier than the paraventricular nucleus. The result is a predominance of AVP over oxytocin in fetal life. It is reasonable to wonder whether this reflects greater functional importance of AVP during life _in utero_.

The Nature of the Fetal Neurohypophysial Principles

The unusual biological activities of the fetal neural lobes could be explained by differential production of AVP and oxytocin through development. However, there still remained the possibility that the ratios reflected the presence of new neurohypophysial peptides. Therefore, chromatographic studies were started to establish the identity of the fetal principles (Vizsolyi and Perks, 1969, 1976b; Perks and Vizsolyi, 1973; Perks, 1977). This work showed that midterm fetal pituitaries contained an extra principle, arginine vasotocin, AVT.

Studies at Midterm. Initial work used paper chromatography to separate the principles of fetal sheep and seals, mainly to determine whether any oxytocin was presnt at that stage (Vizsolyi and Perks, 1969, 1976b; Perks and Vizsolyi, 1973). The youngest fetuses investigated chromatographically were fetal seals just before midterm (0.39 - 0.43 of term). Extracts from 13 neural lobes were chromatographed on Whatman 3 MM paper, using butanol: acetic acid: water, 4:1:5. There were two main results: 1) small amounts of oxytocin appeared to be present, since a fast-running peak corresponded with control synthetic oxytocin; 2) the slow-running peak at Rf 0.2 - 0.3 showed unusual biological activities. It corresponded in position to AVP, but contained 15.2 times too much oxytocic activity to be AVP alone. The presence of a basic peptide with relatively high oxytocic activity suggested the presence of AVT, a principle formerly thought to be confined to submammalian vertebrates. This principle could be detected by its enormous ability in promoting water reabsorption by the frog bladder; therefore the eluates were assayed on this preparation (Sawyer, 1960). The assay gave a value 587 times too great for AVP. Closely similar results were found for slightly older sheep fetuses (0.55 - 0.62 of term). Again, the slow-running vasopressor peak showed oxytocic and frog-bladder activities 10.3 and 452 times too great, respectively, to be AVP alone. This suggested the presence of AVT in slightly lower amounts than in the fetal seals.

More conclusive evidence for AVT needed purification and chemical analysis (for details, see Perks and Vizsolyi, 1973; Perks, 1977). Accordingly, studies turned to the fetal seal, since the adult neural lobe showed activities higher than in any mammal studied, perhaps related to the marine environment (5260 ± 536 mU/mg vasopressor activity; 4085 ± 543 mU/mg oxytocic activity). Extracts from 42 lyophilised neural lobes from midterm fetal seals (0.56 - 0.68 of term) were passed through a large (200 × 2.5 cm), slowly built G-15 Sephadex column. Three peaks of activity were eluted. The first peak, of oxytocic activity, was further purified on CM-Sephadex, and gave a single peak with the biological activities and amino acid analysis of oxytocin. The last peak of vasopressor activity, was purified on phosphocellulose, and the single peak eluted gave the amino acid analyses of AVP. The middle peak showed high frog-bladder activity. It was further purified on an IRC-50 column, and yielded a single peak which showed pharmacological properties in exact parallel with AVT, when assayed directly against the synthetic peptide by different methods. The amino acid analysis was that of AVT (Asp, 1.03; Glu, 1.00; Pro, 1.00; Gly, 1.20; 1/2 Cys, Cys acid, 1.95; Ileu, 0.69; Tyr, 0.43; Arg, 0.80).

Studies near to Term. Studies of near-term sheep fetuses showed that AVT was almost entirely lost from the pituitary.

Extracts from 8 neural lobes from fetal sheep were passed through a specially-built G-15 Sephadex column, made by a method which had been shown to give a remarkable purification of adult sheep material (Vizsolyi and Perks, 1976b). The column gave a low oxytocic peak which appeared to consist of oxytocin itself. The peak eluted in the position of oxytocin, showed no potentiation with magnesium (method of Munsick, 1960), and gave the amino acid analyses of oxytocin. There was also a slower vasopressor peak. This peak gave the amino acid analysis of AVP alone. However, a trace of AVT may have remained, since one fraction (one out of eight within the vasopressor peak), showed twice the oxytocic and 80 times the frog-bladder activity of AVP. Nevertheless, this represented almost complete disappearance of AVT by late gestation.

Chemical purification was not attempted in late fetal seals, but bioassay data suggested that little or no AVT remained in their pituitaries.

Studies in the Adult. An extract from 16 neural lobes from adult sheep were purified by the same G-15 Sephadex system used for late-term fetal sheep. Remarkably pure preparations were obtained (Vizsolyi and Perks, 1976b). Chromatographic mobilities and amino acid analyses suggested the presence of oxytocin and AVP. There was no excess oxytocic activity in the AVP peak, and no evidence for AVT in the adult glands. Similar studies in adult seal glands gave the same results.

Changes in AVT through Fetal Life. Once the presence of AVT had been established by chemical studies, the frog-bladder assay was used to follow changes in its level in the neural lobe through fetal development. The most comprehensive study was in the seal. In early gestation, frog-bladder activity due to AVT predominated over all other activities (FB = 500, V = 115.7, O = 30.3 mU/mg dry powder; 0.19 of term). It rose to 4500 mU/mg just after midterm (0.68 of term), then declined towards birth. Levels were low just after 0.70 of term, and close to birth almost all the frog-bladder activity could be accounted for by the intrinsic activities of the oxytocin and AVP present. Later, the virtual absence of AVT at birth was confirmed by immunological assays (Dogterom et al., 1980). However, this result must be taken with reserve, since the seal tissues had undergone prolonged storage before immunoassay.

Studies in the sheep gave similar results. Calculations based on the purification data showed that frog-bladder activity again predominated at midterm (FB = 12259, V = 616, O = 30 mU/mg dry powder; 0.55 - 0.62 of term). Again, there was a fall before late gestation, but the exact timing was not clear. By 0.94 - 0.96 of term no more than a trace of AVT could be detected.

Two species differed from the sheep and seal in opposite ways. The fetal guinea pig seemed to retain AVT into late gestation (FB/O ratio = 13.4, at 0.92 of term). This may reflect the slow maturation of the system, pointed out earlier. Despite this, the excess frog-bladder activity had fallen greatly by birth, and there was no clear evidence for AVT in the adult. The pig showed the opposite situation. There was no evidence for AVT at any stage of development (0.57 of term - adult). This is probably related to the production of lysine vasopressin (LVP) in the adult gland.

Discussion. The work presented here gave the first evidence for AVT in the mammalian pituitary. It suggested the recapitulation of the basic principle of lower vertebrates (Haekel, 1866). Later, confirmation of its presence came from biological and immunological studies in other laboratories.

Biological studies, depending mainly on frog-bladder activity, have supported the presence of AVT, particularly in early fetal pituitaries. In unpublished observations, Pickering found evidence for AVT in fetal sheep. Pavel (1975) examined the chromatographic mobility, enzymic susceptibility, and frog-bladder activity of principles in early human neurohypophyses. His important conclusion was that AVT was the only principle present in early fetal life (0.20 - 0.23 of term; term = 280 days). Later, at 0.46 - 0.55 of term, the pituitaries appeared to contain oxytocin, AVP and AVT. However, after prolonged tissue culture, only AVT entered the outer medium, and this led Pavel to suggest that AVT was produced by the ependyma.

Although the biological criteria have often proved reliable in identifying peptides, there must be reservation until chemical structure has been established. This is important here, since the pineal has been shown to contain a larger AVT-like molecule, with frog-bladder activity (Neacsu, 1972). However, the amino acid analysis of the seal principle, in our laboratory, suggests that the peptide detected in the human fetal pituitaries was probably AVT.

Recent rapid progress in immunological methods has produced further confirmation of the presence of AVT. Skowsky and Fisher (1977) studied human and sheep pituitaries by a double antibody method. This compared results from an antibody specific to AVP with those from antiserum which reacted to both AVP and AVT. They confirmed Pavel's (1975) finding of AVT in human fetal pituitaries before midterm (0.28 - 0.48 of term). Again, AVT predominated over AVP, although the absolute values for the peptides were low compared to those given by other workers, and there were problems in allowing for the oxytocin present. Despite some irregular values, perhaps due to postmortem changes or to variable depletions from the stress of abortion, the percentage of AVP tended to rise towards midterm. Late-term fetal sheep pituitaries (0.74 - 0.93 of term) also contained distinct levels of AVT, although they were low compared with AVP (ratio AVP/AVT = 7.1, for average values in mU/mg wet weight). However, Alexander et al. (1973), using a similar double antibody method, found no more than a slight excess of activity due to AVT in fetal sheep at 0.91 and 0.95 of term. This agress with our conclusions in near-term sheep and seals, and with the immunological estimates on our late seal material by Dogterom et al. (1980). The apparent differences between different workers are probably due to differences in gestational age. In addition, there may be variations in the precise timing of the late decline in AVT in different individuals, or in different varieties of sheep. Possible support for individual variations may be found in tables shown by Skowsky and Fisher (1977).

Perhaps the best immunological evidence for AVT comes from work on rat fetuses by Lederis et al. (1980). A single antibody with high specificity for AVT showed large quantities present as early as 0.38 of term (term = 21 days). There was a pronounced decline in the final stages of gestation, but AVT persisted, and even increased in the first days after birth. In assessing these results, it must be remembered that the rat is born remarkably immature, and this fact extends to the neurohypophysis (Yakovleva, 1965). Therefore, the more rapid decline in AVT close to birth may correspond to that seen after midgestation in other species. In addition, these workers also concluded that the AVP-producing neurons "matured" before those producing oxytocin.

The presence of AVT in the fetal pituitary raised three important points: (1) It showed that we cannot assume that the hormones of the fetus are the same as those of the adult; (2) it raised the concept of recapitulation of a "lower vertebrate" hormone in fetal life; and (3) it seemed possible that this apparent recapitulation was not a developmental oddity, but important in fetal function.

The work reported here was not designed to investigate function, but to explain unusual characteristics of the fetal neurohypophysis and to establish the nature of its principles. Studies of function would have been better served by estimating blood levels, although plasma assays were not well developed when this work was started. However, the studies of the pituitary showed that possible functions of AVT must be considered. In addition, they suggested that an early dominance of AVT was superceded by that of AVP in later fetal life, and especially around birth. Therefore, further work turned to possible functions of these peptides in fetal and perinatal life.

THE ACTIONS OF AVT AND AVP ON FETAL TISSUES

In searching for a function for AVT in fetal life, help came from studies of lower vertebrates. The most notable effect of AVT was a powerful stimulation of the movement of water through the skin and bladder of the frog (Brunn, 1921; Sawyer, 1960). Since AVT appeared to promote the passage of water through membranes, but was only present during fetal life, it seemed possible that it might act on the passage of water through those membranes which were limited to fetal life - for example, the amniotic membrane. Accordingly, the first studies were carried out on the isolated amniotic membrane of the guinea pig.

Studies of the Amniotic Membrane

The Effects of AVT. Thirty-six amniotic membranes from midterm guinea pigs (0.46 of term) were set up *in vitro*, in salines which reproduced the electrolytes of the natural fluids, and which maintained small osmotic and hydrostatic gradients from amniotic to maternal solutions. The movement of water was followed gravimetrically, and the bulk flow was recorded on a continuous basis (Vizsolyi and Perks, 1974).

Under the conditions of the experiment, water usually passed slowly from the fetal to the maternal side (average rate, 3.93 $mg/cm^2.min$). However, when synthetic AVT at 8-100 mU/ml of amniotic saline was placed on the fetal surface, the flow of water slowed ($n = 28$) and usually reversed ($n = 15$), so that in 21 tests there was a net uptake into the potential amniotic cavity, against prevailing gradients (maximum reversed flow, 10.4 $mg/cm^2.min$). The linear relationship between the log-dose and the response indicated a threshold of 6.4 mU/ml, which was considered relatively high. Controls, and tests with oxytocin (100 mU/ml), showed no response. These were the first physiological studies to show that the amnion could move fluid by an active process and that it could respond to hormones (Perks and Vizsolyi, 1973; Vizsolyi and Perks, 1974).

The Effects of AVP. Ten tests with synthetic AVP (2 - 100 mU/ml) showed closely similar effects, but the threshold dose was lower (approx.

1.0 mU/ml). This suggested that AVP, not AVT, was more likely to be of potential importance to the amnion. Therefore all later work on the amnion concentrated on AVP.

It seemed necessary to check the results by an independent method. Therefore the studies were repeated using measurements of diffusional flux (Holt and Perks, 1977). In 18 experiments, amniotic membranes from guinea pigs between 0.62 of term and term (some overdue) were set up in an Ussing-type, continuous flow, double-perfusion cell, which allowed the membrane to be bathed by the same amniotic and maternal salines used in the bulk flow experiments. The continuous flow system reduced problems due to dead spaces and unstirred layers. Addition of partially purified AVP/LVP to the fetal surface of the membrane (50 - 500 mU/ml) increased the unidirectional flux of tritiated water by up to 12.3 ± 1.5%, in the maternal-fetal direction. The reverse flux only increased 1.6%. This suggested a net uptake of water into the amniotic compartment. The responses showed a similar latency to those seen in bulk-flow studies (5-10 min.), but, in contrast, they seemed to occur in a series of 'spikes'. When quantitated, they showed a linear relationship with the log-dose of vasopressin, but the threshold indicated was high (46 mU/ml). This was not unexpected, since AVP is known to have little effect on diffusional flux. Maetz (1968) has shown that these principles can increase water movement across amphibian skin by 160%, yet have no clear effect on unidirectional flux. Therefore, it was probably surprising to see any responses by this method. In consequence, the technique was abandoned in further studies, in favor of the bulk-flow system. However, it had confirmed the basic observations.

The perfusion cell served a second purpose. It showed, as expected, that the movement of water against gradients was probably linked to ions, since AVP produced a closely similar movement of $^{22}Na^{+}$ ions towards the amniotic saline (9 tests; average increase in flux, 13.6 ± 6.3%).

Although the studies gave reasonable log-dose/response relationships for AVP, there was considerable variability between different membranes. It was possible that this was due to age. Therefore, the maximum capacity of the amnion to move water in response to AVP was studied at different gestational ages (Pisani and Perks, unpublished observations). Thirty-nine amniotic membranes from guinea pigs between 0.44 of term and term were studied by the in vitro, gravimetric method. Synthetic AVP was added to the fetal surface at 100 mU/ml, to produce maximal responses, with recovery. In general, responses increased 10-fold between 0.45 and 0.85 of term. After this, the effect declined rapidly, and was effectively lost close to term (one weak response from 13 membranes, at 0.94 of term). Controls never showed any responses.

These changes in response seemed to reflect changes in the structure of the membrane. Ten membranes, divided into three age groups, were examined by transmission electron microscopy (Pisani and Perks, unpublished observations). Early membranes (0.41 - 0.56 of term), where small responses to AVP were expected, showed characteristics of an epithelium capable of moderate transport activity. The cells were thin and flattened, with no overlap, but with many intracellular organelles and large nuclei. More important, the apical surfaces, which contacted the amniotic fluid showed microvilli, and there were distinct, if narrow, intercellular spaces. Mature membranes (0.74 - 0.91 of term), where responses to AVP were large, had a complex structure. Their overlapping, cuboidal cells retained the many organelles and large nuclei, but had doubled in thickness. Three changes were striking; 1) there was a prominent surface coating (glycocalyx) on the apex of the cells, similar to that seen in the gallbladder and some other transporting epithelia (Fawcett, 1965, 1981); 2) the apical microvilli had become abundant and formed elaborate networks; and 3) the intercellular spaces had become wider and convoluted, with complex intrusions of microvilli. Clearly, the various microvilli could increase surface area for transport. In contrast, the near-term group of membranes (0.94 of term - term, and one overdue), which failed to respond to AVP, showed widespread degeneration. The cells were thin, with almost complete loss of cytoplasmic organelles, and flattened nuclei. The cytoplasm was full of filaments, and there were similarities to keratinizing skin (Parakkal and Alexander, 1972; Montagna and Parakkal, 1974). Apical microvilli were short, thick, and scarce, and the intrusions into the lateral spaces were lost. The spaces themselves had become wide, with open connections to the connective layer below. This may be related to the sudden increase in permeability to Na^+ and Cl^- seen at that time (Holt and Perks, 1977b; Goh and Perks, unpublished observations). This dramatic degeneration was sudden, between 0.91 and 0.94 of term (62-64 days of gestation). This agreed well with the disappearance of the response to AVP. Structural changes of this type have been seen before by King (1978) although the exact timing was not clear. They occur also in other species, but over different time frames. In the cat, the whole epithelium sloughs off a few days before birth (Tiedmann, 1979), In the human, there are increasing numbers of degenerating cells towards term, but uniform degeneration is seen only in overdue fetuses (Hoyes, 1975).

The electron microscope was also used to show the effects of AVP upon the membrane at different stages of development. Studies were carried out on four amniotic membranes, ranging from midterm to overdue. In the early group (0.56 of term), the amnion was divided into three sections. The first was supported in the gravimetric apparatus used for the physiological studies, and treated for 15 min with AVP at 100 mU/ml; this would produce a maximal response in the same conditions used for studies of bulk flow. The second was treated similarly, but received only control AVP-carrier solution. The third section was studied directly, without incubation. Electron microscopy showed that the AVP-treated membrane responded with a dramatic dilation of the intercellular spaces. Their areas increased 260% over those of the control incubated membrane (change significant, $P < 0.001$; Kontron MOP Videoplan Image Analyser). The spaces in the control incubated membrane were only slightly larger than in the unincubated control. No other effects of AVP could be seen. The same effect was found in a mature membrane, at 0.76 of term, treated with AVP. An incubated control of the same age group (0.8 of term) showed no dilations. However, AVP had no observable effect on the lateral spaces of an overdue membrane, which showed the usual degenerative changes.

The changes in the response to AVP, and the structure of the membrane appeared to correlate with changes in the volume of amniotic fluid. The volume was estimated by direct collection in 76 guinea pigs from 0.44 of term to term. Over the period when the response to AVP was increasing, the volume rose steadily by almost 4-fold (approx. 0.47 to 0.82 of term). It then declined, and at the time the response to AVP was lost, reached its lowest point (0.94 of term). After this, when the membrane was permeable and degenerate, the volume rose again, but with wide individual variations. These results suggest a possible physiological relationship between the response of the amnion to AVP and amniotic volume. However, the evidence is only circumstantial, and the changes may not be in any way direct, but relate to an underlying common cause.

The Effects of Fetal Pituitary Extracts. In five experiments, extracts from neural lobes of fetal seals at 0.44 of term were tested on isolated amniotic membranes from guinea pigs at 0.46 of term, by the gravimetric method (extracts: 0.012 - 0.02 ml, extracted at 3 mg/ml in 0.25% acetic acid; neutralized with 2N NaOH). Powerful responses were produced (average change in flow towards the amniotic saline, +9.11 $mg/cm^2 \cdot min$). These were greater than could be accounted for by the biological activities present (final concentrations, FB:V:O = 54:20:5 mU/ml). This raised the possibility of synergism between the peptides, or the presence of further factors.

The Effects of Prolactin. The apparent - if fortuitous - success of phylogenetic reasoning in suggesting that AVT might affect the amnion, also suggested that prolactin should be studied. Prolactin is not only known for its effects on reproduction, but also for those on osmoregulation in lower vertebrates. In fish, it is the principal hormone concerned in freshwater adaptation (Hirano, 1977). In general, it decreases osmotic permeability of membranes, and increases sodium transport in the direction most needed in a particular species (Hirano, 1977). In addition, large amounts of prolactin are found in amniotic fluid (Friesen et al., 1972; Parke, 1973; Josimovich et al., 1974). Accordingly, the effects of prolactin were tested on the amniotic membrane.

Prolactin was able to reduce the passage of water through the amnion, in the fetal-maternal direction (Holt and Perks, 1975). Ten experiments were carried out by the gravimetric method, on guinea pig amnia (0.92 - 1.00 of term; one overdue). In 5 experiments, ovine prolactin (20 μg/ml, amniotic saline) caused small increases in the fetal-maternal flow in the first hour, followed by profound and significant falls, which averaged almost 60% reduction by 3 h. One overdue membrane showed an unusually large reduction. Five control membranes showed no such changes.

These results were checked by measurements of unidirectional fluxes of ^3H_2O in the double-circulation cell used for AVP. In this case, amniotic saline was used on both sides of the membranes, and the work included younger amnia, from 0.62 of term to term, with one overdue. In 8 experiments, ovine prolactin at 10 μg/ml (fetal surface) caused a fall in fetal-maternal flux which reached 9.7 - 17.0% by 3 h. Younger membranes reacted in the same way as near-term preparations, and, again, one overdue amnion showed an unusually large reduction. In contrast, 5 control experiments showed small increases in flux, which averaged 5.1% by 3 h. Therefore, the probable effect of prolactin was to reduce fetal-maternal flux by 15-22%.

Later studies with the double-circulation cell showed that prolactin also caused a net movement of sodium outwards, in the fetal-maternal direction (Holt and Perks, 1977b). Studies were carried out on 24 amniotic membranes between 0.47 and 0.87 of term. Ovine prolactin, at 10 μg/ml (fetal surface) increased unidirectional flux of $^{22}Na^+$ by 53.6 ± 10.1% in the fetal-maternal direction by 3 h. One extra membrane, which was overdue, failed to respond. There were no effects on the reverse flux of sodium, and no responses from controls.

Therefore, it seems clear that prolactin is also capable of affecting the amnion, and the amniotic fluid.

Discussion. Our original observation that AVP could cause the amnion to move fluid into the potential amniotic cavity was confirmed directly by Manku et al. (1975). The effect was not merely a rise in permeability, since movement was against osmotic and hydrostatic gradients. The exact mechanism is still not known. The studies of sodium suggested that movements of ions were involved, but these could involve Cl^- ions as readily as sodium. In fact, Mellor and Slater (1971) have suggested the possibility of a Cl^- pump in the amnion. The remarkable dilations of the lateral spaces produced by AVP suggest that these structures are involved, but their exact role is not clear. Similar dilations have been seen during maximal transport in the gallbladder of the rabbit (Tormey and Diamond, 1967), and in *Necturus* (Spring and Hope, 1978, 1979). If the effects should prove to be physiological, the increase in response with age might reflect the increase in volume of the fluid, either directly, or because the fall of surface area/volume during growth would require greater activity of the amnion to produce the same net change in the composition of the fluid. The failure of any responses, structural or physiological, in the amnion at term are in agreement with the brief comments of Seeds (1967) and Leontic and Tyson (1977). These authors found that vasopressin had no effect on water movement through the human amnion at term (although Leontic and Tyson used the insensitive method of diffusional flux). The apparent degeneration of the guinea pig membrane close to birth probably relates to the rise in permeability to ions and nonelectrolytes seen in our experiments and by North and Segal (1976). The changes near to term may reflect the reduced need for controlling mechanisms at a time when the fetus is less vulnerable to loss of amniotic fluid (Adolph, 1967). They could also be linked to the mechanisms of parturition.

Although these studies shifted from AVT to AVP, because AVP was more potent on older amnia, more studies are needed before we can discount an importance for AVT in early fetal life.

After our initial demonstration that prolactin could reduce water movement out of the amniotic space, the first workers to look for similar effects were Manku et al. (1975). Although they used closely similar methods, a discrepancy resulted. Manku et al. found an increase in fetal-maternal flow. However, the disagreement was not real. Manku et al. incubated with prolactin for only one hour, and failed to see the later fall, which became profound in the third hour. Our original experiments also show an early increase in water movement, but this was not significant (Holt and Perks, 1975). Recent additional experiments have resulted in statistical significance, so there is no further disagreement (Goh and Perks, unpublished observations). It is possible that both responses might be useful in different situations, but most workers accept that a long latency of action is typical of physiological effects of prolactin (Johnson et al., 1974; Bern, private communication). In fact, the lag in responses is useful in suggesting that the effects are biological, and not merely osmotic artifacts.

Some years later, our results were confirmed by Leontic and Tyson (1977) and Leontic et al. (1979). These workers showed a delayed fall in diffusional permeability to water when human or ovine prolactin was placed on the fetal surface of human amnia, recovered after delivery. The effect was blocked by prolactin antibody and by ouabain. Human placental lactogen and growth hormone had no action. Three preterm membranes (0.35 - 0.40 of term) reacted similarly, although the results were not conclusive. It seems remarkable that, despite degenerative changes seen in term membranes of both guinea pigs and humans (Hoyes, 1975), prolactin acts similarly, if not more strongly, close to birth. This raises many questions on its mechanism of action.

These results, taken together, suggest that AVP (AVT) is capable of moving water into the amniotic sac, while prolactin can slow its return, while removing sodium ions towards the mother. These actions of prolactin might influence the low osmotic pressure of amniotic fluid in late gestation. Whether these effects are physiological remains uncertain, although the mechanisms, and presumably receptors, are clearly present.

Studies of the Yolk Sac

AVP is also capable of directing fluid through the yolk sac, which lies outside the amnion of the guinea pig. In preliminary experiments on bulk-flow, by the gravimetric method, 40 mU/ml of AVP on the fetal surface increased passage of fluid inwards towards the amnion and fetus (n = 2). Again, a fetal neurohypophysial extract (FB = 54, V = 20, O = 5 mU/ml; 0.44 of term) produced a powerful change (n = 2). At this time, AVT itself has not been investigated.

Studies of Fetal Skin

The Effects of AVT and AVP. Once again, phylogenetic considerations suggested that AVT or AVP might act on fetal skin. AVT, in particular, is well known for the Brunn effect, in which it stimulates uptake of water through amphibian skin (Brunn, 1921). Morphology also supported this possibility, since the amnion, which reacted to these peptides, is continuous over the umbilical cord, and with the fetal skin.

Studies with the electron microscope showed a close similarity between the temporary outer periderm layer of the fetal skin, and the amnion of midterm guinea pigs (0.51 of term) (Pisani and Perks, unpublished observations). In particular, microvilli with filamentous coats contacted the amniotic fluid and the large nuclei, mitochondria, and rich stores of glycogen suggested a metabolically active, transporting epithelium. However, by 0.76 of term there had been striking changes. The periderm was lifting away, its cells had lost their microvilli and intracellular organelles, and the cytoplasm was full of filaments. These changes were strongly reminiscent of those in the near-term amnion. It is possible that both membranes represent skin programmed with different time-scales for keratinization. The clear parallels between early fetal skin and the amnion suggested that AVP and AVT should be tested on the skin.

Pilot studies showed that the midterm fetal skin could respond to neurohypophysial peptides (Holt and Perks, 1977a). Fetal skin from guinea pigs (0.46 - 0.51 of term) was set up in the double-circulation cells used to measure diffusional fluxes through the amnion. Because of the insensitivity of the method, large doses of AVP/LVP (500 - 1000 mU/ml) were tested on the amniotic surface. There were large responses, with up to 30.4 ± 13.7% increase in flux of ^3H_2O towards the fetus.

Later, these pilot studies were repeated with better methods (Pisani and Perks, unpublished observations). These showed that both AVT and AVP stimulated water uptake (bulk flow) of water into the fetus, but that AVT was almost three times more effective. The work used the more sensitive gravimetric system, but with maternal saline on both sides of the skin. Synthetic peptides were applied to the serosal surface, which would normally receive hormones in the blood. Studies of AVT and AVP were done side by side on skin from sibling fetuses; only the first response was quantitated. Skin was taken from 35 fetal guinea pigs of 0.49 - 0.70 of term. AVT (22 tests) and AVP (18 tests), at 5 - 100 mU/ml, produced similar responses after a latency of 3 - 9 min. Fluid moved inwards towards the fetus, and the maximum rate (0.86 ± 0.06 $mg/cm^2.min$) was similar to that of the amnion. Log-dose/response graphs showed linear relationships which were closely parallel for the two peptides, with AVT always the more effective. The threshold showed that AVT was 2.7 times more potent than AVP. Controls gave no effects, and there were no changes with gestational age.

Discussion. In the search for a function for AVT, the fetal skin appears to be the most likely target organ, at this time. This raises the possibility of 'ontogeny recapitulating phylogeny' (Haekel, 1866) not only on the molecular level of the AVT molecule, but also on the functional level. AVT acts similarly on the skin of frogs (Brunn, 1921).

Just after our preliminary studies of diffusional fluxes were completed, France (1976) published interesting results on the skin of fetal sheep (0.32 - 0.67 of term) and pigs (0.51 of term). Again, the relatively insensitive methods of diffusional flux were used. She found that LVP (200 mU/ml) increased the diffusional permeability to water in fetal sheep, but there was no net movement. In addition, she found that LVP increased the natural uptake of sodium by the skin. However, AVT was much more effective. Fetal pig skin was less reactive. LVP had no effect on Na^+ or water fluxes. However, AVT, which is not present in the pig pituitary, had the strange effect of reducing the flux of Na^+ towards the fetus. Prolactin was also tested on the pig. As in the amnion, short exposure to prolactin (1 IU/ml; 25 min) increased water fluxes in both directions, but reduced that of Na^+. Clearly, long-term effects on bulk flow are needed. However France's studies support the possible influence of hormones, particularly AVT and prolactin, on fetal skin.

Our electron microscope studies have not yet extended to the structural effects of AVT or AVP on fetal skin. However, recent work by Riddle (1985), on human fetal skin (0.23 - 0.35 of term) has shown that the potential second messenger, cAMP (as db-cAMP), produces aggregates of intramembranous particles in the apical membrane of the peridermal cells. Such changes are associated with transepithelial water flow in other tissues (Brown et al., 1983). Riddle suggests that before keratinization fetal skin may function as an osmoregulatory organ; our results, and those of France, indicate AVT (AVP) and prolactin as the possible controls.

Studies of the Fetal Bladder

The Effects of AVP/LVP. The ability of AVT to cause water reabsorption from the frog bladder (Sawyer, 1960) suggested that this function might be retained during ontogeny. This possibility was suggested also by the fact that urine in the bladder of the fetal guinea pig was known to be hypertonic to that in the kidney (Kleinmann, 1970). Therefore a pilot study was started on the isolated bladder of the guinea pig (Holt and Perks, 1977a).

In four experiments, bladders from late fetal guinea pigs (0.75 - 0.91 of term) were set up in the double-perfusion cell used for diffusion fluxes through the amnion. The mucosal surface contacted amniotic saline, the serosal, fetal saline. Again, because of the insensitivity of the diffusional flux method, relatively high doses of AVP/LVP (100 mU/ml) were placed on the serosal (blood) surface. Four tests showed large increases, up to 85% in the movement of tritiated water out of the bladder (average: 49.4 ± 17.8%). A rough approximation indicated a response 12 times greater than in the amnion.

Discussion. The experiments on the bladder were only preliminary in nature. However a phylogenetic approach seemed to have indicated a possible fetal function for vasopressin in the guinea pig. The large response made the bladder more likely to be a physiological target organ than the other tissues, especially since it was shown by the least sensitive method. Clearly, bulk flow studies were needed. However, the results of other workers on other species were less clear. France et al. (1976) have carried out more extensive studies on isolated bladders from fetal sheep (0.80 - 0.96 of term) and pigs (0.86 of term). AVP/LVP (200 mU/ml) caused only slight increases in fluxes of ^3H_2O in both directions, and there was no net change. Bulk flow from the lumen of the bladders increased, but the change could not be shown to be significant. In fetal pig bladders, AVP/LVP was without effect on ^3H_2O fluxes or bulk flow, which was surprising since it enhanced the enlargement of intercellular spaces during osmotic gradients. However, AVT had strong effects increasing fluxes of ^3H_2O, but again, with no net change. It also failed to affect bulk flow from the lumen of the bladders, but increased loss after they had been rendered almost waterproof by prolactin (1 IU/ml). This powerful effect of prolactin resembles its effect on the amnion, and was the most notable change recorded. AVP/LVP and AVT also changed Na^+ fluxes, to produce reabsorption by the fetus.

Although a number of aspects are not yet clear, these studies suggest that the fetal bladder may be controlled by AVP/LVP, AVT, and prolactin.

Studies of the Fetal Small Intestine

The Effects of AVP. Rapid drinking by the fetus (Jeffcoate and Scott, 1959; Bradley and Mistretta, 1973) suggests that the gut could be a major area for fluid control, and investigations of the effects of hormones were badly needed. However, our recent studies of the isolated small intestine, by the everted sac technique (Mainoya et al., 1974), have failed to show any stimulation of water absorption by AVP (Nielsen and Perks, unpublished observations).

The small intestine was removed from 8 fetal guinea pigs (0.91 - 1.00 of term). Each gave 4 sections, with average lengths close to 9 cm. The sections ranged from one close to the stomach to one close to the caecum. Each section was everted, ligated at one end, and attached to an open plastic funnel at the other. Krebs-Henseleit saline was placed on both surfaces of the sac (37°C), and the outside was oxygenated with 95% O_2 and 5% CO_2. Movement of fluid was determined by weighing at 5- or 10-min intervals. Experiments were of the A-B-A design. During the middle hour, saline containing 10 or 50 mU/ml synthetic (or natural) AVP was placed on both surfaces of the gut.

Of 26 segments used, all but one showed rapid uptake of fluid. This was strongest closest to the stomach (average, 26.65 ± 4.0 mg/cm length.hr; n = 6), and declined steadily towards the caudal segment (14.67 ± 2.24 mg/cm length.hr. This was in contrast to comparable data from 15 segments from 5 pregnant adult guinea pigs, where the average uptake was 17 times lower (1.25 ± 3.93 mg/cm length.hr). This was partly due to less consistent behavior of the adult tissues, where 8 segments reabsorbed, but 7 secreted fluid. In 26 tests of AVP (10 and 50 mU/ml), there was no stimulation of fluid uptake by the fetal gut. In fact, there were small reductions, which recovered after removal of AVP at the lower dose level. Results did not change with the position along the gut.

Discussion. The isolated fetal small intestine showed a consistent and unusually high absorption of fluid. Unexpectedly, this was highest close to the stomach. The average absorption for the whole fetus can be calculated at 14.7 ± 2.1 ml/kg body weight.hr (n = 6). However, the rate *in vitro* may not reflect that *in vivo*.

Despite the ability of AVP to direct fluid towards the fetus in all other tissues studied, it failed to increase uptake through the intestine. Although effects in the fetus might well be different, this agrees with its lack of effect on the adult rat jejunum, in a similar preparation, when given alone (Mainoya, 1975a). However, AVP does inhibit the uptake produced by prolactin in the adult preparation (Mainoya, 1975a, b). This could explain the small depression produced by AVP in the fetal intestine, if prolactin was already stimulating uptake. Since prolactin does act on the adult guinea pig jejunum (Mainoya et al., 1974), it might prove to be the main hormonal control of the fetal intestine, in this species; more work is needed.

Although we must conclude that AVP fails to direct water towards the fetus by the intestinal route, work was restricted to the near-term fetus, when the amnion and skin also fail to respond. It is still possible that stimulation by AVP or AVT could occur in earlier gestation.

Studies of the Fetal Lungs

The last fetal tissue to be studied was the lung. During fetal life the lungs produce large quantities of fluid, comparable to those from the kidneys (Jost and Policard, 1948; Barnes, 1976; Strang, 1977; Olver, 1977, 1983). The production is probably important for correct development of the lung, and it makes a substantial contribution to the amniotic fluid. However, it is vital to stop production and clear the lungs at birth. At the onset of our experiments, nothing was known of any controlling mechanisms. Since AVP directed fluid inwards, towards the fetus, in many fetal tissues, it seemed possible that it might help to drain the lungs at birth. Studies were carried out in three preparations, initially in the acute fetal goat, later in the chronic sheep, and recently in the isolated lungs from fetal guinea pigs.

Factors which reduce Lung Liquid Production

1. The effects of AVP.

a) The acute goat. Initial studies were carried out on 22 late fetal goats (0.92 - 1.00 of term), which were delivered by Caesarean section under chloralose anesthesia, and laid beside the mother with their umbilical circulations carefully preserved. Lung liquid secretion was measured by a dye dilution method, in which impermeant Blue Dextran 2000 (Pharmacia) was placed in the lungs; after one hour of equilibration, 0.5 ml samples were withdrawn at 10-min intervals, and the rate of secretion determined by the rate of dilution of the dye (Beckman 25 Spectrophotometer; 620 nm). Rates were calculated over 1-h intervals, from the slope of total volume against time (for details, see Perks and Cassin, 1982; Cassin and Perks, 1982). This method avoids the irregularities of early 'drip' techniques, which were grossly affected by body movements. It also allows statistical analyses of short-term changes, and the direct detection of reabsorption, not possible in later 'bag' collection methods.

Intravenous infusion of synthetic AVP (1.6 - 32.2 mU/kg.min) into 16 goat fetuses of 0.89 of term or older produced prolonged periods of lung liquid reabsorption (13 fetuses), or reduced secretion (3 fetuses) (Perks and Cassin, 1982). The effect often persisted beyond the time of infusion. However, the response did not arise until about 131 days of gestation (0.89 of term); five fetuses below that age failed to respond. Once established, the response increased steadily towards term, and a large effect was seen after a low infusion rate at that time (1.8 mU/kg.min). A linear log dose/response curve suggested a threshold at about 1.3 mU/kg.min. This would correspond to a blood level of about 50 μU/ml (Rurak, 1978; private communication), a value one-fifth of the average level in sheep fetuses at birth (Stark et al., 1977, 1979), and one-tenth that of the human baby at vaginal delivery (Chard et al., 1971; Pohjavuori and Fyhrquist, 1980). This suggests the effect could be physiological.

b) The chronic sheep. Since the acute preparation may be affected by anesthesia or stress, further studies were carried out ón chronic fetal sheep (Perks and Cassin, 1985; Cassin, Gause and Perks, unpublished observations). The method was essentially the same as that used for the fetal goats; however, the fetuses were delivered under sterile conditions with halothane anesthesia, provided with a tracheal loop, carotid and jugular cannulae, and returned to the uterus for continued development. Five days were allowed for recovery. The loop cannula allowed fluid to take its normal course to the buccal cavity between studies, but provided direct access to the lungs during experiments. Fluid production was determined by the rate of dilution of both Blue Dextran 2000 and albumen-I^{125}.

Studies on 29 sheep fetuses at 0.87 - 0.97 of term showed that synthetic AVP could depress lung secretion greatly and occasionally cause reabsorption. In 18 tests on fetuses above 0.91 of term (134 days), AVP was infused into the jugular vein in four dose ranges (28.5 ± 0.01, 20.2 ± 0.4, 10 ± 0.9, and 4.4 ± 0.03 mU/kg.min, for 2 h). All three upper ranges produced clear reductions in secretion, which averaged 66% and 43% reduction in the first hour in the two top ranges, and reached 39% in the second hour in the remaining group. Effects persisted beyond infusion. The lowest dose range tested gave no clear effects. Saline controls (0.027 ± 0.003 ml/kg, min; n = 4) produced a slight increase in secretion (Cassin and Perks, 1982). Fetuses left without infusion showed no significant changes over 4 h (0.95 - 0.96 of term; n = 3). The effect of AVP arose at about 0.91 of term (134 days); 3 fetuses below this age failed to respond.

The effects of AVP seemed less striking than in the acute goat. However, both the combined data, and sequential studies in the same fetus, showed that the effect increased with developmental age, as far as studies were possible (up until 141 days). Linear extrapolation suggested that strong reabsorption could be expected at normal term (147 days). This pin-points an important disadvantage of the chronic preparation. The 'chronic' fetus is not entirely normal, since, in the hands of many groups, delivery occurs early (usually at approximately 142 days, in our laboratory); therefore, simple developmental effects arising close to natural term, not linked directly to the mechanics of delivery, may not be seen. Similarly, the sensitivity of the preparations increased towards natural term. Log-dose/response curves at specific ages showed a fall in threshold to 5.2 mU/kg.min by 0.95 of term (139 days), and extrapolation predicted a threshold sensitivity at 3.7 mU/kg.min at term. This corresponds to blood levels of 0.15 mU/ml (Rurak, 1978; private communication). This is below the average value reported at delivery in sheep (Stark et al., 1977), and far below that of a fetus in respiratory distress (Stark et al., 1979); therefore the effect may be physiological.

c) The guinea pig lung in vitro. The in vitro preparation offers a method for avoiding many complicating influences arising in the fetus, and allows tests with drugs which may be fatal to the whole animal. Recently, we have developed an isolated preparation of the fetal guinea pig lung (Dore, Dyer, Thom, Marshall and Perks, unpublished observations). Fetal lungs were supported in Krebs-Henseleit saline (37°C), supplied with 95% O_2 and 5% CO_2. A tracheal cannula led to a 1.0 ml tuberculin syringe; this allowed gentle withdrawal of lung fluid, and 10 µl samples could be taken through a side arm with a fixed-volume Hamilton syringe. Secretion rates were measured by a scaled-down version of the Blue Dextran 2000 method, as used in the acute goat, with a similar protocol (Cassin and Perks, 1982). After 30 min equilibration, samples were collected at 10-min intervals for 3 h. An A-B-A design was used. Test solutions were applied to the serosal/pleural surfaces by changing the external solution for the middle hour. Isolated lungs secreted steadily. The average rate at term was 2.05 ± 0.21 ml/kg body weight.hr (n = 78), comparable to the average for chronic fetal sheep (3.1 ± 0.3 ml/kg.hr), and above that for near-term fetal sheep (0.87 ± 0.42 ml/kg.hr) (Perks and Cassin, 1985a).

AVP was shown to reduce secretion or promote reabsorption in the isolated fetal lung (Marshall and Perks, unpublished observations). In studies of 38 isolated lungs (0.94 - 1.00 of term), 24 received synthetic AVP (0.6, 0.3, 0.1 or 0.01 mU/ml), and 14 acted as external controls. At all three higher dose levels, AVP produced reabsorption or reduced secretion. At the highest dose, 3 lungs showed reabsorption, 3 reduced secretion; the average reduction was 90%, and, unlike lower doses, the effect persisted beyond treatment. AVP at 0.01 mU/ml had no effect. Controls showed a slight stimulation. When responses were averaged and quantitated, the linear log dose/response curve showed a threshold of 0.048 mU/ml, well below the blood levels reported at birth in sheep and humans, as discussed earlier.

2. The effects of epinephrine. During our early experiments on AVP, Walters and Olver (1978) showed that epinephrine could produce reabsorption of lung liquid in fetal sheep. This was confirmed in our acute fetal goats.

Doses of 0.3 - 6.72 μg/kg.min (i.v.) given to 10 fetuses (0.80 - 1.00 of term) showed the same effect, although it was weak in the youngest fetus (118 days) (Perks and Cassin, 1982). Similar results were obtained in our chronic fetal sheep (Perks and Cassin, 1982; unpublished observations). In 17 fetal sheep (0.90 - 0.98 of term), epinephrine infused into the jugular vein at 0.18 - 0.29 μg/kg.min changed secretion into reabsorption (although reabsorption was sometimes delayed until the second hour of infusion). Unlike the effect of AVP, that of epinephrine reversed rapidly after infusion. However, as with AVP, the responses increased greatly between 0.90 and 0.97 of term. The effect of epinephrine probably involves sodium transport, since amiloride placed in the lung fluid (10^{-4} M or above) slowly reversed the responses (n = 11). Epinephrine-treated controls continued to reabsorb (n = 6). This confirms the findings of Olver et al. (1981).

Recently, we have shown the effects of epinephrine in the isolated guinea pig lung (Cassin et al., 1986; Dyer and Perks, unpublished observations). Studies were made on 30 preparations (0.90 - 1.00 of term). Six lungs treated with epinephrine (1.8×10^{-5} M) changed their average secretion rate of 1.65 ml/kg.hr to a reabsorption of -0.38 ml/kg.hr; secretion returned after epinephrine was removed. Controls showed no significant change (n = 6). Ouabain at 10^{-5} M in the outer saline failed to inhibit normal secretion (n = 6), or the effect of epinephrine (n = 6). At a higher concentration (10^{-4} M), it appeared to inhibit secretion, but this may be an artifact, since ouabain at this level produced difficulties in withdrawing fluid.

The effects of epinephrine were also tested on the fluid uptake by the isolated lung (Dore and Perks, unpublished observations). Uptake from the outer medium was measured gravimetrically in lungs of 0.85 - 1.00 of term. The average uptake was 1.20 ± 0.06 g/kg body weight.hr (n = 83); this was not statistically different from the secretion rate (t-test for unequal variances, Parkes, 1973). Uptake was stopped by combined sodium iodoacetate (10^{-3} M) and cyanide (10^{-4} M) (n = 5). In 20 tests of epinephrine (1.8×10^{-5} M), there was an immediate loss of fluid from the preparations (average -0.64 ± 0.38 g/kg.hr). This continued until epinephrine was removed, after 2 h. The change was blocked by propranolol (8.4×10^{-7} M; n = 4), and reduced by amiloride (10^{-3} M; n = 6), so that β-receptors and sodium transport were probably involved. Controls showed continuous uptake of fluid for 4 h (n = 20).

3. <u>The effects of somatostatin</u>. Preliminary studies suggest somatostatin may produce reabsorption of fluid. During studies on 3 acute, near-term fetal goats, intravenous infusion at 125 μg/kg.hr promoted reabsorption, but 1.39 μg/kg.hr had no effect. Further studies are needed.

4. <u>The effects of lung expansion</u>. During the earlier work on the acute fetal goats, it seemed possible that the effects of AVP might be accentuated by the expansion of the lungs at birth; this might open channels in the epithelia, or expose receptors, to increase the sensitivity of the membranes to AVP. Therefore, infusion of AVP were tested after the lungs had been expanded with a volume of saline equivalent to that of the first breath (approx. 25 ml) (Perks and Cassin, 1985b). The saline contained Blue Dextran 2000 matched to the optical density of the fluid in the lungs, to avoid artifacts from mixing. Pressure changes were monitored and kept low. Results from 26 acute fetal goats (0.79 of term - overdue) showed that expansion did not change responses to AVP, but that expansions themselves produced reduction of secretion or reabsorption when over 18.6%. Secretion changed to reabsorption at about 50% expansion. The linear log stimulus/response curve between the % fall in secretion and the % expansion suggested that small respiratory movements in utero would not change secretion, but that those at birth would promote removal of fluid. Movements of the ions suggested that expansion activated a Na^+/K^+ ATP-ase pump, which removed Na^+

ions, and with them water. However, this potential system needs further study.

5. Effects of drugs. The lung liquid is probably secreted by the alveolar Type II cells, by a system based on chloride transport (sodium-chloride cotransport) (Olver and Strang, 1974; Olver, 1977, 1983). A number of drugs, such as the "loop" diuretics bumetanide and furosemide, are known to inhibit this type of system (Imai, 1977; Kokko, 1984). However, bumetanide had not been tested on fetal lungs. This was surprising, in light of the importance of treating neonatal problems such as "wet lung" and R.D.S. Therefore, we tested these drugs on the chronic sheep preparations (Cassin et al., 1986). Thirty-four experiments were carried out in 20 fetal sheep (0.87 - 0.98 of term). Bumetanide placed in the lung fluid at $2.2 \pm 0.5 \times 10^{-4}$ M produced immediate, long-term reabsorption of fluid. At $1.1 \pm 0.17 \times 10^{-5}$ M, it reduced secretion significantly. Furosemide was active, but less effective.

These results were checked on the guinea pig lung *in vitro* (n = 31; 0.99 of term - term) (Thom and Perks, unpublished observations). Bumetanide at 10^{-3} M on the serosal/pleural surface produced immediate reabsorption (-0.41 ml/kg.hr; n = 6). At 10^{-4} M, it reduced secretion significantly (n = 6). There was no effect at 10^{-5} M (n = 7), or in the ethanol-carrier or untreated controls (n = 6; each group). In 18 similar studies on furosemide, reductions were produced at 10^{-3} and 10^{-4} M, but the effect of the top dose was less than that of bumetanide. Piretanide also inhibited secretion (10^{-3} M; n = 6).

These studies confirm the importance of Na^+/Cl^- cotransport as the basis of lung secretion, and suggest the possibility that bumetanide, rather than furosemide, might find a use in neonatal respiratory problems.

Factors which increase Lung Liquid Production

1. The effects of saline infusions. Studies in 10 acute fetal goats (0.80 of term - overdue) showed that rapid infusions of saline stimulated lung liquid production (Perks and Cassin, 1982). Rapid intravenous infusion of 0.9% NaCl (0.084 - 0.24 ml/kg.min, for 2-3 h; n = 3) increased fluid production as much as 40-fold, and could even reverse a spontaneous reabsorption. There was also an increase in the Na concentration of the fluid. A moderate infusion (0.053 ml/kg.min, for 2 h) did not increase fluid production, but the secretion rate and concentration of Na^+ ions rose. Slow infusions (0.028 - 0.42 ml/kg.min, for 2 h; n = 3) had no significant effect, and untreated fetuses showed no changes (n = 3). Calculations on those preparations which increased secretion showed that 60-94% of the sodium load was passed out through the lungs by the end of the experiment.

2. The effects of prolactin. Six experiments on acute fetal goats (0.80 of term - overdue) suggested that prolactin could stimulate fluid production (Cassin and Perks, 1982). Three fetuses received ovine prolactin placed directly in the lungs (10 μg/ml). This produced a slow increase in fluid production, which reached a rise of 2.5-fold in the third hour. Na^+ and Cl^- showed parallel increases, but K^+ behaved independently. There were no significant changes in three control fetuses.

3. The effects of pentagastrin. Since gastrin stimulates chloride secretion in the stomach, it seemed possible that it would do the same in the fetal lungs, which are derived from the upper gut. However, infusions of synthetic pentagastrin produced no stimulation of fluid production (0.12 - 2.5 μg/kg.min, i.v., for 40-120 min; goats, 0.94 of term, n = 4) (Cassin and Perks, 1982).

Discussion. These studies suggest that the fetal lungs may be concerned in the control of plasma sodium levels during fetal life, when the kidneys are poor in this regard (Alexander and Nixon, 1961; Capek et al., 1968; Kleinman, 1975). This could imply the long-term regulation of blood volume and pressure in the fetus.

At term, the vital draining of the lungs may be carried out by multiple controls; these include the effects of AVP, epinephrine and expansion of the lungs. The effects of AVP in chronic sheep have been confirmed by Ross et al (1984). Although their drainage technique did not allow detection of reabsorption, but only falls in flow, they found reductions with both AVP and AVT. Of these two, AVT is unlikely to act at birth, since it is lost from the pituitary by late gestation; whether it could control the lungs at earlier stages is not known. The ability of the lungs to respond to AVP arises relatively late (131 days, goat; 134 days, sheep). This seems to reflect the increasing predominance of AVP in later fetal life. Together with the increase in response with age (goat, sheep), and the increasing sensitivity of the lungs (sheep), this suggests that AVP may find its use in lung functioning at birth. Despite large variations in published values for plasma AVP at birth, results from the acute goat, the chronic sheep and the isolated guinea pig lung all suggest thresholds below those found in the stress of vaginal delivery (Chard et al., 1971; Stark et al., 1977, 1979; Pohjavuori and Fyhrquist, 1980; Leffler et al., 1985). This may be another example of the importance of stress in "normal" adaptation to the outer world (Lagercrantz and Slotkin, 1986). In addition, the effects of epinephrine, also released in the stress of birth, could add to those of AVP (Brown et al., 1983; Cassin and Perks, 1982; unpublished observations). The effects of epinephrine also arise late in gestation, and increase with age (Brown et al., 1983, Cassin et al., as above). However, AVP and epinephrine have slightly different effects. Those of epinephrine reverse rapidly. Since plasma epinephrine levels tyipcally decline 30 min after delivery, and return to resting levels within 2 h of birth (Lagercrantz and Slotkin, 1986), it would act only in the early stages after delivery. AVP has a slow but prolonged action, beyond its period of contact, so that it would be well suited for a long-term "mopping-up" of newborn lungs.

The exact mechanisms by which these factors produce reabsorption are not clearly understood. Originally reabsorption was thought to involve the halt of secretion, followed by reabsorption by colloid osmotic forces (Olver, 1977). Recently use of agents such as amiloride have suggested an active reabsorption, involving sodium transport (Olver et al., 1981; Olver, 1983). Our studies in chronic sheep support this contention, although the effects of amiloride were unusually slow. The studies of lung expansion in the goat also suggested activation of a Na^+/K^+ ATP-ase. However, the work on the isolated guinea pig lung may be the strongest evidence yet. Reabsorption was produced by AVP, epinephrine and bumetanide. This took place in the absence of a functioning circulation, and without colloid osmotic forces, since the outer saline was free of large molecules.

CONCLUSIONS

This study has not attempted to review the field of fetal water metabolism, but to integrate the work from our laboratories. Many conclusions are still speculative, but some generalizations seem to be taking shape. In a number of places, the old concept of recapitulation seems to apply to both molecular structure and to functions. There are indications of a transition from an early predominance of AVT, the "lower vertebrate" principle, to that of AVP. During fetal and perinatal life, these peptides are able to direct water inwards, towards the fetus through a variety of routes - through the yolk sac, amnion, skin, bladder, and (probably at birth) the lungs. To these we may add the kidney. Despite relative immaturity in its responses to AVP (Alexander and Nixon, 1961), and a low glomerular filtration rate which limits its ability to excrete sodium and water (Kleinman, 1975), an ability to respond to AVP does increase towards late gestation (Robillard et al., 1979). However, the AVP receptors may still be less functional at birth than in the adult (Robillard and Weitzman, 1980). Oddly, AVP does not seem to affect the important intestinal route, which may prove to be influenced by prolactin. Prolactin is often partly antagonistic to AVT and AVP in this fetal system, as in the amnion and bladder, where it limits water movement.

Although our studies concentrated on the less obvious extraplacental routes, we must not forget the placenta. There is some evidence that AVP and AVT may also influence this pathway in fetal sheep. Firstly, AVP may affect placental permeability (Thornberg et al., 1979). Secondly, it produces a net gain in fetal water, and in common with AVT, slows the passage of water towards the mother, when a gradient operates in that direction (Leake et al., 1981, 1983). Leake et al. thought that this could be a direct effect on placental transfer. However, their second explanation, a possible fall in fetal/placental blood flow, is unlikely, since Rurak and Gruber (1984) found that AVP increased umbilical flow in chronic fetal sheep. In fact, this rise in blood flow could also favor water uptake by the placental route. In contrast, prolactin appears to have little effect on water transfer through the placenta, and acts primarily to inhibit water flow through the chorio-amnion (Ross et al., 1983; Leake et al., 1985).

Despite the growing evidence for a complex hormonal system to direct water towards the fetus, a number of difficulties remain. For example, it is hard to see how AVP could reach the avascular amnion. However, a route through the fetal urine and amniotic fluid is possible (Vizsolyi and Perks, 1974), and both AVP and AVT have been found in human amniotic fluid (Artman et al., 1984). In addition, AVP might accumulate in the fluid, in certain circumstances, since it may be recycled, apparently intact, through the fetal gut and kidneys (Ervin et al., 1986).

Another difficulty surrounds the physiologic significance of the system. In some studies, the doses required for activity are high. This is not true for prolactin, which often acts below the levels found in human amniotic fluid (Holt and Perks, 1975). However, AVP needs relatively high doses to influence the amnion and skin. The problem may lie in the use of in vitro preparations, where the conditions may not be ideal. The supporting salines may not be optimal, and the effects of AVP on the toad bladder are known to be highly sensitive to changes in sodium and calcium concentrations (Roy and Ausiello, 1981; Davis et al., 1981). This could be particularly critical for a single-celled epithelium, such as the amnion. In addition, specialized regions, such as the placental or umbilical amnion could be the true sites of action. The salines could also lack substances with synergistic actions; ACTH and TSH enhance the effects of AVT on the skin of newts (Brown and Brown, 1982). In fetal skin, and larger preparations, penetration of hormones to the active sites could raise problems, in vitro. However, the fetal bladder seems to retain a sensitivity which suggests possible physiologic control.

In considering this problem, the large release of AVP by the fetus must also be remembered. An extensive literature supports this fetal ability. For example, sheep fetuses near term can show serum levels almost 40 times greater than those of the mother, and have a production rate 100 times higher (Skowsky et al., 1973). Release is great in stress and delivery (Chard et al., 1971; Challis et al., 1976; Stark et al., 1977; Pohjavuori and Fyhrquist, 1980). Urine AVP levels in dehydrated human neonates can reach 1.25 mU/ml (Ames, 1953), a level sufficient to affect almost all fetal tissues which respond. The burst of AVP at birth, combined with our results from the fetal lungs, suggest that this peptide may act physiologically to help drain the lungs at birth.

Our basic in vitro studies are finding steadily increasing support from work on chronic sheep. The complex AVT/AVP/prolactin system may well give the fetus access to its surrounding stores of fluid. If this is so, and we can regard water and salts as part of its nutrition, we are returning surprisingly close to the ideas of William Harvey in 1651.

REFERENCES

Acher, R., and Fromageot, C., 1957, Relationship of oxytocin and vasopressin to active proteins of posterior pituitary origin, in: "The Neurohypophysis," H. Heller, ed., pp. 11-16, Butterworths, London.

Adolph, E. F., 1967, Ontogeny of volume regulations in embryonic extracellular fluids, Quart. Rev. Biol., 42:1.

Alexander, D. P., Britton, H. G., Forsling, M. L., Nixon, D. A., and Ratcliffe, J. G., 1973, Adrenocorticotrophin and vasopressin in foetal sheep and the response to stress, in: "The Endocrinology of Pregnancy and Parturition," G. Pierrepoint, ed., pp. 112-125, Tenovus Library Series, Alpha Omega Alpha, Cardiff.

Alexander, D. P., and Nixon, D. A., 1961, The foetal kidney, Brit. Med. Bull., 17:112.

Ames, R. G., 1953, Urinary water excretion and neurohypophysial function in full term and premature infants shortly after birth, Pediatr., 12:272.

Arthur, G. H., 1969, The fetal fluids of domestic animals, J. Reprod. Fertil. Suppl. 9, 45.

Artman, H. G., Leake, R. D., Weitzman, R. E., Sawyer, W. H., and Fisher, D. A., 1984, Radioimmunoassay of vasotocin, vasopressin and oxytocin in human neonatal cerebrospinal and amniotic fluid, Dev. Pharmacol. Ther., 7:39.

Bain, A. D., and Scott, J. S., 1960, Renal agenesis and severe urinary tract dysplasia. A review of 50 cases with particular reference to the associated anomalies, Brit. Med. J., 1:841.

Barnes, R. J., 1976, Water and mineral exchange between maternal and fetal fluids, in: "Fetal Physiology and Medicine," R. W. Bear and P. W. Nathanielsz, eds., pp. 194-215, W. B. Saunders, Philadelphia.

Barnes, A. C., and Seeds, A. E., 1972, The water metabolism of the fetus, pp. 32-47, Charles C. Thomas, Springfield, Illinois.

Benirschke, K., and McKay, D. G., 19853, The antidiuretic hormone in fetus and infant, Obstet. Gynecol., 1:638.

Bourne, G., 1962, The Human Amnion and Chorion, pp. 276, Year Book Medical Publishers, Inc., Chicago.

Bradley, R. M., and Mistretta, C. M., 1973, The sense of taste and swallowing activity in foetal sheep, in: "Foetal and Neonatal Physiology," R. S. Comline, K. W. Cross, G. S. Dawes, and P. W. Nathanielsz, eds., pp. 77, Cambridge University, Cambridge.

Brown, D., Grosso, K. A., and DeSousa, R. K., 1983, Correlation between water flow and intramembrane particle aggregation in toad epidermis, Am. J. Physiol., 245:C334.

Brown, M. J., Olver, R. E., Ramsden, C. A., Strang, L. B., and Walters, D. V., 1983, Effects of adrenalin and of spontaneous labour on the secretion and absorption of lung liquid in the fetal lamb, J. Physiol. (Lond.), 344:137.

Brown, P. C., and Brown, S. C., 1982, Effects of hypophysectomy and prolactin on the water-balance response of the newt, Taricha torsa, Gen. Comp. Endocrinol., 46:7.

Brunn, F., 1921, Beitrag zur Kenntnis der Wirkung von Hypophysenextrakten auf der Wasserhaushalt de Frosches, Z. Ges. Exp. Med., 25:170.
Burton, A. M., and Forsling, M., 1972, Hormone content of the neurohypophysis in foetal, newborn and adult guinea pigs, J. Physiol. (Lond.), 221:6 P.
Capek, K., Dlouha, H., Fernandez, J., and Popp, M., 1968, Regulation of proximal tubular reabsorption in early postnatal period of infant rats: micropuncture study, Proc. Int. Union Physiol. Sci., 7:72.
Cassin, S., Dyer, R., Gause, G., and Perks, A. M., 1986, The effects of loop diuretics and ouabain on lung liquid secretion in fetal sheep and guinea pig, J. Physiol. (Lond.), 371; 240 P.
Cassin, S., and Perks, A. M., 1982, Studies of factors which stimulate lung fluid secretion in fetal goats, J. Devel. Physiol., 4:311.
Challis, J. R. G., Robinson, J. S., Rurak, D. W., and Thorburn, G. D., 1976, The development of endocrine function in the human fetus, in: "The Biology of Human Fetal Growth," D. F. Roberts and A. M. Thompson, eds., pp. 149-194, Taylor and Francis, London.
Chard, T., Hudson, C. N., Edwards, C. R. W., and Boyd, N. R. H., 1971, Release of oxytocin and vasopressin by the human foetus during labour, Nature (Lond.), 234:352.
Danforth, D. N., and Hull, R. W., 1958, The microscopic anatomy of the fetal membranes with particular reference to the detailed structure of the amnion, Am. J. Obstet. Gynecol., 75:536.
Davis, W. L., Jones, R. G., Hagler, H. K., Goodman, D. B. P., and Knight, J. P., 1981, Intracellular water transport in the actions of ADH, in: "Hormonal Regulation of Epithelial Transport of Ions and Water," W. N. Scott and D. B. P. Goodman, eds., Vol. 372, pp. 118-130, Annals of the New York Academy of Sciences.
Diamond, J. M., 1971, Standing-gradient model of fluid transport in epithelia, Fed. Proc., 30:6.
Dicker, S. E., and Tyler, C., 1953a, Vasopressor and oxytocic activities of the pituitary glands of rats, guinea-pigs and cats and of human foetuses, J. Physiol. (Lond.), 121:206.
Dicker, S. E., and Tyler, C., 1953b, Estimation of the antidiuretic, vasopressor and oxytocic hormones in the pituitary glands of dogs and puppies, J. Physiol. (Lond.), 120:141.
Dogterom, J., Snidjewint, F. G. M., Pevet, P., and Swaab, D. F., 1980, Studies on the presence of vasopressin, oxytocin and vasotocin in the pineal gland, subcommissural organ and foetal pituitary gland: failure to demonstrate vasotocin in mammals, J. Endocrinol., 84:115.
Elliot, P. M., and Inman, W. H. W., 1961, Volume of liquor amnii in normal and abnormal pregnancy, Lancet, 2:835.
Ervin, M. G., Ross, M. G., Leake, R. D., and Fisher, D. A., 1986, Fetal recirculation of amniotic fluid arginine vasopressin, Am. J. Physiol., 250:E253.
Fawcett, D. W., 1965, Surface specializations of absorbing cells, J. Histochem. Cytochem., 13:75.
Fawcett, D. W., 1981, The Cell, 2nd edition, W. B. Saunders Co., Philadelphia.
Flexner, L. B., and Gellhorn, A., 1942, The transfer of water and sodium to the amniotic fluid of the guinea pig, Am. J. Physiol., 136:757.
France, V., 1976, Active sodium uptake by the skin of fetal sheep and pigs, J. Physiol. (Lond.), 258:377.
France, V. M., Stanier, M. W., and Wooding, F. B. P., 1976, The effect of hormones and of an osmotic gradient on the structure and properties of mammalian foetal urinary bladder in vitro, J. Physiol. (Lond.), 258:393.
Friesen, H., Hwang, P., Guyda, H., Tolis, G., Tyson, J., and Myers, R., 1972, A radioimmunoassay for human prolactin, in: "Prolactin and Carcinogenesis," A. R. Boyne and E. Griffiths, eds., pp. 64-80, Alpha Omega Alpha, Cardiff.
Haeckel, E, 1866, Generelle Morphologie der Organismen, Berlin.

Harvey, W., 1651, De uteri membranis et humoribus, in: "De Generatione Animalium," pp. 527-568, Ludovicum Elzevirium, Amsterdam. Translation quoted: Robert Willis, "The Works of William Harvey, M. D.," 1965, The Sources of Science, 13:551, Johnson Reprint Corp., New York.
Heller, H., 1961, Occurrence, storage and metabolism of oxytocin, in: "Oxytocin," R. Caldeyro-Barcia and H. Heller, eds., pp. 3-23, Pergamon, New York.
Heller, H., 1966, Hormone content of the hypothalamo-neurohypophysial system, Brit. Med. Bull., 22:227.
Heller, H., and Lederis, K., 1959, Maturation of the hypothalamo-neurohypophysial system, J. Physiol. (Lond.), 147:299.
Hirano, T., 1977, Prolactin and osmoregulation. Prolactin and hydromineral metabolism in the vertebrates, Gunma Symp. Endocrinol., 14:45.
Holt, W. F., and Perks, A. M., 1975, The effect of prolactin on water movement through the isolated amniotic membrane of the guinea pig, Gen. Comp. Endocrinol., 26:153.
Holt, W. F., and Perks, A. M., 1977a, The influence of vasopressin on the passage of tritiated water through the isolated amniotic membrane and other tissues from the fetal guinea pig, Can. J. Zool., 55:1393.
Holt, W. F., and Perks, A. M., 1977b, The effect of prolactin on sodium flux through the isolated amniotic membrane of the guinea pig, Can. J. Zool., 55:1468.
Holton, P., 1948, A modification of the method of Dale and Laidlaw for standardization of posterior pituitary extract, Brit. J. Pharmacol., 3:328.
Hoyes, A. D., 1975, Structure and function of amnion, Obstet. Gynec. Annual, 4:1.
Hutchinson, D. L., Gray, M. J., Plentl, A. A., Alvarez, H., Caldeyro-Barcia, R., Kaplan, B., and Lind, J., 1959, The role of the fetus in the water exchange of the amniotic fluid of normal and hydramniotic patients, J. Clin. Invest., 38:971.
Imai, M., 1977, Effect of bumetanide and furosemide on the thick ascending limb of Henle's loop of rabbits and rats perfused in vitro, Eur. J. Pharmacol., 41:409.
Jeffcoate, T. N. A., and Scott, J. S., 1959, Polyhydramnios and oligohydramnios, Can. Med. Assn. J., 80:77.
Johnson, D. W., Hirano, T., Sage, M., Foster, R. C., and Bern, H. A., 1974, Time course of response of starry flounder (Platichthys stellatus) urinary bladder to prolactin and salinity transfer, Gen. Comp. Endocrinol., 24:373.
Josimovich, J. B., Weiss, G., and Hutchinson, D., 1974, Sources and disposition of pituitary prolactin in maternal circulation, amniotic fluid, fetus and placenta in the pregnant rhesus monkey, Endocrinology, 94:1364.
Jost, A., and Policard, A., 1948, Contribution experimentale a l'étude du developpement prenatal du poumon chez le lapin, Arch. Anat. Microscop. Morphol. Exp., 37:323.
Kerpel-Fronius, E., 1970, Electrolyte and water metabolism, in: "Physiology of the Perinatal Period," U. Stave, ed., pp. 643-678, Appleton-Century-Crofts, New York.
King, B. F., 1978, A cytological study of plasma membrane modifications, intercellular junctions and endocytotic activity of amniotic epithelium, Anat. Rec., 190:113.
Kleinman, L. I., 1975, Fetal renal function and water and electrolyte balance, in: "The Mammalian Fetus," E. Hafez, ed., pp. 120-153, Charles C. Thomas, Illinois.
Kokko, J. P., 1984, Site and mechanism of action of diuretics, Am. J. Med., 77:11.
Lagercrantz, H., and Slotkin, T. A., 1986, The "stress" of being born, Sci. Amer., 254:100.

Leake, R. D., Ervin, M. G., Ross, M. G., Polk, D. H., Lam, R., and Fisher, D. A., 1985, Ovine fetal-maternal water transfer is independent of fetal prolactin levels, Pediat. Res., 19:986.
Leake, R. D., Palmer, S., Oakes, G. K., Artman, H., Morris, A. M., and Fisher, D. A., 1981, Arginine vasotocin inhibits ovine fetal/maternal water transfer, Pediat. Res., 15:483.
Leake, R. D., Stegner, H., Palmer, S. M., Oakes, G. K., and Fisher, D. A., 1983, Arginine vasopressin and arginine vasotocin inhibit ovine fetal/ maternal water transfer, Pediat. Res., 17:583.
Lederis, K., Fisher, A. W., Geonzon, R. M., Gill, V., Ko, D., and Raghavan, S., 1980, Arginine vasotocin in fetal, newborn and adult mammals, in: "Hormones, Adaptation and Evolution," S. Ishii et al., eds., pp. 71-77, Japanese Scientific Scoeity Press, Tokyo/Springer Verlag, Berlin.
Leffler, C. W., Crofton, J., Brooks, D. P., Share, L., Hessler, J. R., and Green, R. S., 1985, Changes in plasma arginine vasopressin during transition from fetus to newborn following minimal trauma delivery of lambs and goats, Biol. Neonate, 48:43.
Leontic, E. A., Schruefer, J. J., Andreassen, B., Pinto, H., and Tyson, J. E., 1979, Further evidence for the role of prolactin on human fetoplacental osmoregulation, Am. J. Obstet. Gynecol., 133:435.
Leontic, E. A., and Tyson, J. E., 1977, Prolactin and fetal osmoregulation: water transport across isolated human amnion, Am. J. Physiol., 232:R124.
Levina, S. E., 1968, Endocrine features in development of human hypothalamus, hypophysis, and placenta, Gen. Comp. Endocrinol., 11:151.
Lind, T., 1969, Biochemical changes in human liquor amnii during gestation, J. Reprod. Fertil. Suppl., 9:53.
Lind, T., and Hytten, F. E., 1972, Fetal control of fetal fluids, in: "Physiological Biochemistry of the Fetus," A. A. Hodari and F. G. Mariona, eds., pp. 54-65, Charles C. Thomas, Springfield, Illinois.
Maetz, J., 1968, Salt and water metabolism, in: "Perspectives in Endocrinology; Hormones in the Lives of Lower Vertebrates," E. J. W. Barrington and G. B. Jørgensen, eds., pp. 47-162, Academic Press, New York.
Mainoya, J. R., 1975a, Further studies on the action of prolactin on fluid and ion absorption by the rat jejunum, Endocrinology, 96:1158.
Mainoya, J. R., 1975b, Analysis of the role of endogenous prolactin on fluid and sodium chloride absorption by the rat jejunum, J. Endocrinol., 67:343.
Mainoya, J. R., Bern, H. A., and Regan, J. W., 1974, Influence of ovine prolactin on transport of fluid and sodium chloride by the mammalian intestine and gallbladder, J. Endocrinol., 63:311.
Manku, M. S., Mtabaji, J. P., and Horrobin, D. F., 1975, Effect of cortisol, prolactin and ADH on the amniotic membrane, Nature (Lond.), 258:78.
Mellor, D. J., and Slater, J. S., 1971, Daily changes in amniotic and allantoic fluid during the last three months of pregnancy in conscious, unstressed ewes, with catheters in their foetal fluid sacs, J. Physiol. (Lond.), 217:573.
Montagna, W., and Parakkal, P. F., 1974, The Structure and Function of Skin, 3rd Edition, Academic Press, New York.
Munsick, R. A., 1960, Effect of magnesium ion on the response of the rat uterus to neurohypophysial hormones and analogues, Endocrinology, 66:451.
Neacsu, C., 1972, The mechanism of antigonadotrophic action of a polypeptide extracted from a bovine pineal gland, Rev. Roum. Physiol., 9:161.
North, P. M., and Segal, M. B., 1976, A study of the permeability properties of the guinea pig amniotic membrane, J. Physiol. (Lond.), 256:245.
Olver, R. E., 1977, Fetal lung liquids, Fed. Proc., 36:2669.
Olver, R. E., 1983, Fluid balance across the fetal alveolar epithelium, Amer. Rev. Resp. Dis., 127:S33.

Olver, R. E., Ramsden, C. A., and Strang, L. B., 1981, Adrenaline-induced changes in net lung liquid volume flow across the pulmonary epithelium of the fetal lamb: evidence for active sodium transport, J. Physiol. (Lond.), 319:38.
Olver, R. E., and Strang, L. B., 1974, Ion fluxes across the pulmonary epithelium and the secretion of lung liquid in the foetal lamb, J. Physiol. (Lond.), 241:327.
Parakkal, P. F., and Alexander, N. J., 1972, Keratinization: A Survey of Vertebrate Epithelia, Academic Press, New York.
Parke, L., 1973, Detection of prolactin activity by bioassay in human amniotic fluid, J. Endocrinol., 58:137.
Parker, R. E., 1973, Introductory Statistics for Biology, pp. 19-20, Edward Arnold, London.
Pavel, S., 1975, Vasotocin biosynthesis by neurohypophysial cells from human fetuses. Evidence of its ependymal origin, Neuroendocrinology, 19:150.
Perks, A. M., 1977, Developmental and evolutionary aspects of the neurohypophysis, Amer. Zool., 17:833.
Perks, A. M., and Cassin, S., 1982, The effects of arginine vasopressin and other factors on the production of lung fluid in fetal goats, Chest, 81S:63S.
Perks, A. M., and Cassin, S., 1985a, The effects of arginine vasopressin on lung liquid secretion in chronic fetal sheep, in: "The Physiological Development of the Fetus and Newborn," C. T. Jones and P. W. Nathanielsz, eds., pp. 253-257, Academic Press, London.
Perks, A. M., and Cassin, S., 1985b, The rate of production of lung liquid in fetal goats, and the effect of expansion of the lungs, J. Develop. Physiol., 7:149.
Perks, A. M., and Vizsolyi, E., 1973, Studies of the neurohypophysis in foetal mammals, in: "Foetal and Neonatal Physiology," R. S. Comline, K. W. Cross, G. S. Dawes and P. W. Nathanielsz, eds., pp. 430-438, Cambridge University Press, Cambridge.
Perks, A. M., Vizsolyi, E., Holt, W. F., and Cassin, S., 1978, Hormonal influences on the movement and composition of amniotic fluid, in: "Comparative Endocrinology," P. J. Gaillard and H. H. Boer, eds., pp. 231-234, Elsevier/North Holland Biomedical Press, New York.
Pohjavuori, M., and Fyhrquist, F., 1980, Hemodynamic significance of vasopressin in the newborn infant, J. Pediat., 97:462.
Reynolds, W. A., 1972, Fetal sources of amniotic fluid: an enigma, in: "Physiological Biochemistry of the Fetus," A. A. Hodari and F. Mariona, eds., pp. 3-31, Charles C. Thomas, Springfield, Illinois.
Riddle, C. V., 1985, Intramembranous response to cAMP in fetal epidermis, Cell Tissue Res., 241:687.
Robillard, J. E., Matson, J. R., Sessions, C., and Smith, F. G., 1979, Developmental aspects of renal tubular reabsorption of water in the lamb fetus, Pediat. Res., 13:1172.
Robillard, J. E., and Weitzman, R. E., 1980, Developmental aspects of the fetal renal response to exogenous arginine vasopressin, Am. J. Physiol., 238:F407.
Ross, M. G., Ervin, G., Leake, R. D., Fu, P., and Fisher, D. A., 1984, Fetal lung liquid regulation by neuropeptides, Am. J. Obstet. Gynecol., 150:421.
Ross, M. G., Ervin, M. G., Leake, R. D., Oakes, G., Hobel, C., and Fisher, D. A., 1983, Bulk flow of amniotic fluid water in response to maternal osmotic challenge, Am. J. Obstet. Gynecol., 147:697.
Roy, C., and Ausiello, D. A., 1981, Regulation of vasopressin binding to intact cells, in: "Hormonal Regulation of Epithelial Transport of Ions and Water," W. N. Scott and D. B. P. Goodman, eds., Vol. 372, pp. 92-105, Annals of the New York Academy of Sciences.
Rurak, D. W., and Gruber, N. C., 1984, The effect of vasopressin on fetal oxygenation in sheep, Can. J. Physiol. Pharmacol., 62:27.

Sawyer, W. H., 1960, Increased water permeability of the bullfrog (Rana catesbiana) bladder in vitro in response to synthetic oxytocin and arginine vasotocin and to neurohypophysial extracts from non-mammalian vertebrates, Endocrinology, 66:112.

Sawyer, W. H., 1961, Biologic assays for oxytocin and vasopressin, Methods Med. Res., 9:210.

Seeds, A. E., 1965, Water metabolism of the fetus, Am. J. Obstet. Gynecol., 92:727.

Seeds, A. E., 1967, Water transfer across the human amnion in response to osmotic gradients, Am. J. Obstet. Gynecol., 98:568.

Seeds, A. E., 1972, Amniotic fluid, in: "The Water Metabolism of the Fetus," A. C. Barnes and A. E. Seeds, eds., pp. 48-73, Charles C. Thomas, Springfield, Illinois.

Skowsky, W. R., Bashore, R. A., Smith, F. G., and Fisher, D. A., 1973, Vasopressin metabolism in the foetus and newborn, in: "Foetal and Neonatal Physiology," R. S. Comline, K. W. Cross, G. S. Dawes and P. W. Nathanielsz, eds. pp. 439-447, Cambridge University Press, Cambridge.

Skowsky, W. R., and Fisher, D. A., 1977, Fetal neurohypophysial arginine vasopressin and arginine vasotocin in man and sheep, Pediat. Res., 11:627.

Spring, K. R., and Hope, A., 1978, Size and shape of the lateral intercellular spaces in a living epithelium, Science, 200:54.

Spring, K. R., and Hope, A., 1979, Fluid transport and the dimensions of cells and interspaces of living Necturus gallbladder, J. Gen. Physiol., 73:287.

Stark, R., Hussain, K., Daniel, S., Milliez, J., Morishima, L., and James, L. S., 1977, Characteristics of vasopressin (AVP) release during adrenocorticotrophin (ACTH) induced parturition in the lamb, Pediat. Res., 11:412.

Stark, R. I., Daniel, S. S., Hussain, K. M., James, L. S., and Vande Wiele, R. L., 1979, Arginine vasopressin during gestation and parturition in sheep fetus, Biol. Neonate, 35:235.

Strang, L. B., 1977, Neonatal Respiration; Physiological and Clinical Studies, p. 316, Blackwell Scientific Publications, Oxford.

Thornburg, K. L., Binder, N. D., and Faber, J., 1979, Diffusion permeability and ultrafiltration-reflection-coefficients of Na^+ and Cl^- in the near-term placenta of the sheep, J. Develop. Physiol., 1:47.

Tiedman, K., 1979, The amniotic, allantoic and yolk-sac epithelia of the cat: SEM and TEM studies, Anat. Embryol., 158:75.

Tormey, J., and Diamond, J. M., 1967, The ultrastructural route of fluid transport in rabbit gallbladder, J. Gen. Physiol., 50:2031.

Vizsolyi, E., and Perks, A. M., 1969, New neurohypophysial principle in foetal mammals, Nature (Lond.), 223:1169.

Vizsolyi, E., and Perks, A. M., 1974, The effect of arginine vasotocin on the isolated amniotic membrane of the guinea pig, Gen. Comp. Endocrinol., 52:371.

Vizsolyi, E., and Perks, A. M., 1976a, Neurohypophysial hormones in fetal life and pregnancy. I. Pharmacological studies in the sheep (Ovis aries), Gen. Comp. Endocrinol., 29:28.

Vizsolyi, E., and Perks, A. M., 1976b, Neurohypophysial hormones in fetal life and pregnancy. II. Chromatographic studies in the sheep (Ovis aries), Gen. Comp. Endocrinol., 29:41.

Vosburgh, G. J., Flexner, L. B., Cowie, D. B., Hellman, L. M., Procter, N. K., and Wilde, W. S., 1948, The rate of renewal in women of the water and sodium of the amniotic fluid as determined by tracer techniques, Am. J. Obstet. Gynecol., 56:1156.

Wagner, G., and Tygstrup, I., 1963, Oligohydramnios and urinary malformations in early human pregnancy, Acta Path. Micro. Scand., 59:273.

Walters, D. V., and Olver, R. E., 1978, The role of catecholamines in lung liquid absorption at birth, Pediat. Res., 12:239.

Watson, B. P., 1906, Withdrawal of the liquor amnii, J. Obstet. Gynecol. Brit. Emp., 9:15.

Yakovleva, I. V., 1965, Neirosekretornaga gipotalmo-gipofizarnaya sistema v rannem ontogeneze pozvonochnukha zhivotnykh i cheloveka, Arkhiv. Anat. Gist. i Embryol., 48:79.

THE ORCHESTRATION OF PARTURITION:

DOES THE FETUS PLAY THE TUNE?

G. D. Thorburn

Department of Physiology
Monash University
Clayton, Victoria, Australia 3168

INTRODUCTION

The debate about the relative importance of mother and fetus in the initiation of parturition can be traced back to as early as 460 BC. At this time, Hippocrates proposed that the fetus initiated its own delivery by placing both feet on the fundus of the uterus and pushing. Many years later, the great Italian anatomist, Fabricius ab Aquapendente, argued that the mother played the primary role in initiating delivery of the baby and this was achieved by the contraction of the uterine muscle.

In 1882, Spiegelberg first enunciated the concept that the maturity of the fetus may be linked to the initiation of its delivery. He wrote in a midwifery textbook that "the reason why labor occurs at a definite time must be sought for, not in the uterus and its changes, but in the fetus. It is the maturity of the latter that gives the signal." This theme linking fetal maturity to the initiation of parturition was continued by another obstetrician, Percy Malpas (1933), who investigated a series of cases of prolonged pregnancy associated with fetal anencephaly in humans. He concluded that "the time of the onset of labor is determined by the fetus" and "the fetal adrenal, pituitary, or nervous system, perhaps in combination, are suggested as tissues possibly concerned in the actual excitation of the neuromuscular expulsive mechanisms."

In this paper, I re-examine the proposition that the signal comes from the fetus to initiate parturition and that it does so when the fetus is sufficiently mature to survive in the outside world. This link between fetal maturity and parturition will be explored using a comparative approach. I shall limit my discussion to birds, monotremes, marsupials and eutherians, to demonstrate that as the major site of fetal maturation changes, the fetus requires different strategies to effect birth. Therefore, in birds and monotremes the fetus needs to escape from an egg outside the mother's body whereas in marsupials and eutherians the fetus needs to activate maternal changes in the uterus and cervix.

FETAL MATURITY AND THE INITIATION OF PARTURITION

Our understanding of the link between fetal maturation and the initiation of parturition has increased as a result of intense research on the growth and development of the sheep fetus. In this section, I shall use the fetal sheep as a model to illustrate the link between fetal maturation and the initiation of parturition. I shall then discuss the possible relationship between fetal maturation in other species and the timing of birth.

Cortisol and the Maturation of Fetal Organ Systems

Although the lamb is not fully independent for several months, it is still sufficiently mature at birth to stand, suck and withstand severe climatic conditions. A maturation of the fetal cardiovascular, respiratory and neuroendocrine systems must occur to allow the lamb to make the successful transition from intra- to extra-uterine life. It has become apparent that the maturation of many fetal organ systems is dependent on and coordinated by a prepartum increase in circulating cortisol concentrations (Liggins et al., 1970; Thorburn et al., 1981).

The ability of any newborn animal to maintain adequate ventilation is a good index of its maturity and of its likely survival. Lambs delivered prematurely by cesarean section before 140 days' gestation are unable to establish satisfactory respiration and develop severe respiratory distress. Liggins (1969) observed that when premature parturition was induced at day 135 after intrafetal glucocorticoid infusion, the lambs were able to expand their lungs at delivery. This suggested that the glucocorticoids had accelerated lung maturation and also the production of pulmonary surfactant. Surfactant production increases progressively from about day 120 to term and this correlates with an increase in the fetal concentrations of cortisol (Platzker et al., 1975). Administration of synthetic glucocorticoid to the fetal lamb after day 125 increased the rate of surfactant production.

Another important cause of neonatal mortality is hypoglycemia. During the first 24 h of life, the newborn lamb is at risk of hypoglycemia if it fails to establish regular sucking.

Studies of Comline and Silver (1961) demonstrate that the release of catecholamines in response to asphyxia increased progressively with gestational age. There was an increase in the proportion of adrenaline released in the older fetuses and this may be the result of activation of the enzyme phenylethanolamine-N-methyl transferase (PNMT) by cortisol. The physiological role of adrenaline at term has been reviewed recently by Largercrantz and Slotkin (1986). It prepares the lungs for respiration by stimulating the release of surfactant stored in type II pneumocytes. It inhibits secretion of fetal lung liquid and facilitates reabsorption of fluid from the lung increasing its compliance. It also mobilizes glucose and thus prevents the onset of neonatal hypoglycemia.

In the fetal lamb, liver glycogen accumulates during the last 15 days of pregnancy and this accumulation is prevented by fetal hypophysectomy or adrenalectomy (Barnes et al., 1978). Cortisol infusion into the intact fetus and also into the term adrenalectomized fetuses stimulated a marked increase in liver glycogen. The response of the hypophysectomized fetus was more variable and this suggests that fetal pituitary hormones, such as prolactin or growth hormone, may be required for glycogen accumulation in the fetal lamb.

Insulin secretory mechanisms in the fetal lamb begin to mature just before birth and the adult pattern is achieved in the few days immediately post partum (Bassett and Thorburn, 1971; Bassett et al., 1973). Since the switch to the adult secretory response can be induced prematurely after intrafetal infusion of synthetic ACTH, it has been suggested that the maturation of the insulin response is a consequence of the increasing cortisol secretion in the fetus before delivery.

The abrupt transition from the neutral thermal environment of the uterus to the extra-uterine environment poses problems for the newborn in the maintenance of its body temperature. Brown fat and glycogen accumulated in late pregnancy contribute to the thermogenesis but there is a need for increased production of triiodothyronine (T_3) (Fisher and Odell, 1969). During fetal life, the concentrations of thyroxine (T_4) and reverse T_3, the biologically inactive isomer of T_3, are high whereas the level of T_3 are low. This appears to result from a preferential conversion of T_4 to RT_3 rather than T_3 in the fetus (Chopra et al., 1975).

The increase in the fetal plasma concentrations of cortisol is temporally related to a decrease in the fetal T_4 and RT_3 concentrations (Thorburn and Hopkins, 1973) and an increase in T_3 concentrations (Klein et al., 1978). Intrafetal infusion of synthetic ACTH or dexamethasone caused a decrease in the circulating concentrations of T_4, and RT_3 and a three-fold increase in T_3 concentrations (Thomas et al., 1978). It is thought that the prepartum increase in fetal cortisol stimulates the monodeiodination of T_4 to T_3 rather than RT_3 in the fetal tissues and the biological activity is increased without altering the secretion rate of T_4 (Liggins et al., 1979; Fisher et al., 1977).

Another interesting role for the prepartum cortisol increase in the preparation of the fetus for extra-uterine life is in the control of erythropoiesis.

During fetal development in sheep there is a progressive change in the hemoglobins synthesized by fetal red blood cells. The two embryonic hemoglobins are replaced early in gestation by fetal hemoglobin (Bard et al., 1972; Wood et al., 1976). From 125 days the γ chains of fetal hemoglobin are gradually replaced by the β chains of adult hemoglobin. The 'switch' from fetal to adult hemoglobin production occurs synchronously in all erythropoietic tissues and coincides with the loss of the liver as an erythropoietic organ. As the timing of this 'switch' coincided with the increase in circulating cortisol concentrations, we suggested that cortisol may initiate the β chain synthesis by the fetal reticulocytes (Wood et al., 1976). We have also demonstrated that fetal hypophysectomy slowed the rate of switching from the fetal to the adult form of hemoglobin and also that intrafetal infusion of dexamethasone after day 130 caused a rapid switch in hemoglobin synthesis.

Therefore it can be seen that the development of the lamb to the stage where it can maintain physiological homeostasis at birth is absolutely dependent on the actions of high circulating cortisol concentrations. It is also now well established that the increase in fetal plasma concentrations of cortisol during late gestation is essential for birth to occur.

Experimental Evidence implicating the Fetal Pituitary-Adrenal Axis in the Initiation of Parturition in Sheep

In 1963, cases of prolonged pregnancy in sheep were reported, which were associated with a cyclopian-type fetal malformation. The teratogenic defect was due to the maternal ingestion of skunk cabbage (*Vertrum californicum*) on day 14 of gestation. Although the fetal lambs had pituitaries, the neural connections to these pituitaries were either missing or abnormal (Binns et al., 1963).

Liggins developed the technique of fetal hypophysectomy in sheep (Liggins et al., 1967). When more than 70% of the pituitary gland was destroyed, pregnancy was prolonged. Liggins and his co-workers also showed that hypothalamic destruction or pituitary stalk section prevented the initiation of parturition (Liggins, 1969). However, these studies did not establish whether the fetus played an active or permissive role in the birth process. Liggins (1969) demonstrated that intrafetal infusion of adrenocorticotrophic hormone (ACTH), but not other pituitary hormones, caused premature labor and that intrafetal infusion of glucocorticoids, but not mineralocorticoids, was also effective in inducing premature delivery. These experiments established that an intact fetal pituitary-adrenal axis was essential for the initiation of parturition in sheep and suggested that activation of this pathway would cause premature labor. However, these experiments still did not exclude the possibility that the fetal pituitary-adrenal axis had a permissive role in the initiation of parturition.

Definitive evidence that activation of this axis is involved in the initiation of parturition was presented by Bassett and Thorburn (1969). We showed that there was a gradual increase in fetal plasma glucocorticoids from around 130 days' gestation and that there was a marked increase in fetal corticosteroids in the last 3 to 4 days before birth. This finding was consistent with the observations that growth of the fetal lamb adrenal increased markedly in the last 10 days of intra-uterine life (Comline and Sliver, 1961) and that fetal hypophysectomy prevented and intrafetal infusion of ACTH enhanced this hypertrophy (Comline et al., 1970).

Since these initial observations, it has become apparent that the increase in plasma cortisol concentrations in the fetal lamb represents the link between the fetus and the mother and starts a train of endocrine events leading to parturition. The increase in the fetal plasma cortisol activates placental enzymes that metabolize progesterone to estrogen leading to a dramatic increase in the estrogen:progesterone ratio followed by the release of PGs and the onset of parturition (Fig. 1). These changes will be discussed in detail in a later section.

Therefore, it would appear that the mechanisms which control fetal development also control the timing of birth. I would like to develop the thesis that the fetus dictates the timing of its delivery to coincide with the developmental stage appropriate for its extra-uterine environment and level of parental care. I will pursue this argument through several species including birds, monotremes, marsupials and finally the human.

In this discussion, I define gestation as the length of intra-uterine development. Parturition occurs at the end of gestation and is the delivery of young from the uterus. 'Birth' will be defined as delivery of young into the outside world, be it by expulsion from the uterus or hatching from an egg laid some time previously.

Birds

In eutherians, we have seen that preparation of the fetus for entry into the external environment involves many maturational processes brought about by secretion of cortisol by the fetus. In avian species, such as the domestic hen (Gallus domesticus), fetal maturation occurs within a hard-shelled egg. It is clear that the events leading to oviposition are governed by the mother (Hertelendy, 1983) and are entirely divorced from the events leading to fetal maturation and subsequent hatching. At the time of oviposition (egg-laying), the embryo is less than 24 h old, and obviously unable to survive outside its shell. The hatching of a chick occurs 21 days later. I would suggest that hatching is the counterpart of 'birth' in eutherians. The timing of hatching appears to be dictated by the stage of maturation of the fetus. The chick starts air breathing a couple of days before hatching by pecking its way into the air sac. Whether the synthesis of lung surfactant is increased at this time in preparation for air breathing and whether

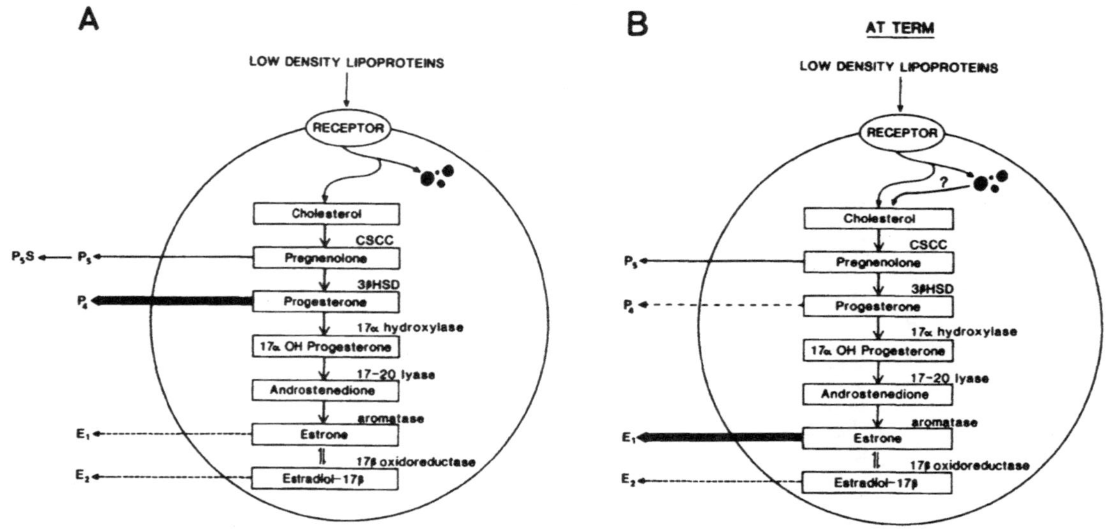

Fig. 1. Steroidogenic pathway in the ovine placenta during gestation (A) and at term (B). At term in response to increased fetal cortisol secretion, the activity of 17α-hydroxylase and C17-20 lyase is increased resulting in a switch from progesterone to estrogens.

this is induced by cortisol, as in eutherians, is unknown. It is interesting that the activity of many enzymes known to be induced by cortisol is increased in the chick just before hatching.

Monotremes

The echidnas, Tachyglossus aculeatus and Zaglossus bruijnii, and the platypus, Ornithorhynchus anatinus are the living representatives of the order Monotremata of the mammalian subclass Prototheria (see Griffiths, 1984 for detailed description). These fascinating creatures are the oldest extant order of Mammalia and are limited to Australia and New Guinea. They are called monotremes because they have only one opening for the genital, urinary and digestive tracts, like birds. They have a particular relevance in any comparative discussion on parturition as they are the only mammals that lay eggs. Little is known about the length of gestation in monotremes since they have not been bred in captivity. Estimates, made from animals brought in from the bush (wild), range from 9-17 days.

This period of gestation is considerably longer than in the domestic hen (about 24 h) and the extent of maturation of the embryo at the time of egg-laying is much greater. As with the hen the time of egg-laying (parturition) appears to be determined by the mother.

When the fertilized egg reaches the uterus it is 4-5 mm in diameter and as it grows to ca. 12 mm, the third layer of the shell becomes apparent. The egg continues to absorb nutritive fluid secreted by the cells of the uterine glands until it reaches its final size of 16.5 × 15 mm and the embryo has formed 18 somites (platypus) or 19 somites (echidna). Egg-laying has never been observed in monotremes and so details are not available.

After parturition in the echidna, the egg is incubated in a primitive ventral pouch and hatching has been observed after a 10.5 day incubation period. Organogenesis occurs in the embryo during this incubation period. The hatchling breaks through the shell, presumably using its egg tooth and caruncle, and struggles with a writhing motion of its head and forelimb to get out of the egg (Griffiths, 1984). The hatchling, although very immature, is, however, sufficiently rugged to survive, and, apparently, takes an active role in escaping from its shell. Whether the adrenal steroids play a role in preparing the hatchling for survival in the external environment is unknown. It is unknown whether the pituitary is functional at the time of hatching but the adrenal is well developed and could therefore be functional.

In monotremes, the emergence of lactation and suckling appears as a developmental initiative.

Marsupials

Although uniformly small related to body weight, the weight of marsupials at birth can vary enormously ranging from 5 mg for Tarsipes spencerae (Honey possum) to 800 mg in the newborn of large kangaroos such as Macropus rufus. As in the case of the monotreme hatchling, one cannot help but be impressed by the marsupial neonate at the time of birth. In both cases, they are able to breathe air, find their way to the teat, and to ingest milk and excrete the byproducts via their gut and mesonephric kidneys. Although the alveolae to not develop until some time after birth in the marsupial (Gemmell and Little, 1982), the lungs contain surfactant at the time of birth in the tammar wallaby and presumably other marsupials (Nicholas, personal communication).

Both the newborn marsupial and monotreme have relatively enormous forelimbs, rudimentary hindlimbs, and anatomically well-developed olfactory organs. These developmental modifications are needed to survive entering the world in a near-fetal condition and having to move about and find, unaided, the mother's mammary gland (Griffiths, 1984). All the organs concerned with vital functions must have reached a sufficient stage of maturity at birth to ensure the survival of the newborn. It does not seem unreasonable, therefore, that the immature marsupial should activate its own birth at a time when its organ systems have reached a sufficient stage of development.

The development of both monotremes and marsupials during early gestation is slow (Griffiths, 1984; Tyndale-Biscoe and Renfree, 1986). For example, the blastocyst of Macropus eugenii, the tammar wallaby, which has an active gestation of 27 days* from the diapause blastocyst stage, takes approximately 17-18 days RPY to achieve 15-20 pairs of somites (Renfree, 1973) whereas in monotremes, the egg at laying has reached a similar stage of development exhibiting 19-20 pairs of somites. This would be consistent with the suggestion that the length of gestation in monotremes is about 20 days (Griffiths, personal communication). As the incubation period in monotremes is about 10.5 days, and so both groups of mammals have reached a similar stage of development at parturition, or at hatching. Monotremes are clearly oviparous but Luckett (1977) has proposed that marsupials should be considered ovoviviparous. The embryo spends the early part of uterine life inside a shell and 'hatches' out of the shell around day 17 of gestation before completing the rest of gestation outside the shell but inside the uterus (Griffiths, 1984).

Therefore in marsupials, a similar stage of development to that in the monotreme egg is achieved within the uterus. Given that the monotreme initiates its own birth by hatching from an egg, it does not seem unreasonable to assume that the marsupial young, on reaching a similar level of maturity, may also initiate its own birth. However, in order to achieve this, the marsupial embryo must activate maternal mechanisms since parturition and hatching or birth into the external environment are no longer separate events. As far as mammals are concerned, we see in marsupials for the first time a synchrony of the fetal and maternal components at birth.

The short gestation and size of the embryo make it difficult to prove experimentally a fetal role for the initiation of parturition. It is known that the adrenals of some marsupials are capable of producing corticoids at birth (Gemmell et al., 1982) and in some, the pituitary of the newborn has cells containing secretory granults (Walker and Gemmell, 1983; Leatherland and Renfree, 1982, 1983) but the functional capacity of the pituitary-adrenal axis is not known. Evidence is available, from embryo transfer experiments in two species of kangaroo, the red and the grey, where it was found that the fetal genotype determines the length of gestation (Poole and Catling, 1974; Poole, 1975). Furthermore, if the young did not determine gestation length, it might be expected that the length of gestation would equal the length of the estrous cycle. If the fetus did play an active role in the timing of birth, one would expect either a shortening or a lengthening of the life of the corpus luteum depending upon the maturational needs of the fetus. In marsupials, the length of gestation is characteristically shorter than the estrous cycle, indicating that the embryo plays an active role in determining the length of gestation.

---

*The length of gestation in the tammar wallaby is 29 days after estrus including ovulation and uninterrupted by an embryonic diapause or 364 days including the diapause, or 27 days after reactivation of the diapause blastocyst by removal or loss of the suckling pouch (RPY) (Tyndale-Biscoe and Renfree, 1986).

Humans

In the human the premature infant may not survive due to the immaturity of the 'fetal' organ systems. The premature infant is at particular risk of respiratory distress syndrome which is due to both a deficiency of lung surfactant and an inability of the baby to inflate its lungs. Therefore, it is vital that birth should not occur until the fetal organ systems that are essential for postnatal survival are mature.

In the sheep, activation of the fetal adrenal induces enzymatic changes in the placenta resulting in a dramatic change in the maternal estrogen: progesterone ratio and a marked stimulation of PG production.

In the human, it has not been possible to demonstrate that the fetus per se is responsible for the initiation of labor. Whereas the fetal zone of the fetal adrenal, via the secretion of estrogen precursors plays an important role by increasing the concentration of estrogen in mother and fetus, this is a gradual process over the last trimester and does not provide the precise signal seen in the sheep.

Therefore, in the human, there is no dramatic change in the estrogen: progesterone ratio before parturition which appears to be initiated by an increased production of PGs by the fetal membranes, decidua and placenta.

In the sheep there may be a mechanism by which an inhibitor of PG synthesis is switched off enabling the uterus to become responsive to the changes in the steroidal hormones (Leach et al., 1983). The way in which the fetus switches off the production of this putative inhibitor(s) is unknown but such a mechanism could be important in the human. In the human switching off of a synthesis and/or release of PG inhibitors and the switching on of stimulators of PG release has been implicated in the initiation of parturition (Olson et al., 1983a; Casey et al., 1983; Billah and Johnston, 1983; Liggins, 1985). It is presently fashionable to consider that the trigger in initiating human parturition resides in the fetal membranes rather than the fetus per se (Mitchell et al., 1978; Olson et al., 1983b; Okazaki, 1981). However, the membranes are, of course, fetal in origin and thus have the same genotype.

If we assume that increased PG production in the reproductive tract initiates delivery, then a failure of inhibitory mechanisms which regulate PG synthesis may be the cause of premature labor. An increase in the circulating level of stress hormones (eg. adrenal glucocorticoids, catecholamines, opioids, etc.) may affect the synthesis and/or release of PG synthesis inhibitors leading to increased PG release and parturition (or abortion). Thus in adverse conditions, labor may be initiated without a fetal trigger and when the fetus is still immature. A similar mechanism is observed in goats where increased levels of adrenal glucocorticoids in the maternal circulation resulting from maternal stress leads to the transplacental movement of cortisol and to the activation of placental enzyme changes resulting in premature parturition.

SUMMARY

I have reviewed the available evidence that the timing of birth is controlled by the fetus and is intimately linked to the appropriate degree of maturity required for neonatal survival. In the domestic hen (*Gallus domesticus*) the chick is mature at birth and is able to seek its own food and water. In altricial species, the young are dependent on the foraging skills of the parents who directly feed the young sometimes after predigestion of the food by the parent (pigeon's milk and penguin's milk). In many ways, this may be regarded as the forerunner of suckling in Prototheria and

Eutheria. The ability of the young to be self sufficient is only gained after varying lengths of time. Nevertheless, I believe that the hatchling in all avian species is responsible for initiating its own entry into the world. Similarly, in Eutherian species there is a considerable variation in the degree of maturation at birth. This varies from the immature young of rodents, such as rats, to the mature young of hystricomorph rodents, such as guinea pigs. The degree of maturity is appropriate for the survival of each species depending upon the level of parental care and the environmental conditions in which the young are born.

It would seem completely illogical to have young born when they are immature and cannot survive, indeed it would spell the end of any species. As a corollary, it follows that if conditions (the external environment) change drastically for any species, then a pattern which was adequate previously may be completely inappropriate and would lead inevitably to the demise of the species unless some appropriate adaptation is made.

THE MATERNAL CONTROL OF THE FETAL CLOCK

Whereas the length of gestation is related to fetal maturity, in many species the photoperiod determines the precise time at which birth occurs on the last day of pregnancy (Lincoln and Porter, 1979). Mitchell and Yochin (1970) found that 87% of rats receiving 14 h of light per day delivered their young during the light (inactive) period. Boer et al. (1975) observed that a majority of rats delivered between 1200 and 1800 h on day 22 and a smaller group gave birth during the early hours of day 23. They noted that with the approach of the dark period at 2000 h on day 22, the animals became increasingly active and those who failed to deliver ceased showing signs of impending parturition. Lincoln and Porter (1979) showed in rats that an 8-h advance in the timing of the light-dark cycle before midpregnancy delays parturition by about 12 h. In this study, the effect of photoperiod was eliminated by lesions placed in the anterior hypothalamus, whereas pinealectomy had no effect. The decrease in plasma progesterone, the rise in prolactin and the appearance of 20α-hydroxysteroid dehydrogenase (HSD) activity in the ovaries were correlated with the time of parturition and were all delayed in rats subjected to the 8 h advance in the photoperiod.

These studies have been interpreted as indicating that the mother plays an important role in the precise timing of parturition. However, another interpretation is that fetal circadian rhythms may be entrained by those in the mother and that the precise timing of events in the fetus which determine parturition can therefore be reset by an altered photoperiod.

Reppert (1984) has demonstrated that the circadian rhythm of activity in the fetal suprachiasmatic nucleus is entrained to the circadian rhythm of activity in the maternal suprachiasmatic nucleus. Whether the activity of the fetal suprachaismatic nucleus is linked to the stimulation of the fetal pituitary-adrenal axis and hence to the initiation of parturition is unknown although it is an intriguing possibility.

Thus the length of gestation is related to fetal maturity but the precise hour of birth may be finally determined by the environmental light/dark cycle acting via the maternal and fetal circadian systems.

THE ROLE OF THE MOTHER AND THE REGULATION OF PROSTAGLANDINS

In Eutherians, it is clear that there are mechanisms to prolong the length of gestation beyond that of estrous or menstrual cycle. I shall briefly examine these mechanisms in both the sheep and human before discussing the maternal role in the events which lead to parturition.

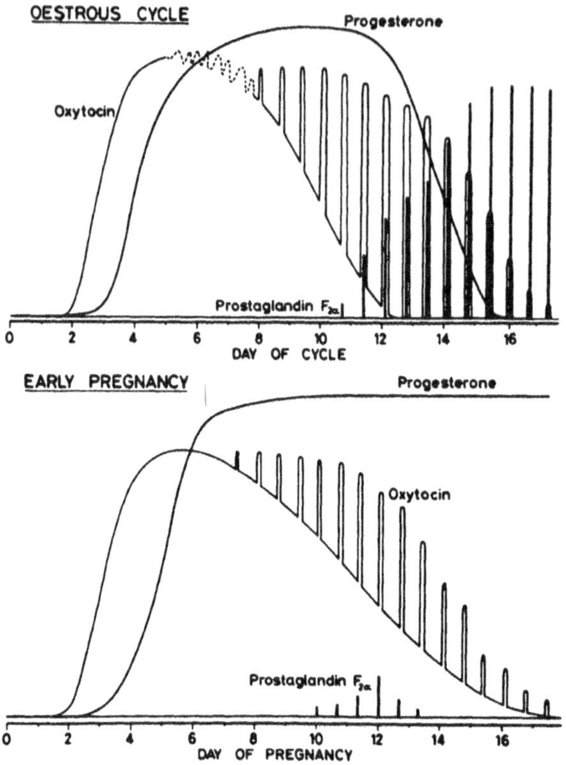

Fig. 2. Utero-ovarian vein concentrations (relative) of oxytocin, $PGF_{2\alpha}$ and progesterone during the estrous cycle and early pregnancy (adapted from Hooper et al., 1986).

Sheep - Early Pregnancy

In the sheep, the pulsatile release of oxytocin from the CL and posterior pituitary causes the pulsatile release of $PGF_{2\alpha}$ from the endometrium. The $PGF_{2\alpha}$, in turn, acts on the CL to cause luteal regression and a decrease in circulating progesterone levels. The action of estrogen on the endometrium is to induce oxytocin receptors which gradually appear towards the end of the luteal phase of the cycle.

In early pregnancy, we found that the pulsatile release of $PGF_{2\alpha}$ was inhibited (Thorburn et al., 1972, 1973) and suggested that this was the cause of luteal maintenance. Subsequently we have shown that the pulsatile release of oxytocin is still present during days 13-16 of early pregnancy (Hooper et al., 1986) but that the pulses of oxytocin fail to elicit PG release (Fig. 2). Others have shown (McCracken, 1980; Philpott et al., 1985) that the population of oxytocin receptors in the endometrium is markedly decreased during early pregnancy. It would seem that the ovine conceptus acts on the endometrium to inhibit the synthesis of oxytocin receptors and possibly also estrogen receptors. This would indirectly block the release of uterine PGs and prevent luteal regression. The mechanism by which the conceptus achieves this is not known, although the ovine conceptus is known to secrete a number of proteins. One of these proteins, possibly OTP_1, described by Godkin and his colleagues (Godkin et al., 1984), may be involved.

I believe that in all eutherians, some mechanism has been evolved to regulate uterine PG synthesis and to permit a prolongation of gestation. A majority of eutherian mammals are dependent on the corpus luteum as a source of progesterone throughout pregnancy (eg. goat, cow, rat, rabbit), and the preparturient decrease in progesterone is associated with regression of the corpus luteum of pregnancy (see Thorburn and Challis, 1979). The luteolytic mechanisms involve the release of $PGF_{2\alpha}$, and are generally similar to those responsible for inducing luteal regression at the end of the estrous cycle (in the case of the goat see Currie and Thorburn, 1976). The withdrawal of the PG inhibitory mechanisms, together with the release of factors such as estrogen and oxytocin that stimulate PG synthesis and release, would then lead to increased PG release, luteolysis and a fall in progesterone concentrations.

In sheep, where progesterone is produced by the placenta, an alternative mechanism is required to achieve the preparturient decrease in progesterone.

Parturition in Sheep

The increase in the plasma concentration of cortisol in the fetal lamb represents an early link in the chain of endocrine events leading to parturition. It has been demonstrated that this increase activates (or induces) placental enzymes that metabolize progesterone to estrogens (Anderson et al., 1975) resulting in a decrease in progesterone in maternal plasma and an increase in estrogens (see Thorburn and Challis, 1979). Recently, Nathanielsz and coworkers have described the hormonal changes before parturition in the sheep in detail (Nathanielsz et al., 1982). They found that the concentration of pregnenolone (Δ5P) and pregnenolone sulphate (Δ5PS), in maternal plasma, decreased in parallel with progesterone during the last 4 days of gestation. The increase in maternal estrone and estrone sulphate followed a similar time course (Magyar et al., 1981). In an earlier study, these workers reported that there were very high levels of Δ5P and Δ5PS in maternal and fetal sheep plasma during the last third of pregnancy. Recent data from our laboratory (McKay et al., 1983) showed that the ovine placenta secretes little Δ5PS into the maternal circulation. These data suggest that pregnenolone is secreted by the placenta and is sulphoconjugated elsewhere. Nathanielsz and coworkers (1982) raised the possibility that Δ5PS may be used as a precursor for estrogen biosynthesis.

It may be, however, that the decrease in plasma levels of Δ5P and Δ5PS observed before parturition may be due to an increased activity of the placental 3βHSD. This would facilitate the conversion of Δ5P through to estrogen rather than it being secreted into the maternal and fetal circulation. It seems possible, therefore, that fetal cortisol may increase the activity of at least three enzymes in the metabolic pathway of estrogen and progesterone.

These preparturient changes in the concentration of progesterone and estrogens in maternal plasma have important consequences. PG production by intra-uterine tissues is increased, oxytocin release from the maternal pituitary is facilitated and the action of oxytocic agents on the myometrium is augmented either by influencing appropriate receptor concentrations and/or provoking ultrastructural changes that enhance electrical conductivity.

In the sheep, the concentration of $PGF_{2\alpha}$ in the utero-ovarian vein increases in parallel with the preparturient increase in estrogen production. Administration of estradiol to pregnant ewes causes a marked increase in PGF release into the utero-ovarian vein and premature delivery. This occurs in the absence of a change in progesterone concentrations in either peripheral or utero-ovarian venous plasma (Currie, 1974). Premature delivery and PGF release can also be induced by progesterone 'withdrawal' following the administration of 3β-hydroxysteroid dehydrogenase inhibitors (Jenkin et al., 1985). Thus a decrease in the circulating levels of progesterone may be a sufficient stimulus to increase PG output in these animals. Undoubtedly, the combination of an increase in estrogen concentration and a decrease in progesterone production has a potent effect on the release of PGs. Recent studies in sheep indicate the major sites of PG synthesis at parturition are the chorio-allantoic membrane and placenta (Mitchell et al., 1976; Evans et al., 1982; Olsen et al., 1985).

The earlier studies of Mitchell and Flint (1978), using a perifusion technique, showed that the late pregnant placenta produced mainly PGE_2. Recently Fowden et al. (personal communication) have shown increasing levels of PGE_2 in maternal and fetal arterial and utero-ovarian venous plasma during the last third of pregnancy. This would indicate that the placenta produces significant amounts of PGE_2. They have suggested that this PGE_2 plays a significant metabolic role and provides a mechanism by which the placenta can dictate metabolic events in the mother and fetus. The PGE_2 produced by the placenta may regulate placental blood flow. Moreover, the significant concentration of PGE_2 measured in maternal arterial plasma (4.8 nmol/l at day 120 increasing to 13.0 nmol/l at parturition) may be important in the evolution of uterine activity during the latter part of pregnancy.

Most investigators have interpreted the regulation of PG biosynthesis as an action of steroid hormones on one or more rate-limiting enzymes of the arachidonic acid cascade. The release of $PGF_{2\alpha}$ at delivery may be the result of stimulation and/or removal of an inhibitor of PG synthesis and release. Recently a number of endogenous inhibitors of PG synthesis have been described. We have reported the presence of a substance in ovine allantoic fluid before 140 days gestation that will inhibit the synthesis of PGs by an ovine endometrial microsomal preparation (Leach et al., 1983). Shemesh and his colleagues (1984) have also isolated a protein from bovine cotyledons (120-240 days' gestation) which inhibit PG biosynthesis. In both instances, the inhibitory substances were not detected around the time of delivery.

Recently, using an in vitro dispersed cell technique, we have shown that the PG synthetic activity of placental cells was low. This may explain the very low circulating concentrations of PGs measured during midpregnancy (50-100 days' gestation) in the sheep (Risbridger et al., 1984). The PG production of these cells increased markedly as term approached.

Estrogens are thought to exert their stimulatory action on PG biosynthesis by increasing the activity of both phospholipases and cyclo-oxygenase. The mechanism by which progesterone exerts its inhibition of PG release is unclear but it has been proposed that it may induce the synthesis of a protein that inhibits phospholipase activity (Thorburn, 1985). Indeed, the action of all the steroidal hormones on PG biosynthesis may be manifest through the action of inhibitors or stimulators of the PG synthetic enzymes.

In sheep, the steroid hormones also influence the release of oxytocin from the posterior pituitary. Oxytocin release is greater in response to vaginal stimulation as parturition approaches (Flint et al., 1975); the release of oxytocin increases with the estrogen:progesterone ratio.

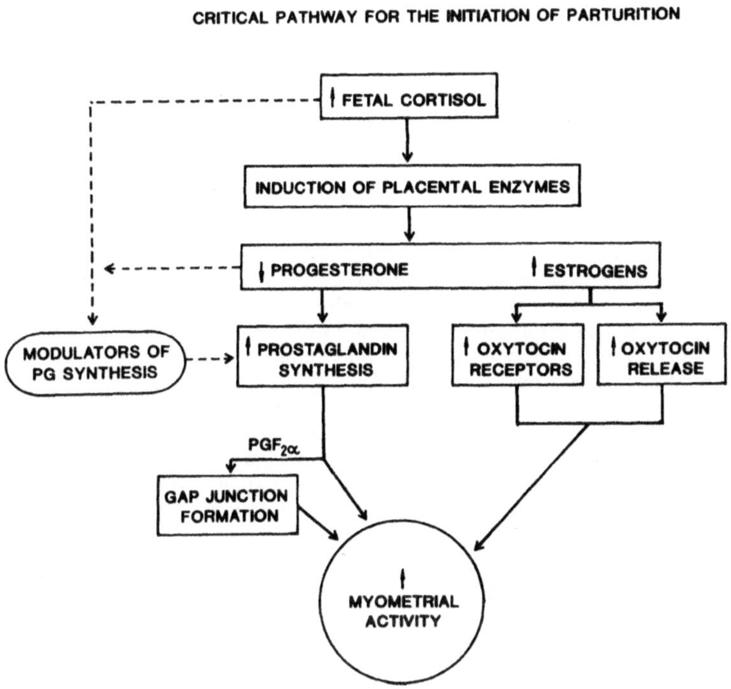

Fig. 3. A model of the regulation of ovine myometrial activity during pregnancy.

In sheep, during late gestation, there is a good correlation between changes in gap junction formation in the myometrium and the estrogen: progesterone ratio (Garfield et al., 1979). In the rat, Garfield and his coworkers have shown a low frequency of gap junctions in the myometrium from animals not in labor. There was a marked increase in the frequency of gap junction in the myometrium during labor which correlated with the change in the estrogen:progesterone ratio. They found that progesterone, in the presence of estrogen, blocked the formation of gap junctions whereas estrogen treatment alone stimulated gap junction formation (Garfield et al., 1982). In vitro, indomethacin (an inhibitor of PG synthesis) and carbacyclin (a PGI_2 analogue) inhibited the formation of gap junctions. Garfield et al. (1982) proposed that estrogen regulates the formation of gap junctions by stimulating protein synthesis. PGs may then influence the insertion of gap junctions into the myometrial cell membrane. This is consistent with the suggestion that PGs facilitate the communication between myometrial cells and enhance propagation in the myometrium resulting in the synchronized contractions seen in labor (Thorburn, 1985). Not only are PGs involved in the synchronous contractions of the myometrium during labor, they are also involved in cervical ripening and softening (Fitzpatrick and Liggins, 1980; Fitzpatrick and Dobson, 1981).

Therefore in the sheep, parturition is dependent upon a sequence of endocrine events culminating in the synthesis and release of PGs which exert their action on the myometrium and cervix (Fig. 3). As we shall see from a comparative viewpoint, the PGs predate the steroid hormones in playing a key role in parturition. Studying the evolution of parturition, I am enormously impressed with how the PGs have retained their primary role.

Birds

Oviposition, the laying of hard-shelled egs, can be regarded as analogous to parturition in eutherian mammals. In birds it is possible to separate the maternal events of parturition, namely, increased uterine activity and changes in the birth canal, from the fetal events which occur in the egg after delivery.

Modern hybrid hens ovulate once every day, 15-30 min after oviposition of the preceding egg. At ovulation, the ovum is released from the ruptured follicle and is picked up by the infundibulum and passed via the oviduct into the uterus. During its passage through the tubular region of the oviduct, albumin is deposited around the yolk and the membranes are formed. These functions of the oviduct are controlled by the steroids. The soft-shelled egg enters the uterus or shell gland, where, after the uptake of water, the hard calcareous shell is deposited. Every 15 min during this 18-h period of active shell formation the animal removes the equivalent of the total blood calcium from the circulation. The uterus is lined by a thick layer of secretory cells which are responsible for the rapid rate of calcium removal and its secretion, mainly as calcium carbonate (Hertelendy, 1983).

Experimental evidence implicating PGs in oviposition was first obtained in the Japanese quail (Hertelendy, 1983). Intra-uterine injection of a low dose of PGs caused premature oviposition. Arachidonic acid (AA), the precursor for PGs, induced oviposition and both indomethacin and aspirin, inhibitors of a key enzyme in PG synthesis, cyclo-oxygenase, blocked this action. Injection of indomethacin several hours before the expected time of oviposition also delayed spontaneous oviposition for up to 48 h in chicken and turkey hens. Phospholipase A_2, the rate-limiting enzyme in the release of AA from phospholipids, induced premature oviposition, and this effect was blocked by indomethacin. These studies clearly demonstrated that the avian uterus is capable of producing PGs and they have a major role in oviposition.

There are a number of other uterotonic agents which will induce premature oviposition (e.g. other oxytocin and vasotocin) that β-mimetic drugs have been shown to inhibit both the action of PGs and the other uterine stimulants.

In eutherians, PGs not only increase uterine activity but also soften the cervix facilitating cervical dilation and the birth of the fetus. In chickens at the uterovaginal sphincter, PGE has been shown to relax the avian vagina. Thus in birds, as in eutherians, PGs both stimulate uterine contractions and relax the vagina. These coordinated actions will facilitate the expulsion of the egg.

Monotremes and Marsupials

Nothing is known concerning the role of PGs in the process of egg-laying in monotremes. In contrast to birds, monotremes have functional corpora lutea which secrete progesterone (Carrick et al., 1975). The precise timing of luteal regression in relation to egg-laying is uncertain (Tyndale-Biscoe and Renfree, 1986, Chapter 10). Hill and Gatenly (1926) argued that luteal regression commenced well before the egg was laid but Griffiths (1984) questioned this conclusion. In many CL-dependent species, for example the goat (Currie and Thorburn, 1977), uterine PGs initiate luteal regression, decrease progesterone concentrations and facilitate uterine PG release thus stimulating uterine activity. In monotremes, PGs also may cause luteal regression and a decrease in progesterone concentrations before oviposition. However, it is not known whether PGs are involved in the actual process of egg-laying.

On the other hand, there is considerable evidence that the PGs are important in parturition in marsupials (Tyndale-Biscoe and Renfree, 1986). A metabolite of $PGF_{2\alpha}$ has been found in the peripheral plasma of three marsupials - the bandicoot (Isoodon macrourus, Gemmell et al., 1980); the tammar wallaby (Macropus eugenii, Tyndale-Biscoe et al., 1983; Shaw, 1983a; Lewis et al., 1986); and the red necked wallaby (Macropus rufogriseus, Walker and Gemmell, 1983).

There is a marked increase in the concentrations of $PGF_{2\alpha}$ in the endometrium and myometrium of the tammar wallaby (M. eugenii) between days 25 and 27 of gestation after R.P.Y. (term = 26.5-27.0 days; Shaw 1983a, b). At the time of delivery, the plasma concentrations of PGFM were only transiently elevated with peak concentrations (2.5 ng/ml) observed immediately before or at birth. Furthermore, minutes after delivery the plasma PGFM concentrations declined to low levels and were undetectable at 2 h post partum (Lewis et al., 1986). The progesterone concentration decreased rapidly during the 24 h before delivery and in 5 of 10 animals in this study progesterone decreased after peak PGFM concentrations. Therefore in the tammar wallaby, PGs may be involved in luteal regression. The decrease in circulating progesterone concentrations coincided with an increase in the concentration of estradiol resulting in a marked change in the ratio of the two hormones before delivery (Tyndale-Biscoe and Renfree, 1986). A transient increase in the plasma concentrations of prolactin has been measured, the peak preceded the fall in progesterone by 4 h in pregnant but not in nonpregnant animals. Therefore, Tyndale-Biscoe and his coworkers (1983) have suggested that the conceptus may initiate luteal regression and parturition by triggering the release of prolactin from the maternal pituitary. Lewis et al. (1986) have concluded that in the tammar wallaby, the essential stimulus for parturition is prostaglandin whereas the prepartum prolactin pulse is important for the initiation of lactogenesis.

In the M. eugenii, during days 13-18 of pregnancy after R.P.Y., little spontaneous activity was observed in either the pregnant or nonpregnant uterine horn and the uterus was unresponsive to oxytocic agents. By day 20 R.P.Y., the gravid uterus responded to oxytocin and the PG analogue, cloprostenol, and the response to both increased as term approached (Renfree and Young, 1979; Shaw, 1983). The myometrium was refractory to both oxytocin and PGs on day 1 post partum. Therefore, the high circulating concentrations of PGs measured before delivery may be important in stimulating myometrial activity.

In M. eugenii, the embryo develops in one or other of the uterine horns. At parturition, the embryo is delivered through the cervix and then the medium vaginal or birth canal. It is probable that the PGs act on the connective tissue of the urogenital strand leading to the formation of the pseudovaginal or birth canal. At birth the histological changes in the connective tissue of the cervix and that surrounding the medium vaginal canal are strikingly similar to those seen in the human and ovine cervix during ripening and softening at parturition.

In the bandicoot (Isoodon macrourus), the CL is retained following delivery and progesterone levels remain elevated for 20-30 days (Gemmell et al., 1980). These authors observed a well-defined peak of PGFM, during the 24 h before and after parturition. Therefore, in this species, the PGs are not luteolytic but probably stimulate uterine activity leading to delivery of the young and possibly influence the development of the birth canal.

A characteristic of the eutherians is the ability of the conceptus to inhibit PG release in the reproductive tract. In some species this leads to maintenance of the CL, and in others to the inhibition of menstruation, and in all to the suppression of uterine activity and failure of the cervix to dilate. By these means, the length of the estrous or menstrual cycle is increased and the myometrium is maintained quiescent permitting the continued development of the fetus within the uterus and an extended gestation. In eutherian species, it is only when this inhibition of PG release is withdrawn, and PG synthesis and release is stimulated, that parturition occurs.

In marsupials, the short gestation may result from a failure of the conceptus to inhibit endometrial and myometrial PG synthesis and release. The above evidence indicates that even in the marsupial which has a unique niche in evolution, PGs may have a central role in parturition. This crucial role is preserved in the primate.

Human - Early Pregnancy

In the human, the end of the menstrual cycle results from regression of the CL and a subsequent decrease in the plasma concentrations of progesterone. This is followed by menstruation and the associated increase in uterine activity. High levels of PGs have been found in menstrual fluid and in the endometrium at menstruation and PGs are considered to be responsible for both menstruation and the associated uterine activity. Clearly both are incompatible with the maintenance of a pregnancy and the human conceptus overcomes this problem in at least two ways. Firstly, it secretes human chorionic gonadotropin, a luteotropic hormone that stimulates the production of progesterone which in turn inhibits uterine activity and may also inhibit PG synthesis and release. The conceptus also inhibits endometrial PG synthesis thus preventing menstruation and increased uterine activity. In contrast to the high concentrations of PGs found in the late luteal endometrium, the decidua (endometrium) of early pregnancy has been found to contain low amounts of PGs (Abel and Baird, 1982). Even in the presence of a tubal pregnancy, Abel and Baird found that the endometrium of the body of the uterus contained low amounts of PGs. This suggested that the conceptus may release a circulating inhibitory factor. The nature of this putative PG-synthetic inhibitory factor is unknown. Clearly, in the human the conceptus acts at an early stage to maintain pregnancy by inhibiting menstruation and uterine activity. This allows it to develop within a quiescent uterus. At the end of pregnancy, the mature term fetus then initiates parturition.

Human Parturition

In the sheep high concentrations of progesterone act to inhibit uterine activity and to render the myometrium insensitive to exogenous PGs. The level of the progesterone 'block' in this species, although gradually decreasing, is significant throughout pregnancy. However, the administration of 3βHSD inhibitors, which cause a decrease in progesterone concentrations, will generally initiate parturition and increase the sensitivity of the myometrium to PGs (Jenkin et al., 1985; Thorburn, 1983, 1985). The preparturient withdrawal of progesterone is also needed for normal cervical dilatation.

Similarly, in humans, progesterone exerts a blocking action on the myometrium which is more profound early in pregnancy and decreases markedly as pregnancy progresses. In early pregnancy, the administration of 3βHSD inhibitors or progesterone receptor antagonists (RU486) make the uterus sensitive to exogenous PGs. A combination of these treatments can induce abortion. The decrease in the progesterone-block on the human myometrium means that endogenous PG production and release must be kept at a low level if premature parturition is to be averted. Thus the maintenance of pregnancy in the human depends on inhibition of PG release. Another factor is the progressive increase in maternal estrogen concentration due to the progressive increase in the production of estrogen precursors by the fetal adrenal. Estrogen has a direct stimulatory effect on the activity of the PG synthetic enzymes (phospholipases, cyclo-oxygenase) and increases the number of oxytocin receptors. This accounts for the progressive increase in oxytocin sensitivity of the human myometrium during the latter part of pregnancy. The increase in estrogen production as a result of fetal adrenal activity is a clear example of the active role of the human fetus in the initiation of parturition.

The extremely low concentrations of PGs in the maternal circulation and amniotic fluid during pregnancy is clear evidence that the PG inhibitory mechanisms are effective. It is controversial whether there is an increase in PG concentrations either in maternal plasma or amniotic fluid before labor (Keirse, 1984). However, there is clear evidence of a progressive increase in the concentrations of $PGF_{2\alpha}$ and PGE_2 in amniotic fluid during cervical dilatation (Keirse, 1984). The major site of production of the PGs remains a matter for debate (Liggins, 1985) but the amnion, chorion, decidua and placenta as well as the myometrium and cervix are capable of producing PGs. McDonald and his coworkers (1978) have shown large amounts of arachidonic acid are released during labor into the amniotic fluid, presumably from the membranes. Liggins (1985) has suggested that there may be cooperation between tissues such that arachidonic acid released from the amnion may be converted to $PGF_{2\alpha}$ by the decidua and myometrium and to PGE_2 by the amnion itself.

An Endogenous Inhibitor of Prostaglanding Synthesis (EIPS) has been found in high concentrations in maternal plasma during pregnancy but EIPS concentrations do not decrease during labor (Brennecke et al., 1982). Therefore, although this factor may contribute to inhibition of PG synthesis during pregnancy, it would not appear to play a role in the initiation of labor (Liggins, 1985).

Liggins and his coworkers (Wilson et al., 1985) have identified two proteins in human amniotic fluid of approximately 80K and 160K daltons, respectively, that are specific antiphospholipases and inhibit the release of arachidonic acid from dispersed human endometrial cells. Data from co-culture experiments of amnion and endometrial cells indicate that these proteins may come from the amnion. They are present in the supernatant from cultures of amnion cells in tissues collected before but not after labor (Liggins, 1985). The inhibitory proteins are presumably of fetal origin (i.e. either from the fetus or membranes) and their synthesis and/or release would appear to be regulated by the fetus. The nature of the regulatory mechanisms controlling the concentrations of these proteins is unknown.

Casey et al. (1983) have shown that fetal (and adult) urine contains a protein of >10K daltons that stimulates PGE_2 synthesis in human amnion cells. Therefore, the release of PGs that initiate labor in the human is due to the withdrawal of inhibitory factors and the release of stimulatory factors on PG synthesis. By this stage of pregnancy the myometrium is more sensitive to PGs. The PGs also act on the cervix to cause cervical ripening and effacement. The changes in the cervix, together with the gradual increase in uterine activity, cause the rupture of lysosomes in the decidua and membranes releasing more phospholipases and further increasing the release of PGs.

There is still controversy about the role of oxytocin in the initiation of human labor (Fuchs and Fuchs, 1984; Soloff, 1983). A marked increase in the number of oxytocin receptors has been measured in human uterine tissues, particularly in the decidua during early term labor and also during premature labor. The increase in oxytocin receptors is probably due to the increase in estrogen concentrations. It is unclear whether PGs can increase oxytocin receptors. This would act to amplify the events at term as oxytocin is known to stimulate PG synthesis. Therefore, oxytocin may either play a primary role or an important permissive role in the initiation of labor.

CONCLUSION

In this paper, I have developed the argument that it would be inappropriate for the survival of the species, for the mother to initiate delivery except in life-threatening circumstances where premature delivery may save the life of the mother. I have proposed that the signal to initiate parturition must come from the fetus when it is sufficiently mature to survive as a neonate.

A comparative approach uniquely illustrates the relative roles of the mother and fetus in the initiation of parturition. In birds, the mother determines the timing of oviposition but the fetus clearly determines the timing of its 'birth' into the environment. In monotremes, the immature hatchling makes its way out of an egg, and we see the emergence of lactation and suckling as the immature young requires prolonged suckling before independence is achieved.

In marsupials, there is for the first time limited intra-uterine development followed by a prolonged period of development in the pouch. In the absence of definitive evidence, it would seem probable that the immature fetus initiates its own delivery from the uterus and makes its own way to the pouch. Finally, the eutherians have developed mechanisms to prolong their intra-uterine existence through the suppression of PG synthesis and release in the reproductive tract of the mother. At the end of pregnancy, the mature fetus initiates a train of complex and coordinated endocrine events which will ensure its safe delivery into the outside world.

ACKNOWLEDGEMENTS

I wish to gratefully acknowledge the help of Drs. I. C. McMillan, G. Jenkin and G. E. Rice in the preparation of this manuscript and talk. I would also like to thank Professor R. V. Short, Drs. Marilyn Renfree, M. Griffiths and R. T. Gemmell for reading the manuscript, offering constructive suggestions and the loan of some beautiful slides for my presentation.

This work was supported by a Program Grant to the author from the National Health and Medical Research Council of Australia.

REFERENCES

Abel, M. H., Smith, S. K., and Baird, D. T., 1980, Suppression of concentration of endometrial prostaglandin in early intra-uterine and ectopic pregnancy in women, J. Endocrinol., 85:379.

Anderson, A. B. M., Flint, A. P. F., and Turnbull, A. C., 1975, Mechanism of action of glucocorticoids in induction of ovine parturition: effect on placental steroid metabolism, J. Endocrinol., 66:61.

Bard, H., Battaglia, H. C., Makowski, E. L., and Meschia, G., 1972, The synthesis of fetal and adult hemoglobin in sheep during the perinatal period, Proc. Soc. Exp. Biol. Med., 139:1148.

Bassett, J. M., and Thorburn, G. D., 1969, Foetal plasma corticosteroids and the initiation of parturition in sheep, J. Endocrinol., 44:285.

Bassett, J. M., and Thorburn, G. D., 1971, The regulation of insulin secretion by the ovine fetus in utero, J. Endocrinol., 50:59.

Bassett, J. M., Thorburn, G. D., and Nicol, D. H., 1973, Regulation of insulin secretion in the ovine foetus in utero. Effects of sodium valerate on secretion and of adrenocorticotrophin-induced premature parturition on the insulin secretory response to glucose, J. Endocrinol., 56:13.

Billah, M. M., and Johnston, J. M., 1983, Identification of phospholipid platelet-activating factor in human amniotic fluid and urine, Biochem. Biophys. Res. Commun., 113:51.

Billah, M. M., Di Renzo, G. C., Ban, C., Truong, C. T., Hoffman, D. R., Anceschi, M. M., Bleasdale, J. E., and Johnston, J. M., 1985, Platelet-activating factor metabolism in human amnion and the response of this tissue to extracellular platelet-activating factor, Prostaglandins, 30:841.

Binns, W., Lynn, J. F., Shupe, J. L. et al., 1963, A congenital cyclopian-type malformation in lambs induced by maternal ingestion of a range plant, Veratrum californicum, Am. J. Vet. Res., 4:1164.

Bleasdale, J. E., Okazaki, T., Sagawa, N., Di Renzo, G. C., Okita, J. R., MacDonald, P. C., and Johnston, J. M., 1983, The mobilization of arachidonic acid for prostaglandin production during parturition, in: "Initiation of Parturition: Prevention of Prematurity," Report of Fourth Ross Conference on Obstetric Research, p. 129, Ross Laboratories, Columbus, Ohio.

Boer, K., Lincoln, D. W., and Swaab, D.F., 1975, Effects of electrical stimulation of the neurohypophysis on labor in the rat, J. Endocrinol., 65:163.

Brennecke, S. P., Bryce, R. L., Turnbull, A. C., and Mitchell, M. D., 1982, The prostaglandin synthase inhibiting ability of maternal plasma and the onset of human labor, Eur. J. Obstet. Gynaecol. Reprod. Biol., 14:81.

Carrick, F. N., Drinan, J. G., and Cox, R. I., 1975, Progesterone and estrogens in peripheral plasma of the platypus Ornithorhynchus anatinus, J. Reprod. Fertil., 43:375.

Casey, M. L., MacDonald, P. C., and Mitchell, M. D., 1983, Stimulation of prostaglandin E_2 production in amnion cells in culture by a substance(s) in human fetal and adult urine, Biochem. Biophys. Res. Commun., 114:1056.

Chopra, I. J., Sack, J., and Fisher, D. A., 1975, 3,3',5'-triiodothyronine (reverse T_3) and 3,3',5'-triiodothyronine (T_3) in fetal and adult sheep, Endocrinology, 97:1080.

Comline, R. S., and Silver, M., 1961, The release of adrenaline and noradrenaline from the adrenal glands of the foetal sheep, J. Physiol., 156:424.

Comline, R. S., Nathanielsz, P. W., Paisey, R. B., and Silver, M., 1970, Cortisol turnover in the sheep fetus immediately prior to parturition, J. Physiol. (London), 210:141P.

Currie, W. B., 1974, Fetal and maternal hormone interactions controlling parturition in sheep and goats, Ph.D. thesis, Macquarie University, Sydney, Australia.

Currie, W. B., and Thorburn, G. D., 1976, Release of prostaglandin F, regression of corpora lutea and induction of premature parturition in goats treated with estradiol-17β, Prostaglandins, 12:1093.

Currie, W. B., and Thorburn, G. D., 1977, The fetal role in timing the initiation of parturition in the goat, in: "The Fetus and Birth," J. Knight and M. O'Connor, eds., p. 49, Ciba Foundation Symposium 47 (New Series), Excerpta Medica, Elsevier, New York.

Ellwood, D. A., Mitchell, M. D., Anderson, A. B. M., and Turnbull, A. C., 1980, The in vitro production of prostanoids by the human cervix during pregnancy: preliminary observations, Br. J. Obstet. Gynaecol., 87:210.

Evans, C. A., Kennedy, T. G., and Challis, J. R. G., 1982, Gestational changes in prostanoid concentrations in intra-uterine tissues and fetal fluids from pregnant sheep, and the relation to prostanoid output in vitro, Biol. Reprod., 27:1.

Fisher, D. A., and Odell, W. D., 1969, Acute release of thyrotropin in the newborn, J. Clin. Invest., 48:1670.

Fisher, D. A., Dussalt, J. H., Sack, J., and Chopra, I. J., 1977, Ontogenesis of hypothalamic-pituitary-thyroid function and metabolism in man, sheep and rat, Rec. Prog. Horm. Res., 33:59.

Fitzpatrick, R. J., and Liggins, G. C., 1980, Effects of prostaglandins on the cervix of pregnant women and sheep, in: "Dilatation of the Uterine Cervix," F. Naftolin and P. G. Stubblefield, eds., p. 287, Raven Press, New York.

Fitzpatrick, R. J., and Dobson, H. 1981, Softening of the ovine cervix at parturition, in: "The Cervix in Pregnancy and Labour," D. A. Ellwood and A. B. M. Anderson, eds., p. 40, Churchill Livingston, London.

Flint, A. P. F., Forsling, M. L., Mitchell, M. D., and Turnbull, A. C., 1975, Temporal relationship between changes in oxytocin and prostaglandin F levels in response to vaginal distention in the pregnant and puerperal ewe, J. Reprod. Fertil., 43:551.

Fuchs, A-R., and Fuchs, F., 1984, Endocrinology of term and preterm labor, in: "Preterm Birth," F. Fuchs and P. G. Stubblefield, eds., p. 39, McMillan Publishing Co., New York.

Garfield, R. E., Rabideau, S., Challis, J. R. G., and Daniel, E. E., 1979, Hormonal control of gap junction formation in sheep myometrium during parturition, Biol. Reprod. 21:999.

Garfield, R. E., Puri, C. P., and Csapo, A. I., 1982, Endocrine, structural and functional changes in the uterus during premature labor, Am. J. Obstet. Gynecol., 142:21.

Gemmell, R. T., Jenkin, G., and Thorburn, G. D., 1980, Plasma concentrations of progesterone and 13,14-dihydro-15-keto prostaglandin $F_{2\alpha}$ at parturition in the bandicoot, Isoodon macrourus, J. Reprod. Fertil., 60:253.

Gemmell, R. T., Singh-Asa, P., Jenkin, G., and Thorburn, G. D., 1982, Ultrastructural evidence for steroid-hormone production in the adrenal of the marsupial, Isoodon macrourus, at birth, The Anatomical Record, 203:505.

Gemmell, R. T., and Little, G. J., 1982, The structure of the lung of the newborn marsupial bandicoot, Isoodon macrourus, Cell and Tissue Research, 223:445.

Godkin, J. D., Bazer, F. W., and Roberts, R. M., 1984, Ovine trophoblast protein 1, an early secreted blastocyst protein, binds specifically to uterine endometrium and affects protein synthesis, Endocrinology, 114:120.

Griffiths, M., 1984, Mammals: monotremes, in: "Marshall's Physiology of Reproduction, Vol. 1, Reproductive cycles of vertebrates," G. E. Lamming, ed., 4th Edition, p. 351, Churchill Livingston, New York.

Hertelendy, F., 1983, Regulation of oviposition, in: "Initiation of Parturition: Prevention of Prematurity," Report of Fourth Ross Conference on Obstetric Research, p. 79, Ross Laboratories, Columbus, Ohio.

Hill, J. P., and Gatenly, J. B., 1926, The corpus luteum of the monotremata, Proc. Zool. Soc. London, 47:715.

Hooper, S. B., Watkins, W. B., and Thorburn, G. D., 1986, Oxytocin, oxytocin associated neurophysin and prostaglandin $F_{2\alpha}$ concentrations in the utero-ovarian vein of pregnant and nonpregnant sheep, Endocrinology, in press.

Jenkin, G., and Thorburn, G. D., 1985, Inhibition of progesterone secretion by a 3β-hydroxysteroid dehydrogenase inhibitor in late pregnant sheep, Can. J. Physiol. Pharmacol., 63:136.

Keirse, M. J. N. C., 1984, Prostaglandins during human parturition, in: "Initiation of Parturition: Prevention of Prematurity," Report of the Fourth Ross Conference on Obstetric Research, p. 137, Ross Laboratories, Columbus, Ohio.

Klein, A. H., Oddie, T. H., and Fisher, D. A., 1978, Effect of parturition on serum iodothyronine concentrations in fetal sheep, Endocrinology, 103:1453.

Largercrantz, H., and Slotkin, T. A., 1986, The 'stress' of being born, Scientific American, April, p. 92.

Leach Harper, C. M., and Thorburn, G. D., 1984, Inhibition of prostaglandin synthesis by ovine allantoic fluid: acute reduction in inhibitory activity during late gestation, Can. J. Physiol. Pharmacol., 62:1152.

Leatherland, J. F., and Renfree, M. B., 1982, Ultrastructure of the nongranulated cells and morphology of the extracellular spaces in the pars distalis of adult and pouch-young tammar wallabies (Macropus eugenii), Cell and Tissue Research, 227:439.
Leatherland, J. F., and Renfree, M. B., 1983, Structure of the pars distalis in pouch-young tammar wallabies (Macropus eugenii), Cell and Tissue Research, 230:587.
Lewis, P. R., Fletcher, T. P., and Renfree, M. B., 1986, Prostaglandin in the peripheral plasma of tammar wallabies during parturition, J. Endocrinol., in press.
Liggins, G. C., Kennedy, P. C., and Holm, L. W., 1967, Failure of initiation of parturition after electrocoagulation of the pituitary of the fetal lamb, Am. J. Obstet. Gynecol., 98:1080.
Liggins, G. C., 1969, Premature delivery of fetal lambs infused with glucocorticoids, J. Endocrinol., 45:515.
Liggins, G. C., Kitterman, J. A., and Forster, C. S., 1979, Fetal maturation related to parturition, Anim. Reprod. Sci., 2:193.
Liggins, G. C., 1985, The paracrine system controlling human parturition, in: "The Endocrine Physiology of Pregnancy and the Periparal Period," Serono Symposium, Vol. 21, p. 205, Raven Press, New York.
Lincoln, D. W., and Porter, D. G., 1979, Photoperiodic dissection of endocrine events at parturition, Anim. Reprod. Sci., 2:97.
McCracken, J. A., 1980, Hormone receptor control of prostaglandin $F_{2\alpha}$ secretion by the ovine uterus, Advances in Prostaglandins and Thromboxane Research, 8:1329.
McDonald, P. C., Porter, J. C., Schwarz, B. E., and Johnston, J. M., 1978, Initiation of parturition in the human female, Seminars in Perinatology, 2:273.
Malpas, P., 1933, Postmaturity and malformation of the foetus, J. Obstet. Gynaecol. Br. Emp., 40:1046.
Magyar, D. M., Elsner, C. W., Eliot, J., Glatz, T., Nathanielsz, P. W., and Buster, J. E., 1981, A combined radioimmunoassay for the measurement of unconjugated and sulfocongjugated pregnenolone, 17-hydroxypregnenolone, dehydroepiandrosterone, and estrone applied to fetal and maternal ovine plasma, Steroids, 37:423.
Mitchell, M. D., and Flint, A. P. F., 1978, Prostaglandin production by intra-uterine tissues from periparturient sheep: use of a superfusion technique, J. Endocrinol., 76:111.
Mitchell, M. D., Bibby, J. C., Hicks, B. R., and Turnbull, A. C., 1978, Specific production of prostaglandin E_2 by human amnion in vitro, Prostaglandins, 15:377.
Mitchell, J. A., and Yochim, J. M., 1970, Influence of environmental lighting on duration of pregnancy in the rat, Endocrinology, 87:472.
Nathanielsz, P. W., Elsner, C., Magyar, D., Fridshal, D., Freeman, A., and Buster, J.E, 1982, Time trend analysis of plasma unconjugated and sulfoconjugated estrone and 3β-Δ^5-steroids in fetal and maternal sheep plasma in relation to spontaneous parturition at term, Endocrinology, 110:1402.
Okazaki, T., Casey, M. L., Okita, J. R., MacDonald, P. C., and Johnson, J. M., 1981, Biosynthesis and metabolism of prostaglandins in human fetal membranes and uterine decidua, Am. J. Obstet. Gynecol., 139:373.
Olson, D. M., Skinner, K., and Challis, J. R. G., 1983a, Estradiol-17β- and 2-hydroxyestradiol-17β-induced differential production of prostaglandins by cells dispersed from human intra-uterine tissues at parturition, Prostaglandins, 25:629.
Olson, D. M., Skinner, K., and Challis, J. R. G., 1983b, Prostaglandin production in relation to parturition by cells dispersed from human intra-uterine tissues, J. Clin. Endocrinol. Metab., 57:694.
Olson, D. M., Lye, S. J., Skinner, K., and Challis, J. R. G., 1985, Prostanoid concentrations in maternal/fetal plasma and amniotic fluid and intra-uterine tissue prostanoid output in relation to myometrial

contractility during the onset of adrenocorticotropin-induced preterm labor in sheep, Endocrinology, 116:389.

Philpott, S. D., Doughton, B., Salamonsen, L. A., and Findlay, J. K., 1985, Oxytocin binding in sheep endometrium during the oestrous cycle and early pregnancy, Proc. Aust. Soc. Reprod. Biol. (Abst), 17:93.

Platzker, A. C. G., Kitterman, J. A., Mescher, E. J., Clements, J. A., and Tooley, W. H., 1975, Surfactant in the lung and tracheal fluid of the fetal lamb and acceleration of its appearance by dexamethasone, Pediatrics, 56:554.

Poole, W. E., and Catling, P. C., 1974, Reproduction in the two species of grey kangaroo, *Macropus giganteus* Shaw and *M. fuliginosus* (Desmarest) I. Sexual maturity and oestrus, Aust. J. Zool., 22:277.

Poole, W. E., 1975, Reproduction in the two species of grey kangaroos, *Macropus giganteus* Shaw, and *M. fuliginosus* (Desmarest). II. Gestation, parturition and pouch life, Aust. J. Zool., 23:333.

Renfree, M. B., and Young, I. R., 1979, Steroids in pregnancy and parturition in the marsupial, *Macropus eugenii*, J. Steroid Biochem., 11:515.

Reppert, S. M., 1984, Development of the circadian timing system, in: "Research in Perinatal Medicine. II. Fetal Neuroendocrinology," F. Ellendorff, P. D. Gluckman, and N. Parvizi, eds., p. 179, Perinatology Press, Ithaca, New York.

Risbridger, G. P., Leach Harper, C. M., Wong, M. H., and Thorburn, G. D., 1985, Gestation changes in prostaglandin production by ovine fetal trophoblastic cells, Placenta, 6:117.

Robinson, J. S., Chapman, R. L. K., Challis, J. R. G., Mitchell, M. D., and Thorburn, G. D., 1978, Administration of extra-amniotic arachidonic acid the suppression of uterine prostaglandin synthesis during pregnancy in the rhesus monkey, J. Reprod. Fertil., 54:369.

Shaw, G., 1983a, Effect of $PGF_{2\alpha}$ on uterine activity, and concentrations of 13,14-dihydro-15-keto-$PGF_{2\alpha}$ in peripheral plasma during parturition in the tammar wallaby (*Macropus eugenii*), J. Reprod. Fertil., 69:429.

Shaw, G., 1983b, Pregnancy after diapause in the tammar wallaby, *Macropus eugenii*, Ph.D. Thesis, Murdoch University, Perth, Western Australia.

Shemesh, M., Hansel. W., and Strauss, J. F., 1984, Modulation of bovine placental prostaglandin synthesis by an endogenous inhibitor, Endocrinology, 115:1401.

Soloff, M. S., 1983, The role of oxytocin receptors in parturition, in: "Initiation of Parturition: Prevention of Prematurity," Report of the Fourth Ross Conference on Obstetric Research, p. 160, Ross Laboratories, Columbus, Ohio.

Spiegelberg, 1882, "Textbook of Midwifery."

Thomas, A. L., Krane, E. J., and Nathanielsz, P. W., 1978, Changes in the fetal thyroid axis after induction of premature parturition by low dose continuous intravascular cortisol infusion to the fetal sheep at 130 days of gestation, Endocrinology, 103:17.

Thorburn, G. D., Cox, R. I., Currie, W. B., Restall, B. J., and Schneider, W., 1972, Prostaglandin F concentration in the utero-ovarian venous plasma of the ewe during the oestrous cycle, J. Endocrinol., 53:325.

Thorburn, G. D., Cox, R. I., Currie, W. B., Restall, B. J., and Schneider, W., 1973, Prostaglandin F and progesterone concentrations in the utero-ovarian venous plasma of the ewe during the oestrous cycle and early pregnancy, J. Reprod. Fertil. Suppl., 18:151.

Thorburn, G. D., and Hopkins, P. S., 1973, Thyroid function in the foetal lamb, in: "Foetal and Neonatal Physiology," R. S. Comline, K. W. Cross, G. S. Dawes, and P. W. Nathanielsz, eds., p. 488, Proc. Sir Joseph Barcroft Centenary Symposium, Cambridge University Press, Cambridge.

Thorburn, G. D., and Challis, J. R. G., 1979, Endocrine control of parturition, Physiol. Reviews, 59:863.

Thorburn, G. D., Waters, M. J., Dolling, M., and Young, I. R., 1981, Fetal maturation and epidermal growth factor, Proc. Aust. Physiol. Pharmacol. Soc., 12:11.

Thorburn, G. D., 1985, Prostaglandins and the regulation of myometrial activity: a working model, in: "The Physiological Development of the Fetus and Newborn," C. T. Jones and P. W. Nathanielsz, eds., p. 381, Academic Press, London.

Tyndale-Biscoe, C. H., Hinds, L. A., Horn, C. A., and Jenkin, G., 1983, Hormonal changes at oestrus, parturition and postpartum oestrus in the tammar wallaby (Macropus eugenii), J. Endocrinol., 96:155.

Tyndale-Biscoe, C. H., and Renfree, M. B., 1986, "Reproductive Physiology of Marsupials," 500 pp., Cambridge University Press, Cambridge.

Walker, M. T., and Gemmell, R. T., 1983, Plasma concentrations of progesterone, oestradiol-17 and 13,14-dihydro-15-oxo-prostaglandin $F_{2\alpha}$ in the pregnant wallaby (Macropus rufogriseus), J. Endocrinol., 97:369.

Walker, M. T., and Gemmell, R. T., 1983, Organogenesis of the pituitary, adrenal and lung at birth in the wallaby, Macropus rufogriseus, Am. J. Anat., 168:331.

Young, I. R., 1978, The physiology of parturition in the macropodid marsupial Macropus eugenii (Desmarest), Ph.D. Thesis, Murdoch University, Perth, Western Australia,

AUTHOR INDEX

SUBJECT INDEX

GPSR Compliance

The European Union's (EU) General Product Safety Regulation (GPSR) is a set of rules that requires consumer products to be safe and our obligations to ensure this.

If you have any concerns about our products, you can contact us on ProductSafety@springernature.com

In case Publisher is established outside the EU, the EU authorized representative is:

Springer Nature Customer Service Center GmbH
Europaplatz 3
69115 Heidelberg, Germany

Batch number: 10761599

Printed by Printforce, the Netherlands